AF323316

FleXible Electronics

From Materials to Devices

FleXible Electronics

From Materials to Devices

Guozhen Shen
Chinese Academy of Sciences, China

Zhiyong Fan
The Hong Kong University of Science and Technology, Hong Kong

NEW JERSEY · LONDON · SINGAPORE · BEIJING · SHANGHAI · HONG KONG · TAIPEI · CHENNAI · TOKYO

Published by

World Scientific Publishing Co. Pte. Ltd.

5 Toh Tuck Link, Singapore 596224

USA office: 27 Warren Street, Suite 401-402, Hackensack, NJ 07601

UK office: 57 Shelton Street, Covent Garden, London WC2H 9HE

Library of Congress Cataloging-in-Publication Data
Names: Shen, Guozhen (Electrical engineer), editor. | Fan, Zhiyong (Electrical engineer), editor.
Title: Flexible electronics : from materials to devices / Guozhen Shen
    (Chinese Academy of Sciences, China) & Zhiyong Fan
    (The Hong Kong University of Science and Technology, Hong Kong).
Other titles: Flexible electronics (World Scientific (Firm))
Description: [Hackensack] New Jersey : World Scientific, [2016] |
    Includes bibliographical references and index.
Identifiers: LCCN 2015031962 | ISBN 9789814651981 (alk. paper)
Subjects: LCSH: Flexible electronics.
Classification: LCC TK7872.F54 F554 2016 | DDC 621.381--dc23
LC record available at http://lccn.loc.gov/2015031962

British Library Cataloguing-in-Publication Data
A catalogue record for this book is available from the British Library.

Desk Editor: Kalpana Bharanikumar

Typeset by Stallion Press
Email: enquiries@stallionpress.com

Printed in Singapore by B & Jo Enterprise Pte Ltd

CONTENTS

CHAPTER 1

CARBON NANOTUBE FLEXIBLE ELECTRONICS

Chuan Wang

Department of Electrical and Computer Engineering,
Michigan State University
428 S. Shaw Lane, Engineering Building #2120,
East Lansing, Michigan, USA
cwang@msu.edu

Single-wall carbon nanotubes (SWNTs) possess fascinating electrical properties and offer new entries into a wide range of novel electronic applications that are unattainable with conventional Si-based devices. The field initially focused on the use of individual or parallel arrays of nanotubes as the channel material for ultra-scaled nanoelectronic devices. However, the challenge in the deterministic assembly of SWNTs has proven to be a major technological barrier. In recent years, solution deposition of semiconductor-enriched SWNT networks has been actively explored for high performance and uniform thin-film transistors (TFTs) on both mechanically rigid and flexible substrates. This presents a unique niche for nanotube electronics by overcoming their limitations and taking full advantage of their superb electronic properties. This chapter focuses on the large-area processing and electronic properties of SWNT TFTs. A wide range of applications in flexible electronics including integrated circuits, radio-frequency (RF) transistors, displays, and electronic skins will be discussed. With emphasis on large-area systems where nm-scale accuracy in the assembly of nanotubes is not required, the demonstrations show SWNTs' immense promise as a low-cost and scalable TFT technology for flexible electronic systems with excellent device performances.

1. Introduction

Single-wall carbon nanotubes (SWNTs) can be considered as monolayer graphene sheets with a honeycomb structure that are rolled into seamless, hollow cylinders. Owing to their small size (diameter around 1–2 nm), as well as their superior electronic properties without surface dangling bonds, SWNTs hold great potential for a wide range of applications in solid-state devices and are envisioned as one of the promising candidates for beyond-silicon electronics. SWNTs can be categorized by their chiral vectors defined on the hexagonal crystal lattice using two integers (m and n). The chiral vectors correspond to the direction along which a graphene sheet is wrapped to result in a SWNT. The electronic properties of SWNTs heavily depend on their chiral vectors and the SWNTs can be either metallic ($m = n$ or $m - n$ is a multiple of 3) or semiconducting (all other cases).[1–4] Using this rule of thumb, one can infer from the possible (n,m) values that one third of SWNTs are metallic and the other two thirds are semiconducting. For practical use as the active channel component of electronic devices, semiconducting SWNTs are commonly used. The advantages of semiconducting SWNTs over other conventional semiconductors are multifold. First of all, the charge carriers in carbon nanotubes have long, mean free paths, on the order of a few hundred nanometers for acoustic phonon scattering mechanism. As a result, scattering-free ballistic transport of carriers at low electric fields can be achieved in carbon nanotubes at moderate channel lengths (e.g., sub-100 nm).[5] Second, the carrier mobility of semiconducting nanotubes is experimentally measured to be $> 10{,}000$ cm^{2}V^{-1}s^{-1},[6,7] at room temperature which is higher than the state-of-the-art silicon transistors. Finally, their small diameters enable excellent electrostatics with efficient gate control of the channel for highly miniaturized devices. Thereby, SWNTs have stimulated enormous interest in both fundamental research and practical applications in nano- and macro-electronics.

Researchers have previously demonstrated excellent field-effect transistors (FETs)[5–12] and integrated circuits[13–17] using individual SWNTs. Figures 1(a) and 1(b) depict the transfer ($I_{DS}–V_{GS}$) and output ($I_{DS}–V_{DS}$) characteristics of the state-of-the-art individual SWNT-FET with self-aligned source/drain contacts and near ballistic transport.[11] Impressive

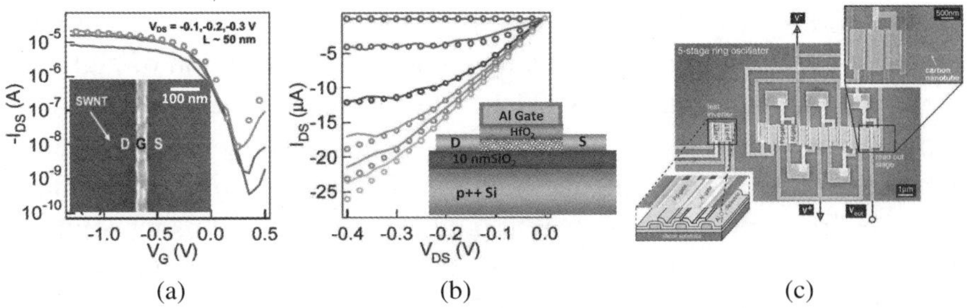

Figure 1. State-of-the-art individual SWNT transistors and circuits. (a) I_{DS}–V_{GS} characteristics of a self-aligned ballistic SWNT-FET with a channel length of 50 nm. Inset: Scanning Electron Microscope (SEM) image of the device. (b) Experimental (solid line) and simulated (open circle) I_{DS}–V_{DS} characteristics of the same device shown in panel (a). Inset: schematic of the device. Reproduced with permission from Ref. 11. (Copyright 2004 American Chemical Society.) (c) SEM image of a 5-stage ring oscillator constructed on an individual SWNT. Reproduced with permission from Ref. 16. (Copyright 2006 The American Association for the Advancement of Science (AAAS).)

performance with subthreshold slope (*SS*) of 110 mV/dec, on-state conductance of $0.5 \times 4e^2/h$ and saturation current upto 25 μA/tube (diameter ~1.7 nm) has been achieved in devices with channel lengths down to 50 nm.[11] Better *SS* of ~70 mV/dec, which is close to the theoretical limit of 60 mV/dec, has also been achieved in transistors with slightly longer channel lengths (500 nm).[10] More recently, SWNT-FETs with sub –10 nm channel lengths have been demonstrated.[12] Such devices exhibit an impressive *SS* of 94 mV/dec, current on/off ratio of 10^4, and on-current density of 2.41 mA/μm, which outperform silicon FETs with comparable channel length. Using SWNT-FETs, integrated circuits with various functionalities have been demonstrated. Notable examples include a five-stage ring oscillator (Fig. 1(c)) and pass-transistor-logic-based integrated circuits (full adder, multiplexer, decoder, D-latch, etc.).[16,17]

Despite the tremendous progress made with individual nanotube transistors and circuits, major technological challenges remain, including the need for deterministic assembly of nanotubes on a handling substrate with nm-scale accuracy, minimal device-to-device performance variation, and development of a fabrication process scalable and compatible with industry standards. Hence, the use of carbon nanotubes for nanoelectronic applications is still long from being realized. On the other hand, the use of SWNT networks, especially based on semiconductor-enriched samples, present a highly promising path for the realization of high performance thin-film transistors (TFTs) for macro- and flexible electronic applications. The most significant advantages of using SWNT random networks for TFTs lie in the fact that the SWNT thin-films are mechanically flexible, optically transparent, and can be prepared using solution-based room temperature processing, all of which cannot be provided by amorphous and poly silicon technologies.[18–20] Compared with organic semiconductors,[21–25] the other competing platform for flexible TFTs, the SWNT thin-films offer significantly better carrier mobility (~2 orders of magnitude improvement). Thereby, large-area TFT applications seem to offer an ideal niche for carbon nanotube based electronics, taking advantage of their superb physical, chemical and electrical properties without being hindered from their precise assembly limitations down to nm-scale.

Numerous research efforts have been devoted to the successful realization of large-scale chemical vapor deposition (CVD) growth of high-density horizontally aligned SWNTs on single crystal quartz or sapphire substrates.[26–34] Transfer techniques have been further developed, enabling the demonstration of high-performance transistors and integrated circuits using the aligned nanotubes on various types of rigid and flexible substrates.[35–42] However, considering the fact that roughly one third of the as-grown nanotubes are metallic, techniques such as electrical breakdown[43] is necessary to remove the leakage-causing metallic paths, which adds complexity, is not scalable, and significantly degrades the device performance due to the high applied fields during the process. Preferential growth of aligned semiconducting SWNTs has been reported recently,[32,44,45] which is an important step forward, however, the purity is not yet high enough to achieve transistors with high on/off current ratio (I_{on}/I_{off}) for digital applications. Therefore, for the purpose of obtaining devices with better I_{on}/I_{off}, it is more attractive to have networks of SWNTs with higher percentage of semiconducting tubes and/or with random orientation where individual nanotubes do not directly bridge the source/drain electrodes, thereby minimizing the metallic pathways.[46–50]

Figure 2 illustrates the most common assembly methods for random nanotube networks including direct CVD growth,[51,52] dry filtration,[53]

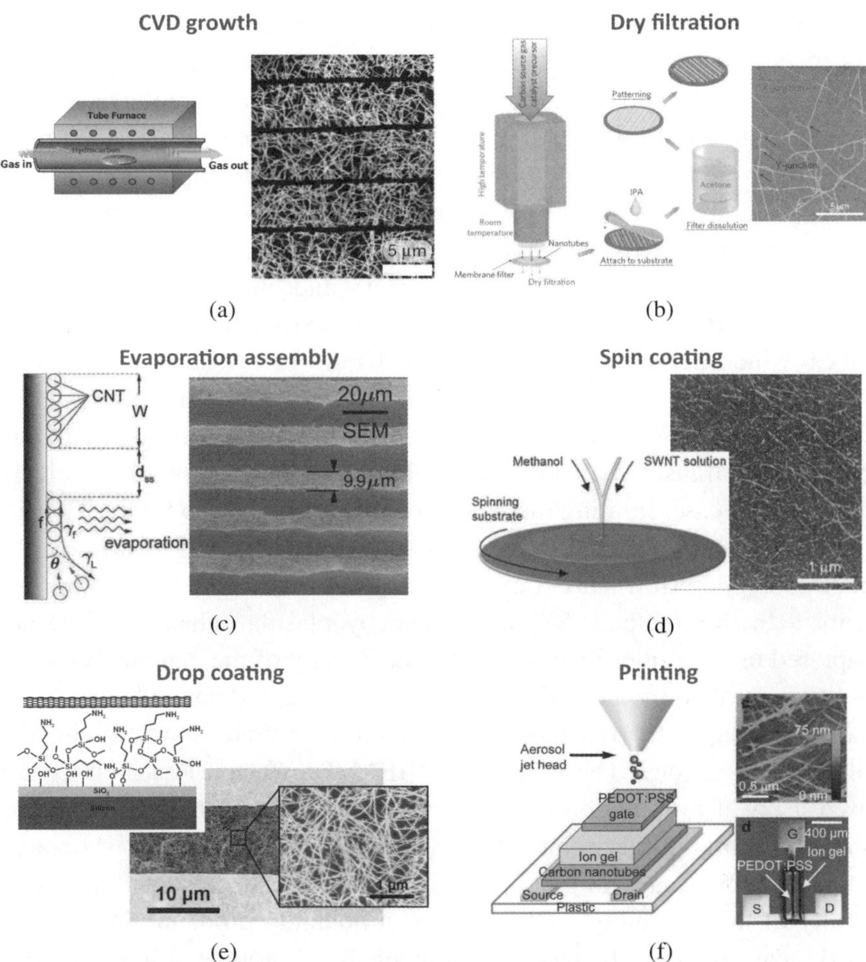

Figure 2. Different methods used for assembling SWNT networks. (a) CVD growth. Reproduced with permission from Ref. 51. (Copyright 2008 Nature Publishing Group.) (b) Dry filtration. Reproduced with permission from Ref. 53. (Copyright 2011 Nature Publishing Group.) (c) Evaporation assembly. Reproduced with permission from Ref. 54. (Copyright 2008 American Chemical Society.) (d) Spin coating. Reproduced with permission from Ref. 55. (Copyright 2004 American Chemical Society.) (e) Drop casting. Reproduced with permission from Ref. 59. (Copyright 2009 American Chemical Society.) (f) Printing. Reproduced with permission from Ref. 64. (Copyright 2010 American Chemical Society.)

evaporation assembly,[54] spin coating[55–57] drop coating[58–63] and printing.[64–68] CVD-grown nanotube networks have been widely explored for TFTs (Fig. 2(a)) and medium scale flexible integrated circuits have been demonstrated by Rogers *et al.*[51] For this method, metal catalysts are typically deposited on the entire substrate using either evaporation or spin coating methods followed by CVD growth using hydrocarbon precursors such as methane, ethylene, ethanol, methanol etc. Despite the tremendous success in making flexible nanotube TFTs and circuits with promising electrical performance, the drawback is the existence of metallic nanotubes, which degrades the current on/off ratio of the devices. Although stripe-patterning (Fig. 2(a) right panel) has been proposed to improve the device on/off ratio by cutting the percolative transport through metallic paths in the transistors,[51] the channel length needs to be made relatively large in this case, limiting the degree of integration in the future. Dry filtration method (Fig. 2(b)) has been used by Ohno and co-workers to achieve high-performance flexible nanotube TFTs and D-flip-flop circuits.[53] In this method, SWNTs grown by plasma enhanced CVD are captured using a filter membrane and the density of the nanotubes can be easily controlled by the collection time. The collected nanotube networks can be subsequently transferred to fabrication substrates by dissolving the filter using acetone. The group at the IBM T.J Watson Research Center used a novel evaporation assembly method to obtain aligned nanotube strips with high purity semiconducting nanotubes (Fig. 2(c)).[54] Although submicron devices with good performance have been achieved, the scalability of this assembly method can be a potential problem. SWNT networks can also be obtained by dropping the nanotube solution onto a spinning substrate (Fig. 2(d)).[55] The drawback for this method is also scalability because the deposited SWNTs often align along different orientations depending on the location on the substrate, preventing wafer-scale fabrication with high uniformity. The other two solution-based SWNT assembly methods — drop coating (Fig. 2(e)) and printing (Fig. 2(f)) — are found to be more promising for large scale applications of nanotube TFTs. For the drop coating method, the substrates are first functionalized with amine-containing molecules, which are effective adhesives for SWNTs. By simply immersing the substrate into the nanotube solution, highly uniform nanotube networks can be obtained throughout the wafer, enabling the fabrication of nanotube TFTs with high yield and small device-to-device variation.[59,60,62,63] Printing (Fig. 2(f)) represents another low-cost approach for fabricating large-scale nanotube TFTs and circuits where SWNT channel, electrodes, and gate dielectric can all be printed using ink-jet[64,65] or gravure printing[66–68] processes. This approach is useful for making cost-effective large-area nanotube circuits requiring only moderate performance as the resolution that can be achieved using printing process is generally lower than the conventional photolithography. Each of these methods discussed above presents unique opportunities and challenges. In this chapter, we will primarily focus on the use of semiconductor-enriched SWNT random networks. We will first discuss the assembly techniques for high density and uniform SWNT networks, and fabrication schemes for large-area, high-performance TFTs on mechanically flexible substrates. Electrical properties study and performance benchmarking will also discussed in detail. Lastly, we review a wide range of potential applications for SWNT TFTs in flexible integrated circuits, RF transistors, displays, and electronic skin.

2. Solution-Processed TFTs using Semiconductor-enriched SWNTs

2.1. *Nanotube Separation and TFT Fabrication*

As discussed in the previous section, one of the major challenges limiting the electronic applications of SWNTs is the coexistence of metallic and semiconducting nanotubes with roughly one third of the as-grown being metallic. The metallic SWNTs cause significant leakage current when the transistors are in the off state, and thus need to be selectively removed. As a result, high-purity semiconducting SWNTs have long been desired in order to achieve devices with high I_{on}/I_{off}. To address this problem, many groups have been actively working on nanotube separation based on electronic types. Hersam *et al.* proposed and demonstrated the use of density gradient ultracentrifugation (DGU) to achieve diameter,[69] electronic type,[70] or even chirality-based[71] separation. By mixing the pristine unsorted SWNTs with surfactants, a density gradient between metallic and semiconducting SWNTs is created and upon ultracentrifugation, the

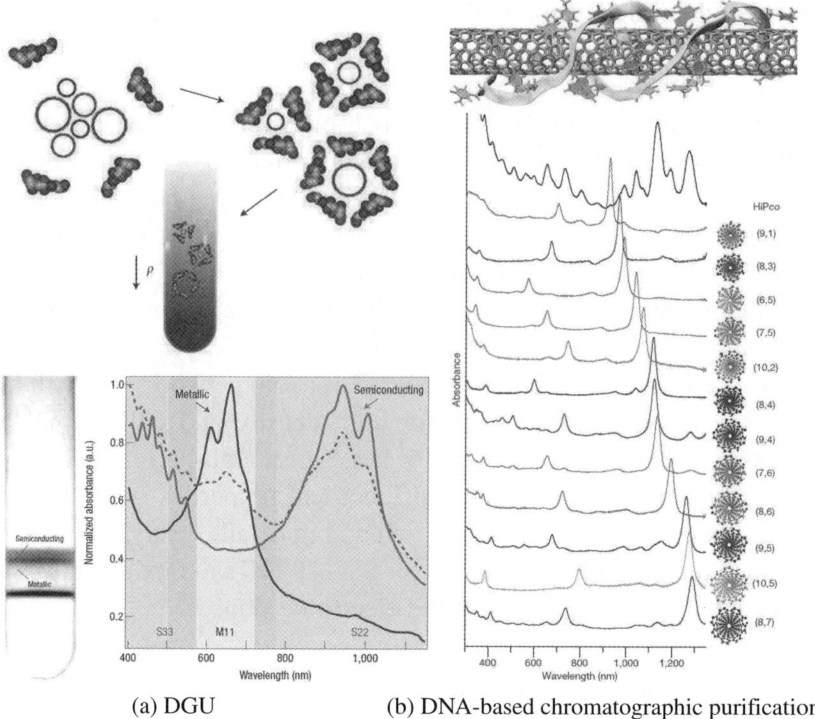

(a) DGU   (b) DNA-based chromatographic purification

Figure 3.   (a) A density gradient between metallic and semiconducting SWNTs is created by mixing the pristine SWNTs with surfactants. Upon ultracentrifugation, the metallic and semiconducting nanotubes can be separated from each other as shown in the UV–Vis–NIR absorption spectra. Reproduced with permission from Ref. 70. (Copyright 2006 Nature Publishing Group.) (b) SWNTs with different chiralities can be separated by chromatographic purification by using DNA with different sequence motifs. Reproduced with permission from Ref. 72. (Copyright 2009 Nature Publishing Group.)

metallic and semiconducting nanotubes are separated from each other as shown in Fig. 3(a). In the photograph of the test tube, the brown band contains primarily semiconducting nanotubes and the green band contains mainly metallic ones. The high purity semiconducting and metallic nanotubes are further evidenced by the UV–Vis–NIR absorption spectra. Noteworthy, the semiconductor-enriched nanotubes are now commercially available, which makes it easy for research groups worldwide to explore device applications based on this purified material system. Even with high purity semiconducting nanotubes, there are still many different possible chiralities, and thus different bandgaps. This factor can negatively affect the device performance and uniformity. In order to achieve chirality-specific purification of SWNTs, DNA modification with different sequence motifs have been used and purified SWNTs with different chiralities have been obtained using chromatographic purification (Fig. 3(b)).[72]

As described in the earlier section, solution-based deposition of the high purity semiconducting nanotubes at macroscales has been reported.[59,60,62,63] Most commonly, the substrates are first functionalized with amine-containing molecules or polymers such as aminopropyltrieth-oxysilane (APTES)[59,60] or poly-L-lysine[62,63] whose molecular structures are shown in Fig. 4(a). By dipping the substrate into APTES or poly-L-lysine, amine-terminated layers are formed on the SiO_2 surface which works as an effective adhesive layer for SWNTs (Fig. 4(b)). The amine-functionalized substrate is subsequently immersed into high-purity (up to 99%) semiconducting nanotube solution followed by deionized (DI) water and isopropanol rinse, and blow dry in nitrogen. Using the method described above, highly uniform SWNT random networks can be achieved at full wafer-scale. As shown in Fig. 4(c), the scanning electron micros-copy (SEM) images taken at different locations on a 3-inch Si/SiO_2 wafer indicate that uniform nanotube deposition is achieved. The importance of substrate functionalization is clearly illustrated in Figs. 4(d) and 4(e). From the images, one can find that the samples with (Fig. 4(d)) and with-out (Fig. 4(e)) an adhesion layer exhibit rather drastic difference in terms of nanotube density and surface coverage.

Using the solution-deposited semiconductor-enriched nanotube thin-films, high performance TFTs can be fabricated on mechanically flexible substrates. A schematic illustration of a representative fabrication process is shown in Fig. 5(a). For the device fabrication on flexible substrates, polyimide is commonly explored, which can be spun coated and cured on a silicon handling wafer.[62,63,73] Polyimide is a high temperature plastic, allowing the dielectric layer and other passive components to be deposited at temperatures as high as 300°C. The fabrication process typically involves gate formation, dielectric deposition, nanotube deposition, source/drain formation, and unwanted nanotube etching (Steps 1–6 in Fig. 5(a).[63,73] Atomic layer deposition (ALD) is used to deposit high-κ oxide layers such as Al_2O_3, HfO_2, or ZrO_2 as the gate dielectric. For the

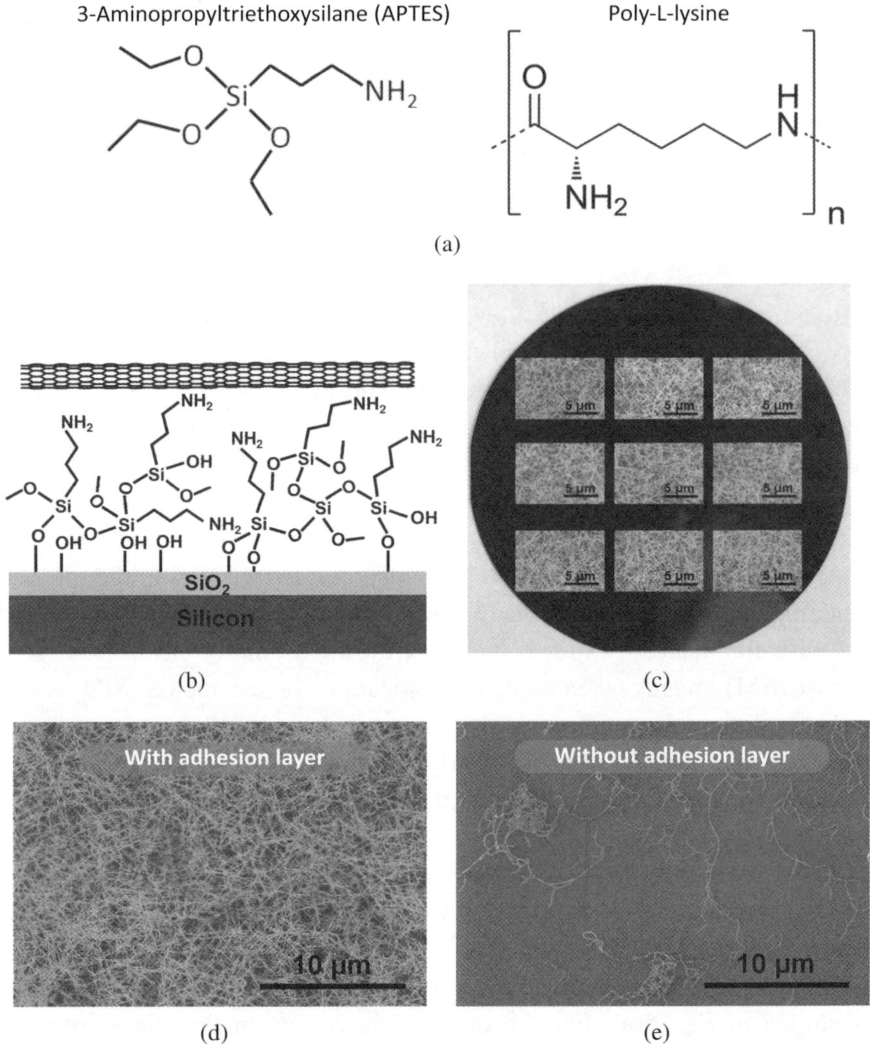

Figure 4. (a) Molecular structure of adhesive layers commonly used for large-scale semiconducting nanotube network deposition. (b) Schematic diagram of APTES-assisted nanotube deposition on Si/SiO₂ substrate. (c) SEM images showing that highly uniform nanotubes are deposited at different locations on a 3-inch Si/SiO₂ wafer. (d, e) SEM images of semiconducting nanotube networks deposited on Si/SiO₂ substrates with (d) and without (e) APTES functionalization. Reproduced with permission from Ref. 59. (Copyright 2009 American Chemical Society.)

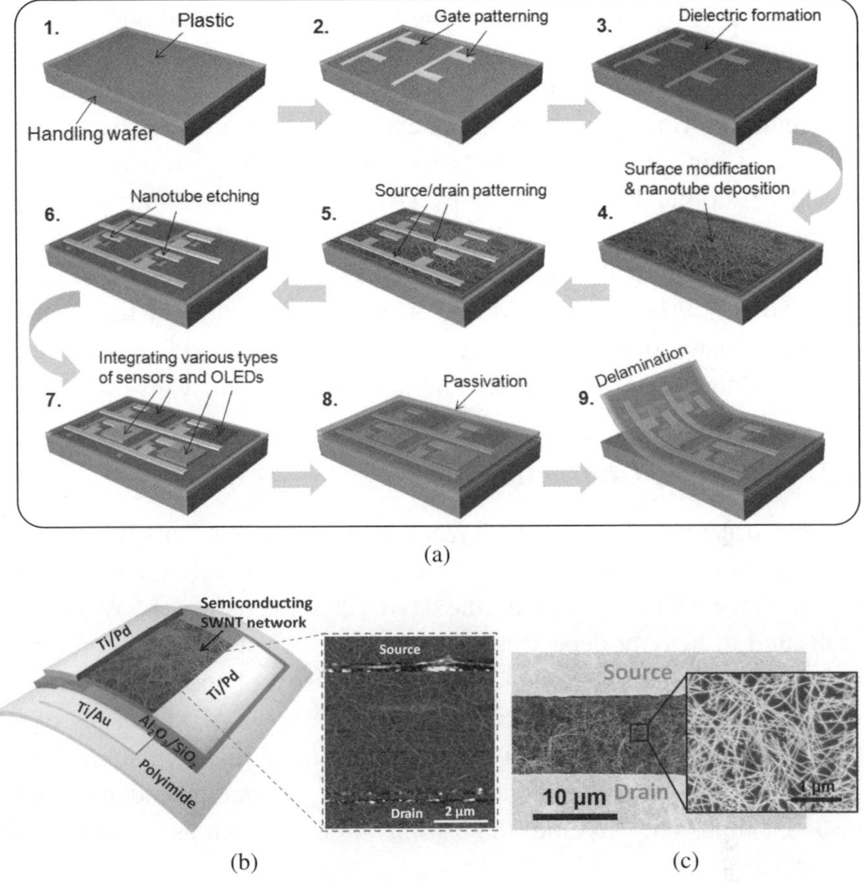

Figure 5. (a) Schematic illustration showing the fabrication of semiconducting nanotube TFTs and integrated circuits on mechanically flexible substrates. Reproduced with permission from Ref. 73. (Copyright 2013 Nature Publishing Group.) (b) Schematic diagram and AFM image of a TFT fabricated on a polyimide substrate. Reproduced with permission from Ref. 63. (Copyright 2012 American Chemical Society) (c) SEM image of a represented carbon nanotube TFT with solution-processed semiconducting carbon nanotube random networks in the channel. Reproduced with permission from Ref. 60. (Copyright 2010 American Chemical Society.)

local back-gated device geometry, a thin layer (thickness, ~2 nm) of SiO_x is often evaporated on top of the high-κ gate dielectric to facilitate a more efficient deposition of SWNTs on the substrate using the amine chemistry described above. In order to precisely define the active channel region, photolithography and oxygen plasma is used to etch away the unwanted nanotubes. For flexible TFTs, the polyimide layer is peeled off from the silicon handling wafer once the fabrication is completed, leaving behind the high performance devices on an extremely bendable plastic substrate (Step 9 in Fig. 5(a). A representative schematic diagram and the corresponding atomic force microscopy (AFM) image of a fully-fabricated nanotube TFTs on flexible polyimide substrates are shown in Fig. 5(b).[63] The SEM image of the channel of a representative device is also shown (Fig. 5(c)).[60] From both SEM and AFM images, one can find that the channel consists of high density, uniform SWNT networks, which is critical for achieving uniform device and circuit performance.

2.2. *Electrical Characteristics*

The electrical properties of the SWNT networks have been systematically studied to evaluate the feasibility of using such material system for high-performance TFTs. First of all, the device characteristics of SWNT TFTs are found to heavily depend on the nanotube density.[61,62] By controlling the duration of nanotube deposition, the density of the network can be controlled as shown in Fig. 6.[62] Figure 6(a) shows AFM images of the semiconducting nanotube networks formed after different durations of deposition, showing the monotonic increase of nanotube density from ~30–65 tubes/μm^2 as the deposition time is increased from 5–90 minutes.[62]

The I_{DS}–V_{GS} characteristics for the devices made with different nanotube densities are shown in Fig. 6(b). Based on the results, histograms of the extracted device on/off current ratio (I_{on}/I_{off}) and unit-width (W) normalized transconductance (g_m/W) are presented in Figs. 6(c) and 6(d). According to these figures, one can find that as the nanotube density increases (i.e., longer deposition time), the g_m/W improves while the I_{on}/I_{off} decreases. This trade-off needs to be taken into consideration when optimizing the device performance for different types of applications. For example, high I_{on}/I_{off} is generally desired for digital logic applications in order to minimize the static power consumption, so it is preferred to use TFTs with lower nanotube density. On the other hand, for high-speed analog or RF applications where g_m is most important, the use of higher

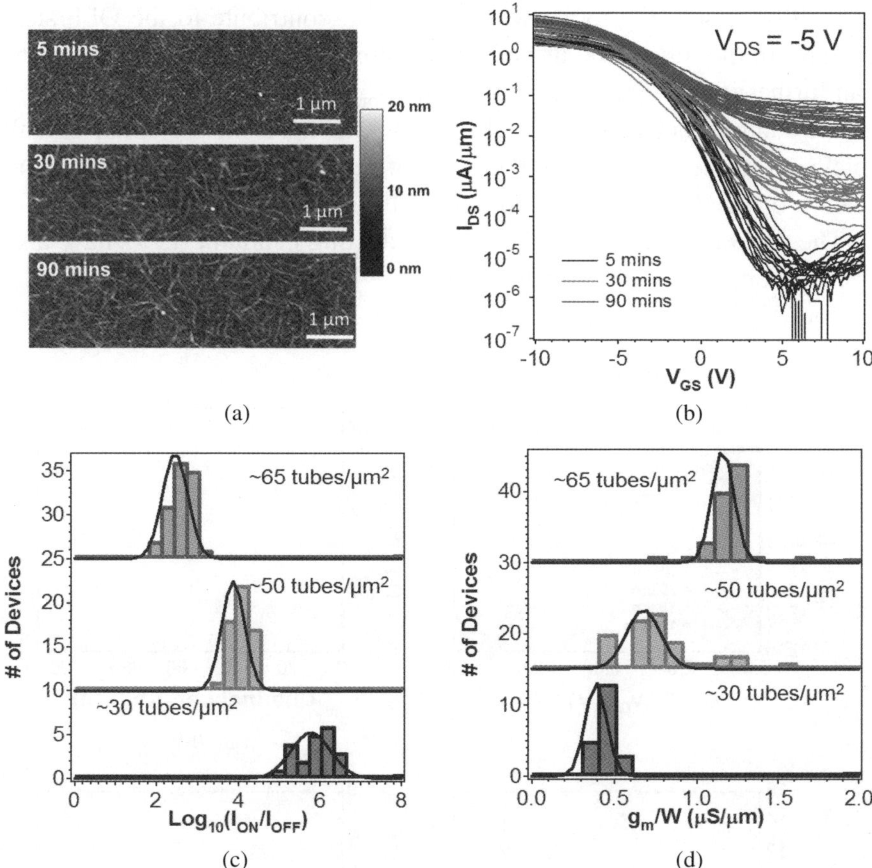

Figure 6. (a) AFM images showing semiconducting nanotube networks formed after different durations of solution deposition. (b) I_{DS}–V_{GS} characteristics of semiconducting nanotube TFTs with different nanotube densities obtained by different deposition times. (c, d) Histogram showing the distribution of on/off current ratio (c) and unit-width normalized g_m (d) for TFTs with different nanotube densities. Reproduced with permission from Ref. 62. (Copyright 2011 American Chemical Society.)

density nanotube networks is more beneficial. The lower I_{on}/I_{off} for high nanotube density networks arises from the higher probability of a direct metallic path between source/drain electrodes given that ~1% of the commercially available semiconductor-enriched nanotubes are still metallic. Furthermore, clear bundling of nanotubes is observed for longer deposition times (Fig. 6(a)) which may also contribute to the OFF state current. In the future, the use of higher semiconductor-enriched solutions can further improve the I_{off} for the same I_{on}. Importantly, the gate control of the random network is clearly strong, arising from the small overall thickness of the network (on the order of a few nm). This is an important feature of the devices, owing to the small diameter of SWNTs.

The scaling properties of the semiconducting nanotube TFTs have also been systematically investigated in Ref. 63. Figure 7(a) shows the $I_{DS}-V_{GS}$ characteristics of representative devices with channel lengths from

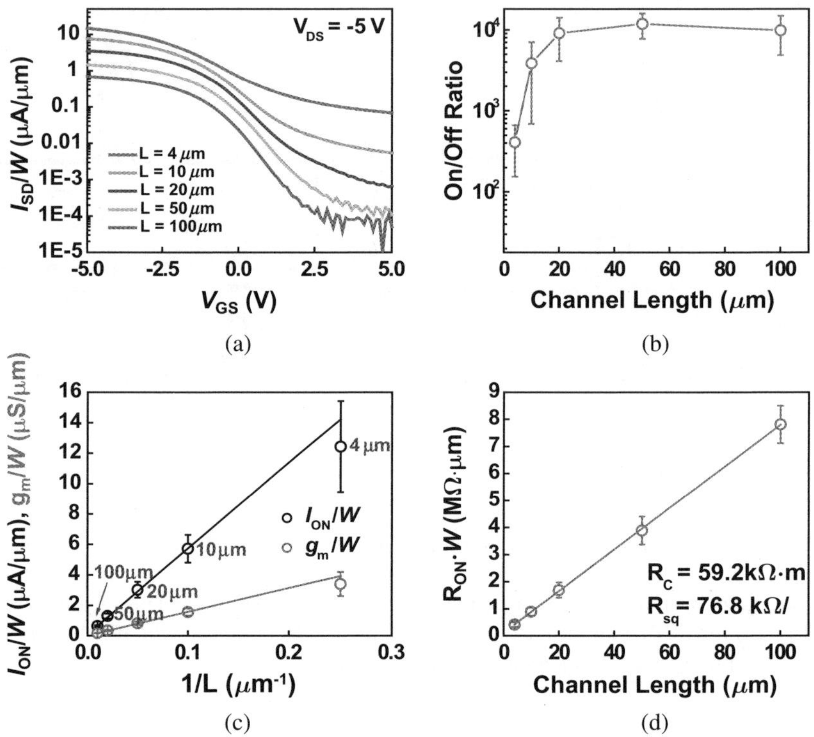

Figure 7.   (a) $I_{DS}-V_{GS}$ characteristics of flexible nanotube TFTs with various channel lengths (4, 10, 20, 50, and 100 μm) measured at V_{DS} = −5 V. (b) Plot of I_{on}/I_{off} as a function of channel length for V_{DS} = −5 V. (c) I_{on}/W at V_{GS} = −5 V and peak g_m/W as a function of $1/L$ for V_{DS} = −5 V. (d) Normalized on-state resistance at V_{GS} = −5 V as a function of the channel length. Reproduced with permission from Ref. 63. (Copyright 2012 American Chemical Society.)

4–100 μm measured at $V_{DS} = -5$ V, from which the device figures of merit such as on-current density (I_{on}/W), g_m/W, and I_{on}/I_{off} are extracted and plotted as a function of channel length (L). Figure 7(b) indicates that as the channel length increases from 4–100 μm, the average on/off current ratio undergoes significant improvement from ~400–10,000, which can be attributed to the decrease in the probability of percolative transport through the metallic nanotubes. For devices with shorter L that is comparable to the length of the SWNTs used for the study (~1–2 μm), the probability of metallic impurities directly bridging the source/drain gets significantly higher, resulting in lower I_{on}/I_{off}. Figure 7(c) shows that the I_{on}/W and g_m/W (at $V_{DS} = -5$ V) of the devices are roughly proportional to $1/L$, consistent with the conventional Metal Oxide Semiconductor FETs (MOSFETs) device models and indicating good uniformity of the transistors. For such solution-processed nanotube TFTs, I_{on}/W and g_m/W as high as 15 μA/μm and 4 μS/μm have been achieved with $L = 4$ μm and $V_{DS} = -5$ V. Interestingly, Figs. 7(b) and 7(c) reveal a similar trade-off between on-current, transconductance, and on/off ratio. The results indicate that for a given nanotube density, longer channel length (e.g., 10 μm) is preferred for digital logic circuits, while smaller channel length (< 4 μm) should be chosen for analog and RF applications. Furthermore, in order to extract the contact resistance, the normalized on-state resistance ($R_{on} \cdot W$) is plotted as a function of L as shown in Fig. 7(d). From the y-axis intercept divided by two, a contact resistance (R_c) of ~59 k$\Omega \cdot \mu$m is extracted. It is worth noting that this R_c value is relatively high because for the nanotube network transistors, only a fraction of the contact area is active for a given unit width. In the future, improving the nanotube density could help to further reduce the overall contact resistance. Nevertheless, the R_c value reported here is low enough for the long channel devices often explored in TFT applications, where the channel resistance is the dominant resistance component.

2.3. *Field-Effect Mobility*

In order to fairly benchmark the performance of SWNT TFTs with other technologies, it is important to examine the field-effect carrier mobility. Carrier mobility is a measure of how fast the charge carriers can travel within the semiconductor material under an electric field. It directly relates to the I_{on}, g_m, and maximum operating speed of the transistors, especially at relatively longer channel lengths where the devices operate in the diffusive regime, and is widely used as a figure of merit for benchmarking semiconductor materials. By knowing the gate capacitance (C_{ox}) and the extracted g_m/W, the field-effect device mobility of the nanotube TFTs can be derived using the following relation

$$\mu_{device} = \frac{L}{V_d C_{ox} W} \cdot \frac{dI_d}{dV_g} = \frac{L}{V_d C_{ox}} \cdot \frac{g_m}{W}.$$

In order to accurately assess the mobility, it is important to directly measure the C_{ox} of the devices. Most published works have relied on either parallel plate model[53] or a more rigorous analytical cylindrical model by considering the electrostatic coupling between the nanotubes.[74] Using the parallel plate model, the C_{ox} would be overestimated, leading to lower extracted device mobility. The analytical model offers better accuracy. However, due to the uncertainty in quantifying the density and diameter for a random network containing hundreds of thousands of SWNTs, the extracted device mobility can still have large uncertainty, depending on the density and diameter values used in the calculations.

Recently, capacitance–voltage (C–V) measurements have been used to directly measure the gate capacitance of the nanotube network TFTs.[51,63] The C–V measurement setup is illustrated in Fig. 8(a),[63] where the gate of the transistor is connected to the HIGH terminal and the source and drain are both connected to the LOW terminal of the C–V analyzer (Agilent B1500A). This measurement setup is similar to the commonly explored technique for the C–V analysis of ultrathin body Si devices. TFTs with slightly underlapped gate are chosen in order to minimize the effect of the parasitic capacitances (Fig. 8(b)). The C–V characteristics of devices with different L are measured under a moderate frequency of ~100 kHz and the C_{ox} values in the accumulation region ($V_{GS} = -5$ V) are extracted (Fig. 8(c)). By using devices with different size, the C_{ox} per unit area can be accurately evaluated from the slope of the linear fit and the field-effect mobilities can be derived using the relation described above. According to Fig. 8(d), one can find that the parallel plate model (blue trace) indeed

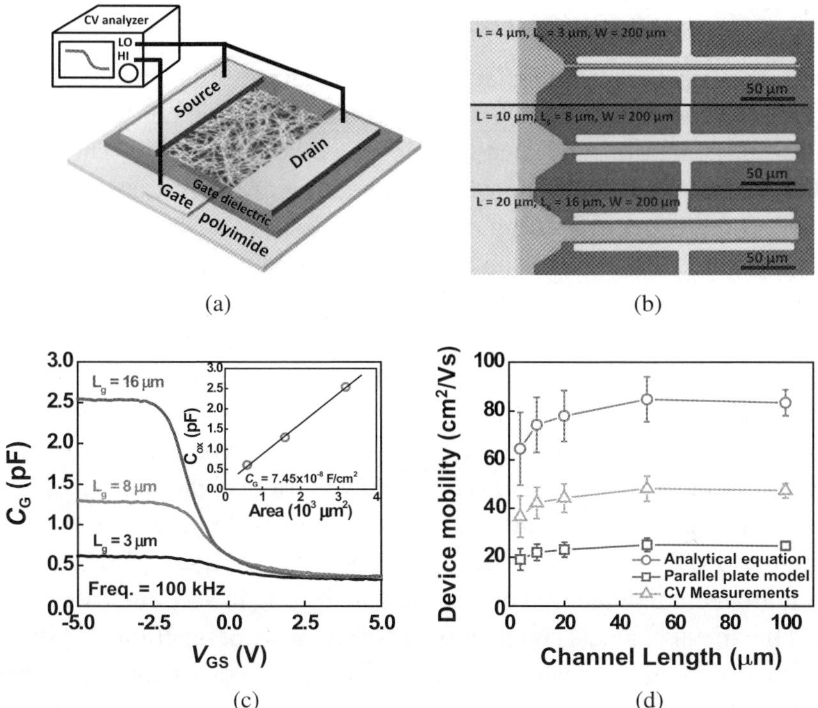

Figure 8. (a) C–V measurement setup used to extract the gate capacitance of nanotube TFTs. (b) Optical microscope images of the underlap-gated devices used for the C–V measurements and C_{ox} extraction. (c) C–V characteristics of the nanotube TFTs measured at 100 kHz. Inset: C_{ox} as a function of the channel area. (d) Field-effect mobility of the semiconducting nanotube TFT as a function of the channel length. Reproduced with permission from Ref. 63. (Copyright 2012 American Chemical Society.)

underestimates the mobility while the analytical model (red trace) overestimates it. The actual device mobility (green trace) for the solution-processed SWNT network is extracted to be ~50 cm²V⁻¹s⁻¹ for 98% semiconducting nanotubes with a density of ~40 tubes/μm². This is close to the device mobility of low-temperature polysilicon (LTPS)[19] and significantly better than amorphous silicon[18] and organic semiconductors.[22–25] The fact that such an impressive mobility can be achieved in devices made using a facile solution-based process makes the nanotube TFTs ideal for a wide range of applications.

2.4. *N-type Nanotube TFTs*

For digital logic applications, it is desirable to have complementary metal-oxide semiconductor (CMOS) operation since it gives rail-to-rail swing, large noise margin, and small static power consumption. The CMOS operation requires a pull-up network consisting of p-type transistors and a pull-down network consisting of n-type transistors. Such circuit configuration ensures that the pull-up and pull-down networks cannot be turned on simultaneously, thereby preventing direct current flow from power supply to ground and reducing the static power consumption. Although CMOS design is widely adopted in silicon-based technologies, it is not straightforward to realize in the SWNT-based platform. Arsenic (As)-fabricated nanotube transistors typically exhibit p-type behavior in ambient environments due to the lower Schottky barrier heights at the nanotube/metal interfaces for most air-stable metal contacts. In order to convert the nanotube FET into n-type, many approaches have been reported in the literature including thermal or electrical annealing in vacuum to desorb O_2 from the metal contacts and thereby lower their work function,[14,15] surface chemical doping,[75–77] electrostatic gating,[78] and metal contact engineering.[79–81] In order to down-select the most practical n-doping method for SWNT TFT integrated circuits, major factors to be considered include air-stability, process reliability, and uniformity. Although promising methods such as using low work function metal contacts (Sc or Y) have been reported for obtaining air-stable n-type SWNT FETs, such methods have not yet been proven to be effective for TFTs with random networks of SWNTs, especially when high device uniformity over large area is required.

Figure 9 illustrates two air-stable methods used to obtain high-performance n-type carbon nanotube TFTs. In Ref. 82, passivation of the nanotube TFTs with high-κ dielectric layer such as HfO_2 deposited by ALD was proposed (Fig. 9(a)) to convert the devices into n-type. The n-type devices obtained using this approach exhibit symmetric electrical performance compared with their p-type counterparts in terms of on-current, on/off ratio, and device mobility (Fig. 9(b)). The mechanism of the carrier type conversion has been attributed to the desorption of moisture and oxygen from the metal contacts during the vacuum baking process in the

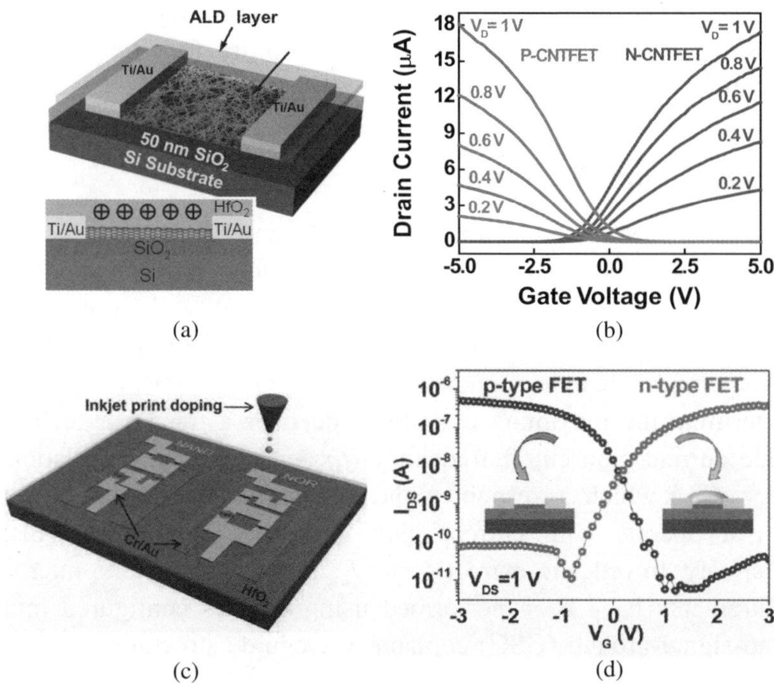

Figure 9. (a) Air-stable n-type semiconducting nanotube TFTs obtained using ALD HfO$_2$ passivation. (b) Symmetric I_{DS}–V_{GS} characteristics have been achieved for the n- and p-type TFTs using the ALD passivation technique. Reproduced with permission from Ref. 82. (Copyright 2011 American Chemical Society.) (c) Air-stable n-type semiconducting nanotube TFTs obtained using surface chemical doping (viologen). (d) I_{DS}–V_{GS} characteristics before and after the viologen doping. Reproduced with permission from Ref. 83. (Copyright 2011 American Chemical Society.)

ALD chamber as well as the deficiency of oxygen atoms in the HfO$_2$ layer deposited at elevated temperatures. This introduces positive fixed charges into the high-κ oxide layer and efficiently shifts the I_{DS}–V_{GS} characteristics of the device from p- to n-type. Chemical doping was also used to demonstrate air-stable nanotube TFTs.[83] In this method, viologen was deposited onto the nanotubes in the channel using inkjet printing (Fig. 9(c)) and TFTs with n-type transfer characteristics were achieved. Using both methods, CMOS inverters with high gain have been demonstrated. It should be noted that despite the success, the reproducibility and uniformity of the n-FETs are still not as good as their p-type counterparts. Therefore, further study is needed to develop n-type TFT platforms with industry standard reproducibility for complex CMOS applications.

2.5. *Frequency Response*

Due to the extremely high carrier mobility[6,7] and ballistic transport,[5] carbon nanotubes hold great potential for applications in RF electronics.[84–88] The frequency response of transistors using networks of high-purity semiconducting nanotubes have been characterized[63,88] for potential use in wireless communication applications. The common figures of merit used to determine the maximum operating speed of a transistor technology include current gain cutoff frequency (f_t) and maximum oscillation frequency (f_{max}), which correspond to the highest frequencies that the transistors can operate with current gain or power gain greater than 1, respectively. In order to extract f_t and f_{max} of the transistors, microwave measurements have been performed using devices configured into the ground-signal-ground (GSG) coplanar waveguide structures (Figs. 10(a) and 10(e)).[63,88] According to the relation $f_t = g_m/(2\pi(C_{gs}+C_{gd}))$, higher g_m and smaller parasitic capacitances lead to better frequency response. For this reason, RF devices are typically designed to have slightly underlapped or self-aligned gate structures in order to minimize the parasitic capacitances.

I_{DS}–V_{GS} and I_{DS}–V_{DS} characteristics of representative RF nanotube transistors with a channel length of ~4 μm on a flexible substrate[63] are presented in Figs. 10(a) and 10(b). The S-parameters, measured using a vector network analyzer (VNA) from 10 MHz to 1 GHz, are plotted in Fig. 10(c). Using the measured S-parameters, the current gain (h_{21}) and maximum available gain (G_{max}) can be deduced as shown in Fig. 10(d). The results are de-embedded using on-chip open and short structures in order to remove the parasitic effects from the probing pads. The f_t and f_{max}, which are defined as the frequency where the h_{21} and G_{max} become 0 dB, are found to be ~170 and 118 MHz for transistors with 4 μm channel length (Fig. 10(d)). This performance is impressive considering the relatively long channel length used and the fact that such devices are fabricated on a mechanically flexible substrate. Such performance is also significantly better than organic semiconductor TFTs whose operating

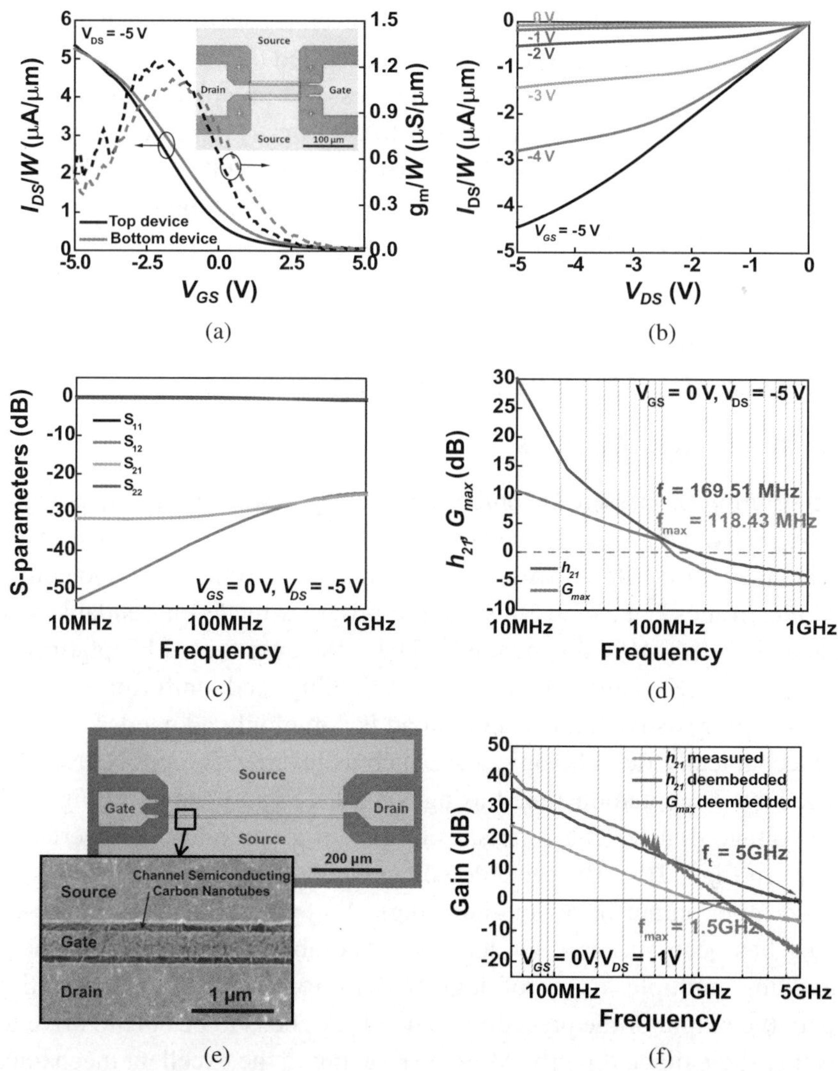

Figure 10. (a) Transfer characteristics of a pair of nanotube RF transistors with $L = 4$ μm, $L_g = 3$ μm, and $W = 100$ μm, measured at $V_{DS} = -5$ V. Inset: Optical microscopy image of the corresponding device. (b) Output characteristics of the same device shown in panel (a). (c) Measured S-parameters for the flexible nanotube RF transistor from 10 MHz to 1 GHz. (d) De-embedded h_{21} and G_{max} of device showing a f_t of 170 MHz and a f_{max} of 118 MHz. Reproduced with permission from Ref. 63. (Copyright 2012 American Chemical Society.) (e) Optical micrograph and SEM image (Inset) of a submicron nanotube RF transistor with $L = 500$ nm. (f) De-embedded h_{21} and G_{max} obtained from the measured S-parameters of the submicron nanotube RF transistor. The f_t is extracted to be 5 GHz at $V_{DS} = -1$ V. Reproduced with permission from Ref. 88. (Copyright 2011 American Chemical Society.)

speed typically lie in the range of kHz or lower. By scaling down the channel length, even faster transistors can be obtained due to the improvement of g_m. RF transistors using solution-processed semiconducting nanotube networks with channel lengths down to 500 nm (Fig. 10(e)) have been obtained on Si/SiO$_2$ substrates.[88] Impressive f_t of ~5 GHz was achieved at a low operating voltage (V_{DD}) of 1 V. This result highlights the practical use of SWNT TFTs for wireless applications.

3. Applications of Solution-Processed Carbon Nanotube TFTs in Flexible Electronics

3.1. *Integrated Circuits*

The solution-processed nanotube TFTs offer a wide range of potential applications such as bendable integrated circuits,[51,53,60,63] active-matrix backplanes for tactile sensors[62,73] and display electronics,[73,89] and conformal electronics.[62] First, we review the progress made on bendable integrated circuits made using nanotube TFTs. Because the *p*-FET platform is currently more mature with higher reliability and uniformity, PMOS design with resistive load or diode-load is commonly adopted.

Mechanically flexible logic gates such as inverter, 2-input NAND, and NOR have been demonstrated using SWNT TFTs as shown in Fig. 11.[63] The voltage transfer characteristics (VTC) of a diode-loaded inverter are shown in Fig. 11(b). Respectable voltage gain of ~30 at V_{DD} = 5 V, symmetric input/output behavior (i.e., single V_{DD} operation), and rail-to-rail swing has been successfully achieved. The above characteristics enable cascading multiple stages of logic blocks for larger scale integration, where the output of the preceding logic block needs to be able to drive the ensuing logic block directly. Moreover, owing to the excellent mechanical flexibility of the carbon nanotube networks, the fabricated inverter circuit is extremely bendable. The measured inverter VTC shows minimal performance change under various curvature radii (down to 1.27 mm) (Fig. 11(c)) and repeated bending tests (2,000 cycles) (Fig. 11(d)). Similarly, 2-input NAND (Fig. 11(e)) and NOR (Fig. 11(f)) logic gates have also been demonstrated with diode load. Both circuits are operated with a V_{DD} of 5 V and input voltages of 5 V and 0 V are treated as logic "1" and "0", respectively. For the NAND gate, the output is "1" when either one of the two

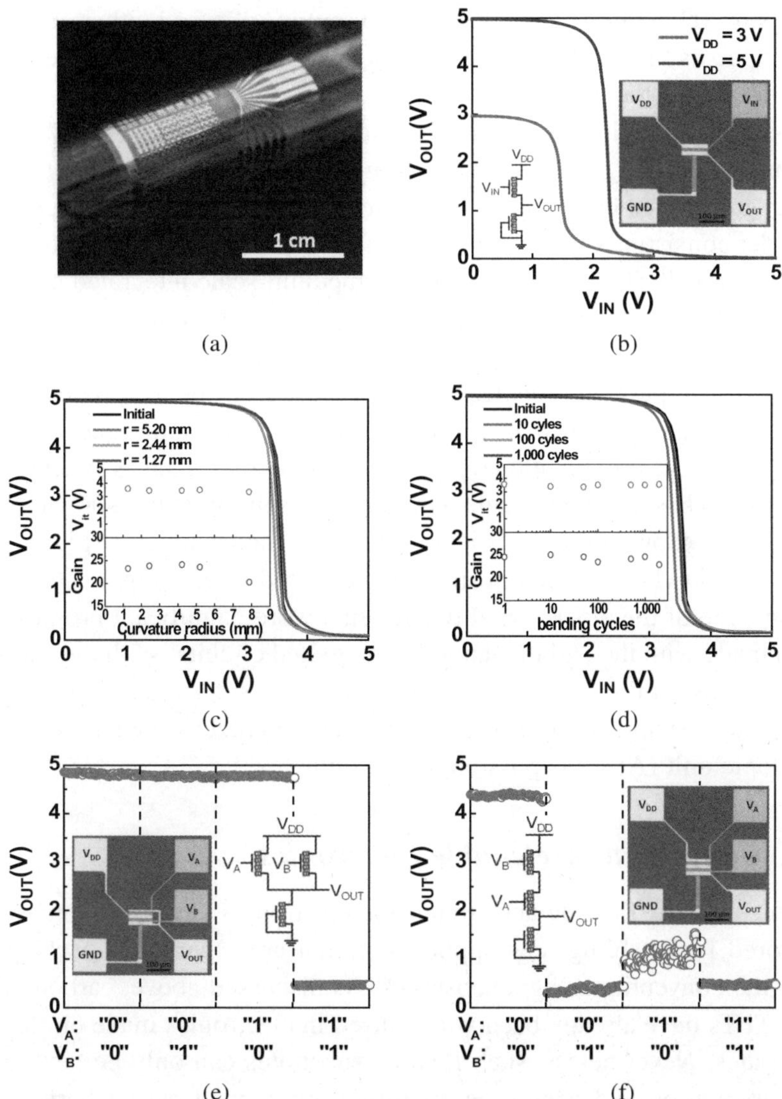

Figure 11. (a) Photograph of a flexible integrated circuit made with semiconducting nanotube TFTs being wrapped on a test tube with a curvature radius of 5 mm. (b) VTC of a flexible nanotube inverter measured with V_{DD} of 3 or 5 V. (c) Inverter VTC measured at various curvature radii. Inset: Inverter threshold voltage and gain as a function of curvature radius, showing minimal performance change even when bent down to 1.27 mm radius. (d) The inverter VTC is measured after various numbers of bending cycles, indicating good reliability after 2,000 cycles. (e, f) Output characteristics of the diode-loaded 2-input NAND (e) and NOR (f) logic gates. Reproduced with permission from Ref. 63. (Copyright 2012 American Chemical Society.)

inputs is "0", while for the NOR, the output is "0" when either one of the two inputs is "1". The measured output characteristics confirm the correct functioning of the circuits.

With the combination of the above-described basic logic blocks, more sophisticated integrated circuits, requiring cascading multiple stages of logic gates, can be readily constructed.[51,53] As an example, a 4-to-16 decoder consisting of 88 carbon nanotube TFTs has been demonstrated by Rogers *et al.*[51] This represents the first medium-scale integrated nanotube circuits. The photograph and optical micrograph of the nanotube decoder are shown in Fig. 12(a). The 4-bit decoder works in a way that only one of the 16 outputs are enabled at a time based on the 16 possible input combinations, as shown in the transient response of the decoder (Fig. 12(b)). Besides combinational logic, sequential logic such as synchronous D-flip-flop has also been made whose optical micrograph is shown in Fig. 12(c).[53] As shown in the transient response of the D-flip-flop (Fig. 12(d)), the input (DATA) can be transmitted and stored at the output (Q) at each rising edge of the clock signal (CLK). In a word, significant progress has been made with the carbon nanotube integrated circuits. With the successful demonstration of both combinational and sequential logic blocks, complete system-level integration for functional blocks such as arithmetic and logic unit (ALU) is possible in the future.

3.2. *Conformal/Stretchable Electronics*

In recent years, flexible and stretchable electronics have been intensively explored for enabling new applications that are otherwise unachievable with the conventional rigid substrates. As discussed above, carbon nanotube TFTs have already been widely used in electronics made on flexible substrates. Nevertheless, such flexible substrates can only be bent along one orientation and cannot be conformally wrapped over spherical surfaces. This limits the applications in wearable electronics where conformal coverage is required.

By cleverly engineering the substrate, a mechanically deformable and stretchable active-matrix backplane containing arrays of SWNT TFTs have been demonstrated.[62] The use of polyimide substrate and the basic fabrication process is the same as the previously-described flexible

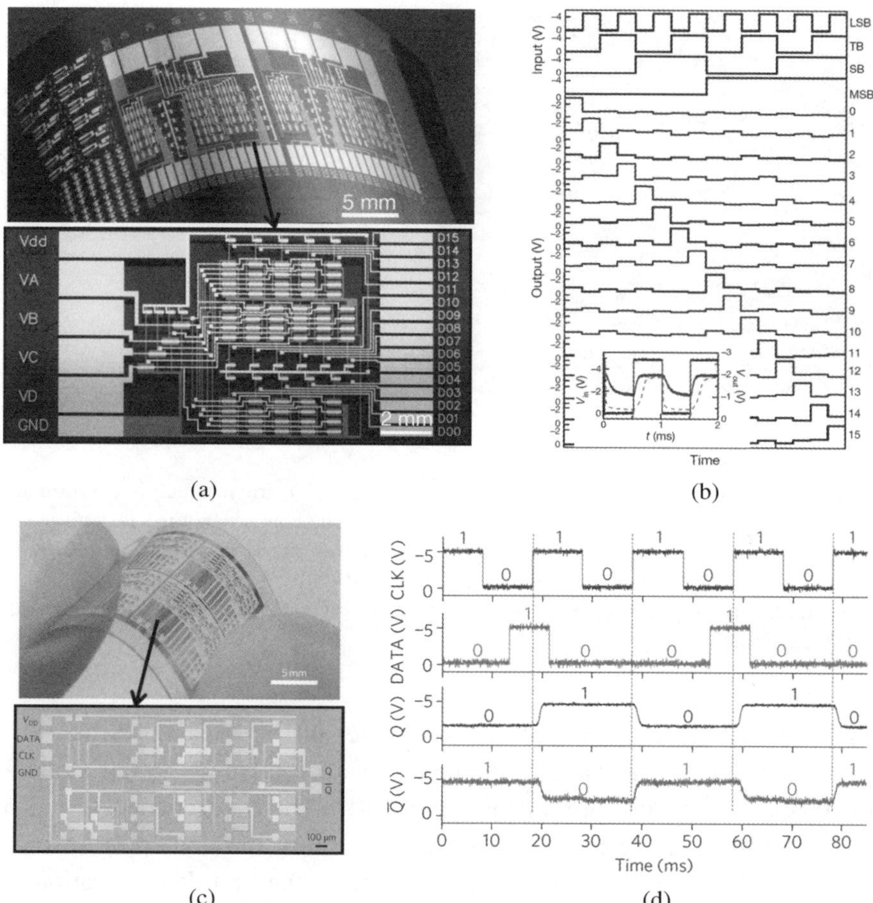

(a)   (b)

(c)   (d)

Figure 12.   (a) Photograph and optical micrograph of a 4-to-16 decoder made with nanotube TFTs. (b) Transient response of the decoder. Reproduced with permission from Ref. 51. (Copyright 2008 Nature Publishing Group.) (c) Photograph and optical micrograph of a D-flip-flop made with nanotube TFTs. (d) Transient response of the D-flip-flop. Reproduced with permission from Ref. 53. (Copyright 2011 Nature Publishing Group.)

nanotube TFTs. However, by laser cutting a honeycomb mesh structure into the substrate, the otherwise non-stretchable polyimide substrate is made stretchable. As shown in Fig. 13(a), the enabled TFT backplane is used to conformally wrap around the surface of a baseball. Nanotube

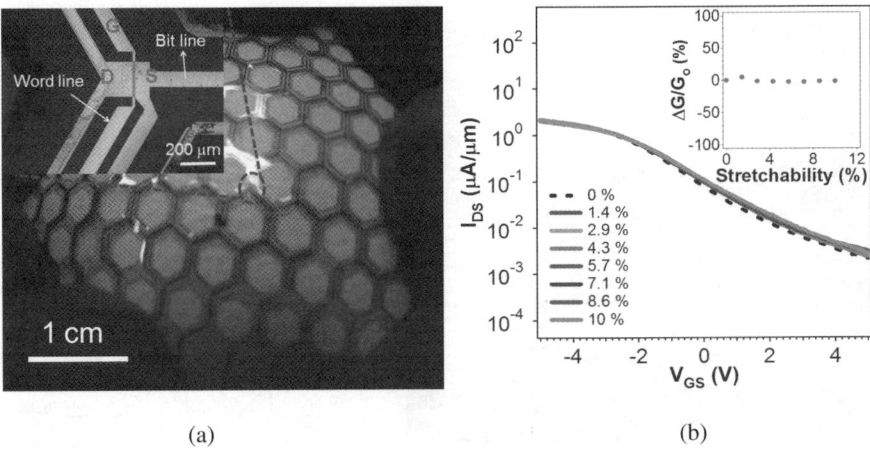

(a)   (b)

Figure 13.   (a) Photograph of a stretchable nanotube TFT array, which is conformally wrapped around a baseball. Inset: Optical micrograph showing a nanotube TFT placed at the corner of a hexagonal opening. (b) I_{DS}–V_{GS} characteristics of the device measured under various stretching conditions. Inset: Normalized on-state conductance as a function of stretchability. Reproduced with permission from Ref. 62. (Copyright 2011 American Chemical Society).

TFTs are strategically placed at each corner of the hexagons where the strain is minimal according to the mechanical simulations.[62] The stretching results of the TFTs are shown in Fig. 13(b). The experiments show that the I_{DS}–V_{GS} characteristics of the devices show minimal performance change even when stretched up to 10%. This is the first demonstration of a novel type of carbon-nanotube-based backplane for stretchable/conformal electronics. Given the minimal thickness of the active channel (i.e., nanotube network thickness), the high mechanical flexibility and stretchability of nanotube TFTs is expected. In future, the use of other stretchable support substrates (e.g., polydimethylsiloxane PDMS) along with stretchable passive components (e.g., electrodes and dielectrics) needs to be explored.

3.3. *Display Electronics*

The pixels in an active-matrix display are controlled by transistors, which act as switches and are typically made from amorphous silicon[18] or polysilicon.[19] While amorphous Si suffers from performance limitations, poly-silicon deposition requires high temperature processing which is often incompatible with plastic substrates. As an alternative channel semiconducting material for TFT applications, solution processed carbon nanotubes offer high mobility, are optically transparent and mechanically flexible.

With the capability of large-area fabrication of high-quality nanotube TFTs with high uniformity, high on/off ratio and excellent mobility,[59,62,63] a monolithically integrated active-matrix organic light-emitting diode (AMOLED) display has been successfully demonstrated on glass substrates (Fig. 14).[89] As one of the most promising candidates for the next generation display technologies, OLEDs hold great advantages compared to liquid crystal displays (LCD) due to their high efficiency, superior color purity, low power consumption, large view angle, low temperature processing, and excellent flexibility.[90–94] The schematic diagram and optical micrograph of one pixel of the nanotube-based AMOLED display are shown in Figs. 14(a) and 14(b), respectively. Each pixel consists of two semiconducting nanotube TFTs, one capacitor, and one OLED. The two-transistors, one-capacitor (2T1C) design (Fig. 14(c) inset) is necessary for output retention in line-by-line scan for displaying dynamic videos. The OLED fabrication involves the sputtering of transparent indium tin oxide (ITO) as the anode, evaporation of hole transporting layer 4-4′-bis[N-(1-naphthyl)-N-phenyl-amino]biphenyl (NPD), emission layer tris(8-hydroxyquinoline) aluminum (Alq3), and LiF and aluminum cathode. As shown in Figs. 14(c) and 14(d), the current flow through the OLED (I_{OLED}) (red line) and light output intensity (green line) can be modulated by more than 10^5 times by changing the voltages applied on the data line (V_{DATA}). The modulation of light output is visually seen in the photographs shown in Fig. 14(d). Figure 14(e) shows a glass substrate containing 7 AMOLED chips. Each chip contains 20×25 pixels driven by a total of 1,000 nanotube TFTs. When all the pixels are turned on ($V_{DATA} = -5$ V, $V_{SCAN} = -5$ V, and $V_{DD} = 8$ V), a yield of ~70 % was achieved (Fig. 14(f)), which is respectable for a proof-of-concept demonstration in the present laboratory-scale experiments. The use of 1,000 nanotube TFTs also represent one of the highest degree of integration to date for carbon-nanotube-based electronics.

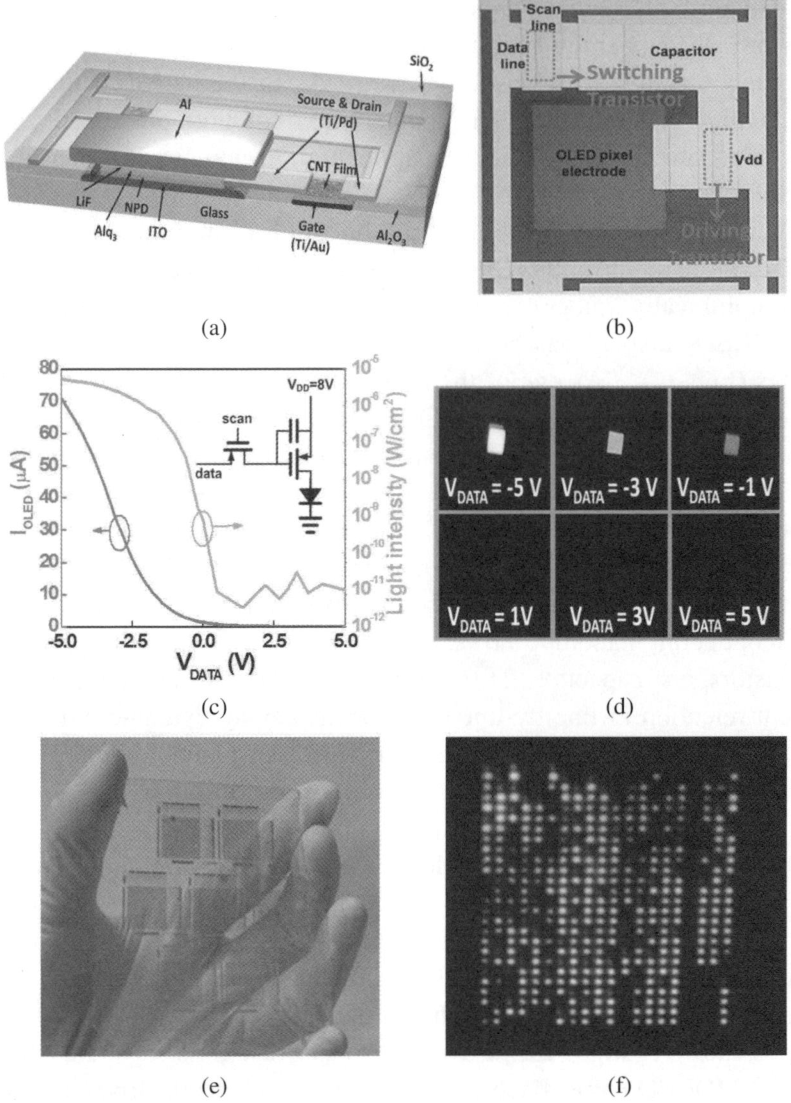

Figure 14.   (a) Schematic diagram showing a pixel of a monolithically integrated AMOLED. Each pixel consists of two semiconducting nanotube TFTs, one capacitor, and one OLED. (b) Optical micrograph of an AMOLED pixel. (c) OLED current (I_{OLED}) (red line) and light intensity (green line) as a function of the voltage on the data line (V_{DATA}). (d) Photographs showing OLED light output under various V_{DATA} voltages. (e) Photograph of seven AMOLED chips fabricated on the glass substrate. Each chip contains 20×25 pixels with a total of 1,000 nanotube TFTs. (f) Photograph of the AMOLED display with all pixels turned on. Reproduced with permission from Ref. 89. (Copyright 2011 American Chemical Society.)

Similarly, the AMOLED display can also be fabricated on a flexible substrate using the same process flow depicted in Fig. 5(a).[73] The photograph of a fully-fabricated flexible AMOLED display driven by carbon nanotube TFTs is shown in Fig. 15(a). Note that the light is emitted through the semi-transparent polyimide substrate, as the opposite side is covered by the aluminum cathodes of the OLEDs. Figure 15(b) shows a

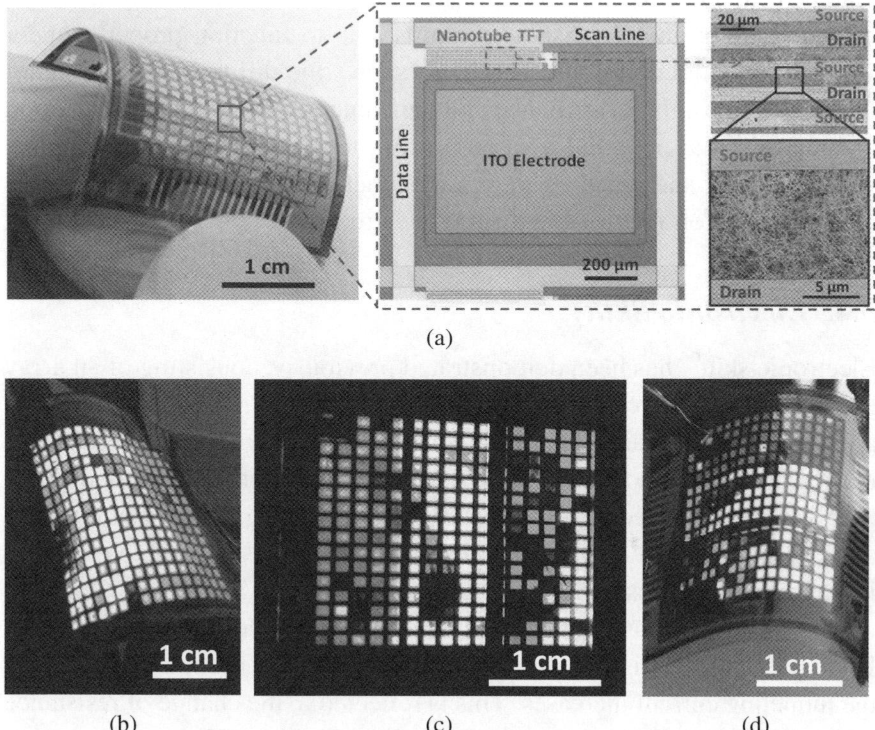

Figure 15.   (a) Left: Optical photograph of a fully fabricated flexible AMOLED display containing 16 × 16 pixels driven by carbon nanotube TFTs with a size of ~3 × 3.5 cm^2. Right: Optical micrograph of one pixel before the OLED deposition step. The drain of the TFT is connected to an ITO electrode, which serves as the anode for the OLED. Inset: SEM image showing the channel of the carbon nanotube TFT. (b) Photo of a single-color (green) AMOLED display being fully turned on and bent. Voltages of −5 V and 10V are applied to all of the scan and data lines, respectively. The pixel yield, as defined by the percentage of OLEDs in the matrix that emit light, is higher than 97%. (c) Photograph of a full-color (red, green, blue) AMOLED display with all pixels being turned on. (d) Photo of the same full-color display shown in (c) being bent. Reproduced with permission from Ref. 73. (Copyright 2013 Nature Publishing Group.)

single-green-color flexible AMOLED display with all of the 16 × 16 pixels being turned on, with a pixel yield of more than 97%. A full-color flexible AMOLED display (that is red, green, blue OLED pixels monolithically integrated on the same substrate) is also demonstrated through a multi-step evaporation of the various emissive layers. Figures 15(c) and 15(d) show the full-color display being fully turned on in the relaxed and bent states, respectively, with a pixel yield of ~85%. Figure 15(d) also confirms that the evaporated organic and metallic thin-films in the display are durable on bending and the display remains able to function properly under various bending conditions. The work shows the utility of carbon nanotube TFTs for high-performance and bendable displays, indicating carbon nanotubes' great potential as a competing technology alongside organic semiconductor and metal-oxide semiconductor as a low-cost and scalable backplane option for flexible display electronics.

3.4. *Electronic Skin*

Electronic skin[95] has been demonstrated previously, consisting of an array of pressure sensors capable of spatial and temporal mapping of pressure applied on the surface. This is done by laminating a pressure-sensitive rubber (PSR) onto a flexible backplane containing arrays of carbon nanotube TFTs.[62] The electronic skin is useful for sensing the pressure stimuli and could find wide range of applications in robotics, wearable electronics, and medical prostheses.

The PSR used has conducting carbon nanoparticles embedded inside. Upon pressing, the distance between the nanoparticles becomes smaller and the tunneling current increases. This is reflected in the change of resistance under various applied pressure.[95] By connecting the PSR in series with a carbon nanotube TFT, one pixel of the active-matrix is formed (Fig. 16(a)).[62] By changing the applied pressure on the PSR, the output conductance of the pixel undergoes significant change as shown in Fig. 16(b). The detection limit of such a configuration is around 1 kPa. As a practical example, the e-skin sensor is used to perform spatial and temporal mapping of the pressure applied on the surface as shown in Fig. 16(c). In the experiment, an L-shaped object is placed on the sensor, and upon electrically scanning through the pixels in the array, a two-dimensional (2D) pressure profile

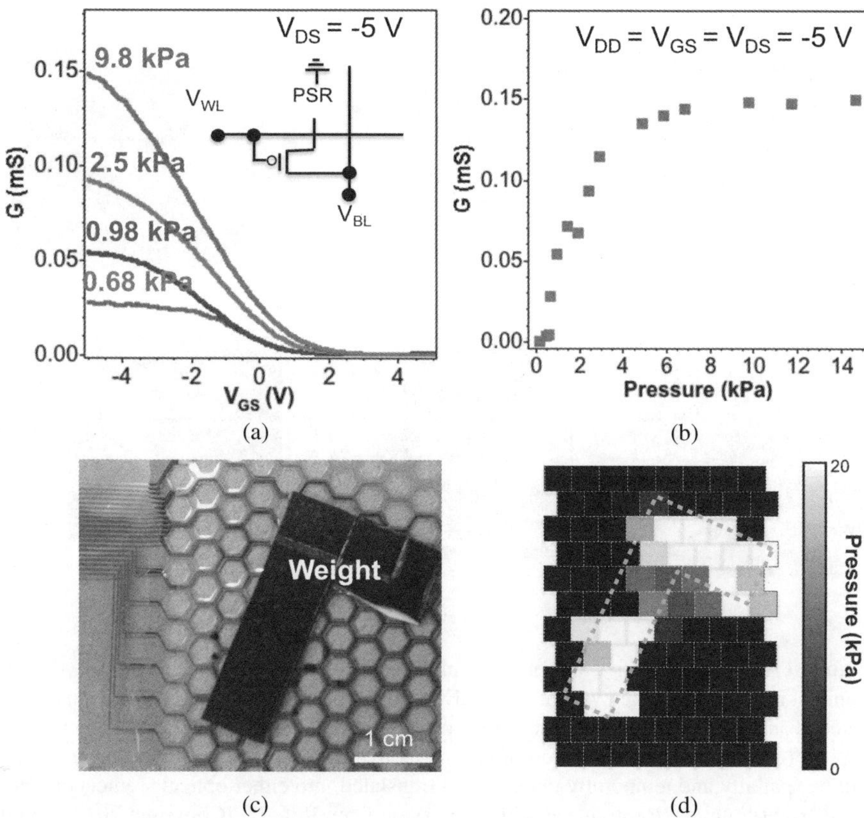

Figure 16. (a) Pressure response of one pixel in the electronic-skin, consisting of a nanotube TFTs active-matrix backplane and a pressure sensitive rubber. The I_{DS}–V_{GS} characteristics of the pixel are measured under various applied pressure. (b) Output conductance at $V_{DD} = V_{GS} = V_{DS} = -5$ V as a function of the applied normal pressure. (c) Photograph showing an electrical skin sensor with an L-shaped object placed on top. (d) The 2D pressure mapping obtained from the L-shaped object in (c). Reproduced with permission from Ref. 62. (Copyright 2011 American Chemical Society.)

is obtained (Fig. 16(d)), which matches well with the shape of the object used.

The Holy Grail for future flexible electronics is to integrate different electronic, optoelectronic components and sensors on the same substrate into a light-weight flexible system with sophisticated sensing and user-interactive

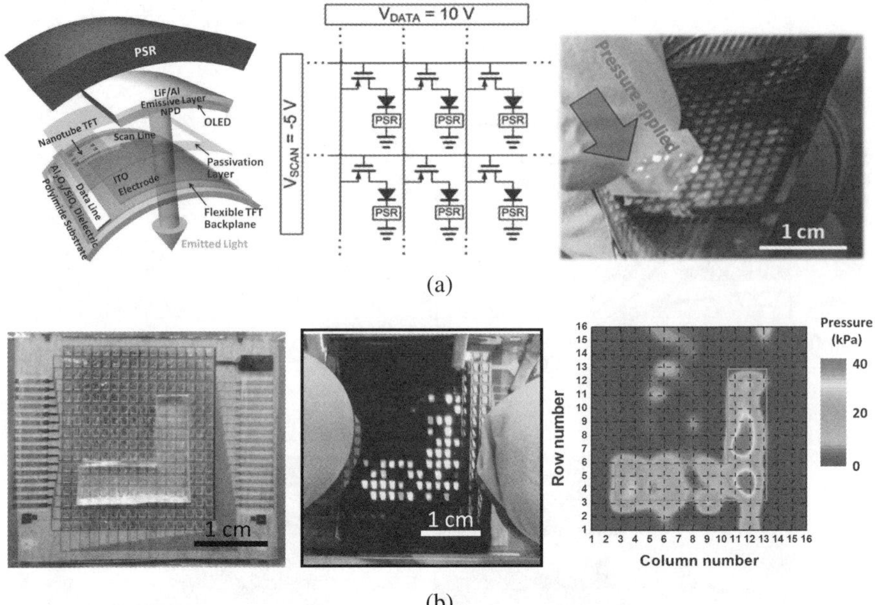

(a)

(b)

Figure 17. (a) Schematics and photograph of the fully fabricated user-interactive electronic skin system. In the system, the OLEDs are turned on locally where the surface is touched and the emitted light intensity corresponds to the magnitude of the pressure applied (right panel). (b) Operation of the interactive electronic skin. The pressure profile can be spatially and temporally mapped and translated into either optical (center) or electrical (right) outputs. Reproduced with permission from Ref. 73. (Copyright 2013 Nature Publishing Group.)

functionalities. This was previously very difficult recognizing the great challenges of integrating a large number of different material systems and device structures in a scalable manner. With the solution-processed carbon nanotube TFT platform, by integrating the concepts of the flexible AMOLED display and the electronic skin, a flexible user-interactive electronic skin that can detect, spatially map the pressure stimuli, and provide instantaneous response through the monolithically integrated OLED display has been demonstrated.[73] As shown in Fig. 17(a), an array of 16×16 pixels were integrated on a polyimide substrate, each pixel consists of a pressure sensor and a TFT that controls the current driving a colored OLED. In the resulting system, the OLEDs are turned on locally where the surface is touched and the emitted light intensity quantifies the magnitude of the applied pressure, allowing the sensed pressure profile to be instantaneously visible without the need of sophisticated data acquisition circuits (Fig. 17(a) right panel and Fig. 17(b)). Immediate applications may be foreseen in interactive control devices, for example in autonomous, reusable touch panels that can be adapted to any curved surface and allow the user to directly see the intensity of their touch on it. The above progress on pressure sensing and visualization is only one example of the possible integrated systems that can be enabled by the fabrication strategy. To take this one step further, it is easy to envision analogous systems for sensing and direct visualization of other environmental stimuli, such as temperature, light, humidity, etc. This can be achieved using a similar platform by simply modifying the sensing element of each pixel and such a system would enable more sophisticated human-surface interfacing, leading to systems with practical functions that are useful for our daily life in the near future.

3.5. *Flexible Visible Light and X-ray Imager*

The electronic skin concept and the use of carbon nanotube TFT backplane can be extended to other applications by simply adding or replacing the sensing components. As an example, by replacing the pressure sensors with light sensors (i.e., photodetectors), a large area flexible imager has been demonstrated for both visible light and X-ray imaging applications.[96] In this system, organic photodetectors (OPDs) are solution processed and monolithically integrated on top of an active-matrix backplane consisting of 18×18 pixels of carbon nanotube TFTs (Fig. 18(a)). The OPDs are fabricated by first thermally evaporating ~9 nm thick MoO_3 hole transportation layer onto the ITO electrode, followed by spin-casting of regioregular poly(3-hexylthiophene) (P3HT) and [6, 6]-phenyl C61-butyric acid methyl ester (PCBM) (80 mg/ml in chlorobenzene, P3HT:PCBM=50:50 weight percent) at 1,000 rpm for 30 second. Al is then evaporated to form the cathode electrode, followed by a thermal anneal at 150°C for 10 minutes in order to form bicontinuous heterojunction network in the organic semiconductor film (Fig. 18(b)). As shown in Fig. 18(d), the fabricated P3HT/PCBM bulk-heterojunction photodetector could detect visible light intensity variation effectively.

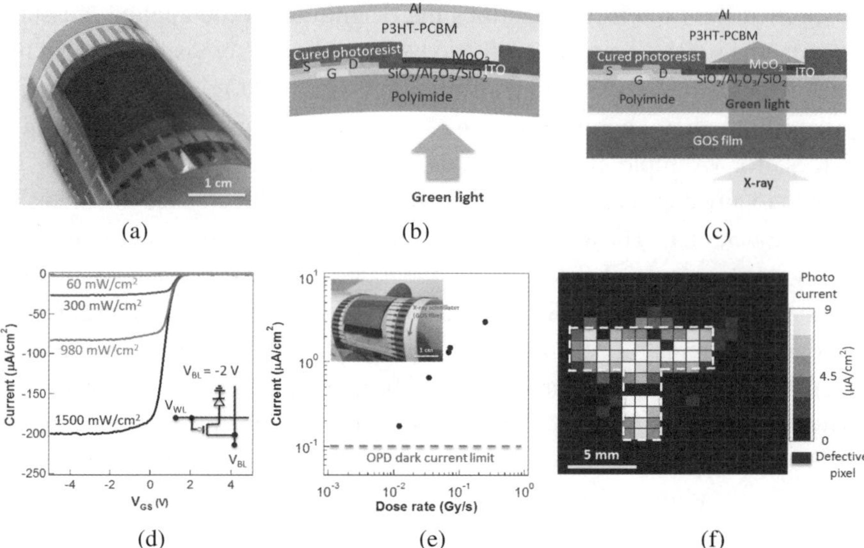

Figure 18. (a) Optical photograph of a flexible imager driven by carbon nanotube TFTs. (b) Cross-sectional schematic showing one pixel of the imager for visible light detection. (c) Cross-sectional schematic showing the imager used for X-ray detection. An X-ray scintillator (GOS film) is used to convert the X-ray irradiation to green light, which is then detected by the imager. (d) Light response of one pixel under various light intensities (λ = 535 nm). (e) Measured electrical current (at a reverse bias of 2 V) as a function of X-ray dose rate. Inset: Optical photograph of the flexible imager with laminated X-ray scintillator. (f) Visible light or X-ray image profile can be obtained by measuring the photocurrent of the 18 × 18 pixels.

Because the absorption spectrum of P3HT:PCBM photodetector covers only the green band of the visible light spectrum, a Gd_2O_2S:Tb (GOS) X-ray scintillator film can be used to enable X-ray detection and imaging (Fig. 18(e)). In this system, the scintillator film converts incident X-ray into visible photons with an energy matching the peak absorption wavelength of the OPDs (Fig. 18(c)). The enabled devices exhibit high performance in terms of imaging sensitivity, response time, and uniformity. By measuring the photocurrent in all the pixels, both visible light and X-ray are successfully imaged (Fig. 18(f)). The work presents a platform for flexible sensor networks for large-area imaging applications.

4. Low-Cost Process Scheme for Flexible Carbon Nanotube Electronics: Printing

Another important feature of solution-processed nanotube TFTs is that they could potentially be fully printed. Printed electronics offer a cost-effective route to achieve mass production of large-area flexible electronic circuits at extremely low cost and high throughput without the need for special photolithography or vacuum-based processes. Numerous research efforts have been devoted in the past to develop a low-cost printed integrated circuit platform using nanotube TFTs. Conventionally, the lack of pure metallic or semiconducting carbon nanotube inks has hindered the fabrication of high performance printed nanotube devices. With the use of commercially available high purity semiconducting nanotube inks, high mobility and high on/off ratio carbon nanotube TFTs have been demonstrated using aerosol-jet printing process.[64] In this process, the gate, source and drain electrodes are first defined using conventional photolithography and lift-off processes. The channel semiconducting nanotube network, ion-gel dielectric, and PEDOT:PSS gate are subsequently defined using aerosol-jet printing process (Fig. 19(a)). Using this method, high-performance transistors, inverters, NAND gates, and ring oscillators have been demonstrated (Fig. 19(b)). Fully-printed transistors and integrated single-pixel OLED control circuits have also been demonstrated by using inks containing silver nanoparticles for the printed electrodes and interconnections and carbon nanotubes as the channel (Fig. 19(b)).[65] The printed OLED driving circuit composes of two top-gated fully printed transistors, which can be used to control the OLED brightness as shown in the right panel of Fig. 19(b). It should be noted that ionic gel has been used in both publications as the gate dielectric for simplified fabrication process. However, due to the unique operating principle of the ionic gel, the devices exhibit ambipolar behavior and the operating speed of the transistors are expected to be slow, limited by the movement of the ions. These problems require further improvements in the future.

For larger scale applications and mass production, gravure (roll-to-roll) printing can be used.[66–68,97] This represents a more cost-effective approach. In this method, the gate, dielectric, source/drain, and carbon nanotube layers are sequentially printed onto a roll of polyethylene terephthalate (PET) film

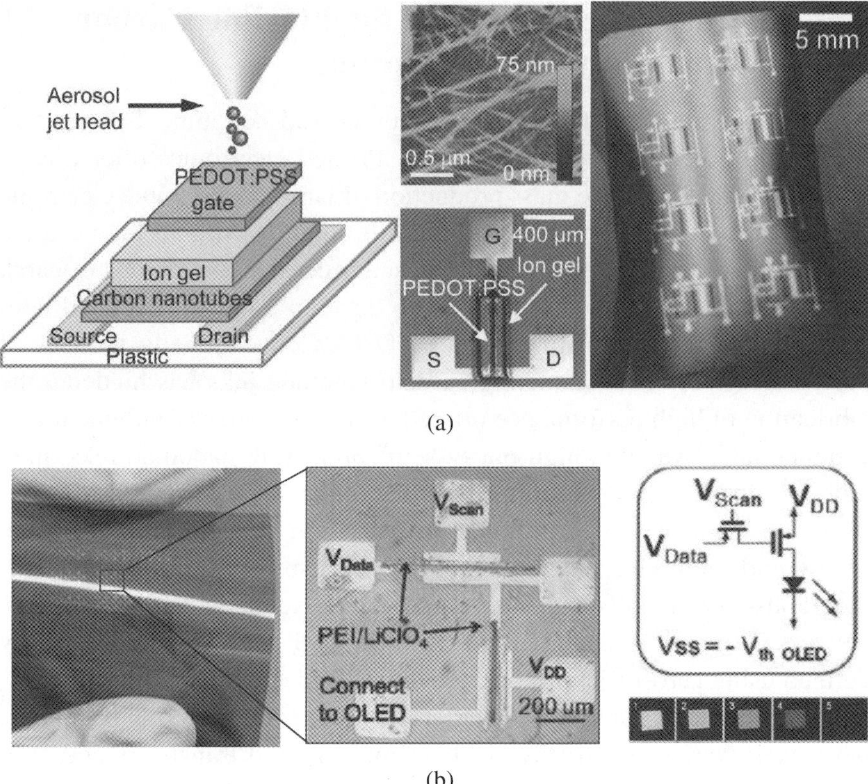

Figure 19.   (a) Nanotube TFTs fabricated using aerosol jet printing where the channel nanotube network, ion-gel dielectric, and PEDOT:PSS gate are defined using a printing process. Left: schematic of the printing process. Center: AFM image and optical micrograph of the printed TFT. Right: photograph of an array of printed NAND logic gates. Reproduced with permission from Ref. 64. (Copyright 2010 American Chemical Society.) (b) Fully printed carbon nanotube TFT (left) and single pixel OLED driving circuitry (right). Reproduced with permission from Ref. 65. (Copyright 2011 American Chemical Society.)

(Fig. 20(a)). Ink formulation and viscosity control are critical in this case in order to achieve uniform printing at large scales. Fully-printed carbon nanotube TFT active matrix backplane (Fig. 20(b)),[96] logic gates, half adder,[68] D-flip-flop,[67] and one-bit RF identification (RFID) tags[66] have been successfully produced. As shown in Fig. 20(d), the devices exhibit excellent performance for a fully printed process, with mobility and on/off current ratio of

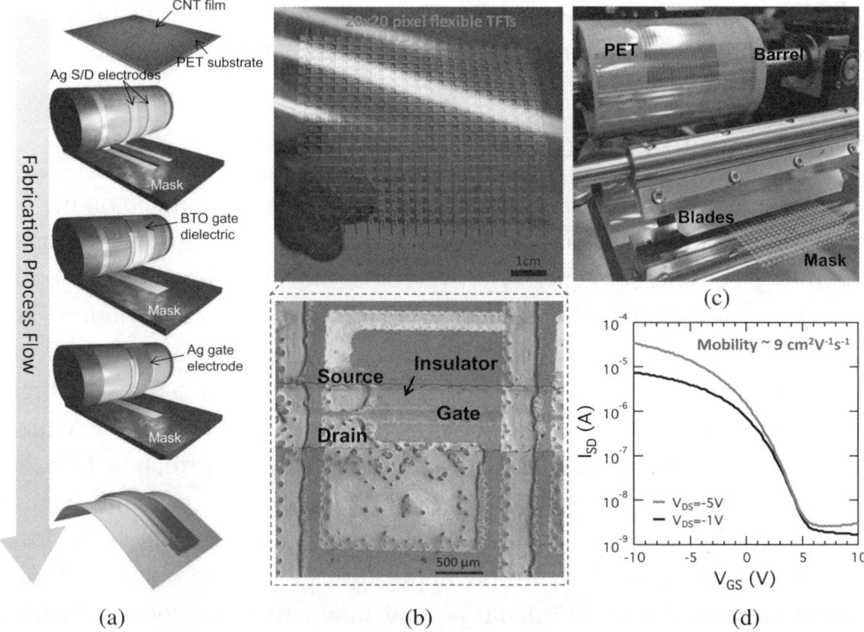

Figure 20. (a) Schematic diagram showing the fabrication process of the fully printed nanotube TFTs on mechanically flexible PET substrates using gravure printing process. (b) Optical photograph of a fully printed carbon nanotube TFT active-matrix backplane with 20 × 20 pixels on a PET substrate. Inset: Optical micrograph of a single TFT showing the printed gate/source/drain electrodes and gate dielectric. (c) Image of the inverse gravure printer used for this work. (d) I_{DS}–V_{GS} characteristics of a representative printed nanotube TFT with a field-effect mobility of ~9 cm²V⁻¹s⁻¹. Reproduced with permission from Ref. 96. (Copyright 2013 American Chemical Society.)

up to ~9 cm²V⁻¹s⁻¹ and 10^5, respectively. Extreme bendability is also observed, without measurable change in the electrical performance down to a small radius of curvature of 1 mm.[96]

Despite the tremendous success in carbon nanotube printed electronics, it should be pointed out that the printing approach is useful for making cost-effective large scale nanotube circuits requiring only moderate performance because the resolution that is currently achieved using printing processes is limited (typically tens of μm or more). For applications where high performance is required and cost is not a major concern, conventional photolithography and metalization is still preferred, primarily adopting the conventional processes utilized for LCD manufacturing, but with a thin layer of polymer cured on the glass handling substrate.

5. Conclusions and Outlook

As a conclusion, this chapter reviews the recent progress made on the use of carbon nanotubes TFTs for scalable, practical, and high performance flexible electronic systems. To overcome the challenges of nanotube assembly, minimizing the device-to-device performance variation, and making the fabrication process scalable and compatible with industry standards, methods have been developed to deposit thin-films of high-purity semiconducting carbon nanotubes at large-scales. This excellent material platform has enabled fabrication of high-performance TFTs on mechanically flexible substrates with superb carrier mobilities of $\sim$50 cm^{2}V^{-1}s^{-1} and high I_{on}/I_{off} >10^4. On the basis of the above achievements, various kinds of electronic applications including conformal integrated circuits, AMOLED displays, and interactive electronic skin have been explored. The successful demonstrations show carbon nanotubes' immense promise for high performance thin-film electronics.

For the future research in this field, predictable and highly uniform device performance is indispensable for more sophisticated nanotube integrated circuits. For instance, precise control of the transistor threshold voltage and subthreshold characteristics is required. As a result, it is important to control not only the electronic type (metal *versus* semiconducting) but also ultimately the chirality of the carbon nanotubes. Since chirality-specific nanotube separation has already been achieved,[71,72] a higher degree of control in the electrical characteristics is expected from TFTs made using such carbon nanotubes. In addition, further lowering of the I_{off}, without sacrificing I_{on} is desirable by increasing the semiconductor purity of the nanotube solutions as well as optimizing the deposition process to minimize the bundling of the assembled nanotubes. From application point of view, the next step is moving toward more sophisticated system-level integrated circuits. For the CMOS operation, platforms that are capable of providing n-type nanotube TFTs with reproducibility and uniformity comparable to their p-type counterparts need to be developed. With the superior electrical performance and the promising progress on system-level integration and development of fully printed fabrication process, one can envision that the solution-processed carbon nanotube TFTs could find tremendous applications in future low-cost, high-performance flexible user-interactive electronics and wearable electronics.

Acknowledgments

C.W. acknowledges support from Michigan State University. Some sections of this book chapter are adapted from a review paper by the author previously published in Chemical Society Reviews,[98] reproduced by permission of The Royal Society of Chemistry (RSC).

References

1. M. S. Dresselhaus, G. Dresselhaus and Ph. Avouris, (Eds.) *Carbon Nanotubes — Synthesis, Structure, Properties, and Applications*, (Springer, 2001).
2. A. Javey and J. Kong, (Eds.) *Carbon Nanotube Electronics*, (Springer, 2009).
3. J. Wildoer, L. Venema, A. Rinzler, R. Smalley and C. Dekker, *Nature*, **391**, 59–62 (1998).
4. T. Odom, J. Huang, P. Kim and C. Lieber, *Nature*, **391**, 62–64 (1998).
5. A. Javey, J. Guo, Q. Wang, M. Lundstrom and H. Dai, *Nature*, **424**, 654–657 (2003).
6. T. Durkop, S. A. Getty, E. Cobas and M. S. Fuhrer, *Nano Lett.*, **4**, 35–39 (2004).
7. X. Zhou, J. Y. Park, S. Huang, J. Liu and P. L. McEuen, *Phys. Rev. Lett.*, **95**, 146805 (2005).
8. S. Tans, A. Verschueren and C. Dekker, *Nature*, **393**, 49–52 (1998).
9. R. Martel, T. Schmidt, H. R. Shea, T. Hertel and Ph. Avouris, *Appl. Phys. Lett.*, **73**, 2447–2449 (1998).
10. A. Javey, J. Guo, D. Farmer, Q. Wang, D. Wang, R. Gordon, M. Lundstrom and H. Dai, *Nano Lett.*, **4**, 447–450 (2004).
11. A. Javey, J. Guo, D. Farmer, Q. Wang, E. Yenilmez, R. Gordon, M. Lundstrom and H. Dai, *Nano Lett.*, **4**, 1319–1322 (2004).

12. A. D. Franklin, M. Luisier, S.-J. Han, G. Tulevski, C. M. Breslin, L. Gignac, M. S. Lundstrom and W. Haensch, *Nano Lett.*, **12**, 758–762 (2012).

13. A. Bachtold, P. Hadley, T. Nakanishi and C. Dekker, *Science*, **294**, 1317–1320 (2001).

14. V. Derycke, R. Martel, J. Appenzeller and Ph. Avouris, *Nano Lett.*, **1**, 453–456 (2001).

15. A. Javey, Q. Wang, A. Ural, Y. Li and H. Dai, *Nano Lett.*, **2**, 929–932 (2002).

16. Z. Chen, J. Appenzeller, Y. Lin, J. S. Oakley, A. G. Rinzler, J. Tang, S. J. Wind, P. M. Solomon and Ph. Avouris, *Science*, **311**, 1735 (2006).

17. L. Ding, Z. Zhang, S. Liang, T. Pei, S. Wang, Y. Li, W. Zhou, J. Liu and L.-M. Peng, *Nat. Commun.*, **3**, 667 (2012).

18. R. A. Street, (Ed.) *Technology and Applications of Amorphous Silicon*, (Springer, 2000).

19. S. Ucjikoga, *MRS Bull.*, **27**, 881–886 (2002).

20. A. J. Snell, K. D. Mackenzie, W. E. Spear, P. G. LeComber and A. J. Hughes, *Appl. Phys. A*, **24**, 357–362 (1981).

21. C. D. Dimitrakopoulos and D. J. Mascaro, *IBM J. Res. Dev.*, **45**, 11–27 (2001).

22. S. R. Forrest, *Nature*, **428**, 911–918 (2004).

23. G. H. Gelinck, H. Edzer, A. Huitema, E. van Veenendaal, E. Cantatore, L. Schrijnemakers, J. B. P. H. van der Putten, T. C. T. Geuns, M. Beenhakkers, J. B. Giesbers, B.-H. Hiusman, E. J. Meijer, E. M. Benito, F. J. Touwslager, A. W. Marsman, B. J. E. Van Rens and D. M. de Leeuw, *Nat. Mater.*, **3**, 106–110 (2004).

24. T. Sekitani, U. Zschieschang, H. Klauk and T. Someya, *Nat. Mater.*, **9**, 1015–1022 (2010).

25. T. Sekitani and T. Someya. *Adv. Mater.*, **22**, 1–19 (2010).

26. A. Ismach, L. Segev, E. Wachtel and E. Joselevich, *Angew. Chem. Int. Ed.*, **43**, 6140 –6143 (2004).

27. A. Ismach, D. Kantorovich and E. Joselevich, *J. Am. Chem. Soc.*, **127**, 11554–11555 (2005).

28. S. Han, X. Liu and C. Zhou, *J. Am. Chem. Soc.*, **127**, 5294–5295 (2005).

29. C. Kocabas, S. Hur, A. Gaur, M. Meitl, M. Shim and J. A. Rogers, *Small*, **1**, 1110–1116 (2005).

30. L. Ding, D. Yuan and J. Liu, *J. Am. Chem. Soc.*, **130**, 5428–5429 (2008).

31. N. Patil, A. Lin, E. R. Myers, K. Ryu, A. Badmaev, C. Zhou, H.-S. P. Wong and S. Mitra, *IEEE Trans. Nanotechnol.*, **8**, 498–504 (2009).

32. L. Ding, A. Tselev, J. Wang, D. Yuan, H. Chu, T. P. McNicholas, Y. Li and J. Liu, *Nano Lett.*, **9**, 800–805 (2009).

33. S. W. Hong, T. Banks and J. A. Rogers, *Adv. Mater.*, **22**, 1826–1830 (2010).

34. C. Wang, K. M. Ryu, L. Gomez, A. Badmaev, J. Zhang, X. Lin, Y. Che and C. Zhou, *Nano Res.*, **3**, 831–842 (2010).

35. X. Liu, S. Han and C. Zhou, *Nano Lett.*, **6**, 34–39 (2006).

36. S. J. Kang, C. Kocabas, T. Ozel, M. Shim, N. Pimparkar, M. A. Alam, S. V. Rotkin and J. A. Rogers, *Nat. Nanotechnol.*, **2**, 230–236 (2007).

37. S. J. Kang, C. Kocabas, H. S. Kim, Q. Cao, M. A. Meitl, D. Y. Khang and J. Rogers, *Nano Lett.*, **7**, 3343–3348 (2007).

38. L. Jiao, B. Fan, X. Xian, Z. Wu, J. Zhang and Z. Liu, *J. Am. Chem. Soc.*, **130**, 12612–12613 (2008).

39. C. Wang, K. Ryu, A. Badmaev, N. Patil, A. Lin, S. Mitra, H.-S. P. Wong and C. Zhou, *Appl. Phys. Lett.*, **93**, 033101 (2008).

40. K. Ryu, A. Badmaev, C. Wang, A. Lin, N. Patil, L. Gomez, A. Kumar, S. Mitra, H.-S. P. Wong and C. Zhou, *Nano Lett.*, **9**, 189–197 (2009).

41. F. Ishikawa, H. Chang, K. Ryu, P. Chen, A. Badmaev, L. De Arco Gomez, G. Shen and C. Zhou, *ACS Nano*, **3**, 73–79 (2009).

42. A. Lin, N. Patil, K. Ryu, A. Badmaev, L. De Arco Gomez, C. Zhou, S. Mitra and H.-S. P. Wong, *IEEE Trans. Nanotechnol.*, **8**, 4–9 (2009).

43. P. G. Collins, M. S. Arnold and Ph. Avouris, *Science*, **292**, 706–709 (2001).

44. G. Hong, B. Zhang, B. Peng, J. Zhang, W. M. Choi, J.-Y. Choi, J. M. Kim and Z. Liu, *J. Am. Chem. Soc.*, **131**, 14642–14643 (2009).

45. Y. Che, C. Wang, J. Liu, B. Liu, X. Lin, J. Parker, C. Beasley, H.-S. P. Wong, and C. Zhou, *ACS Nano*, **6**, 7454–7462 (2012).

46. Q. Cao and J. A. Rogers, *Nano Res.*, **1**, 259–272 (2008).

47. Q. Cao and J. A. Rogers, *Adv. Mater.*, **21**, 29–53 (2009).

48. N. Rouhi, D. Jain and P. J. Burke, *ACS Nano*, **5**, 8471–8487 (2011).

49. C. Kocabas, N. Pimparkar, O. Yesilyurt, S. J. Kang, M. A. Alam and J. A. Rogers, *Nano Lett.*, **7**, 1195–1202 (2007).

50. N. Pimparkar, C. Kocabas, S. J. Kang, J. Rogers and M. A. Alam, *IEEE Electron Device Lett.*, **28**, 593–595 (2007).

51. Q. Cao, H. S. Kim, N. Pimparkar, J. P. Kulkarni, C. Wang, M. Shim, K. Roy, M. A. Alam and J. A. Rogers, *Nature*, **454**, 495–500 (2008).

52. E. S. Snow, J. P. Novak, P. M. Campbell and D. Park, *Appl. Phys. Lett.*, **82**, 2145–2147 (2003).

53. D. Sun, M. Y. Timmermans, Y. Tian, A. G. Nasibulin, E. I. Kauppinen, S. Kishimoto, T. Mizutani and Y. Ohno, *Nat. Nanotechnol.*, **6**, 156–161 (2011).

54. M. Engel, J. P. Small, M. Steiner, M. Freitag, A. A. Green, M. C. Hersam and Ph. Avouris, *ACS Nano*, **2**, 2445–2452 (2008).

55. M. A. Meitl, Y. X. Zhou, A. Gaur, S. Jeon, M. L. Usrey, M. S. Strano and J. A. Rogers, *Nano Lett.*, **4**, 1643–1647 (2004).

56. M. C. LeMieux, M. Roberts, S. Barman, Y. W. Jin, J. M. Kim and Z. Bao, *Science*, **321**, 101–104 (2008).

57. M. Vosgueritchain, M. C. LeMieux, D. Dodge and Z. Bao, *ACS Nano*, **4**, 6137–6145 (2010).

58. E. S. Snow, P. M. Campbell, M. G. Ancona and J. P. Novak, *Appl. Phys. Lett.*, **86**, 033105 (2005).

59. C. Wang, J. Zhang, K. Ryu, A. Badmaev, L. Gomez and C. Zhou, *Nano Lett.*, **9**, 4285–4291 (2009).

60. C. Wang, J. Zhang and C. Zhou, *ACS Nano*, **4**, 7123–7132 (2010).

61. N. Rouhi, D. Jain, K. Zand and P. J. Burke, *Adv. Mater.*, **23**, 94–99 (2011).

62. T. Takahashi, K. Takei, A. G. Gillies, R. S. Fearing and A. Javey, *Nano Lett.*, **11**, 5408–5413 (2011).

63. C. Wang, J.-C. Chien, K. Takei, T. Takahashi, J. Nah, A. M. Niknejad and A. Javey, *Nano Lett.*, **12**, 1527–1533 (2012).

64. M. Ha, Y. Xia, A. A. Green, W. Zhang, M. J. Renn, C. H. Kim, M. C. Hersam and C. D. Frisbie, *ACS Nano*, **4**, 4388–4395 (2010).

65. P. Chen, Y. Fu, R. Aminirad, C. Wang, J. Zhang, K. Wang, K. Galatsis and C. Zhou, *Nano Lett.*, **11**, 5301–5308 (2011).

66. M. Jung, J. Kim, J. Noh, N. Lim, C. Lim, G. Lee, J. Kim, H. Kang, K. Jung, A. D. Leonard, J. M. Tour and G. Cho, *IEEE Trans. Electron Devices*, **57**, 571–580 (2010).

67. J. Noh, M. Jung, K. Jung, G. Lee, J. Kim, S. Lim, D. Kim, Y. Choi, Y. Kim, V. Subramanian and G. Cho, *IEEE Electron Device Lett.*, **32**, 638–640 (2011).

68. J. Noh, S. Kim, K. Jung, J. Kim, S. Cho and G. Cho, *IEEE Electron Device Lett.*, **32**, 1555–1557 (2011).

69. M. S. Arnold, S. I. Stupp and M. C. Hersam, *Nano Lett.*, **5**, 713–718 (2005).

70. M. S. Arnold, A. A. Green, J. F. Hulvat, S. I. Stupp and M. C. Hersam, *Nat. Nanotechnol.*, **1**, 60–65 (2006).

71. A. A. Green and M. C. Hersam, *Adv. Mater.*, **23**, 2185–2190 (2011).

72. X. Tu, S. Manohar, A. Jagota and M. Zheng, *Nature*, **460**, 250–253 (2009).

73. C. Wang, D. Hwang, Z. Yu, K. Takei, J. Park, T. Chen, B. Ma, and A. Javey, *Nat. Mater.*, **12**, 899–904 (2013).

74. Q. Cao, M. Xia, C. Kocabas, M. Shim, J. A. Rogers and S. V. Rotkin, *Appl. Phys. Lett.*, **90**, 023516 (2007).

75. X. Liu, C. Lee, J. Han and C. Zhou, *Appl. Phys. Lett.*, **79**, 3329–3331 (2001).

76. M. Shim, A. Javey, N. W. S. Kam and H. Dai, *J. Am. Chem. Soc.*, **123**, 11512–11513 (2001).

77. C. Klinke, J. Chen, A. Afzali and Ph. Avouris, *Nano Lett.*, **5**, 555–558 (2005).

78. Y. Lin, J. Appenzeller, J. Knoch and Ph. Avouris, *IEEE Trans. Nanotechnol.*, **4**, 481–489 (2005).

79. Z. Zhang, X. Liang, S. Wang, K. Yao, Y. Hu, Y. Zhu, Q. Chen, W. Zhou, Y. Li, Y. Yao, J. Zhang and L. M. Peng, *Nano Lett.*, **7**, 3603–3607 (2007).

80. L. Ding, S. Wang, Z. Zhang, Q. Zeng, Z. Wang, T. Pei, L. Yang, X. Liang, J. Shen, Q. Chen, R. Cui, Y. Li and L. M. Peng, *Nano Lett.*, **9**, 4209–4214 (2009).

81. C. Wang, K. Ryu, A. Badmaev, J. Zhang and C. Zhou, *ACS Nano*, **5**, 1147–1153 (2011).

82. J. Zhang, C. Wang, Y. Fu, Y. Che and C. Zhou, *ACS Nano*, **5**, 3284–3292 (2011).

83. S. Y. Lee, S. W. Lee, S. M. Kim, W. J. Yu, Y. W. Jo and Y. H. Lee, *ACS Nano*, **5**, 2369–2375 (2011).

84. C. Rutherglen, D. Jain and P. Burke, *Nat. Nanotechnol.*, **4**, 811–819 (2009).

85. C. Kocabas, H. Kim, T. Banks, J. Rogers, A. Pesetski, J. Baumgardner, S. Krishnaswamy and H. Zhang, *Proc. Natl. Acad. Sci.*, **105**, 1405–1409 (2008).

86. L. Nougaret, H. Happy, G. Dambrine, V. Derycke, J.-P. Bourgoin, A. A. Green and M. C. Hersam, *Appl. Phys. Lett.*, **94**, 243505 (2009).

87. N. Chimot, V. Derycke, M. F. Goffman, J. P. Bourgoin, H. Happy and G. Dambrine, *Appl. Phys. Lett.*, **91**, 153111 (2007).

88. C. Wang, A. Badmaev, A. Jooyaie, M. Bao, K. L. Wang, K. Galatsis and C. Zhou, *ACS Nano*, **5**, 4169–4176 (2011).

89. J. Zhang, Y. Fu, C. Wang, P. Chen, Z. Liu, W. Wei, C. Wu, M. E. Thompson and C. Zhou, *Nano Lett.*, **11**, 4852–4858 (2011).

90. S. Forrest, P. Burrows and M. Thompson, *Laser Focus World*, **31**, 99–101 (1995).

91. J. Sheats, H. Antoniadis, M. Hueschen, W. Leonard, J. Miller, R. Moon, D. Roitman and A. Stocking, *Science*, **273**, 884–888 (1996).

92. C. Tang, *Dig. Soc. for Information Display Int. Symp.*, **27**, 181–184 (1996).

93. P. Burrows, S. Forrest and M. Thompson, *Current Opinion in Solid State & Material Science*, **2**, 236–243 (1997).

94. G. Gu and S. Forrest, *IEEE J. Sel. Top. Quantum Electron.* **4**, 83 (1998).

95. K. Takei, T. Takahashi, J. C. Ho, H. Ko, A. G. Gillies, P. W. Leu, R. S. Fearing and A. Javey. *Nat. Mater.*, **9**, 821–826 (2010).

96. T. Takahashi, Z. Yu, K. Chen, D. Kiriya, C. Wang, K. Takei, H. Shiraki, T. Chen, B. Ma and A. Javey, *Nano Lett.*, **13**, 5425–5430 (2013).

97. P. H. Lau, K. Takei, C. Wang, Y. Ju, J. Kim, Z. Yu, T. Takahashi, G. Cho and A. Javey, *Nano Lett.*, **13**, 3864–3869 (2013).

98. C. Wang, K. Takei, T. Takahashi and A. Javey, *Chem. Soc. Rev.*, **42**, 2592–2609, (2013).

CHAPTER 2

NANOMATERIAL-BASED FLEXIBLE SENSORS

Kuniharu Takei

Department of Physics and Electronics, Osaka Prefecture University
1-1 Gakuen-cho Naka-ku, Sakai, Osaka 599-8531, Japan
takei@pe.osakafu-u.ac.jp

1. Introduction

To improve the convenience, comfort, and security of daily life, flexible sensor components will play an important role. For example, they can be attached onto infrastructure and vehicles to monitor stress and crack information. Another crucial application is wearable flexible devices; a truly flexible device that attaches onto the human body without pain will provide comfort and convenience by monitoring real-time health conditions. Although conventional solid devices can be used for these applications, considering the target's size (e.g., bridge, water or gas pipes, airplanes, etc.), the sensor must cover macroscale surfaces conformally with an ultra-lightweight device, especially for vehicle applications. Furthermore, the sensor must be economical, including the fabrication, transportation, and installation.

Flexible devices have potential to develop the aforementioned sensor systems. (1) Flexible (plastic) substrates are inexpensive compared to crystal wafers like silicon, which are used in conventional electronics.

45

(2) Transportation and installation costs should be drastically reduced since flexible sensor systems should be very lightweight and compact. In particular, flexible devices should be more eco-friendly as transportation costs, which consume a lot of fuel and energy, should be greatly reduced because flexible devices can be folded like paper. (3) Flexible wearable devices can be attached onto human skin without awareness of the user, enabling monitoring of real-time health conditions and saving lives.

As described above, flexible sensors have diverse applications, and by creating a large-scale flexible sensor tape or intelligent wall paper, these devices can be attached to any surface. To realize flexible devices, many electrical components have been developed (e.g., electrodes,[1–6] thin-film transistors (TFTs),[7–14] sensors,[15–19] and displays[20–23]) using various material systems on flexible substrates. In this chapter, as an important component of flexible electronics, flexible sensors, especially those using nanomaterials, are introduced.

2. Flexible Sensor Applications

For a variety of applications and reasons, flexible sensors composed of inorganic and/or organic materials have been widely studied. As stated in the introduction, a key application is a wearable health monitoring system. Due to the globally aging population, health monitoring without seeing a doctor should be a large market in the near future and should help save costs for patients, doctors, insurance providers, etc. Flexible/wearable devices must incorporate TFTs, sensors, batteries, actuators, etc. In addition, if printable sensor components are developed, an economically fabricated sensor network that can meet user-demand may be realized. Although printable TFTs have been reported,[24,25] sensor components fabricated via printing techniques are not well studied. Figure 1 shows examples reported in 2014, which include strain and temperature sensors using printable nanomaterial composite films.[15,16]

3. Nanomaterial Patterning on Flexible Substrates

Challenges in realizing flexible sensors are how to achieve low-cost, macroscale, and high performance/low power sensor devices on flexible

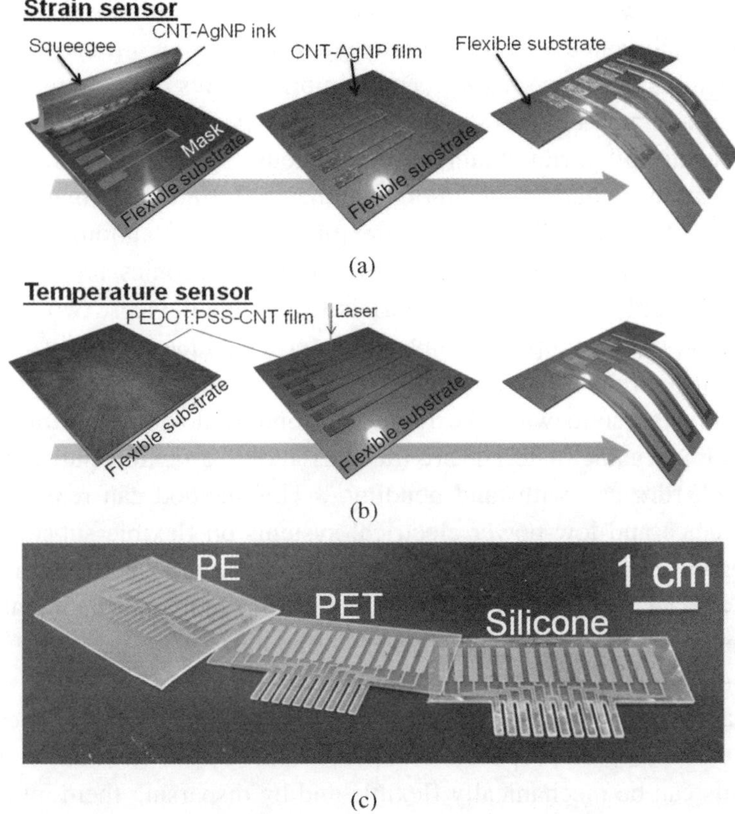

Figure 1. Large-scale and fully printed strain and temperature flexible sensors. Schematic of the fabrication process for (a) strain sensor using a screen printer and (b) temperature sensor using a laser cutter. (c) Printed sensors can be patterned on various plastic substrates (e.g., polyethylene (PE), polyethylene terephthalate (PET), and silicone rubber). Reproduced with permission from Ref. 16. (Copyright 2014, American Chemical Society.)

substrates. One solution toward integrated flexible sensors is macroscale-printing methods. To pattern flexible sensor materials, nanomaterial printing has potential to fabricate device components on flexible substrates.

When organic materials are used as active components, much effort has focused on improving the stability and mobility of transistor applications via solution-based processes. Someya and Sakurai *et al.* developed an ultra-lightweight flexible organic device,[26] and demonstrated a wireless system and a wet sensor on a flexible substrate as a proof-of-concept.[27] Bao *et al.* fabricated a health monitoring system using a transistor-based high sensitive pressure sensor.[28] These improvements of the transistor and flexible integrated circuits should lead to low-power and economic electronics using fully printed fabrication methods.

In addition to innovations in organic materials, inorganic material systems have been developed for a flexible system[7-14] although standard inorganic materials used in the conventional electronics are rigid substrates. To apply to inorganic materials to flexible devices, two methods have been proposed: epitaxial layer transfer techniques and utilization of bottom-up inorganic nanomaterials. In the former, an ultrathin epitaxial layer from the solid wafer is transferred onto a flexible substrate.[8,19,29] Although inorganic materials are transferred onto a flexible substrate, the nanothick film can withstand bending.[30] This method can realize high-performance and low-power electrical systems on flexible substrates by transferring Si integrated circuits[31,32] or III–V semiconductor high speed devices.[8] The latter method utilizes bottom-up inorganic nanomaterials (e.g., nanoparticles,[15,16] nanowires,[7,9,17] nanotubes,[10-14] graphene,[33] and two-dimensional layered transition metal dichalcogenides[34,35]). The advantage of using inorganic nanomaterials is higher performance and sometimes improved properties compared to the bulk.[36] In addition, nanomaterials can be mechanically flexible and by dispersing them into solution, they can be applied to solution-based printing processes, which should eventually lead to low-cost high-performance devices. However, the challenge of using inorganic nanomaterials is how to uniformly pattern them over large flexible substrates. This section mostly focuses on inorganic nanomaterial flexible devices, including their printing techniques.

To fabricate highly uniform patterning of electrical materials on flexible substrates, a chemical surface treatment on either a flexible substrate or patterning onto nanomaterials is an important technique. Because nano inorganic materials usually have a negative charge, there is a strong interaction when the surface of a flexible substrate has a positive charge. In fact, there are reports about the surface chemistry with different self-assembled monolayers (SAMs)[37,38] (Fig. 2). The results[37,39] show that a surface with a poly-L-lysine treatment has a high density of nanomaterial

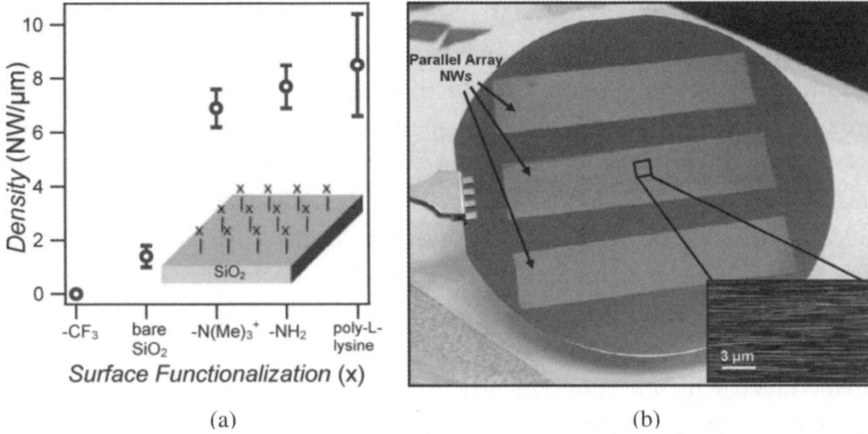

Figure 2. Surface modification for nanowire printing. (a) Printed nanowire density with different surface functionalizations. (b) Picture of a parallel nanowire arrays on a Si substrate with a poly-L-lysine surface treatment. Reproduced with permission from Ref. 37. (Copyright 2008, American Chemical Society.)

transfer for Ge/Si core/shell nanowires because the van der Waals interaction between nanowires and poly-L-lysine, which as a positive charge, is the strongest in the SAMs in the list of Fig. 2.

As demonstrated above, the surface chemistry is critical to realize an optimal surface without sacrificing the electrical properties and uniformity in high-performance electrical materials for active components such as a sensor on a flexible substrate.

4. Flexible Interconnection Layers

A common component in all electronics is the interconnection between active components. Although this chapter describes flexible sensors, flexible interconnection layers are introduced first because they are key components in flexible sensors. Because the resistivity of the interconnection layer must be low to reduce electrical power loss, a metal layer (e.g., Au or Al) is often used in conventional rigid electronics. This requirement is applicable to flexible devices and sensors. In research on flexible devices, a thin metal layer is deposited via an evaporation, sputtering, or other vacuum deposition tools. However, because these metal layers usually have high Young moduli (>10 GPa), the probability of cracks or delamination from the substrate is high when the substrate is bent or folded. Additionally, when economical macroscale flexible devices are considered, the vacuum-based deposition method is unsuited to form an interconnection layer. The costs of the infrastructure and manufacturing throughput for metal deposition are directly related to the device cost. Because flexible devices must be macroscale, the cost per area must be drastically reduced compared to conventional electronics fabricated via semiconductor wafer processes.

Several approaches have been proposed to replace the conventional metal layer with an economically formed flexible electrode. The first approach is to use nano or microsized silver (Ag) particles that can be applied to macroscale printing methods (e.g., gravure printers,[25,40] screen printers,[41] and ink-jet printers,[42] etc.). After printing Ag particles, curing the ink forms a Ag film on a flexible substrate by melting the Ag particles. Because Ag materials have a low resistivity, films can economically pattern a relatively good interconnection layer on macroscale flexible substrates. Although controlling the composition of the Ag particles from micro to nanosize creates a flexible metal layer, for foldable devices designed to be worn or extremely flexible, the mechanical–electrical reliability must be improved.

Nanocarbon materials have been proposed for foldable interconnection layers.[1,43] A foldable electrode using graphene can be patterned by a transfer method from solution to a filter and user-defined flexible substrate. Although graphene foldable electrodes have a high sheet resistance (~200–800 Ω/square)[1] compared to an evaporated metal layer as a foldable electrode (~0.7 Ω/square for Ni, ~0.01 Ω/square for Ag),[2] the mechanical reliability in terms of the electrical resistance change against folding is improved. Developing and optimizing material systems should further improve the sheet resistance and foldability. However, to realize reliable, flexible, and foldable electronics, the balance between sheet resistance, foldability, and scalability is important. In addition, additional printed metal layers need to be developed to improve device performance in terms of the work function to control band offset between metal and semiconductor materials.

5. Flexible Sensors

An economically fabricated sensor network on a flexible substrate will be a challenge for future electronics. Applications of flexible sensors should differ from those of solid-based sensors. Additionally, flexible sensors must be able to discriminate between stimuli and strain because flexible sensors are often bent to conform to different types of surfaces. If the output signal of the sensor changes as a function of substrate bending (i.e., strain), reliable sensing information cannot be achieved. This section introduces different types of sensors on flexible substrates.

5.1. *Optical Sensors*

As demonstrated by the prevalence of digital cameras and cell phones, image sensors or optical sensors are currently one of the most important components in electronics. Optical sensors can operate via different mechanisms (e.g., electron–hole generation at a pn junction in a semiconductor or photoconductor of materials). For an image sensor, charge-coupled device (CCD)[44] or complementary metal-oxide-semiconductor (CMOS) technologies[45] are mainly used as high-speed and high-resolution imagers. However, to apply these technologies to flexible image sensors, it is more important to investigate the material system first due to complicacy of these technologies on a flexible substrate.

First, an optical sensor material needs to be considered. The detection wavelength, which can vary from ultraviolet to infrared light, depends on the bandgap of the sensor materials. The relation between bandgap energy E and wavelength λ can be expressed as

$$E(J) = \frac{hc}{\lambda}, \tag{1}$$

where h is Planck's constant (6.626×10^{-34} m^{2}kg/s (or Js)) and c is the speed of light (3.0×10^8 m/s). It should be noted that energy E in Eq. (1) is calculated in Joules (1 eV $= 1.602 \times 10^{-19}$ J). By converting the unit, the bandgap energy required to detect certain wavelength of light can be expressed as

$$E_g(eV) = \frac{1241}{\lambda(nm)}.$$ (2)

Based on Eq. (2), a material can be appropriately selected for the target wavelength. For example, a wearable digital camera or a wearable skin-care sunlight device may require visible light ranging from 390 nm to 700 nm or ultraviolet light around 300 nm, respectively.

Several parameters characterize optical sensors. The first one is the signal-to-noise ratio (SNR), which is expressed as[46]

$$SNR = \frac{I_{Light} - I_{Dark}}{I_{Dark}},$$ (3)

where I_{Light} and I_{Dark} are the photo current and dark current, respectively. Another is the linear dynamic range (LDR) or photo sensitivity linearity, which is described as[47]

$$LDR(dB) = 20\,Log\left(\frac{J_{Light}}{J_{Dark}}\right),$$ (4)

where J_{Light} is the photocurrent density under a light intensity of 1 mW/cm^2 and J_{Dark} is the dark current density. For example, LDR for a Si photodetector is around 120 dB (Ref. 47). To evaluate the sensor, responsivity is another factor, and can be calculated by

$$responsivity\left(\frac{A}{W}\right) = \frac{J_{Light}}{L_{Light}},$$ (5)

where L_{Light} is the light intensity. By comparing these parameters in an optical sensor, the appropriate applications can be determined.

An X-ray imager on a flexible substrate has been demonstrated as an example of a flexible optical sensor for medical applications.[48,49] The imager consists of organic photodetectors with carbon nanotube active matrix circuitry (Fig. 3). The visible light imager is comprised of regioregular

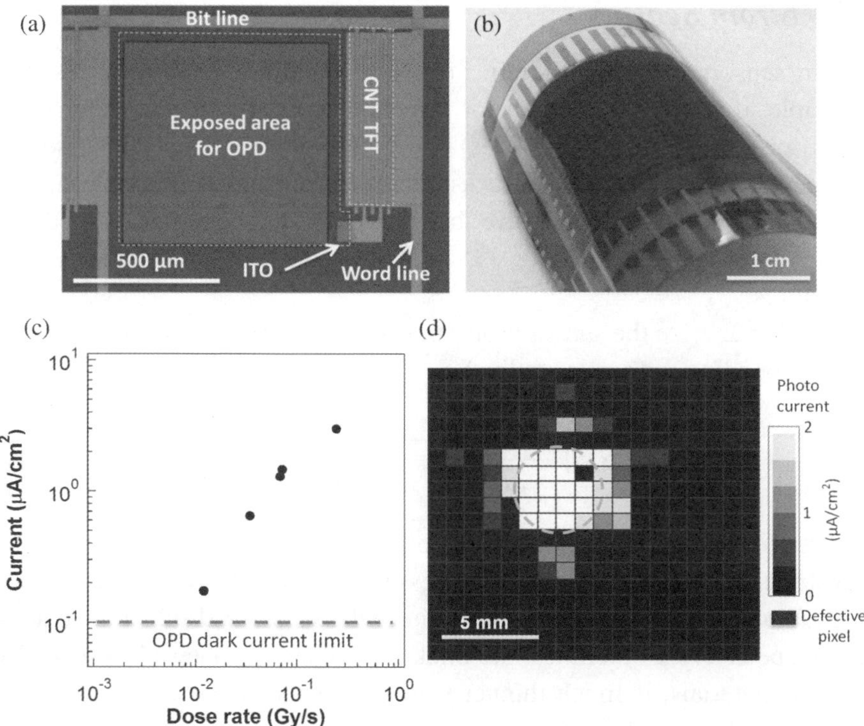

Figure 3. Flexible optical imager. (a) Optical microscope image of one pixel in an 18 × 18 array. Pixel consists of a Carbon-nanowire transistors (CNT) for the active matrix circuitry and organic photodiode (OPD). (b) Picture of a flexible optical imager. (c) Photocurrent response as a function of the X-ray dose rate. (d) X-ray image mapping of a circular object. Reproduced with permission from Ref. 48. (Copyright 2013, American Chemical Society.)

poly(3-hexylthiophene) (P3HT) and [6,6]-phenyl C61-butyric acid methyl ester (PCBM).[50] To acquire X-ray images, a Gd_2O_2S/Tb (GOS) scintillator film[51] can convert X-rays into visible light with ~2.27 eV energy (~547 nm wavelength), and then visible light is detected by the organic photodetector integrated on a flexible substrate. Based on reports for the optical sensors, LDR is around 64 dB at a light wavelength of 535 nm. A flexible optical image sensor that absorbs both visible light (535 nm, intensity 100 μW/cm^2) and X-rays has been demonstrated using a photodetector array (18 × 18 array) (Fig. 3(d)).

5.2. *Strain Sensors*

Strain sensors can detect strain in diverse conditions and surfaces. For example, if strain information for infrastructure (e.g., gas or water pipelines) or vehicles (e.g., airplanes) can be monitored in real-time, severe accidents due to breaks or cracks might be prevented. Flexible and/or stretchable strain sensors have been widely demonstrated, especially using nanomaterials,[15,16,52–54] piezomaterials,[55,56] and transistor structures.[57]

To characterize the sensor properties, strain can be calculated as a function of bending (curvature radius) as[58]

$$S = \left(\frac{d_t + d_s}{2R} \right) \frac{(1 + 2\eta + \chi\eta^2)}{(1+\eta)(1+\chi\eta)}, \tag{6}$$

where d_t and d_s are the thicknesses of the strain sensor material and the substrate, respectively. η is the thickness ratio (d_t/d_s). R and χ are the bending radius and the Young modulus ratio of the sensor and substrate materials, respectively. Sometimes, the thickness of sensor materials, especially in nanomaterials, is much thinner than that of the substrate (i.e., $d_t \ll d_s$) with a similar Young modulus. In this case, Eq. (6) can be simplified to

$$S = \frac{d_t + d_s}{2R}. \tag{7}$$

Strain sensors often characterize the sensitivity of the resistance change as a function of applied pressure (i.e., %/Pa) and the gauge factor based on the applied strain calculated by Eqs. (6) or (7). The gauge factor g is defined by the ratio of the relative resistance change and strain,[15] which is given as

$$g = \frac{d\left(\dfrac{\rho_s}{\rho_o} \right)}{dS}, \tag{8}$$

where ρ_o and ρ_s are the resistivity of the film in the relaxed state and under a given strain, respectively. However, the Young modulus of a new material system is often unknown, and in this case, the sensitivity (%/Pa) can be a suitable parameter to understand the sensor performance. It should be noted that the sensitivity and gauge factor parameters indicate completely different information. Although both the parameters are often used, the gauge factor is a property of only the sensor material, while sensitivity (%/Pa) indicates the properties of the final device structure (i.e., sensor material and substrate with a defined structure). Thus, changing the structure can change the sensitivity but not the gauge factor.

As an example of a printable strain sensor, Ag nanoparticles and CNTs in a binder polymer have been proposed to form a strain sensor on a flexible substrate using a screen printing or painting method.[15,16] Many strain sensors using nanomaterials, including the one mentioned in the earlier system, involve a mechanism where the distance between nanomaterials increases (decreases), resulting in an increase (decrease) in electrical resistance due to tunneling current between nanoparticles upon applying a tensile (compressive) strain (Fig. 4).[59] This tunneling resistance R_T between nanoparticles, which corresponds to R_0 and R_1 in Fig. 4, is expressed[59] as

$$R_T \propto e^{\beta l}, \tag{9}$$

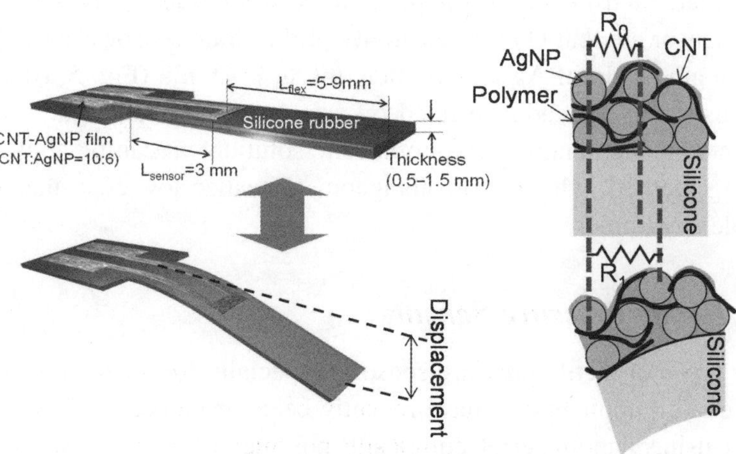

Figure 4.   Strain sensor mechanism. Bending the substrate applies tensile stress into the AgNP–CNT film, increasing the resistance due to the increased AgNP distance. Reproduced with permission from Ref. 16. (Copyright 2014, American Chemical Society.)

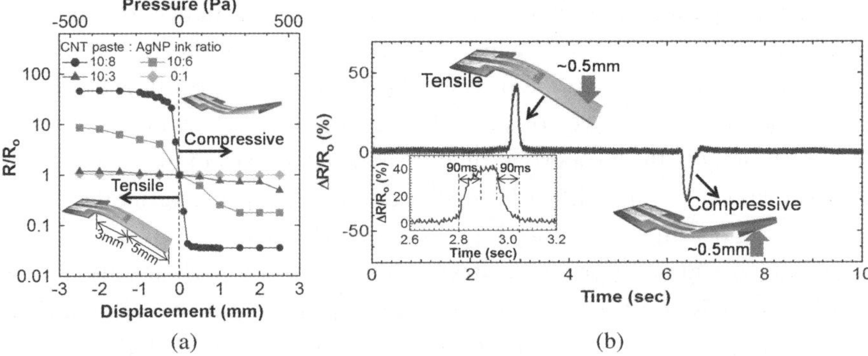

Figure 5.   Characteristics of a printed flexible strain sensor. (a) Resistance change as functions of displacement (bottom *x*-axis) and applied pressure (top *x*-axis). (b) Real-time measurements of the tensile and compressive stress. Response time is about 90 ms. Reproduced with permission from Ref. 16. (Copyright 2014, American Chemical Society.)

where β is a parameter of electron tunneling through organic linker molecules connecting the nanoparticles in the film and l is the distance between nanoparticles. The CNTs in Fig. 4 should achieve a wide dynamic range of strain sensing because the CNTs that entangle surrounding Ag nanoparticles help electrically connect Ag nanoparticles when a large tensile strain is applied to the strain sensor. The keys of this strain sensor are that (1) the sensitivity of the strain sensor is easily tuned by changing ratio of Ag nanoparticle ink to CNT ink (Fig. 5(a)) and (2) the sensor can be readily printed on any surface. Because the sensor is fabricated using nanomaterial-composite solutions, it can be applied to a printing method, which is advantageous to realize low-cost, macroscale flexible electronics.

5.3. *Tactile Pressure Sensors*

Many types of tactile pressure sensors, especially for artificial electronic skin (e-skin) applications, have recently been developed on flexible substrates using nanomaterial composite polymer films,[60] transistor structures,[61] and piezo-materials.[62,63] The mechanism for a tactile pressure sensor using a nanomaterial composite polymer film is similar to that for a strain sensor. By applying the pressure onto the sensor, the nanomaterial

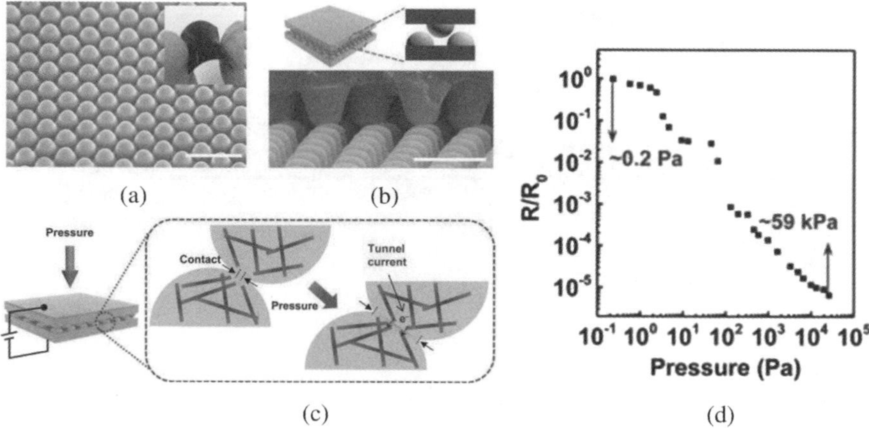

Figure 6.   Microdome-structured high-sensitivity tactile pressure sensor. (a) Microscope image of the microdome structures. (b) Schematic and microscope images of the interface between the microdomes to create the electrical contacts. (c) Schematic of the tactile pressure sensing mechanism. (d) Electrical property as a function of pressure. Minimum detectable pressure is ~0.2 Pa. Reproduced with permission from Ref. 60. (Copyright 2014, American Chemical Society.)

distance typically decreases, resulting in a decrease in the electrical resistance or an increase in the capacitance between nanomaterials. Since nanomaterials are dispersed in the polymer film, tactile pressure sensors are mechanically flexible. To improve the sensitivity toward tactile pressure, microstructures (e.g., a semi-sphere structure with a nanomaterial composite film[60] or a transistor structure[61]) have been proposed.

Figure 6 shows an example using CNTs and polymers with microdome arrays[60] where a small tactile pressure causes the arrays to interlock, increasing the current in between the CNTs in the polymer (Eq. (9) and Fig. (6c)). By reading the current, the tactile pressure can be determined. The report states that the pressure sensitivity is −15.1 kPa^{-1} and the minimum and maximum pressure detections are ~0.2 Pa and ~59 kPa, respectively.

Another approach to realize a highly sensitive tactile pressure sensor is to utilize a transistor structure (Fig. 7(a)).[61,64] The sensor has a polymer film with a microstructure as a gate dielectric layer of a transistor fabricated on a flexible substrate. Applying pressure by squeezing the gate

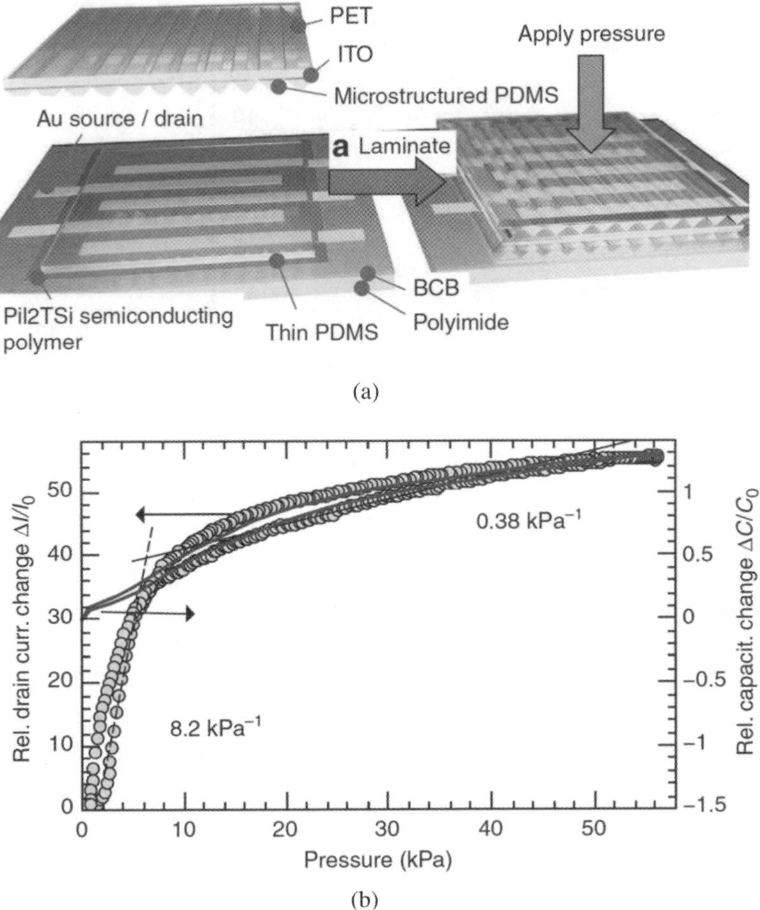

Figure 7. Transistor-type tactile pressure sensor. (a) Schematic of a tactile pressure sensor with a micro-pyramid structure and an organic transistor. (b) Electrical property as a function of pressure. Reproduced with permission from Ref. 64. (Copyright 2013, Nature Publishing Group.)

polymer film decreases the distance between the transistor material and gate electrode, which increases the gate capacitance. Changing the gate capacitance alters the charge density in the semiconductor material (Charge density $Q = C \times V_g$, where C is the gate capacitance and V_g is gate voltage.). Modifying the microstructure on the polymer and using the subthreshold region of transistor operation results in a small applied pressure creating a large change in the drain-source current of the transistor. For example, one report indicated a high sensitivity of ~8.4 kPa^{-1} and <20 Pa minimum pressure (Fig. 7(b)).

Although these tactile pressure sensors can be realized on the flexible substrates, they also detect strain by bending the flexible substrate. Thus, to precisely measure the tactile pressure, the substrate must be fixed on the surface of an object. Studying strain engineering and material/structural engineering should improve the selectivity between tactile pressure and strain and realize truly flexible electronic systems.

5.4. *Temperature Sensors*

Another possibility is a temperature sensor to monitor human health or environmental information for wearable devices or application of smart walls, respectively. Flexible temperature sensors have been demonstrated using a pn junction diode[65] composed of either organic or inorganic materials[16] and the temperature coefficient of resistance (TCR) of a metal layer[66] such as platinum (Pt).

For example, an organic pn junction diode temperature sensor has been formed by a p-type copper phthalocyanine (CuPc) and n-type 3,4,9,10-perylenetetracarboxylic-diimide (PTCDI) deposited by a vacuum sublimation system.[65] The report commented that the sensor is slightly unstable when used in ambient air due to oxygen and moisture. On the other hand, an inorganic pn junction diode for the temperature sensor has also been demonstrated using conventional silicon technology by transferring the Si diode temperature sensor from a wafer onto a flexible/stretchable substrate.[67] Because this sensor is fabricated with silicon, it is stable and reliable.

The current I of the ideal pn junction diode as a function of temperature is calculated as

$$I = I_0(e^{qV/kT} - 1), \tag{10}$$

where I_0 is the reverse bias saturation current, V is the applied voltage at the pn junction, k is the Boltzmann constant, and T is the temperature. It should be noted that I_0 also has temperature dependence due to the excitation of electrons and holes moving to the valence band and conduction band, respectively.

For a TCR temperature sensor, the electrode is often made of Pt due to the relatively high TCR value and high stability of Pt over a wide temperature range. The thermal sensitivity on a plastic substrate patterned by a conventional lithography method is ~0.53%/°C (Ref. 66).

Both the diode and TCR sensors have been fabricated using a vacuum system to deposit each material. For practical wearable electronics, a process that realizes macroscale and economic sensors of flexible substrates using a printing method must be developed.

One example of a temperature sensor fabricated via a printing method has been reported using a mixture of conductive poly(3,4-ethylenedioxythiophene)polystyrene sulfonate (PEDOT:PSS, 1.3 wt% in water) and carbon nanotube ink for the temperature sensor material (Fig. 8).[41] The experimentally determined temperature sensitivity based on TCR for PEDOT:PSS and CNT films are ~0.4%/°C and ~0.18%/°C, respectively. However, mixing solutions with a 10:1 weight% ratio (PEDOT:PSS solution: CNT ink) and using the resistor as the temperature sensor increase the sensitivity up to ~0.61%/°C, which is higher than the values for films of PEDOT:PSS or CNT without a mixture. Thus, the mechanism to detect temperature using a mixture of PEDOT:PSS and CNT is most likely electron hopping at the interface between PEDOT:PSS and CNTs. This electron hopping conductance G as a function of temperature T is often described based on Mott's variable range hopping model as[68]

$$G \propto e^{-1/T^{\alpha}},\tag{11}$$

where $\alpha = 1/4$ and $1/3$ for three dimensions and two dimensions, respectively. A hopping electron moves to the next particle or grain in the material with the shortest distance and lowest energy.

By applying this mixture ink to a shadow mask printing method[41] or etching a film using a laser cutter tool,[16] a highly sensitive temperature sensor can be fabricated on a large-scale flexible substrate without employing expensive infrastructure (e.g., evaporator, sputter, and lithography tools). The hysteresis of the temperature change and resistance change against mechanical bending are relatively small (Figs. 8(b) and 8(c)). In particular, the mechanical reliability is important for the flexible

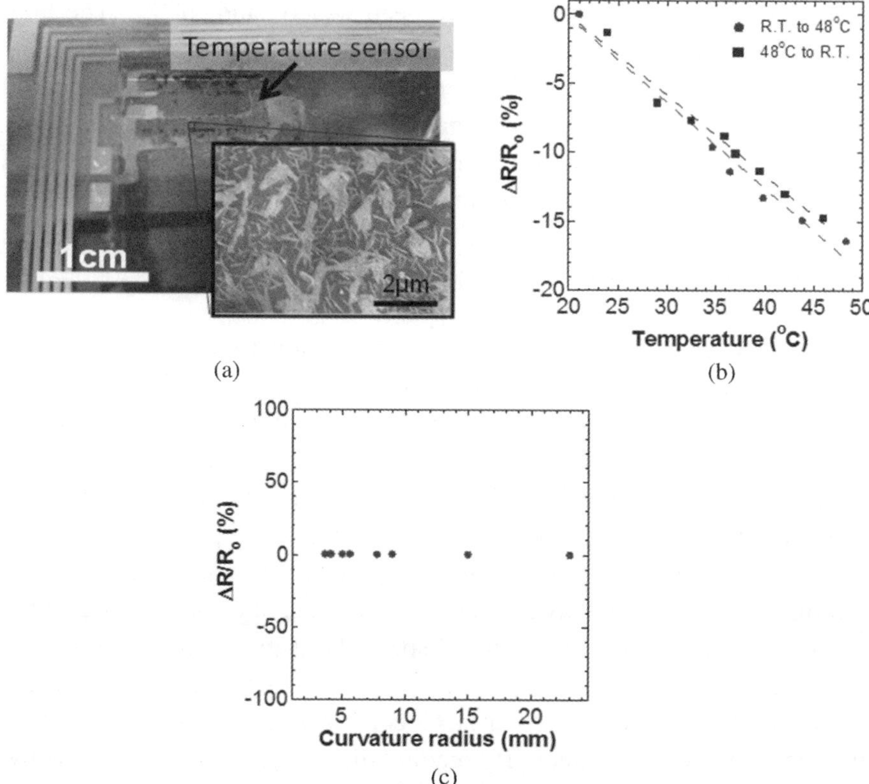

Figure 8.   Printable flexible temperature sensor. (a) Pictures of a temperature sensor on a Kapton substrate. (b) Electrical property as a function of temperature from room temperature (R.T.) to 48°C and vice versa. (c) Mechanical reliability of the temperature sensor upon bending the flexible substrate. Reproduced with permission from Ref. 41. (Copyright 2014, John Wiley and Sons.)

and wearable electronics (Fig. 8(c)), suggesting that the sensor material introduced here is promising for a flexible temperature sensor because it meets the low-cost fabrication and selectivity again strain requirements.

5.5. *Gas Sensors*

Gas detection on a flexible substrate is a beneficial application that can be used in a limited space or dangerous environment. Highly sensitive nanomaterial-based gas sensors have been widely studied.[69–72] The high sensitivity in nanomaterials is most likely due to two reasons. First, the surface area of a high-density patterned nanomaterial is greatly increased compared to bulk. Second, when a gas attaches onto the surface of a nanomaterial, the resistance is drastically changed by moving the sub-band around the Fermi level due to the quantum effect. The quantum effect phenomena have been well studied using two-dimensional InAs materials onto Si/SiO_2 substrates.[73] These phenomena are applicable to flexible substrates without sacrificing performance.

To detect gas using materials such as semiconductors, a catalyst is necessary to absorb or attach the gas ion and modify the surface potential. For example, a palladium (Pd) catalyst is often used for hydrogen sensing because hydrogen absorbs onto Pd to form palladium hydride (PdH_x)[74]; a Schottky contact-type high-sensitivity hydrogen gas sensor with a Pd catalyst[74] can detect hydrogen concentrations below 5 ppm using aligned Si nanowire arrays. Due to the surface potential change upon Pd absorbing hydrogen, the Schottky barrier is changed, causing the conductance of the Si nanowires to change drastically. Especially, if the channel bias can be controlled near the subthreshold region, a small surface potential change by attaching a gas ion shows a large conductance change, realizing an ultrahigh sensitive gas sensor. By reading this conductance difference, the gas concentration can be determined.

Some gases (e.g., N_2O and NH_2S) do not require a catalyst because they are very sticky on a semiconductor surface and easily modify the surface potential of a semiconductor. In fact, a high sensitivity (~20 ppb) for N_2O detection has been achieved by a Si-based flexible sensor without a catalyst (Fig. 9).[70]

One challenge for future gas sensors on both rigid and flexible substrates is to realize high selectivity for different gases. Currently, gas sensors cannot recognize the gas difference when a mixed gas is applied to the sensor because the catalysts or materials usually react with different chemicals, limiting the selectivity of gas detection. However, selectivity should be addressed as a wearable electronics for secure human life. Another challenge is how to imitate animal (e.g., dog) noses, which requires both an ultrahigh sensitivity (ppb level) for every gas and ultrahigh selectivity. Contributions from different research fields (e.g., chemistry,

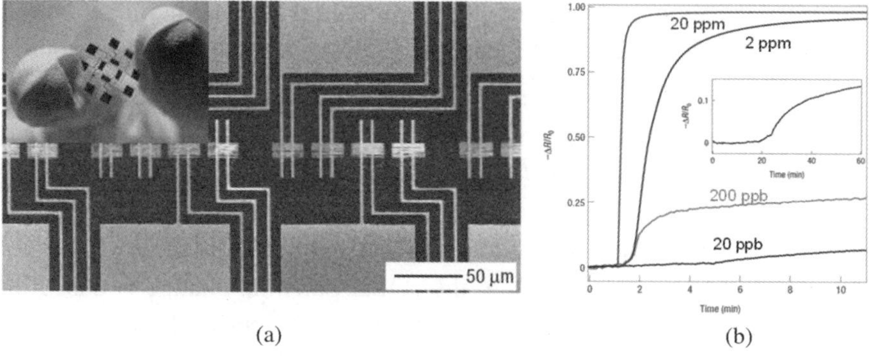

(a)   (b)

Figure 9.   Si nanowire-based gas sensor. (a) Microscope image and picture of a flexible gas sensor array. (b) Electrical response for different concentrations of NO_2 gas up to 20 ppb. Reproduced with permission from Ref. 70. (Copyright 2007, Nature Publishing Group.)

material science, electrical engineering, etc.) will be necessary to realize an electrical nose.

5.6. *pH Chemical Sensors*

It is difficult to determine even if an unknown liquid is dangerous. Usually the first step to identifying a potential threat is to determine the pH. Due to dramatic progress in robotics, robots will play a key role in assessing environments, including identifying unknown liquids in the near future. If a pH chemical sensor can be fabricated on a flexible substrate and has other functionalities, it will help robots to protect from coming into contact with dangerous chemicals.

The pH is determined by measuring the potential difference. An ion sensitive field-effect transistor (ISFET) has been widely developed on rigid and flexible substrates to sense this difference.[75–78] An ion is attached on the gate region of transistor, and the potential is read by the difference between the ion and the reference electrode. Due to the potential on the FET gate, the source–drain current is modulated based on the ion potential to determine the pH of the solution placed onto the FET gate. The concept is very similar to the gas sensor described above except that the target phase is a vapor for a gas sensor and a liquid for a pH sensor.

The mechanism and physics of ISFETs are well studied and understood. The actual reaction potential E for ISFETs can be expressed theoretically by the Nernst equation[79] at 25°C as

$$E = E° - \frac{0.0592}{n} \log Q, \tag{12}$$

where $E°$ is the standard potential of the ion and Q is the reaction quotient. Because $E°$ has a constant value, the actual potential change is based on the 0.0592 factor corresponding to the maximum sensitivity of 59.2 mV/pH. It should be noted that Eq. (12) cannot be applied to gas sensors.

Another factor is that the protection layer for devices from liquid must not sacrifice ion sensitivity. Usually liquids contain ions (e.g., sodium and potassium), which often work as mobile ions that break transistors under the operating bias or cause malfunctions. To prevent this problem, SiN, Al_2O_3, etc. have been proposed as the protecting layers for ISFETs. Because ions cannot diffuse inside these materials and these materials are sensitive to pH, they are suitable protecting layers without sacrificing pH sensitivity even if they are layered over the transistor.

For a flexible pH sensor, different types of materials for the ISFET channel have been reported (e.g., nanotubes,[75] oxide semiconductor materials,[76] and organic materials[77]). As an example, a carbon nanotube ISFET has been prepared using single-walled CNTs (SWCNT) and poly(diallydimethyammonium chloride) (PDDA) as polyelectrolytes for the channel material (Fig. 10). Unfortunately, the paper did not report the sensitivity (mV/pH), but based on the source–drain current as a function of pH level, it is clear that a flexible ISFET using CNT can be turned in a pH sensor. In addition to the pH ISFET sensor, this paper demonstrated a glucose sensor utilizing an ISFET by adding a glucose oxidase layer on top of channel layer.[75] Figure 11 and Eq. (13) depict the mechanism to convert the glucose level into pH level.

$$\beta - D - glucose + O_2 \xrightarrow{GO_x} D - glucono\text{-}\delta\text{-}lactone + H_2O_2$$

$$D - glucono\text{-}\delta\text{-}lactone + H_2O \rightarrow D\text{-}gluconate^- + H^+ \tag{13}$$

$$H_2O_2 \xrightarrow{+0.7V} O_2 + 2H^+ + 2e^-.$$

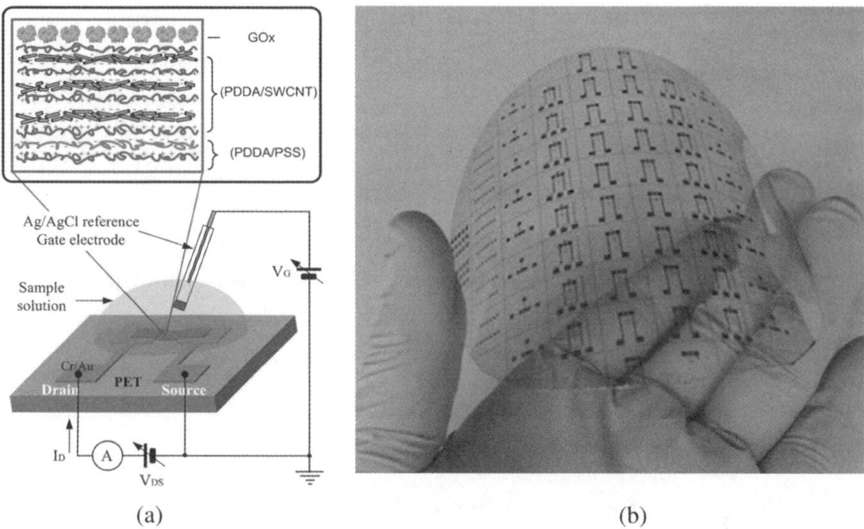

Figure 10. Flexible pH and glucose sensors. (a) Measurement setup and detailed schematic structure of the sensing materials. (b) Picture of flexible pH and glucose sensors. Reproduced with permission from Ref. 75. (Copyright 2010, Elsevier.)

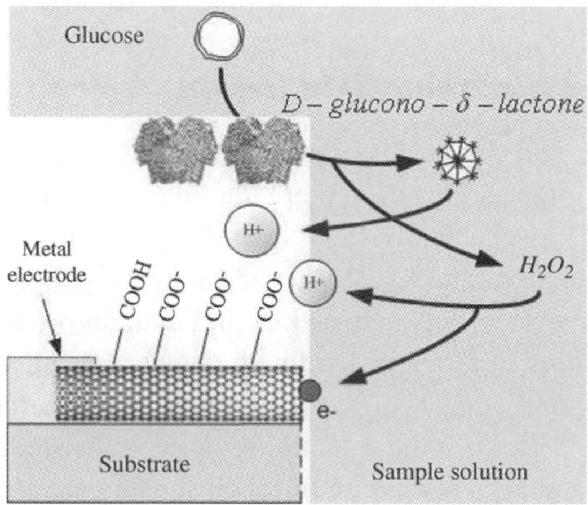

Figure 11. Mechanism for glucose sensing using the sensor structure in Fig. 10. Reproduced with permission from Ref. 75. (Copyright 2010, Elsevier.)

Due to the oxidation of H_2O_2 by the gate bias of the ISFET, glucose generates three hydrogen ions. Thus, the sensitivity of glucose level is relatively high around 28.4 μA/mM.

To identify unknown liquids in a robotic application and to develop a glucose monitoring system (e.g., diabetes health monitoring), the ISFET technique for pH sensing has great potential in future flexible sensors. However, to realize a truly flexible pH sensor using the ISFET method, a monolithically integrated reference electrode with a good stability must be developed on a flexible substrate. The ISFET and glucose sensor described above both used a separately prepared reference electrode.

6. Flexible Sensor Integration

Sensor integration for a sensor network on a flexible substrate has been demonstrated. However, a hurdle for practical flexible devices is advancing integration techniques. In this section, methods and devices with regard to flexible integrated sensors are introduced as examples. It should be noted that the selectivity of each sensor is an important factor to detect the targets correctly and to integrate different types of sensors.

6.1. *Artificial Electronic Skin (e-skin)*

For the robotic and prosthesis applications, e-skin that imitates human skin is of great interest. However, there are several requirements to imitate human skin: (1) mechanically flexible and lightweight, (2) pressure and temperature detection, and (3) low power and low voltage operations. In addition, the interface between e-skin and human/robots is important. E-skin applications in terms of hardware devices are often demonstrated as tactile pressure sensor arrays to detect touch similar to human skin.[17,28,61,63,65,80] Besides these tactile sensors, transistors for future signal processing and high resolution of touch sensing are also integrated on the flexible substrates.[17,63] Both organic and inorganic materials have been proposed for tactile sensors and transistors for e-skin applications. In 2004, Someya *et al.* reported the first e-skin integrated with a tactile sensor array with active matrix circuitry,[80] which was a breakthrough in both e-skin research and organic transistor integration. In 2005, the same group developed temperature sensor arrays integrated with tactile sensors and active matrix circuitry on a flexible substrate.[65] However, the organic transistor had a low field effect mobility (~1.4 cm^2/Vs), and consequently, a high voltage was required to operate e-skin.

New material systems composed of organic and/or inorganic materials have been proposed to improve the transistor mobility. In 2010, two works on e-skins with flexible substrates were reported at the same time. One improved the transistor mobility via an inorganic nanowire array,[17] while the other developed a highly sensitive tactile pressure sensor.[64]

Compared to research on organic materials, fewer works have employed inorganic materials for flexible devices because inorganic materials are usually rigid, making mechanical flexibility difficult. However, the 2010 report used well-aligned high-density inorganic nanowire arrays patterned by a printing method as described above.[17] Using a nanowire film as the transistor channel material realized a low voltage operation for an e-skin with a relatively high field effect mobility of ~20 cm^2/Vs (Fig. 12). When the material size decreases from micro- to nanometers, the material becomes mechanically flexible.[30] Hence, the reported inorganic nanowire array transistors can realize both high performance and mechanical flexibility and may eventually realize low voltage operations on flexible substrates.

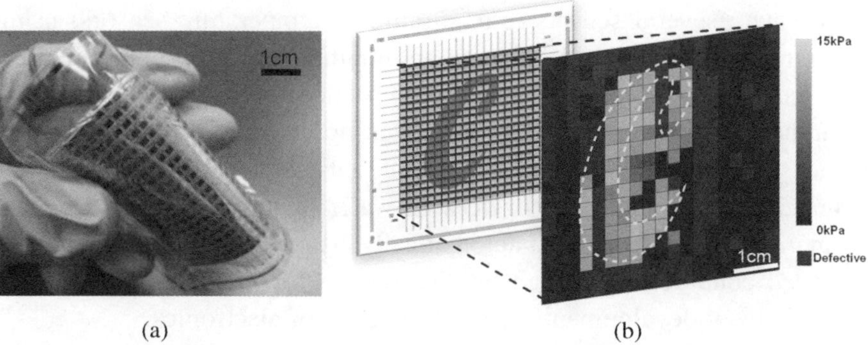

Figure 12. Artificial electronic skin. (a) Picture of an e-skin with an integrated nanowire array transistor for active matrix circuitry and a pressure sensitive rubber. (b) Two-dimensional pressure distribution by placing an object "C" onto the e-skin measured at an operating voltage <5 V. Reproduced with permission from Ref. 17. (Copyright 2007, Nature Publishing Group.)

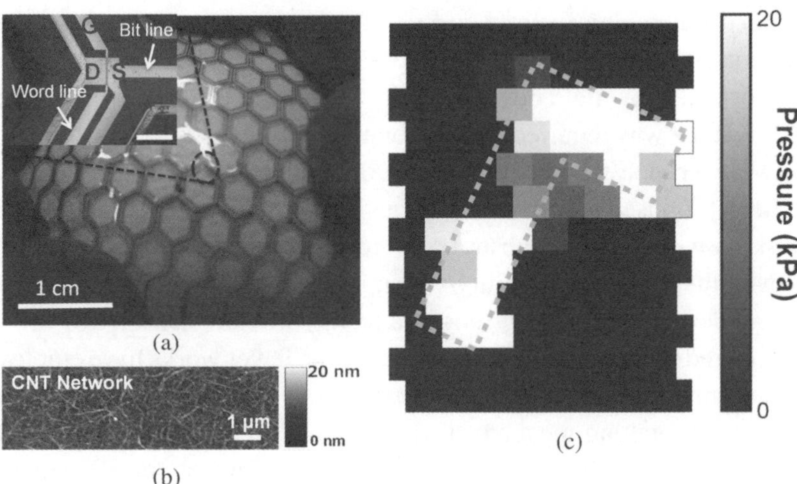

Figure 13.   Stretchable e-skin. (a) Photo of the stretchable e-skin covering a baseball. (b) Atomic force microscopy image of the printed CNT network film. (c) Two-dimensional pressure distribution measured by the stretchable e-skin. Reproduced with permission from Ref. 10. (Copyright 2011, American Chemical Society.)

After these developments, different kinds of e-skins, stretchable (Fig. 13)[10] or the so-called epidermal electronics (Fig. 14),[81,82] have been widely proposed. Rogers *et al.* proposed two key developments: an electrical tattoo[81] and a dissolvable device.[83] An electrical tattoo is a wireless device with several sensors (e.g., strain and temperature sensors) using inorganic micro/nanothick films on an ultrathin substrate (total thickness is ~7 μm).[81] Because it is ultrathin, it can be placed on human skin without awareness of the user. In addition, the group also demonstrated flexible devices using silk, Mg, and MgO materials, which dissolve in water (Fig. 15).[83,84] Because these devices are water soluble, they can be applied as an implantable device that automatically disappears after measuring the intended condition. Although hurdles remain for practical applications, these types of developments open a new class of electronics.

6.2. *Artificial Electronic Whisker (e-whisker)*

One interesting natural phenomenon is the function of an animal whisker or hair on human skin that can sense weak wind flow or spatial

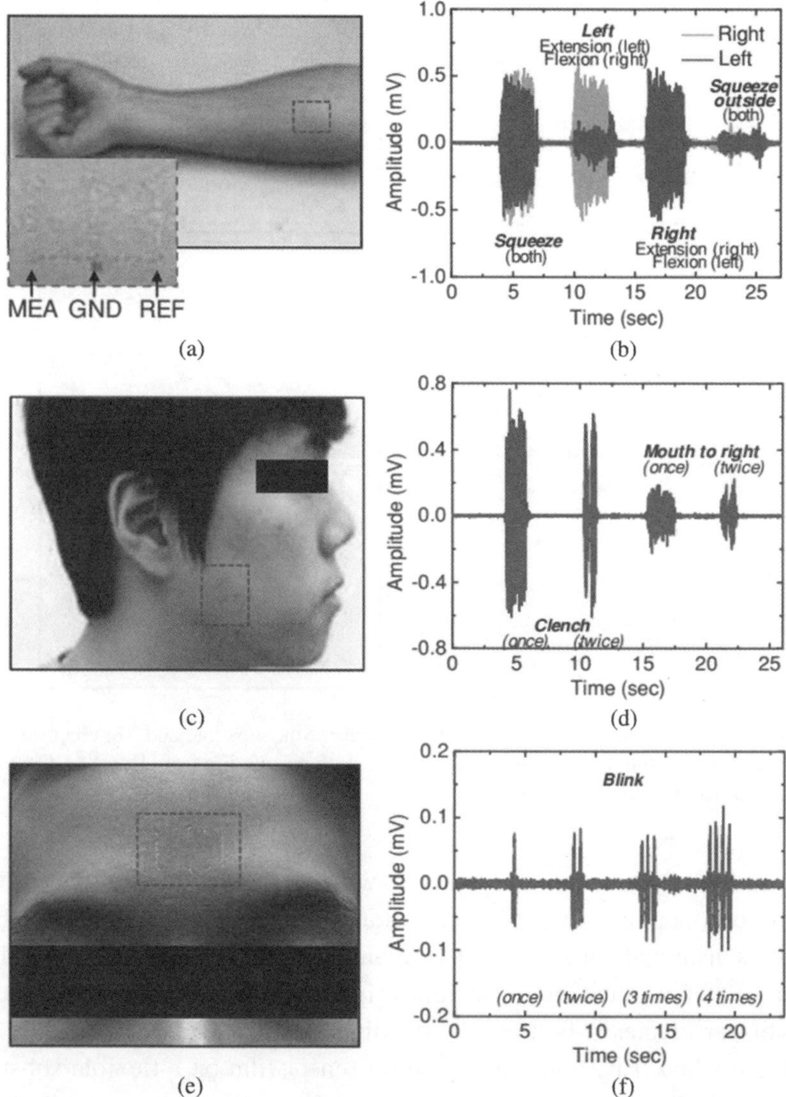

Figure 14. Epidermal electronic device. Surface electromyography to monitor the movement of (a) (b) forearm. (c) (d) face, and (e) (f) forehead. Reproduced with permission from Ref. 82. (Copyright 2013, John Wiley and Sons.)

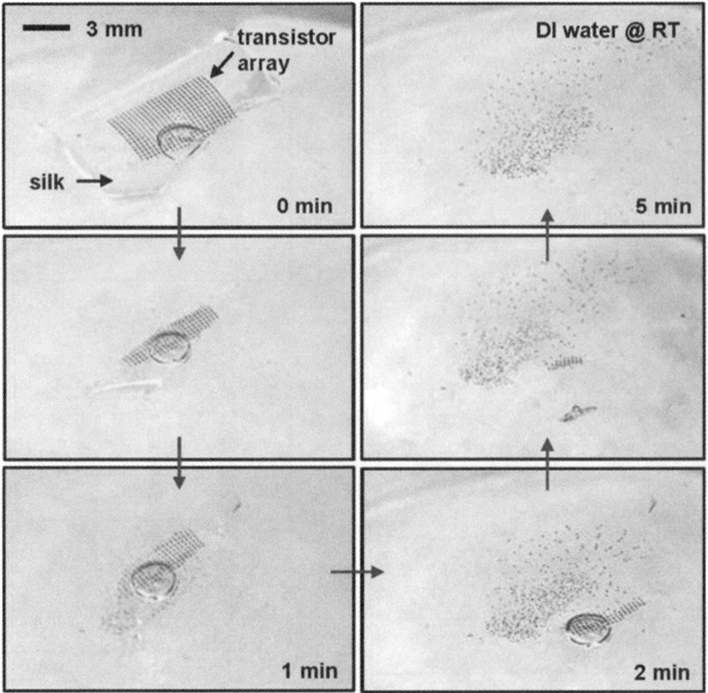

Figure 15.   Physical transient of a flexible device. Silk substrate and Mg electrodes dissolve in water within five minutes. Reproduced with permission from Ref. 84. (Copyright 2013, John Wiley and Sons.)

distribution by touching an object. E-whisker devices have a long history in robotic applications, but conventional e-whiskers are built by a bulk strain sensor and a metal wire to imitate the animal whisker, making them bulky.[85,86] For microrobots or lightweight robotic applications, this bulky e-whisker creates a bottleneck. Flexible strain sensors can replace this bulky e-whisker by patterning a strain sensor film on a flexible substrate with a whisker structure.

By patterning a composite film as described in the section of strain sensors on a whisker-like flexible substrate and mounting on a semispherical object, e-whiskers can detect gas flow mapping with a flow rate of 1 m/s of nitrogen. Similar to an object touching an animal whisker, scanning with an e-whisker array can map the three-dimensional spatial distribution (Fig. 16).[16] Combining e-whiskers with e-skin on a flexible substrate

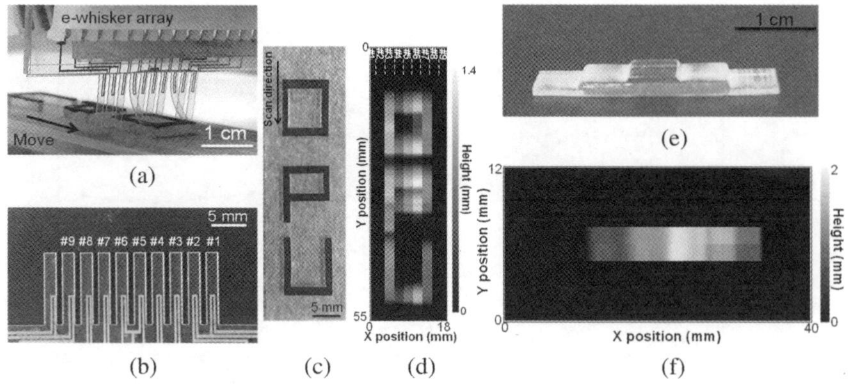

Figure 16. (a) Photo of the measurement setup for the e-whisker scanner. (b) Photo of the e-whisker array. (c) Photo of a scanned target with letters of "OPU" and (d) output result of the height change, which corresponds to the electrical resistance change. (e) Photo of the stair-like object and (f) output result, depicting that the e-whisker can measure at least a 400 μm height difference. Reproduced with permission from Ref. 16. (Copyright 2014, American Chemical Society.)

should realize various applications over a large dynamic range of pressure.

As a proof-of-concept for a multifunctional integrated system, a strain and temperature sensing e-whisker array has been fabricated using the nanomaterial composite films described above.[16] Figure 17 shows the selectivity of the strain and temperature sensors against temperature and strain, respectively. Because each sensor has high selectivity of strain and temperature, the sensor network of strain and temperature sensors can collect both results without artifacts. In addition, laminating the temperature-sensing e-whisker and strain-sensing e-whisker fabricates a multi-functional (temperature and strain) e-whisker (Fig. 18). The printable e-whisker array can map the temperature and shape distribution successfully. Hence, the multi-functionalities of these devices can be beyond natural whiskers, allowing these electronics to collect more information.

6.3. *Clinical Applications*

Flexible devices have also been employed in clinical applications to reduce the burden on patients and doctors and to monitor medical

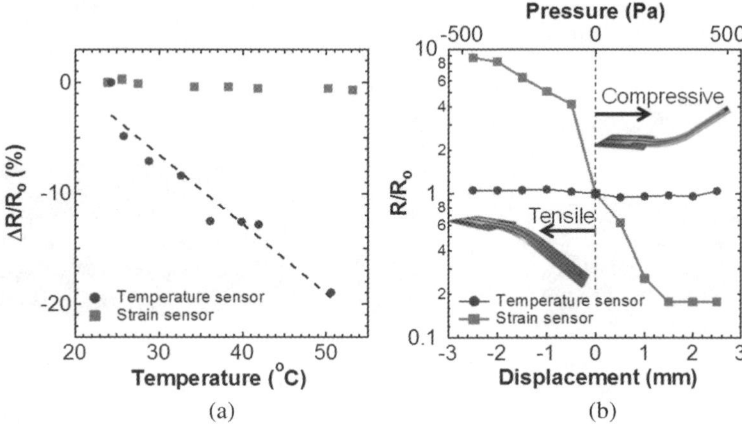

Figure 17.   Resistance changes as a function of (a) temperature and (b) strain for both temperature and strain sensors. Reproduced with permission from Ref. 16. (Copyright 2014, American Chemical Society.)

conditions safely. There are several approaches to replace conventional medical tools to flexible devices (e.g., catheter with several sensors,[66] an implantable brain activity monitoring sensor,[31,87] and cardiac electro-physiology sensors[88,89]).

By arranging the metal layer and studying stretchable devices in the field of strain engineering, a balloon catheter with several sensors has been developed (Fig. 19).[66] In addition to containing a light emitting diode array, this catheter is integrated with temperature, flow, tactile, and elec-trogram sensors. Moreover, epicardial ablation lesions can be created wirelessly using the integrated radio-frequency (RF) ablation electrodes, confirming that flexible sensors and systems with integrated user-defined multifunctional devices can be used in various clinical applications. Specifically, flexible sensors and systems may be able to assist doctors during surgery and reduce the burden on patients.

Another application for clinical and implantable devices is to record brain activity using a flexible actively multiplexed electrode array that conformally covers the brain surface. Demonstrations of *in vivo* brain activity mappings (e.g., sleep spindles, visual evoked responses, and elec-trographic seizures[31,87]) have allowed neuroscientific and practical appli-cation studies (e.g., epilepsy) to move forward.

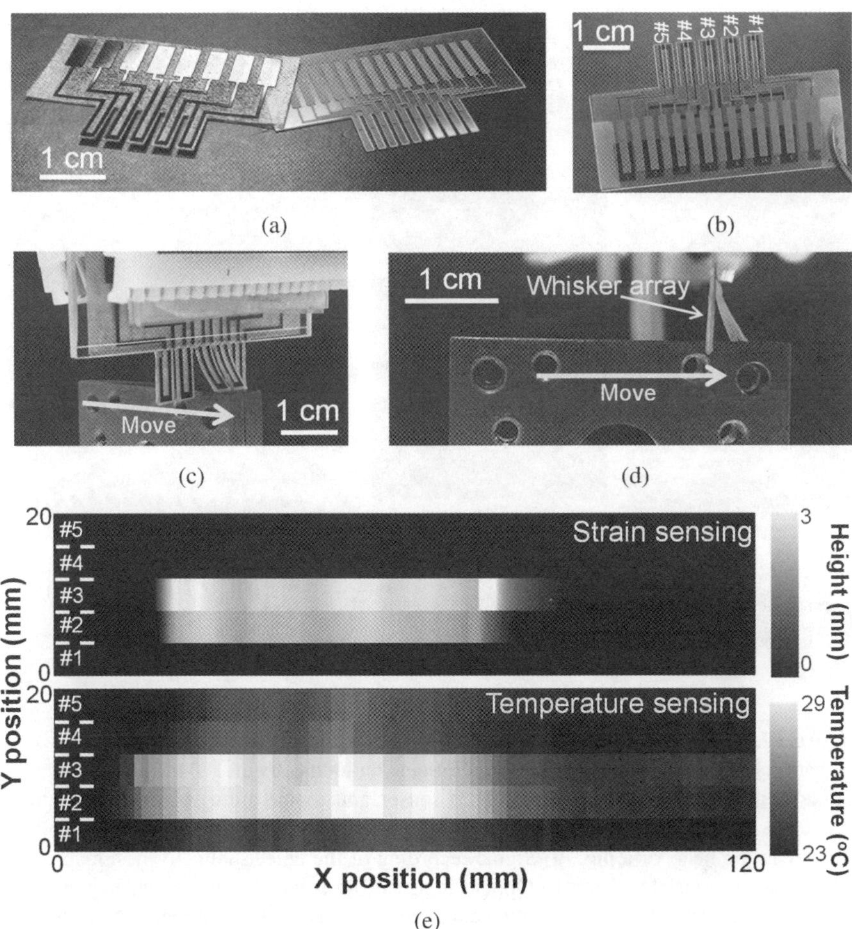

Figure 18. Multifunctional e-whisker array. Photos of (a) each temperature and strain sensor array. (b) laminated multifunctional e-whisker, and (c,d) experimental setup. (e) Scanning results of the three-dimensional mapping and temperature distributions. Reproduced with permission from Ref. 16. (Copyright 2014, American Chemical Society.)

Although flexible devices for clinical applications are in the research and development phase with animal tests by an academic group, more intense studies by medical groups and institutes should make these technologies available as clinical tools for humans as soon as possible. This pioneering work will certainly open the door for new electronics that achieve comfortable, convenient, and safe human lives.

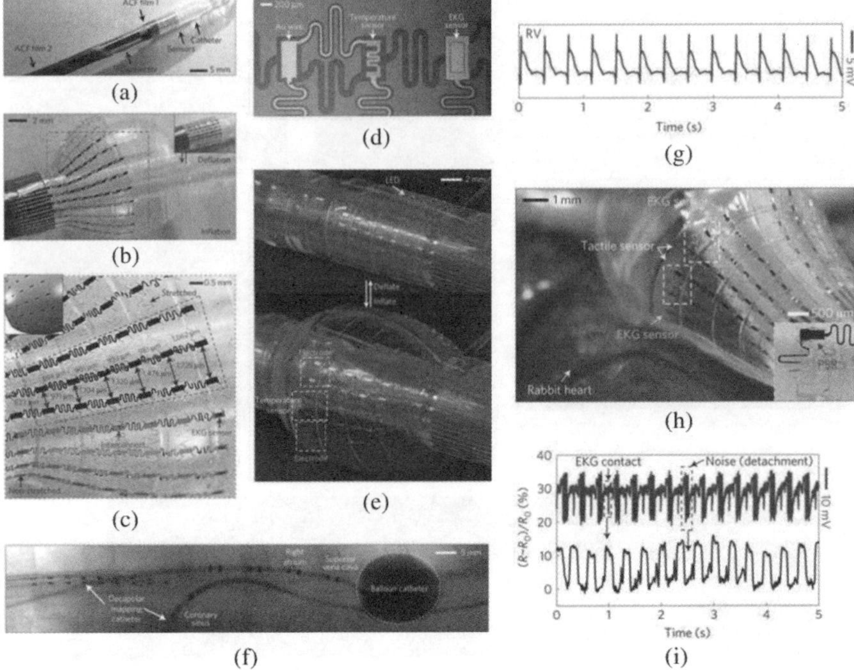

Figure 19. Balloon catheter with several sensors and stretchable electrodes. Photos of (a) a stretchable balloon catheter. (b,c) an inflated balloon catheter. (d) magnified image of a pixel with electrogram mapping (EKG) sensor and temperature sensor. (e) a balloon catheter with a light-emitting diode (LED). (f) X-ray angiography image of a balloon catheter in the heart of a pig. (g) Signal recording of the epicardial activation map of the anterior right ventricle (RV). (h) Optical image of an inflated balloon catheter in direct contact with a rabbit heart. (i) Simultaneous recording of the EKG signal and mechanical beats of a heart. Reproduced with permission from Ref. 66. (Copyright 2011, Nature Publishing Group.)

6.4. *Health Monitoring Devices*

Wearable health monitoring systems that attach onto human skin have attracted much attention to monitor health conditions in real-time. Although conventional wearable health monitoring systems contain bulk sensor components, ultra-lightweight and truly flexible health monitoring systems are required to increase the number of users. If an economic device can be placed on a person without his or her awareness, many people, even those who do not seriously care about their health, will try such a device. As demand increases, the device will become more affordable.

Recently highly sensitive tactile pressure sensors have been widely studied for pulse wave of the radial artery monitoring using a flexible device.[64,90,91] This highly sensitive tactile pressure sensor can detect pulse waves by placing it on the wrist due to heartbeats. The record of long-term pulse waves probably allows us to predict the health condition.

7. Human-Interactive Flexible Sensors

Future electronics should interact with humans more. Conventional electronics can only detect or display some information. However, if electronics and humans interact more, electronics will become more convenient, comfortable, and useful in everyday life. In fact, there are already several reports on human-interactive flexible sensors.

7.1. *Pressure–light Responsive e-skin*

Light-responsive, flexible, interactive devices upon applying pressure onto an e-skin are the first step to demonstrate human-interactive electronics. The concept is that the e-skin responds to a human eye by emitting light that depends on a human touch. The 2013 report provided a proof-of-concept where the device performed like a touch panel display,[92,93] and this light-responsive e-skin can also detect pressure strength by observing the light intensity. In the future, human mind monitoring by observing the touch pressure on a display will likely be demonstrated because the pressure of a human touch often depends on the state of mind (e.g., angry, sad, happy, etc.). This kind of light-responsive flexible device should be applicable to future smart wall displays or can be expanded to other electronics.

There are two types of light-responsive flexible e-skins. The first one integrates active matrix circuitry with organic light emitting diode (OLED) array and pressure sensitive rubber (PSR) (Fig. 20).[92] OLED is connected in series to PSR. Applying pressure onto the PSR drastically decreases its resistance, allowing the OLED to drive current through the circuit, which changes the OLED intensity. In conventional displays, the active matrix

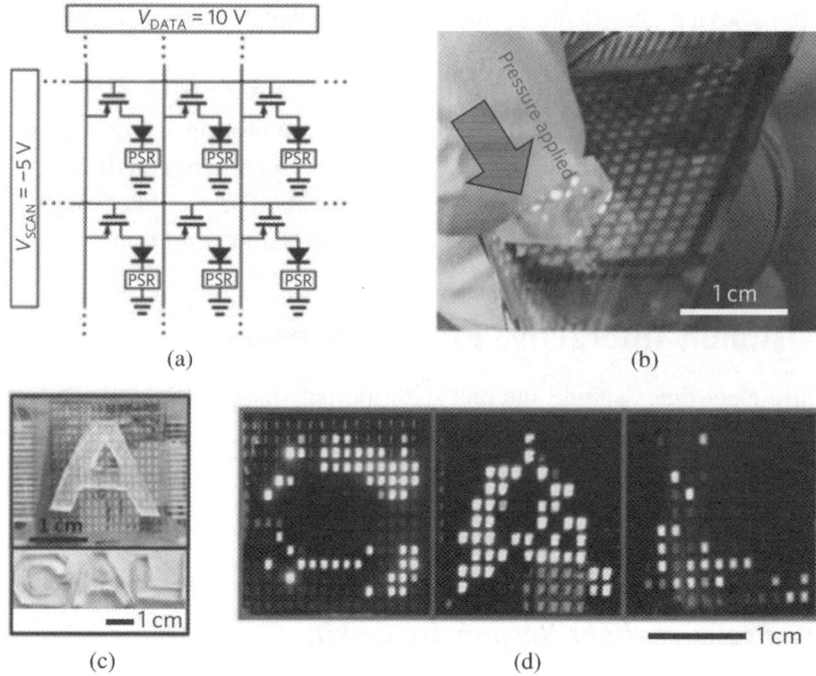

Figure 20.   Light-responsive interactive e-skin. (a) Schematic of the circuit diagram for an interactive e-skin. (b) Photo of the light-response when pressure is applied onto the e-skin. (c) Photos demonstrating the light-response applied pressure with "C", "A", and "L" shapes, and (d) light response results with three color light emitting diode (green, blue, and red from left to right). Reproduced with permission from Ref. 90. (Copyright 2013, Nature Publishing Group.)

circuitry selects which pixels to display in high resolution. The second one utilizes the piezo-phototronic effect using p-GaN/n-ZnO nanowires.[93] Using a p-GaN/n-ZnO nanowire, piezoelectric polarization occurs at the end of nanowire under pressure. This polarization charge works as a gate voltage to modulate the charge separation in a ZnO nanowire. This device can light up under pressure due to charge separation or band bending.

7.2. *Smart Bandages*

Smart bandages are integrated components of wearable health monitoring systems. However, the concept of a smart bandage is worth mentioning in

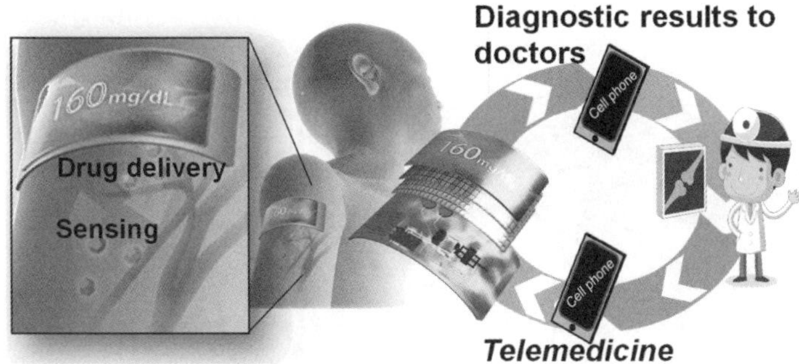

Figure 21. Conceptual image of a human-interactive smart bandage and its wireless communication system. Reproduced with permission from Ref. 41. (Copyright 2014, John Wiley and Sons.)

detail as future human-interactive devices should monitor and record real-time human health conditions by just attaching on the skin (Fig. 21). Patients and doctors can check trends of a health condition through flexible sensors and display. These trends can also help doctors to predict and possibly prevent disease. By integrating a smart bandage and a wireless system, the diagnostic results can be sent to a doctor via a cell phone. Using a cell phone is essential because a flexible wireless system cannot directly send signals due to technology limitation on a flexible substrate. By realizing this wireless real-time diagnostic monitoring system, the number of patients that doctors must physically see can be reduced, allowing doctors to focus on more urgent needs of patients. In addition to doctors, the patient can also view a diagnostic result based on the trend of sensing results on the smart bandage. Moreover, by integrating a drug delivery system or some other curing function, the patient or doctor can also cure or suppress symptoms by delivering a drug.

This is a simple concept of a human-interactive electronic, but depending on the demands, different sensors, actuators, and systems could be integrated easily without drastically increasing the cost. To economically fabricate such devices, the conventional fabrication process for semiconductors cannot be used. One alternative is a printing method. Although the demonstration described below is yet to be integrated with a sensor network system, the first proof-of-concept of a smart bandage has been demonstrated using a printing method.[41]

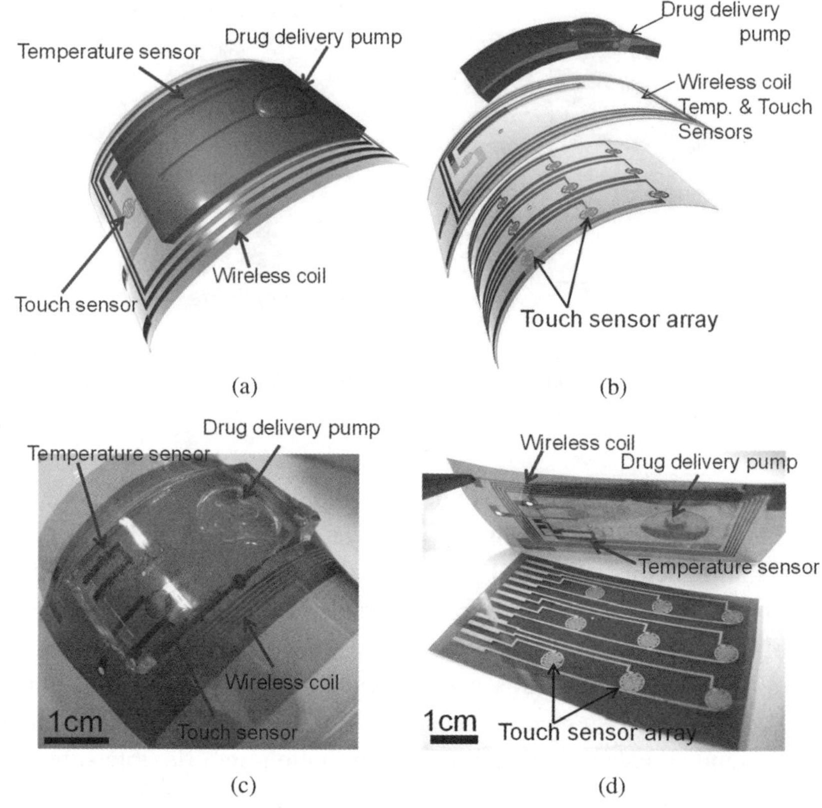

Figure 22.   Prototype of a fully printed smart bandage. (a,b) Schematics of the smart bandage. (c,d) Optical images of a fabricated smart bandage consisting of a temperature sensor, a touch sensor integrated with a wireless coil, and a drug delivery pump for a curing function. Reproduced with permission from Ref. 41. (Copyright 2014, John Wiley and Sons.)

In addition to the sensors, different types of devices using microelectromechanical system (MEMS) techniques are necessary for practical flexible electronics to realize fully functional human interactive devices. As a proof-of-concept, a flexible drug delivery pump and a wireless coil have been integrated. To achieve economical large-scale flexible device systems, these devices have also been fabricated mainly using printing and mold techniques. Figure 22 describes the schematics of an integrated flexible drug delivery pump, temperature sensor, and a capacitive touch

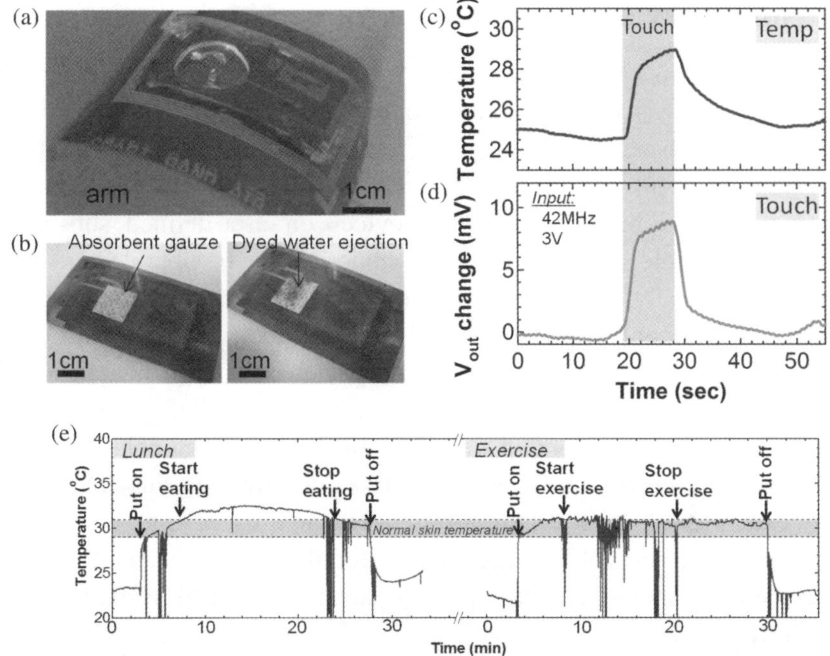

Figure 23.   Photos of (a) a smart bandage on a human arm and (b) before and after water ejection via a smart bandage by applying a pressure onto the drug delivery pump. Real-time measurement of a human touch onto a smart bandage of (c) temperature change and (d) touch detection wirelessly at 42 MHz and 3 V input. (e) Real-time temperature measurement of the skin surface while eating spicy soup and short-time exercise. Reproduced with permission from Ref. 41. (Copyright 2014, John Wiley and Sons.)

sensor connected in series to the sender coil for wireless detection. To realize a human-interactive health monitoring system, the sensing systems may also need to incorporate a curing system such as drug delivery. By attaching a polydimethylsiloxane (PDMS)-based flexible drug delivery pump onto a Kapton substrate, it can be mechanically flexible without delamination or cracks when the device is bent (Fig. 22(c)) or placed on a human arm (Fig. 23(a)). The pump is readily activated with a low applied pressure of ~3.3 kPa, which is equal to a gentle human touch (Fig. 23(b)). Furthermore, temperature and wireless touch detections have been demonstrated as real-time measurements by placing onto a human skin (Figs. 23(c)–23(e)).

8. Summary and Outlook

In summary, this chapter introduces flexible sensors and select examples of high performance printable components using mainly inorganic nano-materials. By dispersing inorganic nanomaterials into solutions, solutions can be used as ink for printing methods, which may realize economical, large-scale, and high-performance devices on user-defined substrates, including flexible substrates. Organic materials for flexible devices have been dramatically developed and improved for practical applications. However, using only organic materials may not realize flexible and wear-able sensor systems. Because each material has its own benefits, both organic and inorganic material systems need to be studied in parallel and heterogeneously incorporated to realize flexible and wearable electronic systems. The sensors on flexible substrates described in this chapter are only examples of the efforts in different research fields (e.g., mechanical engineering, electrical engineering, material science, chemistry, etc.). These efforts should realize practical flexible electrical systems in the near future.

Acknowledgments

The author acknowledges financial support from Osaka Prefecture University.

References

1. W. J. Hyun, O. O. Park and B. D. Chin, *Adv. Mater.* **25**, 4729 (2013).
2. A. C. Siegel, S. T. Phillips, M. D. Dickey, N. Lu, Z. Suo and G. M. Whitesides, *Adv. Funct. Mater.*, **20**, 28 (2010).
3. R. Guo, Y. Yu, Z. Xie, X. Liu, X. Zhou, Y. Gao, Z. Liu, F. Zhou, Y. Yang and Z. Zheng, *Adv. Mater.*, **25**, 3343 (2013).
4. T. Sekitani, Y. Noguchi, K. Hata, T. Fukushima, T. Aida and T. Someya, *Science*, **321**, 1468 (2008).
5. H. Wu, D. Kong, Z. Ruan, P.-C. Hsu, S. Wang, Z. Yu, T. J. Carney, L. Hu, S. Fan and Y. Cui, *Nature Nanotechnol.*, **8**, 421 (2013).
6. J. Du, S. Pei, L. Ma and H.-M. Cheng, *Adv. Mater.*, **13**. 1958 (2014).

7. T. Takahashi, K. Takei, E. Adabi, Z. Fan, A. M. Niknejad and A. Javey, *ACS Nano*, **4**, 5855 (2010).

8. C. Wang, J.-C. Chien, H. Fang, K. Takei, J. Nah, E. Plis, S. Krishna, A. M. Niknejad and A. Javey, *Nano Lett.*, **12**, 4140 (2012).

9. M. Madsen, K. Takei, R. Kapadia, H. Fang, H. Ko, T. Takahashi, A. C. Ford, M. H. Lee and A. Javey, *Adv. Mater.*, **23**, 3115 (2011).

10. T. Takahashi, K. Takei, A. G. Gillies, R. S. Fearing and A. Javey, *Nano Lett.*, **11**, 5408 (2011).

11. C. Wang, J.-C. Chien, K. Takei, T. Takahashi, J. Nah, A. M. Niknejad and A. Javey, *Nano Lett.*, **12**, 1527 (2012).

12. H. Chen, Y. Cao, J. Zhang and C. Zhou, *Nat. Commun.*, **5**, 4097 (2014).

13. C. Wang, K. Takei, T. Takahashi and A. Javey, *Chem. Soc. Rev.*, **42**, 2592 (2013).

14. Q. Cao, H.-S. Kim, N. Pimparkar, J. P. Kulkarni, C. Wang, M. Shim, K. Roy, M. A. Alam and J. A. Rogers, *Nature*, **454**, 494 (2008).

15. K. Takei, Z. Yu, M. Zheng, H. Ota, T. Takahashi and A. Javey, *P. Natl. Acad. Sci. USA*, **111**, 1703 (2014).

16. S. Harada, W. Honda, T. Arie, S. Akita and K. Takei, *ACS Nano*, **8**, 3921 (2014).

17. K. Takei, T. Takahashi, J. C. Ho, H. Ko, A. G. Gillies, P. W. Leu, R. S. Fearing and A. Javey, *Nat. Mater.*, **9**, 821 (2010).

18. X. Liu, L. Gu, Q. Zhang, J. Wu, Y. Long and Z. Fan, *Nat. Commun.*, **5**, 4007 (2014).

19. H. C. Ko, M. P. Stoykovich, J. Song, V. Malyarchuk, W. M. Choi, C.-J. Yu, J. B. Geddes, J. Xiao, S. Wang, Y. Huang and J. A. Rogers, *Nature*, **454**, 748 (2008).

20. J. Liang, L. Li, X. Niu, Z. Yu and Q. Pei, *Nature Photonics*, **7**, 817 (2013).

21. T. Sekitani, H. Nakajima, H. Maeda, T. Fukushima, T. Aida, K. Hata and T. Someya, *Nat. Mater.*, **8**, 494 (2009).

22. L. Zhou, A. Wanga, S.-C. Wu, J. Sun, S. Park and T. N. Jackson, *Appl. Phys. Lett.*, **88**, 083502 (2006).

23. A. Sugimoto, H. Ochi, S. Fujimura, A. Yoshida, T. Miyadera and M. Tsuchida, *IEEE J. Sel. Top Quant.*, **10**, 107 (2004).

24. P. H. Lau, K. Takei, C. Wang, Y. Ju, J. Kim, Z. Yu, T. Takahashi, G. Cho and A. Javey, *Nano Lett.*, **13**, 3864 (2013).

25. M. Jung, J. Kim, J. Noh, N. Lim, C. Lim, G. Lee, J. Kim, H. Kang, K. Jung, A. D. Leonard, J. M. Tour and G. Cho, *IEEE T. Electron Dev.*, **57**, 571 (2010).

26. M. Kaltenbrunner, T. Sekitani, J. Reeder, T. Yokota, K. Kuribara, T. Tokuhara, M. Drack, R. Schwodiauer, I. Graz, S. Bauer-Gogonea, S. Bauer and T. Someya, *Nature*, **499**, 458 (2013).

27. H. Fuketa, K. Yoshioka, T. Yokota, W. Yukita, M. Koizumi, M. Sekino, T. Sekitani, M. Takamiya, T. Someya and T. Sakurai, *Proc. 2014 IEEE Int. Solid-State Circuits Conf.*, 490 (2014).

28. G. Schawartz, B. C.-K. Tee, J. Mei, A. L. Appleton, D. H. Kim, H. Wang and Z. Bao, *Nat. Commun.*, **4**, 1859 (2013).

29. H.-s. Kim, E. Bruechkner, J. Song, Y. Li, S. Kim, C. Lu, J. Sulkin, K. Choquette, Y. Huang, R. G. Nuzzo and J. A. Rogers, *P. Natl. Acad. Sci. USA*, **108**, 10072 (2011).

30. J. A. Rogers, M. G. Lagally and R. G. Nuzzo, *Nature*, **477**, 45 (2011).

31. J. Veventi, D.-H. Kim, L. Vigeland, E. S. Frechette, J. A. Blanco, Y.-S. Kim, A. E. Avrin, V. R. Tiruvadi, S.-W. Hwang, A. C. Vanleer, D. F. Wulsin, K. Davis, D. Contreras, J. A. Rogers and B. Litt, *Nat. Neurosci.*, **14**, 1599 (2011).

32. D. Shahrjerdi and S. W. Bedell, *Nano Lett.*, **13**, 315 (2013).

33. S. Bae, H. Kim, Y. Lee, X. Xu, J.-S. Park, Y. Zheng, J. Balakrishnan, T. Lei, H. R. Kim, Y. I. Song, Y.-J. Kim, K. S. Kim, B. Ozyilmaz, J.-H. Ahn, B. H. Hong and S. Iijima, *Nat. Nanotech.*, **5**, 574 (2010).

34. H. Fang, S. Chuang, T. C. Chang, K. Takei, T. Takahashi and A. Javey, *Nano Lett.*, **12**, 3788 (2012).

35. B. Radisavljevic, A. Radenovic, J. Brivio, V. Giacometti and A. Kis, *Nat. Nanotech.*, **6**, 147 (2011).

36. J. Xiang, W. Lu, Y. Hu, Y. Wu, H. Yan and C. M. Lieber, *Nature*, **441**, 489 (2006).

37. Z. Fan, J. C. Ho, Z. A. Jacobson, R. Yerushalmi, R. L. Alley, H. Razavi and A. Javey, *Nano Lett.*, **8**, 20 (2008).

38. C. Wang, J. Zhang, K. Ryu, A. Badmaev, L. G. D. Arco and C. Zhou, *Nano Lett.*, **9**, 4285 (2009).

39. Z. Fan, J. C. Ho, T. Takahashi, R. Yerushalmi, K. Takei, A. C. Ford, Y.-L. Chueh and A. Javey, *Adv. Mater.*, **21**, 3730 (2009).

40. J. Noh, D. Yeom, C. Lim, H. Cha, J. Han, J. Kim, Y. Park, V. Subramanian and G. Cho, *IEEE T. Electron Pa. M.*, **33**, 275 (2010).

41. W. Honda, S. Harada, T. Arie, S. Akita and K. Takei, *Adv. Funct. Mater.*, **24**, 3299 (2014).

42. Y. Noguchi, T. Sekitani and T. Someya, *Appl. Phys. Lett.*, **89**, 253507 (2006).

43. W. Honda, T. Arie, S. Akita and K. Takei, *Phys. Status Solidi A*, in press (2014) doi 10.1002/pssa.201431481.

44. W. S. Boyle and G. E. Smith, *AT & T TECH. J.*, **49**, 587 (1970).

45. A. E. Gamal and H. Eltoukhy, *IEEE Circuit DEVIC.*, May/June, 6 (2005).

46. Z. Jin and J. Wang, *Sci. Rep.*, **4**, 5331 (2014).

47. X. Gong, M. Tong, Y. Xia, W. Cai, J. S. Moon, Y. Cao, G. Yu, C.-L. Shieh, B. Nilsson and A. J. Heeger, *Science*, **325**, 1665 (2009).

48. T. Takahashi, Z. Yu, K. Chen, D. Kiriya, C. Wang, K. Takei, H. Shiraki, T. Chen, B. Ma and A. Javey, *Nano Lett.*, **13**, 5425 (2013).

49. G. H. Gelinck, A. Kumar, D. Moet, J.-L. van der Steen, U. Shafique, P. E. Malinowski, K. Myny, B. P. Rand, M. Simon, W. Rutten, A. Douglas, J. Jorritsma, P. Heremans and R. Andriessen, *Organic Electronics*, **14**, 2602 (2013).

50. V. D. Mihailetchi, H. Xie, B. de Boer, L. J. A. Koster and P. W. M. Blom, *Adv. Funct. Mater.*, **16**, 699 (2006).

51. C. W. E. van Eijk, *Phys. Med. Biol.*, **47**, R85 (2002).

52. L. Lin, S. Liu, Q. Zhang, X. Li, M. Ji, H. Deng and Q. Fu, *ACS Appl. Mater. Interfaces*, **5**, 5815 (2013).

53. C. Pang, G.-Y. Lee, T.-i. Kim, S. M. Kim, H. N. Kim, S.-H. Ahn and K.-Y. Suh, *Nat. Mater.*, **11**, 795 (2012).

54. D. J. Lipomi, M. Vosgueritchian, B. C-K. Tee, S. L. Hellstrom, J. A. Lee, C. H. Fox and Z. Bao, *Nat. Nanotechnol.*, **6**, 788 (2011).

55. J. Zhou, Y. Gu, P. Fei, W. Mai, Y. Gao, R. Yang, G. Bao and Z. L. Wang, *Nano Lett.*, **8**, 3035 (2008).

56. Q. Liao, M. Mohr, X. Zhang, Z. Zhang, Y. Zhang and H.-J. Fecht, *Nanoscale*, **5**, 12350 (2013).

57. K. Fukuda, K. Hikichi, T. Sekine, Y. Takeda, T. Minamiki, D. Kumaki and S. Tokito, *Sci. Rep.*, **3**, 2048 (2013).

58. Z. Suo, E. Y. Ma, H. Gleskova and S. Wagner, *Appl. Phys. Lett.*, **74**, 1177 (1999).

59. J. Herrman, K.-H. Muller, T. Reda, G. R. Baxter, B. Raguse, G. J. J. B. de Groot, R. Chai, M. Roberts and L. Wieczorek, *Appl. Phys. Lett.*, **91**, 183105 (2007).

60. J. Park, Y. Lee, J. Hong, M. Ha, Y.-D. Jung, H. Lim, S. Y. Kim and H. Ko, *ACS Nano*, **8**, 4689 (2014).

61. S. C. B. Mannsfeld, B. C-K. Tee, R. M. Stoltenberg, C. V. H-H. Chen, S. Barman, B. V. O. Muir, A. N. Sokolov, C. Reese and Z. Bao, *Nat. Mater.*, **9**, 859 (2010).

62. M. Akiyama, Y. Morofuji, T. Kamohara, K. Nishikubo, M. Tsubai, O. Fukuda and N. Ueno, *J. Appl. Phys.*, **100**, 114318 (2006).

63. W. Wu, X. Wen and Z. L. Wang, *Science*, **340**, 952 (2013).

64. G. Schwartz, B. C.-K. Tee, J. Mei, A. L. Appleton, D. H. Kim, H. Wang and Z. Bao, *Nat. Commun.*, **4**, 1859 (2013).

65. T. Someya, Y. Kato, T. Sekitani, S. Iba, Y. Noguchi, Y. Murase, H. Kawaguchi and T. Sakurai, *P. Natl. Acad. Sci. USA*, **102**, 12321 (2005).

66. D.-H. Kim, N. Lu, R. Ghaffari, Y.-S. Kim, S. P. Lee, L. Xu, J. Wu, R.-H. Kim, J. Song, Z. Liu, J. Viventi, B. de Graff, B. Elolampi, M. Mansour, M. J. Slepian, S. Hwang, J. D. Moss, S.-M. Won, Y. Huang, B. Litt and J. A. Rogers, *Nat. Mater.*, **10**, 316 (2011).

67. D.-H. Kim, S. Wang, H. Keum, R. Ghaffari, Y.-S. Kim, H. Tao, B. Panilaitis, M. Li, Z. Kang, F. Omenetto, Y. Huang and J. A. Rogers, *Small*, **8**, 3263 (2012).

68. D. Yu, C. Wang, B. L. Wehrenberg and P. Guyot-Sionnest, *Phys. Rev. Lett.*, **92**, 216902 (2004).

69. Y. S. Kim, *Sens. Actuator B*, **114**, 410 (2006).

70. M. C. McAlpine, H. Ahmad, D. Wang and J. S. Heath, *Nat. Mater.*, **6**, 379 (2007).

71. P.-G. Su, C.-T. Lee, C.-Y. Chou, K.-H. Cheng and Y.-S. Chuang, *Sens. Actuator B*, **139**, 488 (2009).

72. M. A. Lim, D. H. Kim, C.-O. Park, Y. W. Lee, S. W. Han, Z. Li, R. S. Williams and I. Park, *ACS Nano*, **6**, 598 (2012).

73. J. Nah, S. B. Kumar, H. Fang, Y.-Z. Chen, E. Plis, Y.-L. Chueh, S. Krishna, J. Guo and A. Javey, *J. Phys. Chem. C*, **116**, 9750 (2012).

74. K. Skucha, Z. Fan, K. Jeon, A. Javey and B. Boser, *Sens. Actuators B*, **145**, 232 (2010).

75. D. Lee and T. Cui, *Biosens. Bioelectron.*, **25**, 2259 (2010).

76. J. T. Smith, S. S. Shah, M. Goryll, J. R. Stowell and D. R. Allee, *IEEE Sensors J.*, **14**, 937 (2014).

77. A. Loi, I. Manunza and A. Bonfiglio, *Appl. Phys. Lett.*, **86**, 103512 (2005).

78. Y.-L. Chin, J.-C. Chou, T.-P. Sun, H.-K. Liao, W.-Y. Chung and S.-K. Hsiung, *SENSORS ACTUAT B*, **75**, 36 (2001).

79. http://chemwiki.ucdavis.edu/Analytical_Chemistry/Electrochemistry/.

80. T. Someya, T. Sekitani, S. Iba, Y. Kato, H. Kawaguchi and T. Sakurai, *P. Natl. Acad. Sci. U.S.A.*, **101**, 9966 (2004).

81. D.-H. Kim, N. Lu, R. Ma, Y.-S. Kim, R.-H. Kim, S. Wang, J. Wu, S. M. Won, H. Tao, A. Islam, K. J. Yu, T.-i. Kim, R. Chowdhury, M. Ying, L. Xu, M. Li, H.-J. Chung, H. Keum, M. McCormick, P. Liu, Y.-W. Zhang, F. G. Omenetto, Y. Huang, T. Coleman and J. A. Rogers, *Science*, **333**, 838 (2011).

82. J.-W. Jeong, W.-H. Yeo, A. Akhtar, J. J. S. Norton, Y.-J. Kwack, S. Li, S.-Y. Jung, Y. Su, W. Lee, J. Xia, H. Cheng, Y. Huang, W.-S. Choi, T. Bretl and J. A. Rogers, *Adv. Mater.*, **25**, 6839 (2013).

83. S.-W. Hwang, H. Tao, D.-H. Kim, H. Cheng, J.-K. Song, E. Rill, M. A. Brenckle, B. Panilaitis, S. M. Won, Y.-S. Kim, Y. M. Song, K. J. Yu, A. Ameen, R. Li, Y. Su, M. Yang, D. L. Kaplan, M. R. Zakin, M. J. Slepian, Y. Huang, F. G. Omenetto and J. A. Rogers, *Science*, **337**, 1640 (2012).

84. S.-W. Hwang, D.-H. Kim, H. Tao, T.-i. Kim, S. Kim, K. J. Yu, B. Panilaitis, J.-W. Jeong, J.-K. Song, F. G. Omenetto and J. A. Rogers, *Adv. Funct. Mater.*, **23**, 4087 (2013).

85. J. H. Solomon and M. J. Z. Hartmann, *IEEE Trans. Robotics*, **24**, 1157 (2008).

86. D. Kim and R. Moller, *Robot. Auton. Syst.*, **55**, 229 (2007).

87. D.-H. Kim, J. Viventi, J. J. Amsden, J. Xiao, L. Vigeland, Y.-S. Kim, J. A. Blanco, B. Panilaitis, E. S. Frechette, D. Contreras, D. L. Kaplan, F. G. Omenetto, Y. Huang, K.-C. Hwang, M. R. Zakin, B. Litt and J. A. Rogers, *Nat. Mater.*, **9**, 11 (2010).

88. J. Viventi, D.-H. Kim, J. D. Moss, Y.-S. Kim, J. A. Blanco, N. Annetta, A. Hicks, J. Xiao, Y. Huang, D. J. Callans, J. A. Rogers and B. Litt, *Sci. Transl. Med.*, **2**, 24ra22 (2010).

89. J.-W. Jeong, M. K. Kim, H. Cheng, W.-H. Yeo, X. Huang, Y. Liu, Y. Zhang, Y. Huang and J. A. Rogers, *Adv. Healthcare Mater.*, **3**, 642 (2014).

90. X. Wang, Y. Gu, Z. Xiong, Z. Cui and T. Zhang, *Adv. Mater.*, **26**, 1336 (2014).

91. P. Digiglio, R. Li, W. Wang and T. Pan, *Ann. Biomed. Eng.*, doi: 10.1007/s10439-014-1037-1 (2014).

92. C. Wang, D. Hwang, Z. Yu, K. Takei, J. Park, T. Chen, B. Ma and A. Javey, *Nat. Mater.*, **12**, 899 (2013).

93. C. Pan, L. Dong, G. Zhu, S. Niu, R. Yu, Q. Yang, Y. Liu and Z. L. Wang, *Nature Photonics*, **7**, 752 (2013).

CHAPTER 3

GRAPHENE: FROM SYNTHESIS TO APPLICATIONS IN FLEXIBLE ELECTRONICS

Henry Medina, Wen-Chun Yen, Yu-Ze Chen, Teng-Yu,
Yu-Chuan Shih and Yu-Lun Chueh*

*Department of Materials Science, National Tsing Hua University
No. 101, Section 2, Kuang-Fu Road, Hsinchu, Taiwan 30013, R.O.C.
ylchueh@mx.nthu.edu.tw

Graphene has proven to have exceptional electrical and mechanical properties when used in flexible memories, flexible sensors, and field-effect transistors (FET). Despite the outstanding results observed, several issues have to be further improved from synthesis in order to improve the device fabrication yield. In this chapter, the electrical and mechanical properties of graphene will be discussed. Issues related to synthesis will be addressed and recent developments in applications related to flexible electronics will be shown. Finally, summary and conclusions will be given.

1. Introduction

For the last five years, the amount of publications related to flexible electronics has almost doubled moving from 473 papers published in 2008 to 908 papers in 2013, showing a particular interest in its development. On the other hand, graphene has shown astonishing properties that have been described in reviews extensively. In particular, due to its electrical and mechanical properties,[1] graphene is taking a privileged position among other materials in the development of a new series of flexible electronics. After a sustained progress on the mass production of graphene, such as the graphite exfoliation and large area synthesis by chemical vapor deposition (CVD),[2,3] the increasing amount of publications related to graphene flexible electronics has gone from four in 2008 to 110 last year, which is around 10% of the manuscript related to flexible electronics (Fig. 1).

Graphene has shown exceptional optical transparency, suggesting a large amount of applications as transparent electrode in a series of highly flexible electronics.[4–6] High sensitive strain sensors have been developed using graphene with gauge factors (GFs) several times larger than those made of metals.[7] Furthermore, due to its particular ultra-high hole and electron mobility, a new era of radio-frequency (RF) flexible transistor is under development.[8] Despite all the great advances in application development, there are still several challenges to address for graphene to be used in commercial application, starting from reliable synthesis and transfer methods. The most common way to synthesize graphene is by chemical vapor deposition on transition metals owing to the high quality of the films obtained by this method.[4] Nevertheless, the required transfer process became a major bottleneck with several issues to address such as wrinkles, scratches and residues.[9] For this reason, the searches for new

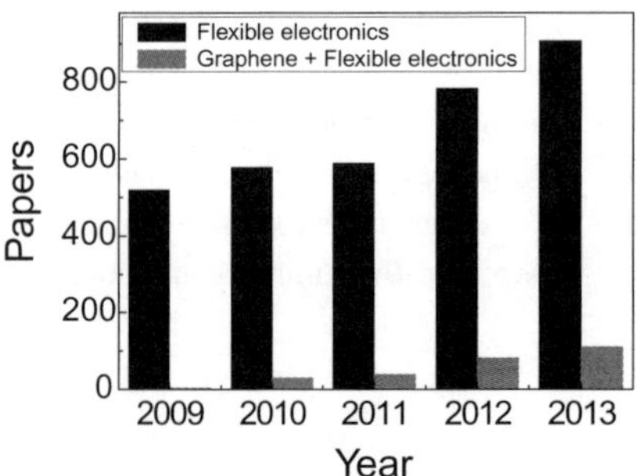

Figure 1.   Statistical data of paper published during the last five years about flexible electronics and graphene flexible electronics. The black bars correspond to the papers published containing the word, "flexible electronics" and the red bars to papers published containing the word, "graphene and flexible electronics". Data obtained from ISI Web of Science.

transfer methods and/or methods for direct graphene synthesis on insulator to avoid the transfer process bottlenecks remain as major research topics. In this chapter, we would like to address the electrical and mechanical properties that make graphene, such a promising material for flexible electronics. Then revise the state-of-art graphene synthesis by CVD, with a special emphasis on those methods for direct synthesis on insulators, limitations and perspectives for its use in flexible applications. Finally, we concentrate on the recent advances of graphene electronics for flexible applications.

2. Mechanical and Electrical Properties of Graphene

2.1. *Mechanical Properties*

The intrinsic strength of the carbon hexagonal lattice was predicted to exceed that of any other material, by utilizing large-scale quantum calculations reported by Zhao *et al.*[10] Experimentally, different methods have been previously applied to determine mechanical properties of various graphitic materials. Some of the first approaches made use of atomic force microscopy (AFM) tips.[11–13] This AFM method can be assumed to be rather accurate on nanotubes due to the uniaxial strain applied by the AFM tip.[14–18] Many researchers have reported assessment of mechanical properties of two-dimensional (2D) materials, such as graphene, using the AFM method. However biaxial strain instead of uniaxial strain has to be taken into account.

Lee *et al.*[19] used an atomic force microscope (AFM) nano-indentation to measure the mechanical properties of monolayer graphene membranes suspended over an open hole with diameters 1.5/1 mm and depth of 500 nm on 300 nm-thick SiO_2. The graphene membranes were scanned by non-contact mode and the AFM tip with the radius of 27.5 nm and 16.5 nm were subsequently positioned to within 50 nm from the center and the mechanical testing was then performed. The force-displacement measurements are shown in Fig. 2. The response of suspended graphene under uniaxial extension can be expressed as a Taylor series in powers of strain, for which the lowest order term leads to a linear elastic response and the third-order term leads to a nonlinear elastic

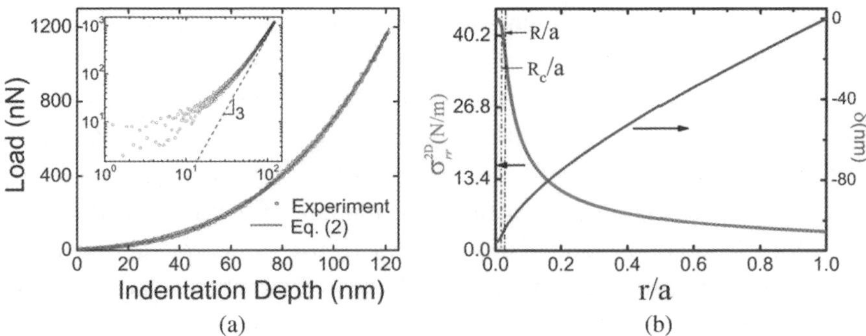

Figure 2. (a) Stress loading curve of suspended graphene by AFM tips. (b) Maximum stress and deflection of graphene membrane versus normalized radial distance at maximum loading. Adapted from Ref. 19.

behavior. Considering the elastic response of the graphene, the equation can be written as

$$\sigma = E\varepsilon + D\varepsilon^2, \tag{1}$$

where E is the Young's modulus and D is the third-order elastic modulus. Due to the real 2D nature of graphene, the force-displacement behavior is given by

$$F = \sigma_0^{2D}(\pi a)\left(\frac{\delta}{a}\right) + E^{2D}(q^3 a)\left(\frac{\delta}{a}\right)^3, \tag{2}$$

where F is the applied force by the tip and δ is the deflection at the center point of graphene membrane.

Figure 3 shows the distribution of the E^{2D} of suspended graphene derived from the above equation. Clearly, the average value of E^{2D} is near to 342 Nm^{-1} and the deviation is only below 10%. These quantities correspond to a high Young's modulus between 0.9 to 1.1 terapascals (TPa). The relation between a maximum stress for a clamped, linear elastic, and circular graphene membrane with an applied load by spherical indenter is given by

$$\sigma_m^{2D} = \left(\frac{FE^{2D}}{4\pi R}\right)^{\frac{1}{2}}, \tag{3}$$

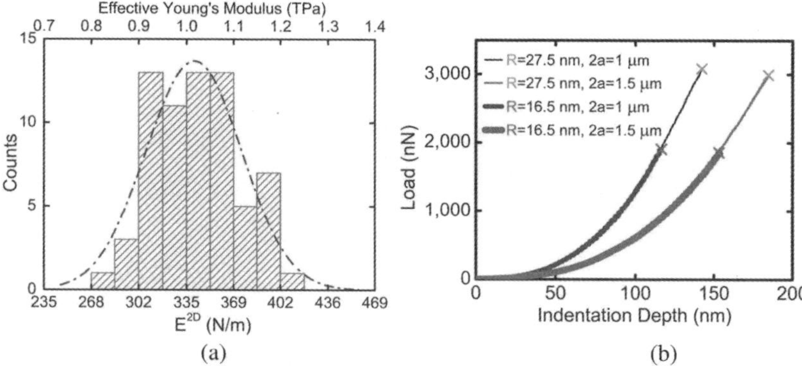

Figure 3. The left panel (a) shows the statistic results of effective Young's modulus of suspended graphene and the right panel (b) displays the stress loading curve of suspend graphene for different tip's diameters and pole sizes. Adapted from Ref. 19.

where σ_m^{2D} is the maximum stress at the central point of the suspended graphene and R is the tip radii pressed on the graphene. Taking the diameter of 16.5 nm for AFM tip and the 1 μm diameter for graphene as an example, the intrinsic strength σ^{2D} is 42 ± 4 Nm^{-1}, which corresponds to $E = 1.0 \pm 0.1$ TPa and $D = -2.0 \pm 0.4$ TPa, respectively. Therefore, the corresponding intrinsic stress of the suspended graphene with the effective thickness assumed to 0.335 nm $\sigma_{graphene}$ is equal to 130 ± 10 GPa at the strain $\varepsilon_{graphene}$ of 0.25.

2.2. *Electrical Properties of Graphene*

Despite the simplicity of the hexagonal graphene structure formed by carbon atoms, the electronic behavior shows fascinating and complex properties, giving high expectation for the possible applications of graphene in electronics. Remarkably, the theoretical calculations predicted several years ago by Wallace[20] show excellent agreement with the experimental results observed after the first experimental isolation of graphene on insulator, allowing the analysis of its electrical properties.

Graphene is formed by a layer of carbon atoms in sp^2 bonding. The orbitals comprise four valence electrons in each carbon atom when bonded on this sp^2 configuration. Two types of orbitals shapes can be differentiated

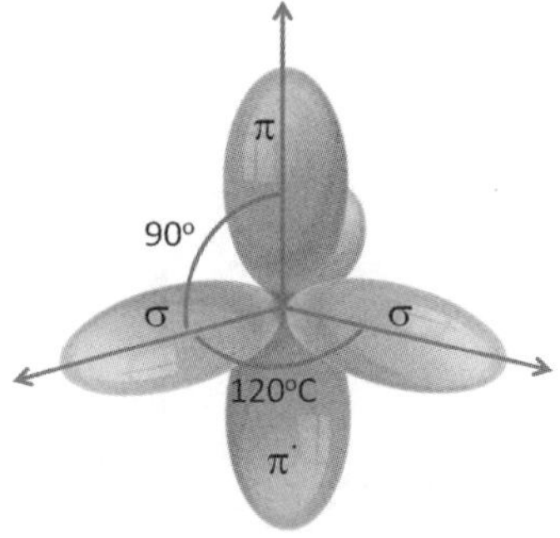

Figure 4.   Schematic of sp^2 orbital hybridization of the carbon atoms in graphene.

to be σ and π, respectively. The σ bond is formed by head-to-head bonding of the in-plane sp^2 orbitals, yielding a strong covalent bond with the nearest carbon atoms while π represents the $2p_z$ electron orbital, which is perpendicular to the graphene sheet that remains unbounded, and overlaps with the π orbital of its nearest neighbor, giving rise to a particular π-conjugated 2D gases (Fig. 4). It is interesting to notice that there is no overlap between σ and π orbitals, which give an advantage for theoretical analysis once each band can be treated from the others, independently. By using tight-binding method as presented by Wallace,[20] the electron band structure for graphene can be estimated.

In the electronic band structure of graphene, except for the π orbital, all orbitals show a very large separation between valence and conduction bands. For the σ orbital, the shortest gap is more than 10 eV around the Γ point, which reveals the strong sp^2 covalent bonding.

On the other hand, the π bands show the largest separation at the Γ point, but at the K point, the valence and conduction band touch at a point with no overlap between π and π^*. Because of that, graphene is described as a semiconductor with zero band gap, or as a semi-metal. Due to this behavior compared with the large gap of the σ bands, the π band is responsible for the particular electronics behavior of graphene. As mentioned before, there is no overlap between σ and π orbitals, simplifying the theoretic analysis and allowing the independent study of the π band from others.

Assuming that the electrical behavior of graphene mainly relays on the π band by using two adjacent carbon atoms as a unit cell, tight-binding

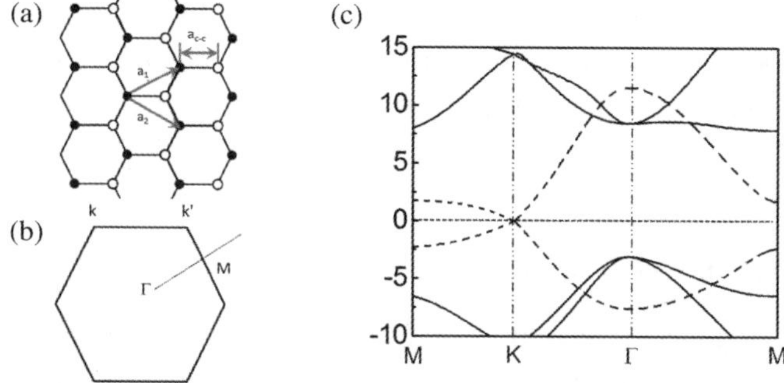

Figure 5. (a) Graphene schematic and the definition of a1, a2, and ac–c used for tight-binding calculations and (b) displays high symmetry points. In (c), the electronic band structure of graphene. The dashed line corresponds to the band for the π orbital while the solid line belongs to the closest σ bands.

method within the first nearest neighbor approximation can be used to deduce the energy dispersion relation close to the k point. For convenience, $a = a_{c-c}\sqrt{3}$ is used (Fig. 5). Then, in the tight binding approximation, the energy dispersion of the π electrons in graphene is given by

$$E = \pm\gamma_0\sqrt{1 + 4\cos^2\left(\frac{1}{2}ak_y\right) + 4\cos\left(\frac{\sqrt{3}}{2}ak_x\right)\cos\left(\frac{1}{2}ak_y\right)}, \qquad (5)$$

where γ_0 is the energy overlap integral between nearest neighbors. The $(-)$ root of the equation represents the valence band, which is fully occupied whereas the $(+)$ root is the conduction band, which is entirely empty. Both bands meet exactly at the k and k' points.

By expanding Eq. (5) around this $k(k0)$ point, a linear relation between E and k can be found given by:

$$E = \frac{\sqrt{3}}{2\gamma_0}a|k|. \qquad (6)$$

Defining $v_f = \left(\sqrt{3}\right)\gamma_0 a/2\hbar$ then E around the k point can be rewritten as

$$E = \hbar v_f |k|. \qquad (7)$$

Interestingly, the experimental results obtained by Novoselov *et al.*[21,22] showed good agreement, unveiling astonishing electrical properties of graphene, such as ultrahigh mobility and low sheet resistance for one atomic layer.

3. Synthesis Methods

Graphene synthesis is usually divided into exfoliate graphene (graphene powder) and graphene films (CVD graphene). In this chapter, we will focus on the synthesis of graphene by CVD methods.

3.1. *CVD synthesis of Graphene on Transition Metals and Transfer Process*

Since 2009,[3] graphene synthesized on copper foils *via* CVD has been the most reliable way for graphene film production due to its ability to synthesize graphene in large area. Despite the high quality of the film obtained after the CVD process, large growth temperatures (~1000°C) are required for the synthesis process. Several attempts have been done to reduce the synthesis temperature. Sun *et al.*[23] proposed a new strategy, using solid carbon source instead of gas-type carbon source, and the growth temperature could be effectively decreased to 800°C by using poly(methyl methacrylate) (PMMA) as a carbon source. The use of binary metal alloys with gold had also been proposed to synthesize graphene at relatively low temperatures, but the quality remained as a concern.[24,25] Further improvement was achieved by Li *et al.*[26] using benzene as carbon feedstock to successfully grow graphene at 300°C. However, the film was not continuous and several safety issues should be taken into account due to the toxic nature of benzene. Zhang *et al.*[27] followed a similar concept utilizing toluene to achieve low temperature synthesis, however continuous film were possible only at 600°C.

Despite the progress to reduce the temperature synthesis by this transition metal catalyst assisted CVD process, graphene needs an extra transfer process to an insulating substrate such as SiO_2 or Al_2O_3 for device. Polymer assisted transfer is the most well-known transfer process and has been reported extensevely.[28–32] Although, several variations of the methods have been proposed, the basic steps were reviewed. First, a protective polymer such as PMMA is deposited on the pre-grown graphene/catalyst metal sample by a spin coating method. Second, the metal catalyst is removed by chemical etching process and the etching solution is chosen accordingly to the metal. Note that $FeCl_3$ or HCl solutions are normally used as etchants to remove Cu or Ni foil. In this step, graphene is protected by the polymer film to avoid the tensile stress to destroy graphene resulting from the surface tension of the etching solution. Due to the low density feature of the polymer/graphene sample, it flows on the surface of the solution. Third, DI water was used to clean polymer/graphene for many times until all the chemical residues were removed. Fourth, the target substrate was then used to support the polymer/graphene. Lastly, the polymer is taken away by chemical etching or high temperature tempering. For example, PMMA can be removed by acetone or annealed in 400°C in H_2 environment. Figure 6 summarizes all the steps required in the procedure.

3.2. *Direct Synthesis of Graphene on Insulators*

Different methods have been proposed in order to synthesize graphene on insulators. Between them, using of remote catalyst, using of an intermediate metal layer as a catalyst, and plasma assisted CVD are three important methods. However, the main drawback of the CVD process is that a further chemical etching process is required to remove the metal thin-film

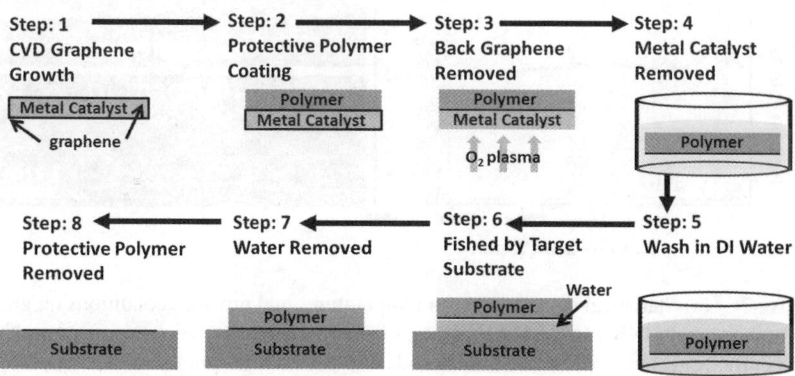

Figure 6.   Schematic process of graphene transfer process.

layer in order to allow the use of graphene inducing residual contaminants and affecting the electrical performance of graphene.

By using a remote catalyst, such as nickel[33] or copper,[34] graphene can be synthesized directly on insulators due to the vapor produced by the metal at temperatures closed to their melting point. In the case of Ni, the temperature and pressure play a critical role on the formation of graphene on oxides.[33] For temperature over 1000°C, the layer growth process becomes more pressure sensitive while for pressures over 100 torr, the formation of graphene spheres become dominant over the formation of graphene films. By further reducing the deposition rate, a conformal coating even on high aspect ratio structures such as silicon nanorods can be precisely controlled.[35] Figure 7 displays Raman and SEM images of the Si nanorods before and after the deposition of graphene. Despite that

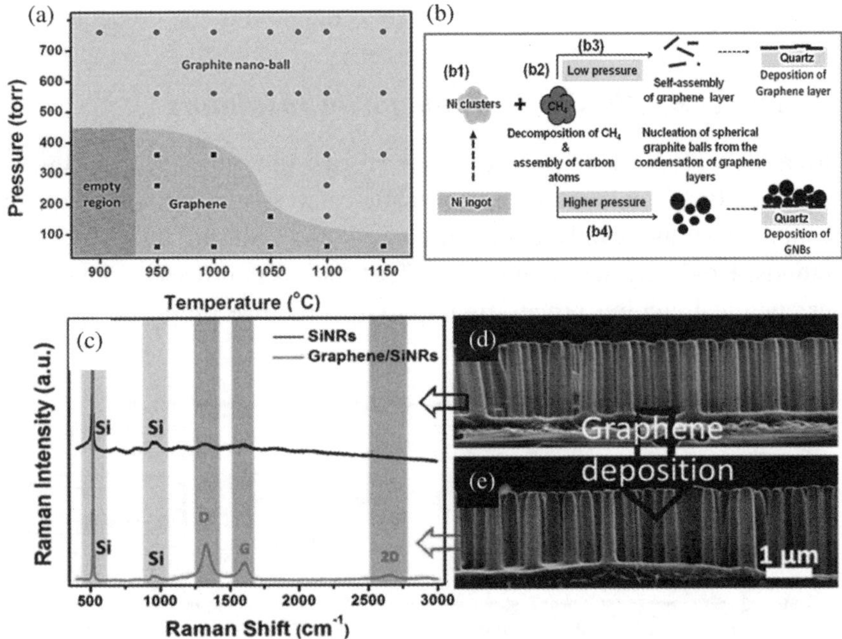

Figure 7. (a) Statistical data at different temperatures and pressure conditions on growth of graphene and GNBs. (b1–b4) Schematic of proposed growth mechanisms of graphene and GNBs. (c) Raman spectra of Si nanorods before and after graphene deposition by Ni vapor method. (d) and (e) are the SEM images of the Si nanorods array before and after graphene deposition. Adapted from Refs. 33 and 35.

this approach allows a precise control of the number of deposited layers even on non-planar surfaces, the temperature required to achieve good quality films is still too high, restricting the amount of substrates suitable for the process.

An alternative method was proposed by Su *et al.*[36] for direct formation of graphene on insulators by carbon atoms diffusion through grain boundaries. In this approach, a metal layer is first deposited on the target substrate. Then, carbon atoms diffuse through the metal layer to reach the substrate during the diffusion process of the carbon atoms, forming sp^2 bonding due to the catalytic properties of the metal layer. Similar approaches have also been reported using Ni as a metal layer.[37,38] In order to reduce the temperature process Yen *et al.*[37] proposes the use of a SiC substrate as a microwave susceptor for rapid annealing. By controlling input power of the microwave, the annealing temperature can be tuned, enabling the controlled segregation of carbon into the target substrate (Fig. 8).

Several variations of CVD systems such as plasma enhanced (PECVD) with RF (13.56 MHz) as reported by Wang *et al.*[39] or CVD (MPCVD) systems with microwave-frequency (2.45 GHz) as reported

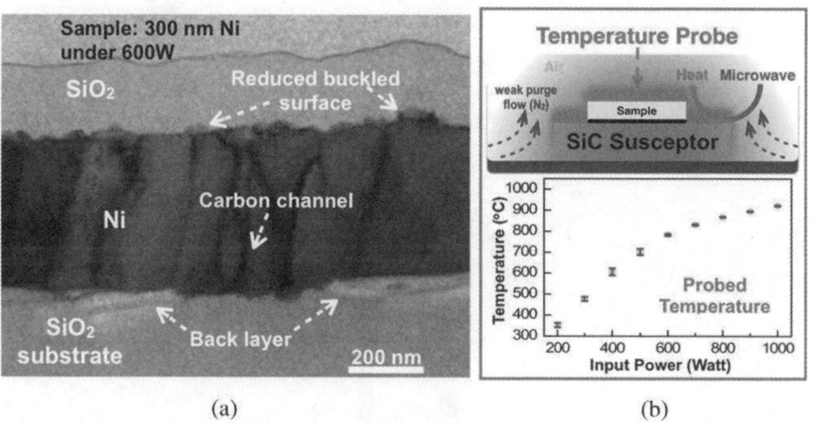

Figure 8. (a) TEM image of the segregated graphene through Ni onto SiO_2 after microwave annealing. In (b) schematic of the rapid heating using SiC substrate as a susceptor for microwave and the corresponding temperature estimated for the microwave input power. Figure adapted from Ref. 37.

by Malesevic *et al.*[40] are used to provide extra energy to the system in order to decrease the process temperature. This significant effect is due to the polarization and ionization of hydrocarbon molecules due to the strong electromagnetic field, increasing the possibility to break the atomic bonds between hydrogen and/or neighboring carbon atoms. Besides, the excited bombardment between the gaseous molecules also help to decompose the hydrocarbon gas. Since the hydrocarbon molecules are decomposed by the plasma, graphene can be deposited on arbitrary substrates. Figure 9 provides a schematic of a typical MPCVD system.

Although this technique provide a possibility to deposit graphene on to the oxide substrate with mono- or bi-layers. Unfortunately, the graphene deposited by PECVD or MPCVD usually exist on a very rough surface. Perhaps due to high nucleation, the graphene deposited by PECVD or MPCVD can grow without following the surface of substrate. Therefore, the formation of nano-flower or nano-wall shapes are often observed.[41–45]

To further improve the roughness, electron cyclotron resonance CVD (ECR-CVD) has been used to deposit graphene onto silica substrate and the process temperature was further decreased to 400°C by using C_2H_4 gas

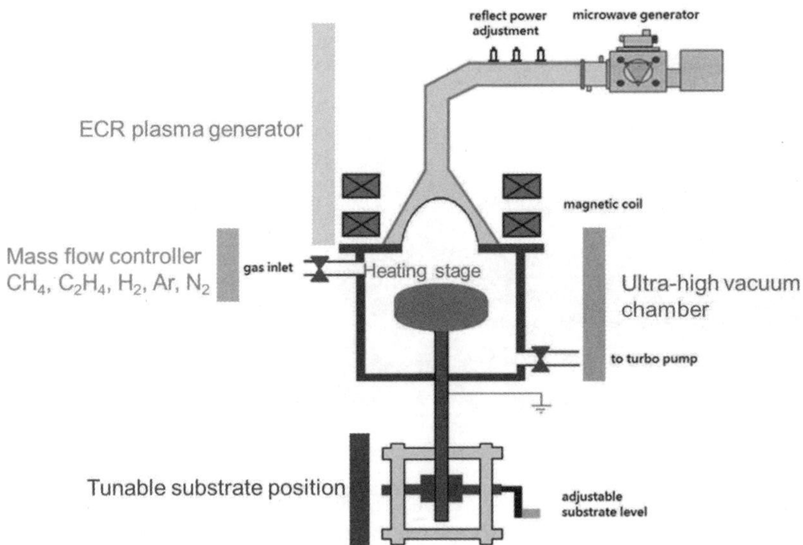

Figure 9.   The schematic diagram for MPCV.

as carbon source.[46] By carefully adjusting the distance between the plasma and the substrate as well as plasma power and the gas flow, large area uniform graphene can be achieved. Besides, a SiC peak is observed in both Raman spectrum and X-ray photoelectron spectrum of graphene deposited by ECR-CVD, indicating that a SiC buffer layer was formed between graphene and silica through the reaction: $3C + SiO_2 \rightarrow SiC + 2CO$ within four minutes. Once the Sic layer is formed, graphene will grow to over all the surface of the sample within one minute. But, in spite of the uniformity and smoothness of the synthesized graphene deposited by ECRCVD, the domain size is relatively small, only 2–3 nm in length Figure 10 shows the graphene deposited on different thickness and its optical and

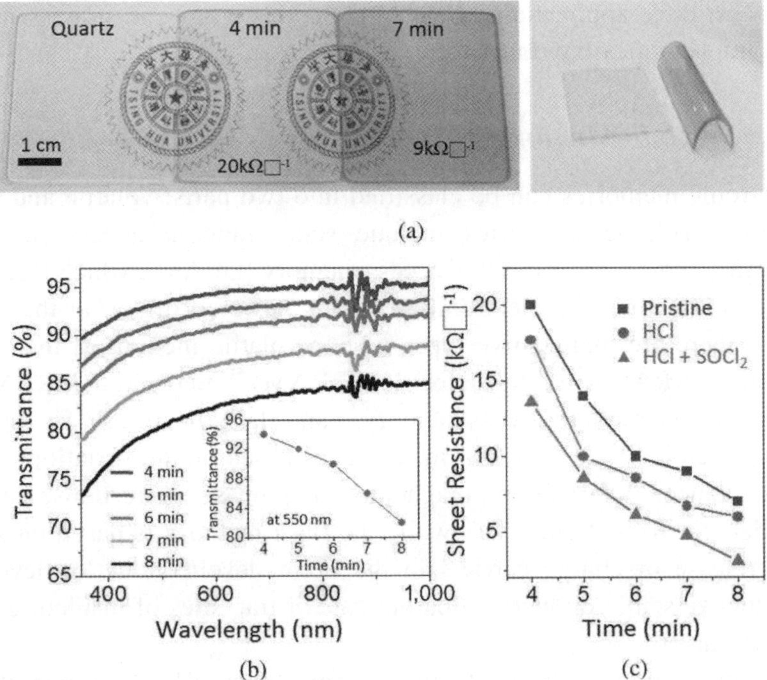

Figure 10. Direct deposition of ECR-CVD nanographene on quartz. (a) Nanographene films with different thicknesses obtained by varying the growth time. (b) Optical transmittance of ECR-CVD nanographene grown on quartz, with different film thickness. The inset compares the transmittance at 550 nm for different growth durations. (c) The sheet resistance decreases as the film thickness increases; it also decreases with increasing doping by chemical functionalization in HCl and $SOCl_2$ aqueous solutions. Adapted from Ref. 46.

electrical performance. In fact, most graphene flakes with few nanometers domain are embedded in the amorphous carbon film. The quality of graphene is limited by the instrument since the energy of the plasma should be high enough to decompose the hydrocarbon molecules and allow them to recrystallize after leaving the plasma region. For that reason, there is a trend between the plasma power and the temperature. In order to further reduce the temperature, the higher plasma power is required.

4. Applications of Graphene in Flexible Electronics

Graphene has shown exceptional performance in several applications in electronics. In particular, for flexible electronics in this chapter, we will focus on three applications: flexible memories, graphene strain sensors, and graphene flexible transistors.

4.1. *Flexible Memories*

Electronic memories can be classified into two parts: volatile and non-volatile. Volatile memories include static random access memory (SRAM) and dynamic random access memory (DRAM), which loses the stored data when the electronic device is powered off.[47] On the other hand, there are mainly five types of non-volatile memories, including resistive (RRAM),[48,49] ferroelectric (FeRAM),[50] magnetic (MRAM),[51] phase change random access memory,[52] and flash memory.[53] The operation principle of the first four memories is based on the variation of the conductance, polarization, magnetism and phase in response to the applied electrical field[49,54–56] while the flash memory is based on store and release of charge carriers in the Fermi level/conduction level of conductive/semi-conductive floating gate or trap sites of insulating data storage layer.[57]

From the previously mentioned memory types, RRAM is one of the most promising candidates for next generation non-volatile memory devices. Its structure is composed of metal/insulator/metal, which is very simple as shown in Fig. 11. Because of the nanometer size of the conduction filament, devices can scale down to nanoscale for future technologies. Furthermore, it has excellent electrical behavior such as high-density

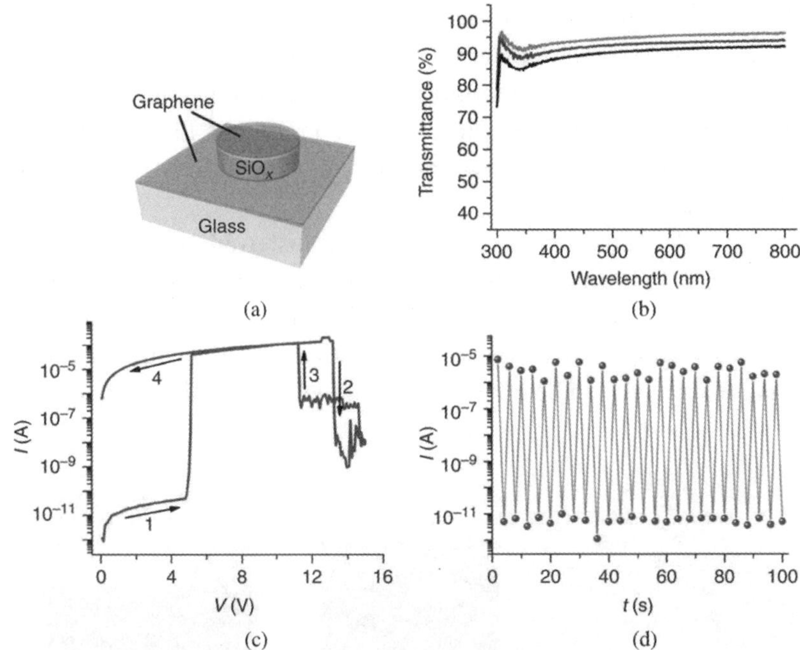

Figure 11.   (a) Schematic of the G/SiO$_x$/G device on glass. (b) Optical transmittance in G/SiO$_x$/G-layered structures with different layer thicknesses of graphene. (c) Characteristic *I–V* curves from an electroformed G/SiO$_x$/G device. The arrows indicate the voltage-sweep directions and the numbers indicate the order. (d) Corresponding memory cycles from the device using +6 V and +15 V as set and reset voltages, respectively. The memory states (current) were recorded at +1 V. Figure reproduced from Ref. 71.

integration, long retention time, and fast switching speed.[58,59] The insulator layer is usually based on transition metal oxide such as, TiO_2,[60] ZnO,[61] HfO,[62] TaO_x[63] or SiO_2,[64] and organic materials.[65,66] For the flexible application, the substrates such as stainless steel[61] and conductive polymers[67,68] are often used. Graphene is also an excellent choice as a conductive layer in flexible memories due to its high transparency, high mechanical strength, and high carrier mobility. Different groups have presented studies making use of graphene in different ways for RRAM applications and showed high flexibility and stable electrical behaviors.[69–71]

In particular, Yao *et al.* used CVD graphene as electrode and SiO_x as an insulator as shown in Fig. 11.[71] Initially, CVD graphene was transferred from Cu foil to glass substrate. Then, a layer of SiO_x film with the thickness of 70 nm was deposited by electron-beam evaporation. The top graphene film was then transferred onto the SiO_x layer and finally the pattern was defined by photolithography. The high transmittance and high on/off ratio (~1,000) is presented in Figs. 11(b) and 11(c), respectively. Furthermore, endurance test was carried out, using pulse sweep and showing good retention.

For flexible operation, graphene was transferred onto high melting point (>280°C) plastic substrate (fluoropolymer, PFA). The corresponding structure and optical view are showed in Fig. 12. The on/off ratio can still reach above 1,000, which is just like the device on glass substrate, meaning that this device can be operated well on arbitrary substrates. However,

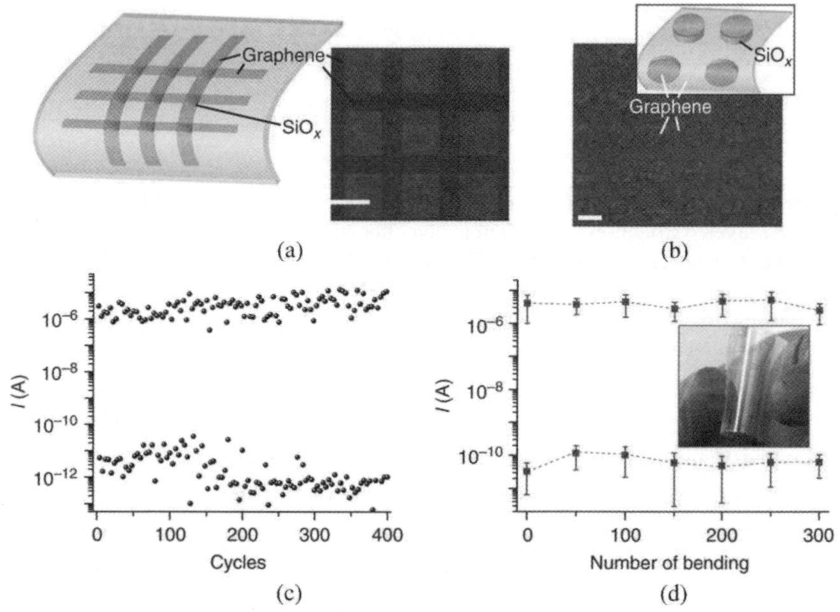

Figure 12.   (a) Schematic of the G/SiO$_x$/G crossbar structures and the optical image of the structures. (b) Optical image of the G/SiO$_x$/G structures with the inset showing the schematic image. (c) Memory cycles from one of the crossbar devices (d) Retention of both on and off memory states from a crossbar device is shown on bending the plastic substrate around a ~1.2-cm diameter curvature. The inset shows the actual transparent memory devices using the pillar structures on the plastic substrate. Adapted from Ref. 71.

the yield of the device on the flexible substrate is lower, which is attributed due to the deformation caused by the contact of the tip and the device. It may cause some defects inside SiO_x, leading to the failure of the device. The bending test is showed in Fig. 12(d). The device still shows great electrical behavior after bended about three hundred times.

4.2. *Strain Sensors*

Recently, graphene was served as the new candidate in strain sensors. Atomic force microscope (AFM) is considered a useful tool to analyze the mechanical properties of different nanomaterial such as nanowires.[11–13] An AFM tip was used in contact mode to deflect a nanowire to measure bending force as the function of the displacements.[11] A different method by attaching two ends of a carbon nanotube between two AFM tips was proposed to access the accurate measurement of Young's Modulus for a single carbon nanotube.[13] However, it is noted that uniaxial strain achieved by the AFM tip is an important step to access mechanical properties. On changing the nanomaterial from nanotubes to 2D materials such as graphene, the assessment of the mechanical properties would also change to biaxial strain instead of uniaxial strain due to one atom layered structure.[19,72,73] Besides, it is possible to form point defects by the AFM tip in contact mode during the measurement.

To solve these problems, a non-contact method to determine mechanical properties of graphene using Raman spectroscopy has been proposed by Mohiuddin and coworkers.[74] Figure 13 shows the corresponding Raman spectra as a function of the uniaxial strain, for which the strain was longitudinally applied to films by stretching the substrate along the basal plane. From the Raman spectra, it should be noted that the shift of peak position would increase with the increase of the strain. The relation between the peak shift and the strain can be fitted with lorentzians and then access the trends for the peaks. The most notable thing is that the G peak would split into two sub-bands G^+ and G^- with larger strain, but it does not happen for 2D peak. This phenomenon can be explained by the effect of uniaxial strain on the optical modes response. The G peak corresponds to the doubly degenerated E_{2g} phonon at the Brillouin-zone

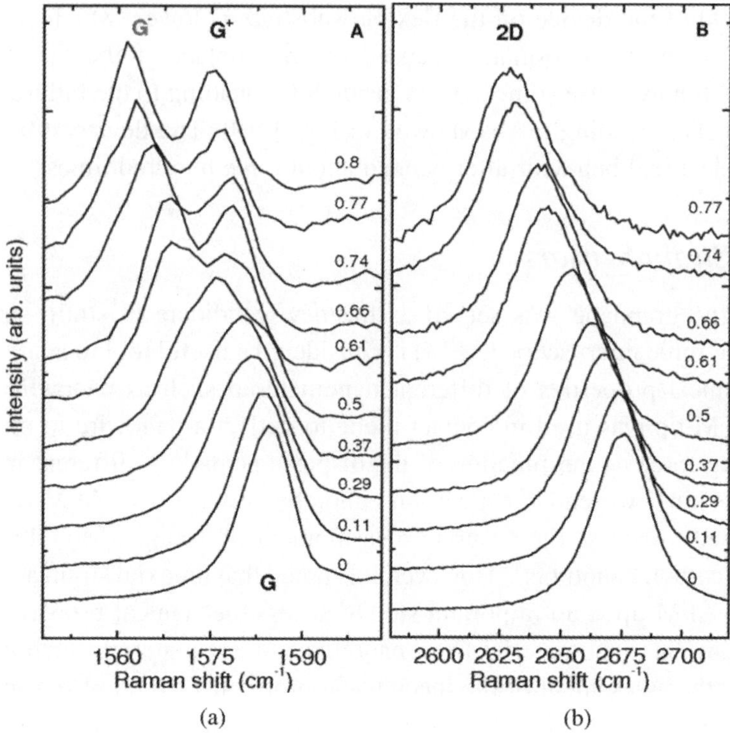

Figure 13. (a) G and (b) 2D peaks as a function of uniaxial strain. The spectra are measured with incident light polarized along the strain direction, collecting the scattered light with no analyzer. Note that the doubly degenerate G peak splits in two Sub-bands G^+ and G^-, while this does not happen for the 2D peaks. The strains, ranging from 0 to 0.8%, are indicated on the right side of the spectra. From Ref. 74.

center. The Gruneisen parameter, γ_{E2g}, and the shear deformation potential, $\beta_{E_{2g}}$, are given by:[74–76]

$$\gamma_{E_{2g}} = -\frac{1}{\omega^0_{E_{2g}}} \frac{\partial \omega^h_{E_{2g}}}{\partial \varepsilon_h}, \tag{8}$$

$$\beta_{E_{2g}} = \frac{1}{\omega^0_{E_{2g}}} \frac{\partial \omega^s_{E_{2g}}}{\partial \varepsilon_s}, \tag{9}$$

where $\varepsilon_h = \varepsilon_{ll} + \varepsilon_{tt}$ is the hydrostatic component of the applied uniaxial strain, l is the longitudinal direction, and t is the direction transverse to the strain. In addition $\varepsilon_s = \varepsilon_{ll} - \varepsilon_{tt}$ is the shear component of the strain and ω^0_{E2g} is the G peak position at zero strain.

Under the uniaxial strain $\varepsilon_{ll} = \varepsilon$ and $\varepsilon_{tt} = -v\varepsilon$, where v is the Poisson's ratio. The strain in these two directions are not the same so that the solution of the secular equation for the in-plane Raman-active E_{2g} phonon would doubly degenerate due to the shear deformation.[75,76]

$$\Delta\omega^{\pm}_{E_{2g}} = -\omega^0_{E_{2g}}\gamma_{E_{2g}}\left(\varepsilon_{ll} + \varepsilon_{tt}\right) \pm \frac{1}{2}\beta_{E_{2g}}\omega^0_{E_{2g}}\left(\varepsilon_{ll} - \varepsilon_{tt}\right). \qquad (10)$$

That is the reason why G peak would split into two peaks G^+ and G^-. Otherwise, under biaxial strain $\varepsilon_{ll} = \varepsilon_{tt} = \varepsilon$; $\Delta\omega^{\pm}_{E2g} = \omega^0_{E2g}\gamma_{E2}(\varepsilon_{ll} + \varepsilon_{tt})$. Then, the shear deformation term is canceled so the G peak does not split. The amount of peak shift of the G peak is the same as that of the 2D peak. This characteristic makes graphene useful to determine whether the strain is uniaxial or biaxial.

Strain sensor application plays an important role in electro-mechanical systems. It is normally used to detect mechanical deformation, such as structural health monitor and force sensors. To investigate the relation between the electrical and mechanical properties of graphene, strain was longitudinally applied to the device to stretch the flexible substrate along the basal plane as shown in Figs. 14(a)[7] and 14(b) presents the electrical measurements for graphene as strain sensor devices. The resistance of graphene increases due to the electron scattering by the lattice distortion when the strain arises before the device fails. Note that the yield point, which indicates the transition point of the elastic and the plastic deformation, was also indicated in Fig. 14(b). With this relationship, the GF can be calculated in order to test the sensitivity of the devices. The GF relates to changes in electrical resistance to the applied strain given by

$$GF = \Delta R/R\varepsilon, \qquad (11)$$

where ΔR is changes in electrical resistance, R is the original resistance at the condition without any strain, and ε is the mechanical strain of the device. By linearly fitting the data in the elastic region, the slopes of

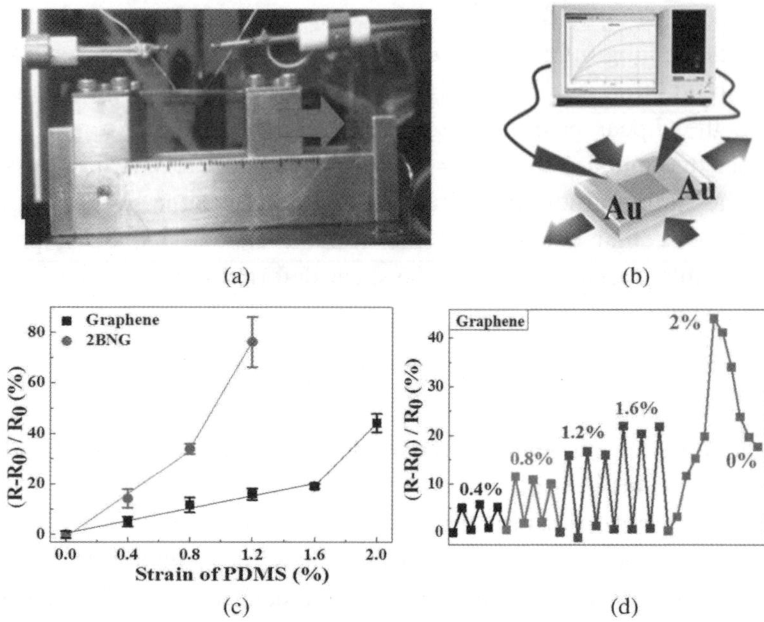

Figure 14. (a) The schematic diagram of the device was placed on a stage used for applying strain to sample/flexible substrate. (b) The normalized resistances of the graphene strain sensor device as a function of strain of flexible substrate. (c) Cyclic measurement for graphene strain sensor device. Adapted from Ref. 7.

graphene is about -0.25 ($1/\varepsilon$), yielding GFs of ~12.2 for the graphene, respectively (Fig. 15(b)). Obviously, graphene exhibits much higher GFs than that of conventional metallic strain gauge. To demonstrate the strain sensor, cyclic measurements of graphene-based strain sensor were measured as shown in Fig. 14(c), respectively. By gradually increasing and then decreasing the strain before reaching the yield point, the change in normalized resistance can be repeated perfectly even after several cycles, provided that the applied strain is under the yield point. Note that once the applied strain exceeds the yield point, namely ~1.2% for the graphene, cyclic measurements fail.

4.3. *Graphene Flexible Transistors*

One of the major difficulties in the development of flexible electronics has been the fabrication of transistors stable enough to work under strain/stress and under a variety of environments. Organic semiconductor materials have been a preferred material, offering flexibility for its role as an active channel for field-effect transistors (FETs), opening up a mixture of promising applications in light-weight and low-power electronics, such as active-matrix elements for plastic displays.[77–80] However, the carrier mobility of organic materials is quite low compared with Si. The typical mobility of organic semiconductors goes from 10^{-3} to 2 cm^2/V·s.[81–84] On the other hand, graphene promises to solve the mobility problem, offering a high-speed, flexible, and transparent FETs on flexible substrates.

Chen *et al.* reported fabricating flexible graphene FETs *via* the transfer printing method to mechanically transfer the exfoliated graphene with pre-defined electrodes to a plastic substrate.[85] For practical applications, large-scale graphene films grown by CVD are desirable. Kim *et al.*[86] prepared flexible FETs using CVD graphene as a channel material exhibited electron mobility of ~90 cm^2/V·s, which is several times larger than those transistors fabricated by organic semiconductors. However, this large intrinsic mobility expected for graphene is not high enough. This is presumably due to the long access lengths (i.e., the ungated channel region between the drain/source contacts and gate), downgrading device performance through the parasitic resistance RA (Fig. 15). Farmer *et al.*[87] showed that self-aligned top gates fabricated using the hydrophobic characteristics of the surface of exfoliated graphene can minimize the access length, yielding a greatly increased carrier mobility and cutoff frequency in the resulting FETs.

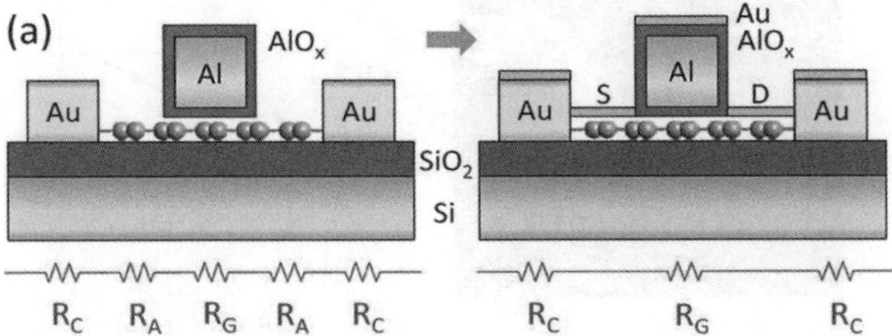

Figure 15.   (a) Schematic diagram of the steps used to fabricate self-aligned graphene FETs on a Si substrate. The lower panel shows the critical resistive components before and after the self-alignment. Reproduced from Ref. 91.

The polymer residues on the surface of the graphene channel due to the transfer process is another problem, which is needed to be addressed.[9] The inhomogeneous coverage of the polymer residue frequently results in pinholes in the stack of high-κ dielectrics (e.g., Al_2O_3, HfO_2, and ZrO_2) grown by atomic layer deposition (ALD). The high temperatures used in the ALD process may also rule out the use of some plastic substrates.[88–90]

To solve this problem Lu *et al.*[91] proposed a method using natural aluminum oxide as a gate dielectric in a self-aligned configuration. This high capacitance product of the thickness of the native aluminum oxide and self-alignment offer several improvements such as reduction of the access resistance, improvement of the on/off ratio, and high electron and hole mobility of 230 cm^2/V·s and 300 cm^2/V·s, respectively. Remarkably, this native aluminum oxide shows to be resistant to mechanical bending, exhibiting self-healing process even upon electrical breakdown. Figure 16 displays the device prepared with native aluminum displaying good performance on the bending test.

Recently, Yeh *et al.*[8] utilized the same technique to fabricate high-frequency transistors on flexible substrates, reaching a maximum oscillation frequency of 32 GHz before bending and 13 GHz after bending

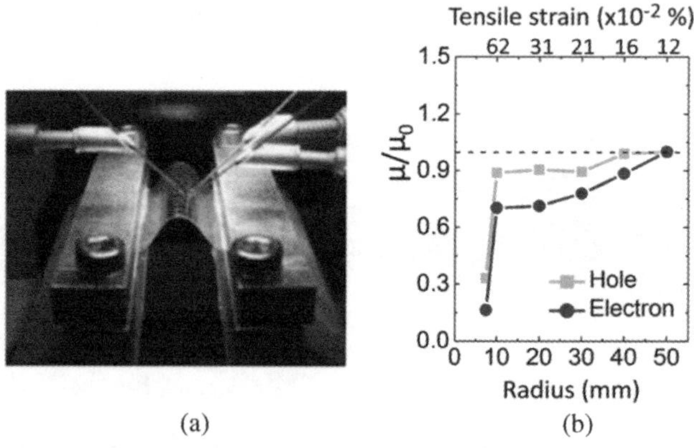

(a)   (b)

Figure 16.   (a) Optical photograph of bending test facility. (b) Normalized mobility as a function of the bending radius. From Ref. 91.

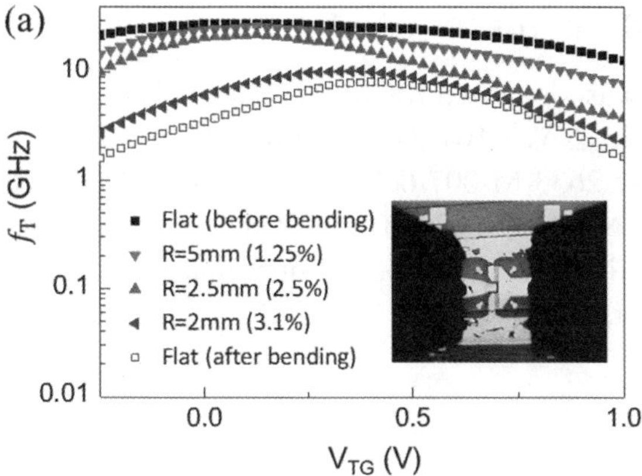

Figure 17.   Extrinsic f_T as a function of the gate bias for a RF G-FET in a flat configuration (before and after bending) and for three different bending states. Inset is the OM image, showing the measurement setup. Design adapted from Ref. 8.

(Fig. 17) probing the exceptional electrical and mechanical properties of graphene as a channel material for flexible transistors.

5. Summary and Conclusions

The quality of CVD graphene has proven to be good enough to be used in practical applications for flexible electronics. However, the low reliability and yield of the transferred graphene raises concerns about graphene use at industrial scale. More work is required to improve or modify the transfer process to achieve higher yield of graphene on good quality graphene transfer. The temperature and quality for direct synthesis of graphene on insulator is yet an issue to overcome. Up-to-date, large area graphene films can be deposited at temperatures as low as 400°C, which is still quite high for flexible substrate. Except for some substrates such as flexible glass that can resist up to 500°C, the temperature should be further reduced to temperatures under 200°C to extend the synthesis to high temperature polymers as a flexible substrate. Nevertheless, graphene has demonstrated to be a versatile material with great potential for applications in flexible electronics, not only as a flexible electrode but also as a highly sensitive strain sensor and as high mobility transistors with outstanding performance even after bending.

Acknowledgments

The research was supported by the Ministry of Science and Technology through Grant Nos. 101-2112-M-007-015-MY3, 101-2218-E-007-009-MY3, 102-2633-M-007-002, and National Tsing Hua University through Grant No. 102N2022E1. Y. L. Chueh greatly appreciates the use of facility at CNMM the National Tsing Hua University through Grant No. 102N2744E1.

References

1. K. S. Novoselov, V. I. Falko, L. Colombo, P. R. Gellert, M. G. Schwab and K. Kim, *Nature*, **490**, 192–200 (2012).
2. K. R. Paton, E. Varrla, C. Backes, R. J. Smith, U. Khan, A. O'Neill, C. Boland, M. Lotya, O. M. Istrate, P. King, T. Higgins, S. Barwich, P. May, P. Puczkarski, I. Ahmed, M. Moebius, H. Pettersson, E. Long, J. Coelho, S. E. O'Brien, E. K. McGuire, B. M. Sanchez, G. S. Duesberg, N. McEvoy, T. J. Pennycook, C. Downing, A. Crossley, V. Nicolosi and J. N. Coleman, *Nat. Mater.*, **13**, 624–630 (2014).
3. X. Li, W. Cai, J. An, S. Kim, J. Nah, D. Yang, R. Piner, A. Velamakanni, I. Jung, E. Tutuc, S. K. Banerjee, L. Colombo and R. S. Ruoff, *Science*, **324**, 1312–1314 (2009).
4. S. Bae, H. Kim, Y. Lee, X. Xu, J.-S. Park, Y. Zheng, J. Balakrishnan, T. Lei, H. Ri Kim, Y. I. Song, Y.-J. Kim, K. S. Kim, B. Ozyilmaz, J.-H. Ahn, B. H. Hong and S. Iijima, *Nat. Nanotechnol*, **5**, 574–578 (2010).
5. A. K. Geim, *Science*, **324**, 1530–1534 (2009).
6. A. K. Geim and K. S. Novoselov, *Nat. Mater.*, **6**, 183–191 (2007).
7. S.-H. Pan, H. Medina, S.-B. Wang, L.-J. Chou, Z. M. Wang, K.-H. Chen, L.-C. Chen and Y.-L. Chueh, *Nanoscale*, **6**, 8635–8641 (2014).
8. C.-H. Yeh, Y.-W. Lain, Y.-C. Chiu, C.-H. Liao, D. R. Moyano, S. S. H. Hsu and P.-W. Chiu, *ACS Nano*, **8**, 7663–7670 (2014).
9. Y.-C. Lin, C.-C. Lu, C.-H. Yeh, C. Jin, K. Suenaga and P.-W. Chiu, *Nano Lett.*, **12**, 414–419 (2011).
10. Q. Zhao, M. B. Nardelli and J. Bernholc, *Phys. Rev. B*, **65**, 144105 (2002).
11. E. W. Wong, P. E. Sheehan and C. M. Lieber, *Science*, **277**, 1971–1975 (1997).

12. J.-P. Salvetat, A. J. Kulik, J.-M. Bonard, G. A. D. Briggs, T. Stöckli, K. Méténier, S. Bonnamy, F. Béguin, N. A. Burnham and L. Forró, *Adv. Mat.*, **11**, 161–165 (1999).

13. M.-F. Yu, O. Lourie, M. J. Dyer, K. Moloni, T. F. Kelly and R. S. Ruoff, *Science*, **287**, 637–640 (2000).

14. G. J. McShane, M. Boutchich, A. S. Phani, D. F. Moore and T. J. Lu, *J. Micromech. Microeng.*, **16**, 1926 (2006).

15. S. Cuenot, S. Demoustier-Champagne and B. Nysten, *Phys. Rev. Lett.*, **85**, 1690–1693 (2000).

16. L. Q. Liu, D. Tasis, M. Prato and H. D. Wagner, *Adv. Mater.*, **19**, 1228–1233 (2007).

17. Z. Yang, R. N. Wang, S. Jia, D. Wang, B. S. Zhang, K. M. Lau and K. J. Chen, *Appl. Phys. Lett.*, **88**, 041913 (2006).

18. B. Lecouvet, J. Horion, C. D'Haese, C. Bailly and B. Nysten, *Nanotechnology*, **24**, 105704 (2013).

19. C. Lee, X. Wei, J. W. Kysar and J. Hone, *Science*, **321**, 385–388 (2008).

20. P. R. Wallace, *Phys. Rev.*, **71**, 622–634 (1947).

21. K. S. Novoselov, A. K. Geim, S. V. Morozov, D. Jiang, M. I. Katsnelson, I. V. Grigorieva, S. V. Dubonos and A. A. Firsov, *Nature*, **438**, 197–200 (2005).

22. K. S. Novoselov, A. K. Geim, S. V. Morozov, D. Jiang, Y. Zhang, S. V. Dubonos, I. V. Grigorieva and A. A. Firsov, *Science*, **306**, 666–669 (2004).

23. Z. Sun, Z. Yan, J. Yao, E. Beitler, Y. Zhu and J. M. Tour, *Nature*, **468**, 549–552 (2010).

24. Y. S. Woo, D. H. Seo, D.-H. Yeon, J. Heo, H.-J. Chung, A. Benayad, J.-G. Chung, H. Han, H.-S. Lee, S. Seo and J.-Y. Choi, *Carbon*, **64**, 315–323 (2013).

25. R. S. Weatherup, B. C. Bayer, R. Blume, C. Ducati, C. Baehtz, R. Schlögl and S. Hofmann, *Nano Lett.*, **11**, 4154–4160 (2011).

26. Z. Li, P. Wu, C. Wang, X. Fan, W. Zhang, X. Zhai, C. Zeng, Z. Li, J. Yang and J. Hou, *ACS Nano*, **5**, 3385–3390 (2011).

27. B. Zhang, W. H. Lee, R. Piner, I. Kholmanov, Y. Wu, H. Li, H. Ji and R. S. Ruoff, *ACS Nano*, **6**, 2471–2476 (2012).

28. X. Li, Y. Zhu, W. Cai, M. Borysiak, B. Han, D. Chen, R. D. Piner, L. Colombo and R. S. Ruoff, *Nano Lett.*, **9**, 4359–4363 (2009).

29. K. S. Kim, Y. Zhao, H. Jang, S. Y. Lee, J. M. Kim, K. S. Kim, J.-H. Ahn, P. Kim, J.-Y. Choi and B. H. Hong, *Nature*, **457**, 706–710 (2009).

30. J. W. Suk, A. Kitt, C. W. Magnuson, Y. Hao, S. Ahmed, J. An, A. K. Swan, B. B. Goldberg and R. S. Ruoff, *ACS Nano*, **5**, 6916–6924 (2011).

31. J. Song, F.-Y. Kam, R.-Q. Png, W.-L. Seah, J.-M. Zhuo, G.-K. Lim, P. K. H. Ho and L.-L. Chua, *Nat Nanotechnol.*, **8**, 356–362 (2013).

32. Y. Han, L. Zhang, X. Zhang, K. Ruan, L. Cui, Y. Wang, L. Liao, Z. Wang and J. Jie, *J. Mater. Chem.*, *C* **2**, 201–207 (2014).

33. W.-C. Yen, Y.-Z. Chen, C.-H. Yeh, J.-H. He, P.-W. Chiu and Y.-L. Chueh, *Sci. Rep.*, **4**, 4739 (2014).

34. P.-Y. Teng, C.-C. Lu, K. Akiyama-Hasegawa, Y.-C. Lin, C.-H. Yeh, K. Suenaga and P.-W. Chiu, *Nano Lett.*, **12**, 1379–1384 (2012).

35. W.-C. Yen, H. Medina, C.-W. Hsu and Y.-L. Chueh, *R. Soc. Chem. Adv.*, **4**, 27106–27111 (2014).

36. C.-Y. Su, A.-Y. Lu, C.-Y. Wu, Y.-T. Li, K.-K. Liu, W. Zhang, S.-Y. Lin, Z.-Y. Juang, Y.-L. Zhong, F.-R. Chen and L.-J. Li, *Nano Lett.*, **11**, 3612–3616 (2011).

37. W.-C. Yen, H.-C. Lin, J.-S. Huang, Y.-J. Huang and Y.-L. Chueh, *Sci. Adv. Mater.*, **6**, 8 (2014).

38. Z. Peng, Z. Yan, Z. Sun and J. M. Tour, *ACS Nano*, **5**, 8241–8247 (2011).

39. J. Wang, M. Zhu, R. A. Outlaw, X. Zhao, D. M. Manos and B. C. Holloway, *Carbon*, **42**, 2867–2872 (2004).

40. A. Malesevic, R. Kemps, A. Vanhulsel, M. P. Chowdhury, A. Volodin and C. Van Haesendonck, *J. Appl. Phys.*, **104**, 8 (2008).

41. P. Hojati-Talemi and G. P. Simon, *Carbon*, **49**, 2875–2877 (2011).

42. P. Hojati-Talemi and G. P. Simon, *Carbon*, **48**, 3993–4000 (2010).

43. K. Shiji, M. Hiramatsu, A. Enomoto, M. Nakamura, H. Amano and M. Hori, *Diamond and Related Materials*, **14**, 831–834 (2005).

44. A. Malesevic, S. Vizireanu, R. Kemps, A. Vanhulsel, C. V. Haesendonck and G. Dinescu, *Carbon*, **45**, 2932–2937 (2007).

45. A. Dato, V. Radmilovic, Z. Lee, J. Phillips and M. Frenklach, *Nano Lett.*, **8**, 2012–2016 (2008).

46. H. Medina, Y.-C. Lin, C. Jin, C.-C. Lu, C.-H. Yeh, K.-P. Huang, K. Suenaga, J. Robertson and P.-W. Chiu, *Adv. Funct. Mater.*, **22**, 2123–2128 (2012).

47. S.-T. Han, Y. Zhou and V. A. L. Roy, *Adv. Mater.*, **25**, 5425–5449 (2013).

48. T. Rueckes, K. Kim, E. Joselevich, G. Y. Tseng, C.-L. Cheung and C. M. Lieber, *Science*, **289**, 94–97 (2000).

49. R. Waser and M. Aono, *Nat. Mater.*, **6**, 833–840 (2007).

50. C. A. P. de Araujo, J. D. Cuchiaro, L. D. McMillan, M. C. Scott and J. F. Scott, *Nature*, **374**, 627–629 (1995).

51. N. Nishimura, T. Hirai, A. Koganei, T. Ikeda, K. Okano, Y. Sekiguchi and Y. Osada, *J. Appl. Phys.*, **91**, 5246–5249 (2002).

52. T. C. Chong, L. P. Shi, R. Zhao, P. K. Tan, J. M. Li, H. K. Lee, X. S. Miao, A. Y. Du and C. H. Tung, *Appl. Phys. Lett.*, **88**, 122114 (2006).

53. R. Bez, E. Camerlenghi, A. Modelli and A. Visconti, *P. IEEE*, **91**, 489–502 (2003).

54. N. Setter, D. Damjanovic, L. Eng, G. Fox, S. Gevorgian, S. Hong, A. Kingon, H. Kohlstedt, N. Y. Park, G. B. Stephenson, I. Stolitchnov, A. K. Taganstev, D. V. Taylor, T. Yamada and S. Streiffer, *J. Appl. Phys.*, **100**, 051606 (2006).

55. J. D. Boeck, W. V. Roy, J. Das, V. Motsnyi, Z. Liu, L. Lagae, H. Boeve, K. Dessein and G. Borghs, *Semicond. Sci. Technol.*, **17**, 342 (2002).

56. S. Tehrani, J. M. Slaughter, E. Chen, M. Durlam, J. Shi and M. DeHerren, *Magnetics, IEEE Transactions on*, **35**, 2814–2819 (1999).

57. J.-S. Lee, *J. Mater. Chem.*, **21**, 14097–14112 (2011).

58. M.-J. Lee, C. B. Lee, D. Lee, S. R. Lee, M. Chang, J. H. Hur, Y.-B. Kim, C.-J. Kim, D. H. Seo, S. Seo, U. I. Chung, I.-K. Yoo and K. Kim, *Nat. Mater.*, **10**, 625–630 (2011).

59. J. J. Yang, M. D. Pickett, X. Li, A. A. OhlbergDouglas, D. R. Stewart and R. S. Williams, *Nat. Nano.*, **3**, 429–433 (2008).

60. K. J. Yoon, M. H. Lee, G. H. Kim, S. J. Song, J. Y. Seok, S. Han, J. H. Yoon, K. M. Kim and C. S. Hwang, *Nanotechnology*, **23**, 185202 (2012).

61. S. Lee, H. Kim, D.-J. Yun, S.-W. Rhee and K. Yong, *Appl. Phys. Lett.*, **95** (2009).

62. Y. S. Lin, F. Zeng, S. G. Tang, H. Y. Liu, C. Chen, S. Gao, Y. G. Wang and F. Pan, *J.f Appl. Phys.*, **113** (2013).

63. Q. Zhou and J. Zhai, *Physica B: Condensed Matter*, **410**, 85–89 (2013).

64. L. Gao, S. B. Lee, B. Hoskins, H. K. Yoo and B. S. Kang, *Appl. Phys. Lett.*, **103**(4), 043503 (2013).

65. Y. Ji, B. Cho, S. Song, T.-W. Kim, M. Choe, Y. H. Kahng and T. Lee, *Adv. Mater.*, **22**, 3071–3075 (2010).

66. Y. Ji, S. Lee, B. Cho, S. Song and T. Lee, *ACS Nano*, **5**, 5995–6000 (2011).

67. S. Jung, J. Kong, S. Song, K. Lee, T. Lee, H. Hwang and S. Jeon, *Appl. Phys. Lett.*, **99**(14), 142110 (2011).

68. J. Liu, Z. Zeng, X. Cao, G. Lu, L.-H. Wang, Q.-L. Fan, W. Huang and H. Zhang, *Small*, **8**, 3517–3522 (2012).
69. H. Y. Jeong, J. Y. Kim, J. W. Kim, J. O. Hwang, J.-E. Kim, J. Y. Lee, T. H. Yoon, B. J. Cho, S. O. Kim, R. S. Ruoff and S.-Y. Choi, *Nano Lett.*, **10**, 4381–4386 (2010).
70. J. Liu, Z. Yin, X. Cao, F. Zhao, L. Wang, W. Huang and H. Zhang, *Adv. Mater.*, **25**, 233–238 (2013).
71. J. Yao, J. Lin, Y. Dai, G. Ruan, Z. Yan, L. Li, L. Zhong, D. Natelson and J. M. Tour, *Nat. Commun.*, **3**, 1101 (2012).
72. I. W. Frank, D. M. Tanenbaum, A. M. van der Zande and P. L. McEuen, *JVST* B **25**, 2558–2561 (2007).
73. M. Poot and H. S. J. van der Zant, *Appl. Phys. Lett.*, **92**(6), 063111 (2008).
74. T. M. G. Mohiuddin, A. Lombardo, R. R. Nair, A. Bonetti, G. Savini, R. Jalil, N. Bonini, D. M. Basko, C. Galiotis, N. Marzari, K. S. Novoselov, A. K. Geim and A. C. Ferrari, *Phys. Rev.* B **79**, 205433 (2009).
75. S. Reich, H. Jantoljak and C. Thomsen, *Phys. Rev. B*, **61**, R13389–R13392 (2000).
76. C. Thomsen, S. Reich and P. Ordejón, *Phys. Rev. B*, 65, 073403 (2002).
77. G. Gelinck, P. Heremans, K. Nomoto and T. D. Anthopoulos, *Adv. Mater.*, **22**, 3778–3798 (2010).
78. L. Zhou, A. Wanga, S.-C. Wu, J. Sun, S. Park and T. N. Jackson, *Appl. Phys. Lett.*, **88**, 083502 (2006).
79. T. Sekitani, Y. Noguchi, K. Hata, T. Fukushima, T. Aida and T. Someya, *Science*, **321**, 1468–1472 (2008).
80. G. H. Gelinck, H. E. A. Huitema, E. van Veenendaal, E. Cantatore, L. Schrijnemakers, J. B. P. H. van der Putten, T. C. T. Geuns, M. Beenhakkers, J. B. Giesbers, B.-H. Huisman, E. J. Meijer, E. M. Benito, F. J. Touwslager, A. W. Marsman, B. J. E. van Rens and D. M. de Leeuw, *Nat. Mater.*, **3**, 106–110 (2004).
81. Y. Li, P. Sonar, L. Murphy and W. Hong, *Energy & Environmental Science*, **6**, 1684–1710 (2013).
82. H. E. Katz, A. J. Lovinger, J. Johnson, C. Kloc, T. Siegrist, W. Li, Y. Y. Lin and A. Dodabalapur, *Nature*, **404**, 478–481 (2000).
83. L.-L. Chua, J. Zaumseil, J.-F. Chang, E. C. W. Ou, P. K. H. Ho, H. Sirringhaus and R. H. Friend, *Nature*, **434**, 194–199 (2005).
84. P. Deng and Q. Zhang, *Polym. Chem.*, **5**, 3298–3305 (2014).

85. J. H. Chen, M. Ishigami, C. Jang, D. R. Hines, M. S. Fuhrer and E. D. Williams, *Adv. Mater.*, **19**, 3623–3627 (2007).

86. B. J. Kim, H. Jang, S.-K. Lee, B. H. Hong, J.-H. Ahn and J. H. Cho, *Nano Lett.*, **10**, 3464–3466 (2010).

87. D. B. Farmer, Y.-M. Lin and P. Avouris, *Appl. Phys. Lett.*, **97**, 013103 (2010).

88. Y.-M. Lin, C. Dimitrakopoulos, K. A. Jenkins, D. B. Farmer, H.-Y. Chiu, A. Grill and P. Avouris, *Science*, **327**, 662 (2010).

89. Y.-M. Lin, K. A. Jenkins, A. Valdes-Garcia, J. P. Small, D. B. Farmer and P. Avouris, *Nano Lett.*, **9**, 422–426 (2008).

90. L. Liao, J. Bai, Y.-C. Lin, Y. Qu, Y. Huang and X. Duan, *Adv. Mater.*, **22**, 1941–1945 (2010).

91. C.-C. Lu, Y.-C. Lin, C.-H. Yeh, J.-C. Huang and P.-W. Chiu, *ACS Nano*, **6**, 4469–4474 (2012).

CHAPTER 4

INTEGRATING SEMICONDUCTOR NANOWIRES FOR HIGH PERFORMANCE FLEXIBLE ELECTRONIC CIRCUITS

Ning Han[*] and Johnny C. Ho[†,‡]

State Key Laboratory of Multiphase Complex Systems,
Institute of Process Engineering,
Chinese Academy of Sciences, Beijing
100190, P. R. China
†Department of Physics and Materials Science,
City University of Hong Kong, Tat Chee Avenue, Kowloon,
Hong Kong S.A.R., P. R. China
‡johnnyho@cityu.edu.hk

In recent years, electronic devices fabricated on flexible and stretchable substrates, namely macroelectronics, with special demands on low-cost materials and fabrications have been widely investigated for practical applications in flexible displays, electronic papers and so on. In particular, the performance of macroelectronics depends highly on the active semiconductor materials employed. Among many promising candidate materials, semiconductor nanowire parallel arrayed thin-film with the high intrinsic carrier mobility has been demonstrated with great potency.

In this regard, this chapter summarizes mainly the basic knowledge and requirement moving from silicon (Si)-based microelectronics to flexible macroelectronics, with the emphasis on the preparation methods, alignment technologies as well as transferring approaches onto flexible substrates of one dimensional semiconductor nanowires (NWs). All these show their great promising prospective not only enhancing the performance and flexible circuits, but also contributing to the large-scale applications of macroelectronics.

1. Introduction

In the past decades, silicon (Si)-based semiconductor microelectronics have become a great success in terms of dramatically reducing the device size and enormously increasing the speed, capacity as well as the functionality every 18–24 months, as commonly known as the Mole law. In the meanwhile, in recent years, another kind of electronics, in which the device size and speed are not as important as their required flexibility and stretchability; however, their relatively low-cost characteristics have found their significant utilization in large-scale displays, electronic papers, sensors, actuators, health diagnosis, radio-frequency (RF) identification, and so on. In such applications, the individual device size can be much larger than microelectronics, even in the micro- or milli-meter scale, while the flexibility and stretchability is critical. This new class of electronics is typically branded as "macroelectronics".[1–6]

In order to tackle this new application domain, simply transferring conventional Si electronic devices, such as thin-film transistors (TFTs), directly onto flexible substrates is not feasible, since the stiffness of inorganic active device materials would not be fully compatible with organic substrates; importantly, the strains induced by any bending would significantly degrade or even destroy the devices. It is exactly the reason that organic semiconductors were extensively exploited for all-organic transistors for macroelectronics in early 1990s. Since then, all macroelectronics or so-called "flexible electronics" were all stemmed from using all polymeric materials in conjunction with the traditional low-cost roll-to-roll printing technology.[7–10]

However, one major problem of utilizing organic semiconductors is their intrinsically low carrier mobility in the range of <10 cm^2/Vs, leading to the poor functionality of flexible devices, especially high performance or high speed is also required for the many advanced applications of next-generation flexible electronics. In this regard, when high carrier mobility inorganic nanomaterials, such as carbon nanotubes (CNTs), graphene, transition metal dichalcogenides (TMDs), semiconductor nanowires (NWs) and nanoribbons (NRs), are adopted in the use as active materials for flexible electronics, they greatly nourish the development of macroelectronics in the last decade.[2,11,12] Specifically, the one-dimensional (1D) nanomaterials, including CNTs, NWs and NRs, would be effective in the mechanical strain isolation and endurance, which makes them fully compatible with flexible substrates. Also, NWs would typically encounter a rather small part of the strain (usually less than 0.5%) induced by the flexible substrate during bending or folding.[13] Till now, the high mobility semiconductor NWs in combination with the use of flexible substrates would be one of the ideal choices for macroelectronics and all these have attracted more and more attention worldwide. In this context, this chapter will mainly review the basic fundamentals in individual components of flexible macroelectronics as well as the advances of the integration of 1D semiconductor NWs for flexible integrated circuits (ICs). Other applications such as flexible solar cells, sensors, and artificial skins etc. will be emphasized in other chapters.

2. Requirements for Flexible Circuits

The major functional unit of flexible circuits is typically field-effect transistors (FETs), similar to the one in Si microelectronics. Therefore, the main components of FETs such as the substrate, gate dielectrics, source/drain/gate electrodes, and active channel materials should all be mechanically flexible or at least be compatible with excellent flexibility,[14] which will be reviewed in this section for the requirement of flexible electronic circuits.

2.1. *Flexible Substrates*

In flexible electronics, the primary requirement is the selection of appropriate flexible substrates such as plastics, rubbers, papers, and metal foils, etc.[15,16]

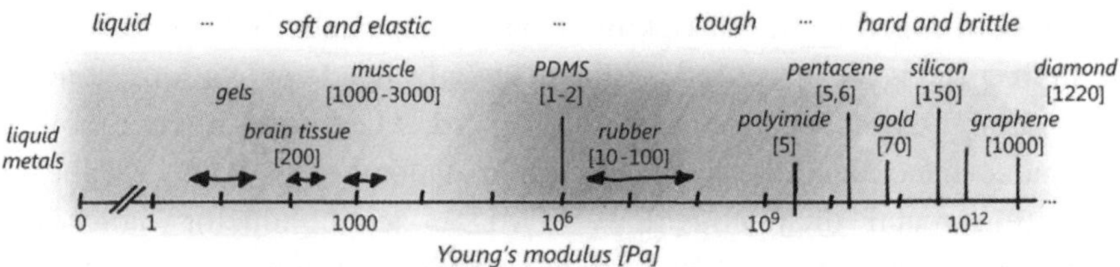

Figure 1. Young's modulus change from liquids, soft materials to hard inorganics. The numbers in brackets are in Pa for soft, MPa for tough, and GPa for hard materials, respectively. Reprinted with permission.[15]

In the literature, one would find many different adjectives for such flexible materials, including elastic, bendable, foldable, stretchable, wearable, conformable, and others. On the other hand, inorganic materials are all rigid, stiff, hard, tough, and brittle. The main difference between flexible and inorganic materials here is their Young's modulus, ranging in kPa to GPa for organics, while >10 GPa for inorganics, which is schematically shown in Figs. 1 and 15 with typical soft to rigid materials exemplified.

In addition, to be an ideal substrate for flexible electronics, the organic polymers should be at least thermally stable, solvent resistant, and high optical transparency for advanced applications.[17,18] To date, there are a number of plastics developed for flexible substrates, such as the frequently used polycarbonate (PC), polyethersulfone (PES), polyimide (PI), polyethylene terephthalate (PET), and polyethylene naphthalate (PEN), etc. as well as their corresponding multilayer composite structures.[19] One major concern in employing these polymer substrates is the requirement of smooth surface for the subsequent processing, and sometimes the surface treatment is essentially needed for the better compatibility with other active device materials.[18] In some cases, especially for the human health monitor, the biodegradability is also important when the devices are implanted into the human body.[20]

2.2. *Flexible Dielectrics*

Typically, the conventional high-k dielectrics commonly used in silicon technology can also be adopted in flexible electronics, since the gate dielectric films are rather thin in the range of ~10 nm in the thickness and

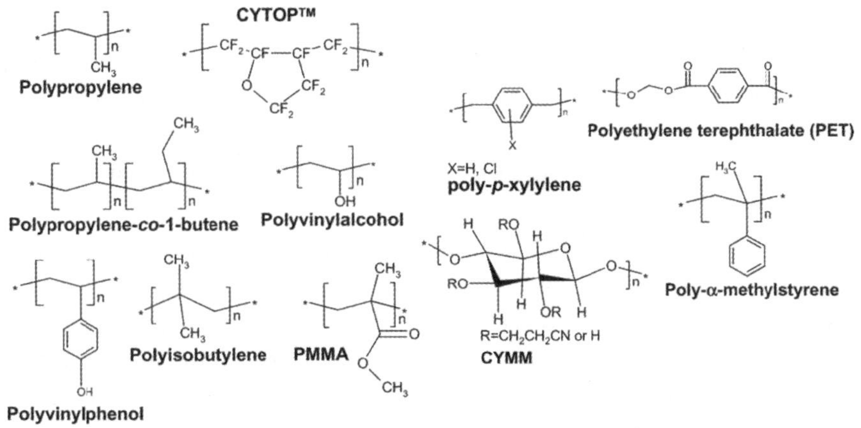

Figure 2.   Molecular structures of the organic polymers commonly used as gate insulators in flexible electronics. Reprinted with permission.[26]

thus they can withhold the required bending strains. In this case, SiO_2, Al_2O_3, and HfO_2 etc. are all usually used and reported in the literature.[21–23] However, these inorganic materials are relatively hard to process, demanding the high vacuum evaporation or atomic layer deposition (ALD) technologies. This way, there would be several associated problems here such as the temperature endurance issue of plastic substrates employed in these processes as well as the resultant high cost of those complex processing techniques. In this regard, several organic materials are generally utilized as gate insulators considering the compatibility with plastic substrates, easy fabrication, low cost, and also because of the high speed induced by high-k dielectrics not being essential for macroelectronics. Consequently, polymers including PI, polyvinylpyrrolidone (PVP), and polymethylmethacrylate (PMMA), etc. with   in the range of 2–3 are frequently used as organic dielectrics in flexible FETs,[9,24–26] as the molecular structures are illustrated in Fig. 2 and properties are summarized in the review of Ref. 26.

2.3. *Flexible Electrodes*

In addition to substrates and dielectrics, the conductive electrodes are another important component in flexible electronics. In microelectronics, metals and highly doped polysilicon are commonly employed for electrodes; however, they are mechanically rigid, and hence are difficult to be compatible with flexible electronics. Generally, there are two kinds of substitutes: (i) shape control in making the metal electrodes pre-deformed to enhance the mechanical flexibility and (ii) adopting the conductive paste such as conductive polymers or inorganic/organic hybrid conductive materials.

For pre-deformed electrodes, Roger's group is a pioneered research team to actively design and develop many different kinds of electrode structures for flexible electronics,[27–30] as representatively depicted in Fig. 3.[27] The main idea is to deposit the metal electrodes in the pre-deformed flexible substrates in order to obtain the convex or concave shape, and then the metal lines would be resulted in the wavy shape after the strain is released (Fig. 3). These wavy electrode structures would be able to even endure the large deformation of substrates with more complicated shape design, and many flexible and stretchable electronics are developed thereof.[31]

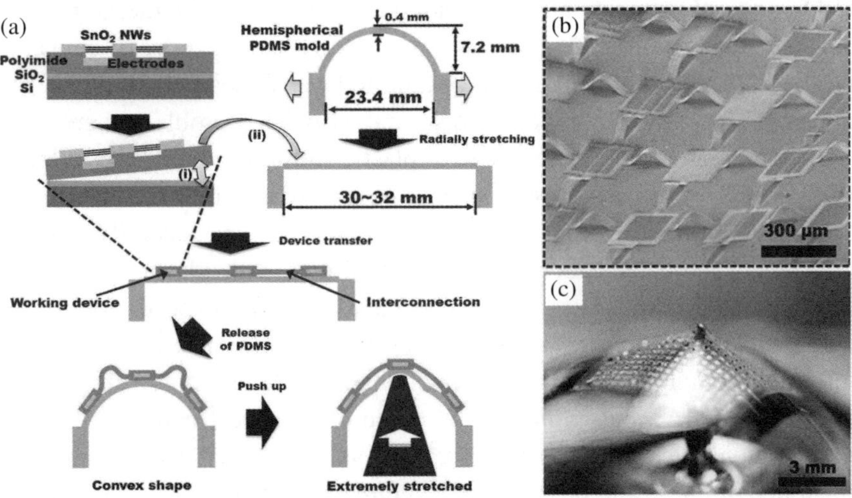

Figure 3. The fabrication schematics and images of the wavy metal electrodes. (a) Fabrication process: (i) the device on PI is first detached from the SiO₂/Si substrate, (ii) the detached device is then placed onto the radial stretched PDMS mold which is originally hemispheric, and after the release of the strain, the PDMS mold retrieves to the hemisphere shape and the metal interconnections become wavy in the shape, (b) corresponding SEM image of the wave shaped metal electrodes, (c) photography of the stretched PDMS mold with devices on it. Reprinted with permission.[27]

At the same time, several organic pastes, including organic conductors or hybrid organic/inorganic conductors are also developed for flexible electrodes.[1,32–34] In fact, the first all-organic FETs reported by Garnier *et al.*[7] employed graphite based polymer ink as the conducting electrodes, and later polyaniline (PANI), PEDOT/PSS are also used for electrodes. For example, the printed PEDOT:PSS electrodes have a conductivity of 100 S/cm, which is high enough for the application of organic FETs.[35] On the other hand, metal NWs and nanoparticles such as Ag and Cu NWs are also synthesized to form the conductive networks, with sheet resistance in the range of tens of ohm/square and transmittance of 80–90% in the visible light, and importantly with little resistance change under the deformation, these nanonetworks can outperform their rigid ITO counterparts, being better suited for flexible and transparent electronics.[34,36–42] Recently, with a more extensive development, the wavy shaped Ag interconnections achieved by the omnidirectional ink printing technology also holds the promise for flexible electrodes.[43]

2.4. *Flexible Active Matrix*

Once all the required components are realized, the first flexible electronics based on all-organic FETs were reported in 1994 by Garnier *et al.* In these prototypes, they have utilized semiconducting polymer α,ω-di(hexyl)sexithiophene as the active channel material,[7] and later pentacene and poly(thienylenevinylene) (PVP) are also be adopted as the active channel and integrated into ring oscillators.[8] Due to the use of organic semiconductors, many kinds of logic circuits are then enabled and fabricated on flexible substrates, such as complimentary metal-oxide-semiconductor (CMOS) inverter,[9,44,45] and microprocessors,[46] etc. However, there is one major problem that the carrier mobility of these organic channels is rather low (~0.06 cm^2/Vs) at the early stage of development, which is even much lower than that of amorphous silicon (a-Si). Recently, the carrier mobility of organic semiconductors is much improved reaching ~10 cm^2/Vs, as summarized in Fig. 4(a).[11,47] Although they can fulfill the low-end requirement for applications such as e-papers and others, these relatively low mobility values, as compared to their crystalline inorganic counterparts, can significantly restrict their further utilizations in advanced

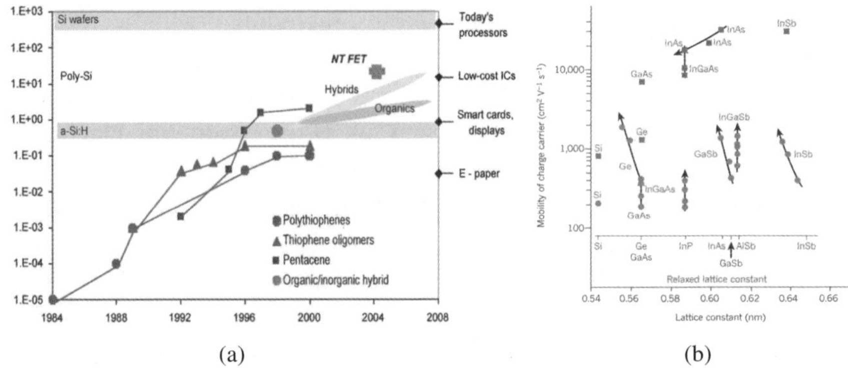

Figure 4. Carrier mobility of typical organic and compound semiconductors. (a) The evolution of the mobility of organic semiconductors,[11] and (b) the electron and hole mobility of III–V semiconductor materials. Reprinted with permissions.[11,55] The fabrication schematics and images of the wavy metal electrodes. (a) Fabrication process: (i), the device on PI is first detached from the SiO_2/Si substrate, (ii) the detached device is then placed onto the radial stretched PDMS mold which is originally hemispheric, and after the release of the strain, the PDMS mold retrieves to the hemisphere shape and the metal interconnections become wavy in the shape, (b) corresponding SEM image of the wave shaped metal electrodes, (c) photography of the stretched PDMS mold with devices on it. Reprinted with permission.[27]

flexible electronics including as flexible ICs and processors. In this regard, inorganic nanostructured semiconductors with high mobility and better performance have become more attractive as the active channel materials here.

Moreover, amorphous silicon (a-Si)-based TFTs are also typically used for flexible electronics if the strain encountered during the operation is not too huge. Notably, the a-Si films are usually deposited at a relatively low temperature by plasma-enhanced chemical vapor deposition (PECVD), and thus the organic substrate would not be degraded significantly at these temperatures.[48,49] In any case, this low-temperature (e.g., 75°C) deposited a-Si TFTs have only a low carrier mobility of ~0.04 cm^2/Vs. An increase of the deposition temperature up to 150°C can enhance the mobility to 12 cm^2/Vs, but a further increase in the temperature would induce the thermal degradation of organic substrates such that the mobility of this type of a-Si cannot reach the ones of ~100 cm^2/Vs in rigid Si technology, as demonstrated in Fig. 4(a).[11] Similar constraints also exist for other better-performed thin-film based compound semiconductors, which mainly composed of amorphous or nanocrystalline materials such as In-Ga-Zn-O (a-IGZO)[50] and CdSe.[51] resulting in the relatively low electron mobility in the order of 10–20 cm^2/Vs .

CNTs networks are attractive candidates for the channel material in flexible electronics because of their easy fabrication and relative high mobility of 10–100 cm^2/Vs.[52,53] In recent times, Roger's group has synthesized perfectly aligned CNTs by CVD on the quartz substrate and subsequently transferred to flexible substrates after the device fabrication, in which the corresponding electron mobility can reach as high as ~1,000 cm^2/Vs.[54] The details of these CNT applications can be found in other chapters. It should as well be noted that the electron mobility of III–V compound semiconductors can easily exceed 10,000 cm^2/Vs as illustrated in Fig. 4(b),[55] and recently, n-type InAs NWs and p-type GaSb NWs posses high electron and hole mobility of ~20,000 and ~350 cm^2/Vs,[56–60] illustrating their promising prospect in flexible electronics. Furthermore, when integrated onto plastic substrates, the NWs would only encounter a rather small proportion (~0.35%) of strain induced by the substrate deformation.[13] More importantly, the NWs would become mechanically flexible themselves at this high aspect ratio as confirmed by the AFM deformation test.[61] Therefore, the inorganic NWs can be prepared into the spring shape, which would further eliminate the influence of the bending strain with no sacrifice to their electronic properties.[62] Based on all these evidence, semiconductor NWs are one of the most promising alternative channel materials in macroelectronics as they are advantageous in both the high carrier mobility as well as the good compatibility with flexible substrates.[63,64] Hence, in this chapter, we would mainly focus on the preparation, integration and application of 1-D NWs for flexible circuits.

3. Progress of NW-Based Flexible Circuits

3.1. *Preparation and Characterization of NWs*

Generally, the preparation methods of semiconductor NWs fall into two distinct categories, which are "top-down" and "bottom-up" approaches.

In the top-down technology, bulk materials are etched with the assistant of well-defined masks and/or by the anisotropic property of the crystallography, leaving the residuals in one dimension forming horizontal or vertical NWs. On the other hand, precursors would assemble in the bottom-up technologies along the preferred directions; as a result, 1D NWs are then grown along certain crystal orientations with the lowest surface energy.

3.1.1. Top-Down Wet Chemical Etching

There are mainly two techniques in the top-down method, one is liquid phase wet chemical etching, and the other one is gas phase reactive ion etching (RIE). In the first case, different anisotropic etching rate is often observed in different crystal planes; this way, NWs in the masked region would only be obtained in the orientation with the minimum chemical etching rate. While in the RIE process, reactive ions would mostly etch materials in the vertical direction; therefore, the resultant orientations of NWs would only depended on the orientation of the single crystalline wafer used.

In practice, as shown in Fig. 5(a), the wet chemical etching method is widely adopted in the preparation of Si NWs and other 1D nanostructures assisted by the Ag nanoparticle catalyst.[65–67] Specifically, this etching process includes two successive steps: firstly, oxidant such as H_2O_2 is reduced at the Ag surface, and the holes are then transferred to the Ag/Si interface, oxidizing Si into SiO_2; following that, HF diffuses into the Ag/SiO_2 interface and dissolves the SiO_2 layer to make the etching process continue and thus NW array is then attained. It should be noted that this etching preferentially occurs on the Si(100) surface; therefore, the <100> oriented NWs can only be obtained. If other oriented Si wafers are used such as Si(110) or Si(111), NWs are still resulted along the <100> direction, tilting to the wafer surface. Lately, researchers can get other morphologies such as nanocones and nanopencils by tailoring the concentration of oxidant/etchant as well as repeating the etching process for multiple times.[68] Also, a horizontal crack layer can be gained by a simple water soaking treatment, which would facilitate NW array harvesting for functional electronics.[67] Similarly, free standing triangular shaped GaAs

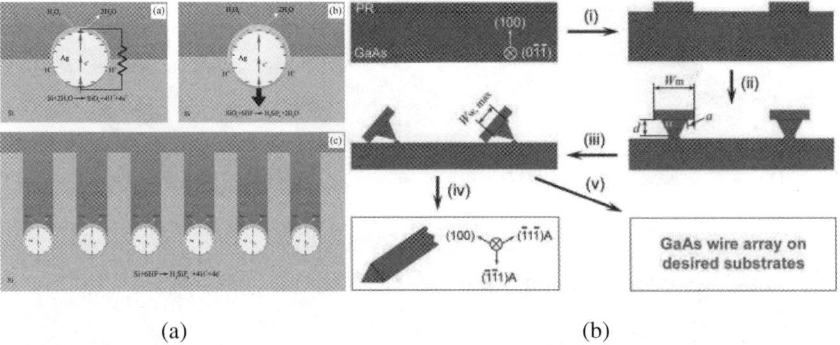

(a)   (b)

Figure 5. Schematics of the wet chemical etching methods for vertical and horizontal NW preparation. (a) Silver nanoparticle assisted etching of vertical Si NWs from mother wafers: (i) Si is electrochemically oxidized into SiO_2 catalyzed by Ag nanoparticles, (ii) the SiO_2 layer is etched by HF and Ag nanoparticle moves deeper to repeat (i) and (ii), and (iii) Si NWs are then finally obtained, (b) anisotropic chemical etching of the horizontal GaAs NWs: (i) patterning the photoresist by lithography on GaAs wafers, (ii) anisotropic etching of the GaAs wafer by H_3PO_4 and H_2O_2 using patterned photoresist as the mask, (iii) free standing GaAs NWs are attained after a complete etching, (iv) photoresist removal by acetone or O_2 plasma, and (v) NWs get transferred to other substrates. Reprinted with permission.[65,69]

NWs,[69,70] and Si ribbons,[71] etc. can be prepared using the wet chemical etching processes with lithography defined photoresist mask.

3.1.2. Top-down RIE

On the other hand, RIE process is successful implementing on most semiconductors because RIE can homogeneously etch the exposed surface and NWs are obtained in the masked region of the wafer. This process is typically shown in Fig. 6, taking Si NWs as an example in combination with the wet chemical etching.[72] In details, the photoresist mask is lithographically defined on the $Si_3N_4/SiO_2/Si(110)$ wafer, and nanochannels are then gained by the XeF_2 RIE. After an etching plane is obtained by the anisotropic chemical etching, the channels are then oxidized to extend such that the cracks are all oxidized and leaving the core area of Si NWs with the doping process adopted for functional electronics. Similarly, Si nanoribbons and NRs can also be obtained by varying the process parameters of

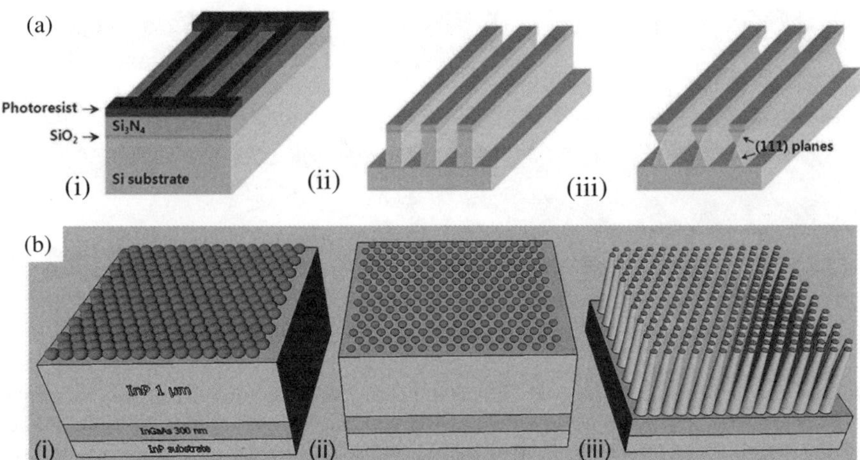

Figure 6.   Schematics of the dry RIE etching methods for vertical and horizontal NW preparation. (a) RIE for the horizontal Si NW preparation: (i) photoresist mask definition, (ii) etching of Si using the Si$_3$N$_4$ layer as a hard mask, (iii) anisotropic wet chemical etching for the formation of NWs, (b) RIE for the construction of vertical InP NW arrays: (i) dispersing SiO$_2$ nanospheres onto the O$_2$ plasma pre-treated InP wafer by spin coating, (ii) reducing the size of nanospheres by RIE (CHF$_3$ gas), and (iii) NW array is then obtained by the selective etching (Cl$_2$/H$_2$/CH$_4$ mixture) on the exposed area of the InP surface. Reprinted with permission.[72,74]

the combined dry and wet etching processes.[73] In contrast, other masks such as silica spheres can be used for the etching mask; this way, the vertical NW arrays, instead of the horizontal one, can be attained and then harvested for the subsequent device fabrication.[74]

In both cases, the as-obtained NWs would inherit the physical, chemical, and electrical properties of their mother wafer materials. In this case, the electrical property can be easily controlled by just tailoring that of the mother wafer. Moreover, the NW diameter and length can also be easily manipulated by the masks employed and the processing conditions. However, there is one disadvantage that single crystalline wafers are required as the sacrifice precursor while the etching process is relatively complicated, especially for the RIE. All these together would make the cost here relatively high.

3.1.3. Bottom-Up Vapor Phase Growth

The bottom-up strategy assembles precursor atoms to the realization of 1D nanomaterials in such a way that the growing crystal plane would have the highest Gibbs energy while leaving the lower energy planes to form the sidewalls. It is noticeable that the surface energy of nanomaterials differs much from their bulk counterparts; therefore, this leads to the growth of NWs along different directions in which sometimes this growth direction depends on the NW diameter. Similarly, this surface energy can be tailored by adding surfactants to make the 1D nanomaterials stacking along their lower energy planes. The atomic precursors can be either assembled in liquid solution or in gas phase, with growth technologies well-known as the hydrothermal/solvothermal or chemical vapor deposition (CVD), respectively.

In the vapor phase growth, metal nanoparticles such as Au, Ni, and Pd are adopted as the catalyst to lead the NW growth, commonly referred to as the vapor–liquid–solid (VLS) process firstly reported in the 1960s.[75] In this process, precursors such as Si, Ge, Ga, and In would be dissolved in the catalyst particles gradually, and then become supersaturated due to the Gibbs–Thomson effect in the nanometer scale, i.e., $\ln(C_d/C_0) = 4\gamma V_m/(dRT)$, where C_d is the concentration of precursor in the catalyst nanoparticle with the diameter of d, C_0 is the equilibrium concentration in flat surface ($d \rightarrow \infty$) materials, γ is the surface energy, V_m is the molar volume of catalyst, R is a constant (8.314 J mol^{-1} K^{-1}), and T is the growth temperature.[76] The precursors are then precipitated from the supersaturated catalyst nanoparticles to induce the NW growth accordingly. This process is believed to occur at temperatures above the eutectic temperature of the binary phase, while it is also sometime observed to happen under the eutectic temperature, namely the vapor–solid–solid (VSS) mechanism.[77–80] One apparent difference between VLS and VSS is that the catalyst shapes are mostly hemispherical in VLS but polyhexagonal in VSS, as given in Fig. 7. A detailed study recently shows that the NW quality, in terms of the growth rate, uniformity of orientation, and crystal defects, are highly dependent on their diameter (Fig. 7), which quantitatively dictates the supersaturation of the catalyst particles.[81] All these results infer that the

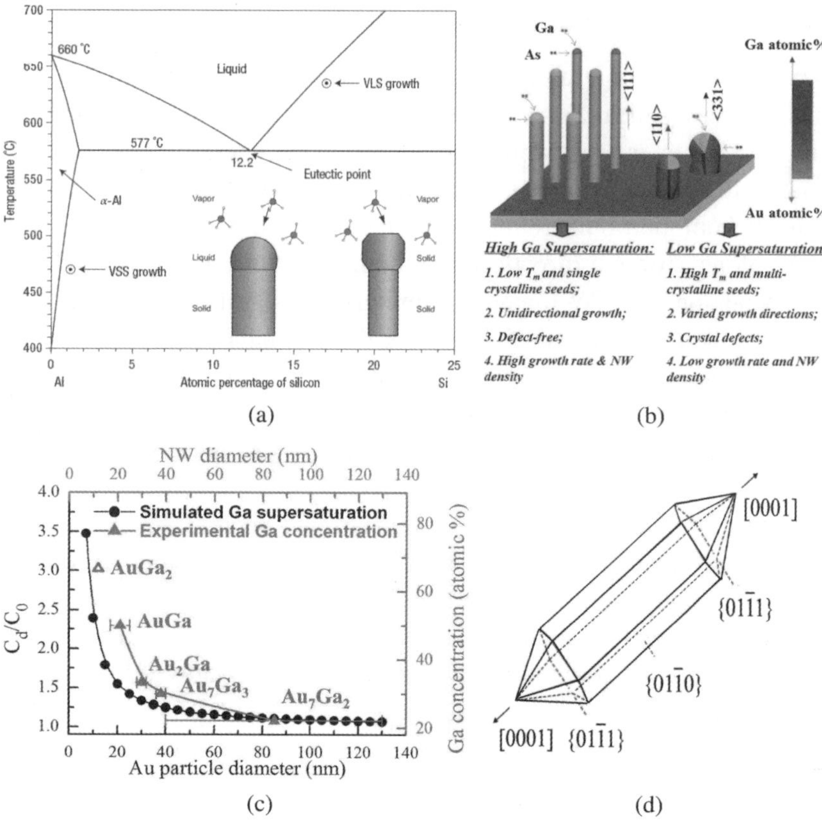

Figure 7. Schematics of vapor phase growth mechanisms of semiconductor NWs. (a) Determination of growth temperature zones of Si NWs in the binary phase diagram of Al–Si, with hemispherical catalyst tips inferring the VLS growth and polygonal tips inferring the VSS growth, (b) crystal quality depending on the Ga supersaturation in Au nanoparticle catalysts with different diameters, (c) simulation of the Gibbs–Thomson effect and observed by HRTEM and EDS on the Ga supersaturation in Au catalyst seeds with different diameters, (d) VS growth mechanism of ZnO NWs along the polar <001> direction with the lowest surface energy. Reprinted with permission.[78,80,88]

NWs are quasi-epitaxial grown from their catalyst nanoparticles. A further control of the catalyst supersaturation by the two-step growth method as well as by the catalyst alloying process is successful in controlling the crystal quality and electronic property of the resultant GaAs, InAs, CdTe, and Ge NWs.[81–84]

In the meanwhile, the metallic component of III–V and II–VI compound semiconductors such as Ga, In, and Cd can also serve as the catalyst for the NW growth, which is known as the self-catalyzed growth. This technique consists of a unique characteristic of the homogenous catalyst employed in the growth system, as compared with the heterogeneous catalyst of Au or Ni. One significant advantage of this self-catalysis is that there is not any foreign element used in the growth, and hence no catalytic atom residual would exist in the NW body, while Au atoms are usually found in the Au-seeded NW body and are believed to be detrimental to the electronic properties of NWs.[85,86] However, the self-catalyzed growth process needs to be well-controlled as the growth window is typically much narrower than that of the heterogeneous growth method. Nowadays, much success has been achieved for the synthesis of high-quality self-catalyzed GaAs NWs.[87] Alternatively, if neither homo- nor heterogeneous catalyst is used, NWs might also grow via a direct assemble of gaseous atom precursors, which is identified as the vapor–solid (VS) mechanism, as illustrated in Fig. 7(d).[88] This VS process is most applicable to metal oxide NWs such as ZnO, but this would as well compete with the VLS/VSS growth of other semiconductor NWs, resulting in a surface coating problem frequently.[89,90] Therefore, special attention should be paid to the growth processes in synthesizing semiconductor NWs, where different growth mechanisms would dominate and compete against each other in order to obtain the process equilibrium.

3.1.4. Bottom-Up Liquid Phase Growth

Liquid phase growth methods are generally divided into two categories based on the medium used: (i) hydrothermal using aqueous solution and (ii) solvothermal utilizing organic solutions. In the hydrothermal method, the process temperature is relatively lower, as compared with the one of solvothermal. Hundreds of degree higher in the temperature is always needed in the solvothermal technique since it requires the supercritical fluid as the organic solvent, and thus the process temperature is highly dependent on their supercritical temperatures. Similar to those in the vapor phase growth, the growth mechanism in the liquid phase synthesis also involves two main strategies, that are the self-assembly and the catalytic growth.

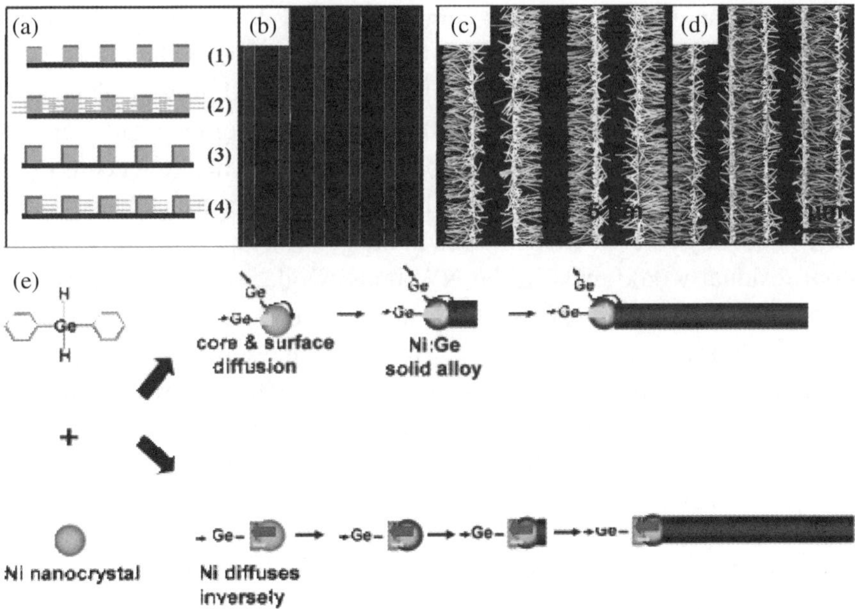

Figure 8.   Growth mechanisms and SEM images of the liquid phase grown NWs. (a)–(d) Hydrothermal grown horizontal ZnO NWs, (a) schematics of the growth process, (b) SEM image of the ZnO pattern shielded on one top and one side by 10 nm thick Cr, (c) and (d) SEM images of the ZnO NWs grown by the hydrothermal process, (e) growth mechanism of Ni catalyzed Ge NWs via the solvothermal process with supercritical fluid, including the Ge surface diffusion to the NW body and the Ni counter-diffusion from the NW body. Reprinted with permission.[91,94]

As depicted in Fig. 8, ZnO NW arrays can be prepared directly on the pre-patterned seeds by the hydrothermal method[91] by self-assembly of the precursors while the seeds are used as the nucleation site for the NW growth. Specifically, the 300 nm ZnO strips are photolithographically defined and magnetron sputtered onto the Si(100) substrate to serve as the active seeding layer and a 10 nm thick Cr layer is deposited on top of it to inhibit the vertical growth. The substrate is then aged at 80°C for 12 hour in the solution of precursor $Zn(NO_3)_2$ and surfactant hexamethylenetetramine (HMTA) with both concentrations of 0.0025 mol/L. After that, the lateral grown ZnO nanorod arrays are finally obtained. Furthermore, by passivating one sidewall of the ZnO strip with Cr or Sn, one side growth can also be attained. It is worth noting that the NW array is grown at a moderate temperature of 80°C, in which plastic materials such as Kapton films are stable and can also serve as the substrate. Therefore, this hydrothermal process is highly applicable to many other metal oxide semiconductor NWs, such as TiO_2[92] and SnO_2[93] etc., but is seldom used for the synthesis of other semiconductor NWs since there is none available and relevant precursor soluble in water.

Instead, group IV, II–VI, and III–V semiconductor NWs can be synthesized by the solvothermal technique employing metal-organic precursors soluble in organic medium, especially in the supercritical fluid. Most solvothermal processes involve heterogeneous catalyst particles; this way, NWs can be synthesized by the VLS/VSS mechanism similar to those in the vapor phase growth. For example, Ge NWs are successfully synthesized by using diphenylgermane (DPG) as the precursor and Ni or Au nanoparticles as the catalyst in anhydrous toluene, hexane and chloroform at 410–460°C and 23.4 MPa, which is in the supercritical fluid status.[94] The growth temperature is far below the eutectic temperature of Ni–Ge and Au–Ge binary alloy, and the non-spherical shaped $NiGe_x$ catalyst tips infer the solid phase diffusion mechanism. Meanwhile, the NW diameter distribution is rather narrow for Ni catalyzed NWs than those of Au-catalyzed NWs, indicating that there is significant seed particle aggregation in the case of Au-catalyzed growth process. All these results show the advantage of Ni over Au as the catalyst, while the growth mechanism is schematically discussed in Fig. 8(e), highlighting the combined effect of Ge precursor diffusion from surface of solid Ni and solid Ni counter-diffusion from Ge NWs. Importantly, the bulk diffusivity of Ni in Ge is extremely fast such that the corresponding reaction kinetics can be simply controlled by the Ge precursor diffusion. Employing the similar processes, Si, GaAs, and GaP NWs can also be prepared successfully.[95–99]

At the same time, there are other methods for the NW preparation through the "bottom-up" strategy, including templated synthesis using anodic aluminum oxide (AAO)[100,101] and electronic spinning.[102–104] 1D nanomaterials synthesized by these methods are more often polycrystalline with low carrier mobility; hence, they are mostly applied in thermo-electronics and batteries, while seldom used in electronic circuits, and thus are not emphasized herein. Anyway, these bottom-up approaches provide the flexibility for control of the morphology, crystal structure, and doping of NWs for different electronic applications. In the meanwhile, several problems are also observed in the non-uniform distribution of NW diameter, length, crystal orientation, and doping profile, etc., which need a further fine tune of the growth parameters for large-scale application in electronics with uniform electronic properties. All these related recent developments will be discussed in the following section.

3.1.5. Characterization of NWs

In order to assess the suitability of NWs for certain applications as well as the effect of different synthesis parameters on the NW characteristics, physical, chemical, optical, and electrical properties of NWs have to be explored systemically. More related information can be found elsewhere in the Refs. 105–107 for details. This section mainly introduces some conventional techniques as well as advanced methods for the fundamental characterization of NWs.

Nowadays, as widely adopted, microscopy methods are commonly engaged for the morphological characterization of NW materials, including scanning electron microscopy (SEM), transmission electron microscopy (TEM), and atomic force microscopy (AFM), etc. All these techniques are robust in determining the diameter distribution, surface roughness and hierarchical structures in the nanometer scale. In addition, the *in situ* (high-resolution, HR) TEM observation of NW growth can provide the solid proofs as well as the determination of the growth mechanism in real-time.[108] Recently, a dark field optical microscopic method is also developed for the NW diameter determination based on the scattering spectra of NWs, which would facilitate the first and fast access of the NW quality right after the synthesis.[109]

The energy dispersive spectroscopy (EDS) and electron energy loss spectroscopy (EELS) associated with the SEM and TEM can as well provide the elemental composition within the resolution of ~1%, while the X-ray photoelectron spectroscopy (XPS) and ultraviolet photoelectron spectroscopy (UPS) can yield the similar elemental information in the outmost ~5 nm surface with the higher resolution. The exact composition of the main components can be determined by inductively coupled plasma atomic emission spectroscopy (ICP-AES) after the NWs are dissolved in certain solutions, and the trace dopants can be detected by the high resolution secondary ion mass spectroscopy (SIMS). In recent times, due to all these advancements in the composition assessment, Au atoms, existing in the NW body via the residual catalyst in the VLS growth, can be directly imaged by the high resolution scanning (STEM), which is believed to induce trap states in NWs deteriorating their physical properties.

The determination of NW crystal structure is rather complicated, especially for the bottom-up grown NWs, which is not as easily confirmed by the mother wafers as in the case of top-down approaches. Since the surface-to-volume ratio is extremely large for nanomaterials, the non-equilibrium crystal structure in the bulk materials would become thermally stable or even preferential in the nanometer scale. For instance, the cubic zinc-blende (ZB) structure is very stable in most III–V semiconductors, while hexagonal wurtzite structure (WT) will always compete with ZB counter-parts down to certain criteria and eventually becomes dominant below the critical diameter such as 10 nm for GaAs NWs and 60 nm for InAs NWs.[110,111] In this context, the conventional structural characterization techniques including X-ray diffraction (XRD) and Raman spectroscopy cannot be sufficient for the crystal structure determination of individual NWs. In this case, the selected area electron diffraction (SAED) and HRTEM of NWs are essential to assess their crystal structure and to determine their growth orientation. Besides, the NW growth mechanism can also be verified by the crystal structural analysis of the catalyst tips and NW body by SAED and HRTEM technologies,[112] and lately by the extremely high energy surface XRD.[113]

Last of all, the electronic bandgap of NWs should be investigated, especially for the thin wires which may associate with the blue-shift in their bandgaps, as well as the composition dependent composite alloys. Photo-luminescence (PL) and cathodoluminescence (CL) are widely used for the NW bandgap assessment if the studied NWs have the good crystal quality. On the other hand, when there are significant amounts of defects acting as non-radioactive recombination centers with no observable luminescence emitted, or when the bandgap is too large/small to be sensed by detectors, ultraviolet–visible absorption/reflection spectroscopy can be hired to deduce the bandgap, with details easily found in the literature.[114,115]

3.2. *NW Alignment Technologies*

Although single NWs demonstrate excellent electric performances, the current density is still relatively low for the practical use, and thus it is necessary to integrate individual NWs into arrays with good registration. It is noticed that most top-down obtained NWs are directly well aligned in vertical/horizontal arrays, as illustrated in the figures in the previous sections; however, bottom-up NWs are always randomly distributed on the substrates. As a result, technologies aligning NWs have been vastly explored in the last decade in order to fabricate complex circuits in a large scale employing bottom-up NWs. In this section, some of the technologies are summarized, such as the directly grown method, Langmuir–Blodgett technique, blown bubble film technology, fluid assisted alignment, electric field assisted alignment, and contact printing technology, etc.

3.2.1. Grown in Arrays Directly

The directly grown ZnO NW array by hydrothermal scheme (Fig. 8) is a representative example of this method; however, no functional electronics have been fabricated by this technique at that stage. In the meanwhile, horizontal aligned GaAs NW arrays directly grown on GaAs substrates,[116] and the knock-down method for vertical NWs to align horizontally[117] are all transferable and promising for flexible electronics. Recently, Kim *et al.* synthesized Si NW arrays directly between two predefined Si channels using Au catalyst by CVD method as shown in Fig. 9.[118] By tuning the process parameter, a p-i-n structured Si NW array is prepared and demonstrated with the good diode property which can be later integrated into functional logic gates. Although the substrate utilized is the rigid SiO_2 in this report due to the high growth temperature of Si NWs, the entire device can still be transferable by technologies introduced in 3.3.3 onto flexible substrates; therefore, it is prospective in flexible electronics. In this context, directly growing high carrier mobility NW arrays, such as GaAs instead of ZnO by the hydrothermal/solvothermal method, would be promising for flexible electronics, because the relatively moderate growth temperature would be endurable for the plastics, and hence no post-transfer step would be needed.

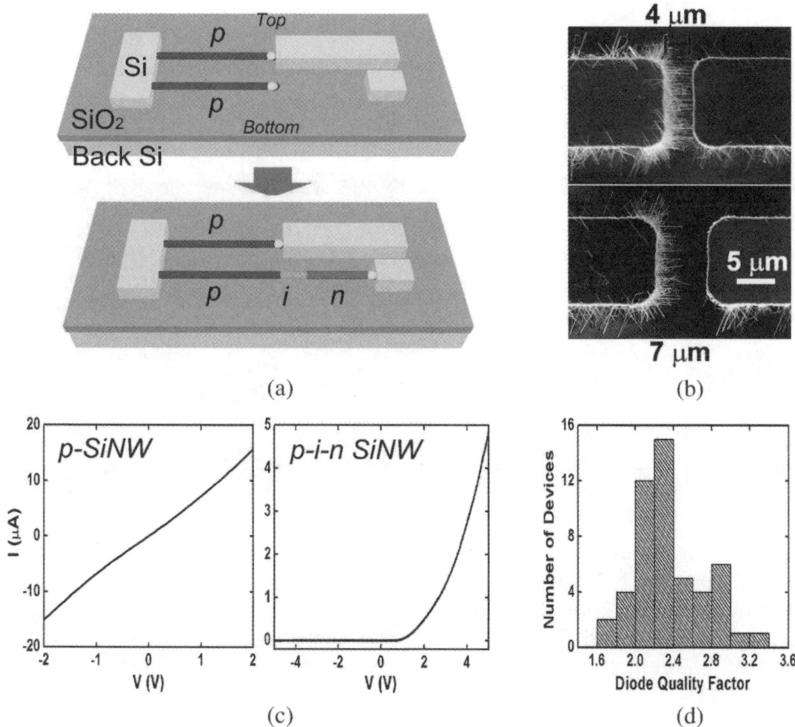

Figure 9.   Schematics, SEM images and electric properties of directly grown Si NW devices. (a) Schematic of the directly grown p-type Si NW resistor (top) and p-i-n Si NW diode (bottom), (b) SEM images of the Si NW devices with different channel gap, (c) IV curves of the Si NW resistor and diode, and (d) the ideality factor of the p-i-n Si diode distribution histogram. Reprinted with permission.[118]

3.2.2. Langmuir–Blodgett Technique

In early 1900s, Langmuir and Blodgett discovered that a highly ordered monolayer of organics (LB film) will form on the surface of a solid when it is immersing into or emerging from the monolayered organic-water interface. Decades later in early 2000s, researchers succeeded in getting an ordered 1D nanomaterial layer such as nanorods and nanowires using similar process, substituting the monolayered organics by 1D nanomaterial/ surfactant monolayers.[119] The process is schematically illustrated in Fig. 10, taking NWs as an example. Firstly, Si NWs are monodispersed onto the surface of water assisted by surfactant 1-octadecylamine in an LB trough.

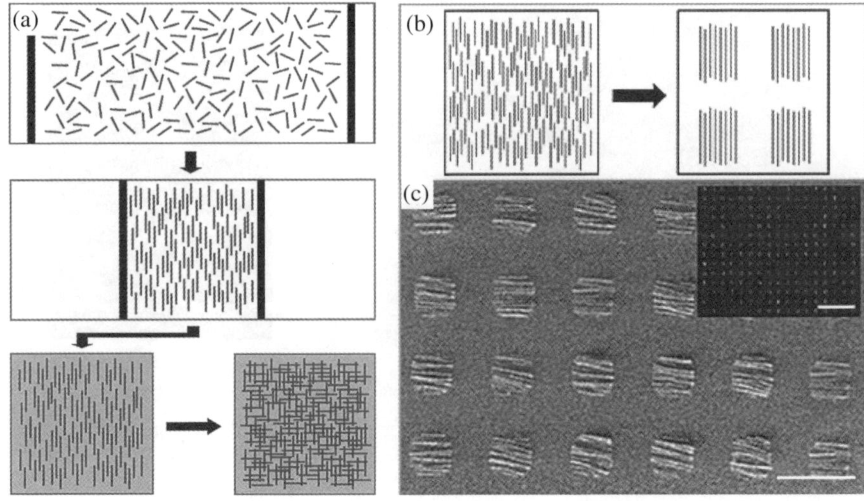

Figure 10. Schematics and SEM images of NW arrays aligned by the LB method. (a) Steps in LB process: NWs are dispersed by surfactant on the surface of water, and then are aligned by the compression of the surface, which are finally transferred onto other substrate, (b) schematics of the patterned transfer by the LB process, and (c) the SEM images of the NWs arrays by patterned LB transfer. Reprinted with permission.[119]

Then the monolayer surface is compressed, and the NWs are aligned along their long axis by the compression. Later, the aligned NW arrays are transferred onto any substrate immersed or emerged through the layer, which can be repeated to get higher density or cross structures. It should be noted that the inter-NW spacing can be easily tuned by controlling the compression from micrometer to close-packed structure. Furthermore, the NWs arrays can be selectively registered onto substrates aided by the lithographically defined areas as shown in Fig. 10(b), showing the versatility of this easy alignment and transferring technology. By choosing different liquids, surfactants, different kinds of NWs such as Si, Ge etc. they have to be successfully aligned into array and integrated into electronic devices.[119,120]

3.2.3. Blown Bubble Film Technology

The blown bubble technology aligns NWs by the sheer force on the expanding bubble films, and then the NW/polymer film can be transferred to any substrate in the large scale. Yu *et al.* first reported the aligned Si NW arrays fabricated by this technology in the following route[121]: Si NWs are modified by 5,6-epoxyhexyltriethoxysilane and combined with an epoxy solution to get a suspension of 0.01–0.22 wt.%; then after the viscosity of the NW solution reached 15–25 Pa·s, the suspension is dropped onto the surface of a die and blown to a bubble by N_2 (150–200 kPa); and finally, the blown bubble film with aligned NWs on it can be transferred to both rigid and flexible substrates for the further construction into electronic devices as illustrated in Fig. 11. Similar to the LB method, the spacing and NW density in the array can be manipulated easily by the loading percentage of NWs in the suspension. Especially, this technology is widely applicable to most 1D nanomaterials, such as the CNTs, and

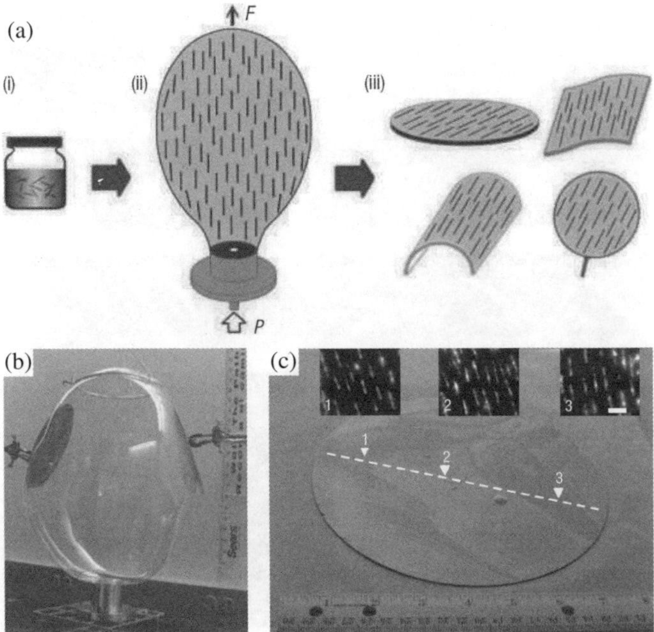

Figure 11.   Schematics and photographs of blown bubble film for aligning NWs. (a) Process schematics including (i) NWs suspended in the polymer, (ii) polymer/NW bubble blown over a circular die by N_2, (iii) NW array on polymer films got transferred onto different rigid and flexible substrates, (b) photography of the blown bubble transferring step, and (c) photography and microscope image of the transferred NW array on Si wafer. Reprinted with permission.[121]

CdS nanowires, etc.[122] Similarly, the strain release process of a stretched substrate will also lead to NWs aligning uniformly on the surface.[123]

3.2.4. Fluid Assisted Alignment

It is well known that the pressure on the front and rear side of a solid in a fluid liquid is different. This way, the force difference would drive the 1D materials to align along the fluid flowline in order to stabilize the system, which is a well-known phenomenon in nature similar to the timbers flowing in the river. In 2001, Huang *et al.*[124,125] reported aligning GaP, InP and Si NWs by the fluid flow in microchannel as shown in Figs. 12(a) and 12(b). Typically, the Si NWs are suspended in ethanol and the suspension is then

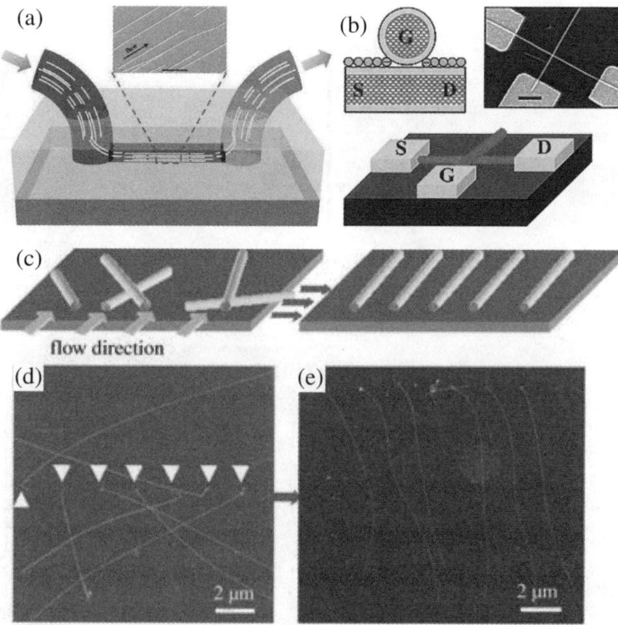

Figure 12. Schematic and SEM images of Si and Ge NW arrays fabricated by the fluid assisted alignment technique. (a) Schematics of the NW array constructed by the microchannel fluid method, (b) configuration and SEM images of the cross NW FET obtained by a repeated cross fluid process, (c) schematics of the fluid assisted alignment of anchored Ge NWs, (d) and (e) are the SEM images of the Ge NWs before and after alignment by fluid. Reprinted with permission.[2,127,128]

driven through the fluidic channel formed between a poly(dimethylsiloxane) (PDMS) mold and the flat substrate. After the liquid flow, NWs are anchored on the substrate by van der Waals force or electrostatic force along the channel as depicted in the SEM image, and NW crosses can be integrated by a second cross fluidic layer. Detailed process parameter control can enable the NW array alignment and the density tailoring. For example, the alignment angle can be controlled by the flowrate. The NW density (spacing) can be tailored by the flow duration and the substrate surface modification. Specifically, the NH_2-terminated monolayers coated on substrates would enhance the NW deposition rate and density than the CH_3-terminated monolayers or bare SiO_2 surfaces, due to the enhanced interaction of positively charged NH_2-group with the NW surface. This method is robust in manufacturing NW network logic gates as summarized in the literature.[126]

Furthermore, the randomly grown Ge NWs on substrates can also be aligned *in situ* by the liquid flown over, as shown in Figs. 12(c)–12(e).[127] This method resembles the phenomenon that grass can be aligned in the streamlet by the flow. In this case, special attention should be taken that the NWs are fixed tightly on the substrate and the flow rate needs to be small in order not to cut the NWs. It should also be noted that the LB method is executed in a static liquid medium, while this fluid method is performed in a flowing liquid. As no surfactant is needed in the fluid assisted technology, which is essential in LB method to suspend the NWs, there would not be any surface modification effect on the NWs which would influence the NW electronic property profoundly as discussed in the following sections.

3.2.5. Electric field assisted alignment

In some cases, if there is any dipole in the NWs such as semiconductor NWs with the axial p–n junction, an electrostatic field can be utilized to align these NWs into arrays by the electrostatic force. If there is no intrinsic dipole in the NWs, an alternative electric field can still be employed to induce a dipole between the two ends of the NWs and align them accordingly. In this case, the electric field assisted NW alignment is usually occurs in a liquid suspension of NWs, which is named as electrophoresis driven by the direct current (DC) and as dielectrophoresis driven by the alternative current (AC).

Electrophoresis can be used to successfully align the axial p-n Si NWs as reported by Lee *et al.*[129], with a high yield of 97.7% in nearly 200 tests as indicated in Fig. 13. The p–n Si NWs are prepared by CVD technology via the VLS mechanism by tuning the dopant (B_2H_6 for p-type and PH_3

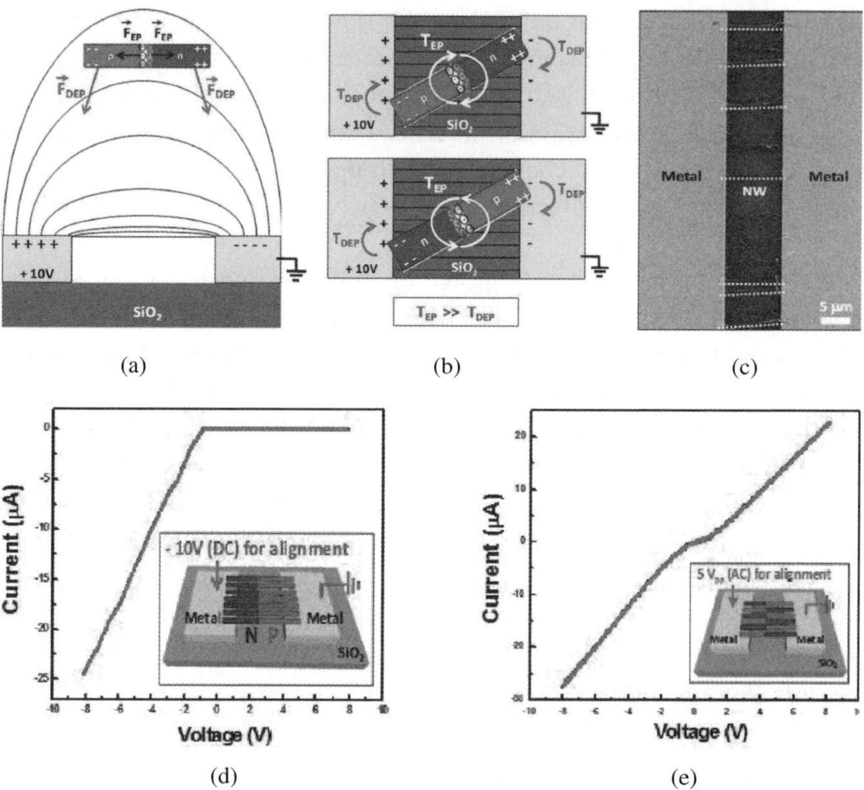

Figure 13.   Schematics, SEM image and electrical properties of NW arrays obtained by the electric field assisted alignment. (a) The origin of electrophoresis force and dielectrophoresis force of a polar NW in the electric field, (b) schematics of the alignment of polar p–n Si NWs by the electrophoresis force, (c) SEM images of the aligned p–n Si NW array between electrodes, (d) IV curve of the p–n Si NW array attained by the DC voltage showing the rectifying effect, and (e) IV curve of the p–n Si NW array by the AC voltage showing no rectifying effect. Comparison of (d) and (e) illustrates the polar alignment as well as the morphology alignment of polar NWs by the DC voltage. Reprinted with permission.[129]

for n-type) as well as the molar ratio of Si/B = 4000 and Si/P = 1500. The p–n Si NWs are then suspended in isopropyl alcohol (IPA) by sonication, and then drop-casted onto the lithographically defined electrodes. After a DC voltage supply of 10 V, the NWs are then aligned between the two electrodes effectively not only in terms of the morphology, but also in terms of the polar direction. In order to serve as a control, an AC voltage of 5 V can also be used to align the p–n Si NWs as well, but with neutral overall dipole (i.e., randomly directed p–n junctions) as illustrated in Fig. 13 no rectifying effect is observed, as one of the DC aligned NW arrays. The kinetic analysis shows that the electrophoretic force is far larger than the dielectrophoretic force in rotating the counter directed NWs and to align the NWs into arrays. On the other hand, the axially homogeneous NWs such as p-type Si obtained by either top-down wet chemical etching or by bottom-up CVD method, can be effectively aligned in patterned electrodes in the large scale and even on insulating plastics by the dielectrophoresis technology.[130,131]

Generally, the advantages of the electric field assisted alignment technologies includes that the NWs can be well aligned due to the spatial distribution of the electric field, which would be better than other technologies, and the NWs can be aligned due to their intrinsic polar rather than only the morphology in other technologies. Furthermore, the NW density can also be controlled by the NW concentration and even by multi-layered processes. In any case, the NW density aligned in the arrays seems not as high as those obtained in other technologies, which might be improved further.

3.2.6. Contact printing technology

The above-mentioned NW aligning technologies all involve a suspension of NWs in liquid or polymer process in order to harvest NWs from the substrate first and then align them in the subsequent step. The NW dispersion would be good for individual NWs, but the obtained NW array would have a limited density due to the NW to NW repulsion. And importantly, as the electrical properties of semiconductor NWs are highly sensitive to their surface states, the surfactants or solutions used in these technologies might have some influence on the corresponding electronic performance of the fabricated devices. Therefore, a direct transfer of NWs from their growth substrate to a device substrate is desired, which can avoid any possible detrimental effects to their electrical properties.

Recently, Javey *et al.* developed a dry transfer technology called contact/roll printing technology, which enables the direct, large-scale, patterned transfer of semiconductor NWs onto various rigid and flexible substrates.[132–134] The materials used are the CVD synthesized semiconductor NWs such as Ge/Si,[134,135] InAs,[135,136] GaAs,[82] and $In_xGa_{1-x}As$[137] etc. typically grown on amorphous SiO_2 substrate, and the prepatterned device substrate obtained by the conventional photolithography. Specifically, the NW donor substrate is first placed upside down onto the receiver device substrate in full contact, and pushed at a certain velocity against the receiver with optimized pressure. After the photoresist lift-off, the NW arrays are then left on the receiver device substrate for the further circuit fabrication. In this dry transfer process, the NW alignment might not be good and an oil lubricant would enhance the alignment greatly with no significant influence on the electrical properties. Furthermore, a positively charged NH_2-functional polymer or poly-L-lysine used for SiO_2 surface modification can also increase the NW density drastically by ~2–5x, while the neutral CF_3–grouped modifier would inhibit NW adhesion on the substrate. It should be further noted that the cross layout of NW arrays as well as the vertically stacked multilayer devices can also be easily fabricated by repeating the printing process in the desired directions, demonstrating its advantages over other NW alignment technologies. In the meantime, NWs can be successfully transferred to flexible substrates such as polyimide for flexible electronics by this contact printing. The only disadvantage of this method is the printed NWs in the arrays are only about half the length as-grown, because NWs would be cracked by the pressure and friction in the contact printing process, which however can be easily surmounted by growing the NWs with sufficient length, e.g., >10 μm.

To further explore the possibility of contact printing technology being compatible with the traditional roll-to-roll printing for low-cost flexible electronics, a roll printing technology is developed by growing semiconductor NWs on cylindrical substrates and mimicking the contact printing process to transfer NWs onto flexible substrates with the high efficiency as shown in Fig. 14.[138] The main difference between the contact and roll

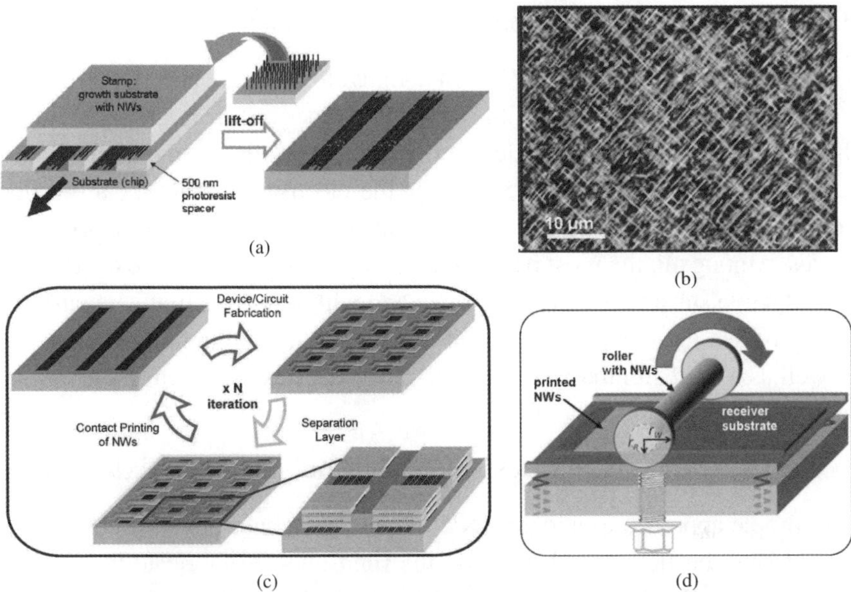

Figure 14. Schematics and photograph of the contact and roll printing technology. (a) Schematic of the contact printing method, including that NWs are flipped away from the prepatterned substrate and then the NW arrays are obtained after the photoresist lift-off, (b) schematics of the repeated contact printing method for the vertically stacked device integration, (c) photograph of two layer printed NWs constructed in a cross layout, and (d) schematics of the roll-printing technology with different radii of roll and wheel. Reprinted with permission.[132,134,138]

printing is the friction induced by the receiver substrate would be low due to the rolling motion of the donor substrate in the roll printing. In order to transfer the NWs efficiently, a differential rolling process is developed by mismatching the roller and wheel radii, which enables a relative sliding motion of the roller with respect to the substrate. Therefore, the roll-printing technology would be the most promising technique in the fabrication of NW arrays for flexible electronics, especially in this direct dry transferring process, the high NW density and multilayer printing can be readily obtained on any rigid or flexible substrates. More importantly, all these are compatible with the existing low-cost roll-to-roll printing technology for the manufacturing of flexible electronics.

3.3. *Integration into Flexible Logic Circuits*

Till now, several strategies have been adopted to integrate NW arrayed films for electronic devices onto flexible substrates. One scheme is simply to attach the flexible substrates onto rigid ones, and then exfoliating them after fabricating the electronic devices. Secondly, the electronics devices can be first fabricated on rigid substrates and then transferred directly onto flexible substrates. Among all, the most promising one is to directly fabricate the devices onto flexible substrates on a large-scale and with low-cost. In this regard, the following session will give a brief introduction to these different technologies as well as a brief summary of the state-of-the-art flexible electronics.

3.3.1. Patterning on Plastics/Rigid Substrate and then Exfoliate

One simple approach to integrate NW arrayed film onto flexible substrates is to first spin the flexible substrate on the rigid ones, to fabricate the devices directly on the flexible substrates and then to peel them off after the fabrication. This method is compatible with most NW alignment technologies, as typically shown in Fig. 15. More importantly, this can make full use of the conventional lithography process as long as the flexible substrates are stable during the entire process scheme. For example, the thermally and chemically stable PI is commonly used as the flexible substrate, which can be spin-coated onto the SiO_2/Si handling wafer at a thickness of 24 μm. Then the InAs NWs (~30 nm in diameter and ~10 μm in length) grown by CVD method are contact-printed on the PI, followed by that top-gated FET devices are fabricated with 50 nm thick Ni S/D electrodes (1.5 μm in channel length), 8 nm thick ALD Al_2O_3 gate dielectrics and 40 nm thick Al gate electrode.[23] This flexible NW arrayed TFTs hold the record of oscillation frequency up to 1.8 GHz and a cutoff frequency of ~1 GHz due to the high electron mobility of InAs NWs. Importantly, this flexible structure is strain-resistant with only 0.5% of the strain endured by the NW arrays as a result of the relatively high Young's modulus of InAs than that of PI as well as due to the relatively small dimension of the NWs.

3.3.2. Transfer Printing of the Fabricated Devices onto Plastics

However, if the flexible substrate is not compatible with the lithography process, the entire device structure can be first fabricated on rigid Si

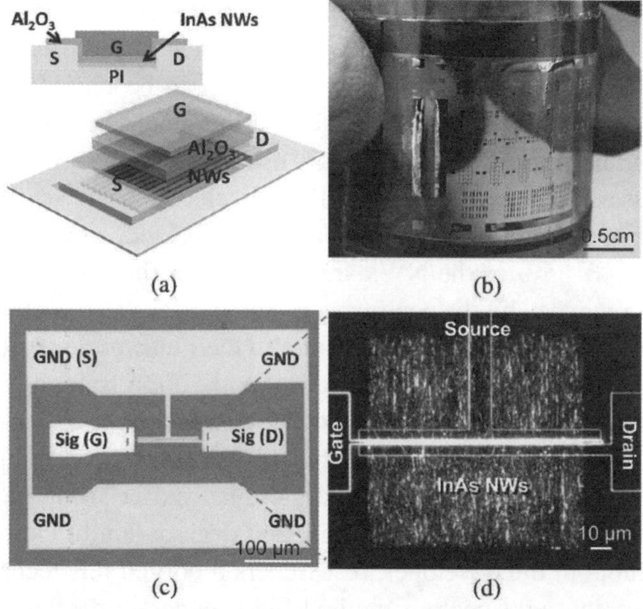

Figure 15. Schematics and photography of InAs NW array FET fabricated on the flexible polyimide substrate. (a) Schematics of the InAs NW array FET layouts, (b) photography of the flexible electronics, (c) microscope image of the FET device with ground-signal-ground electrodes, and (d) dark field microscope image of one individual InAs NW array FET. Reprinted with permission.[23]

substrates and then later get transferred onto flexible substrates such as by PDMS.[139] This technology is purely based on the adhesion difference of the devices (especially the metal electrodes) on rigid donor substrate versus the one on the flexible receiver substrates. Therefore, metals with weak adhesion to Si surface are usually adopted, including Au and Pd, and a top Ti layer can be added in order to increase this adhesion to the flexible substrates after the transfer. Typically, Si NWs grown by CVD method are contact printed on Si wafers and the Au/Ti electrodes are defined by lithography and e-beam evaporation. Then, liquid PDMS was casted onto the whole device wafer and enveloped it after curing at room temperature overnight. Finally, the PDMS film with a thickness of ~5 mm is peeled off with the entire NW arrayed device embedded inside (channel length ranging from 3–25 μm). This technology is advantageous in that, the peeled PDMS shows wavy shaped surface, which would dissipate strains in bending or stretching. Also, the Si wafer can be reused to fabricate devices again for the repeated transfer. Furthermore, the NWs and metal electrodes can be transferred separately using the double or multiple transfer processes,[139,140] facilitating the flexible electronic device integration onto various substrates such as Kapton, taps, and even Petri dishes. However, ~80–85% of the NWs are broken after the peel-off due to the large strain induced by the peeling step. Some cracks also occur in PDMS and electrodes, and become detrimental to their electronic performances, especially when there is gate dielectric in the FET structure. All these necessitate the further optimization of the transfer process.

As the significant advantage of this device transfer technology lies in the compatibility with flexile substrates that suffer from problems in lithography process such as shrinking and degradation in the resist baking or soluble in the developer, researchers modified this technology by substituting the wild peeling method by a moderate peeling called the water assisted transfer as shown in Fig. 16. In this modified process,

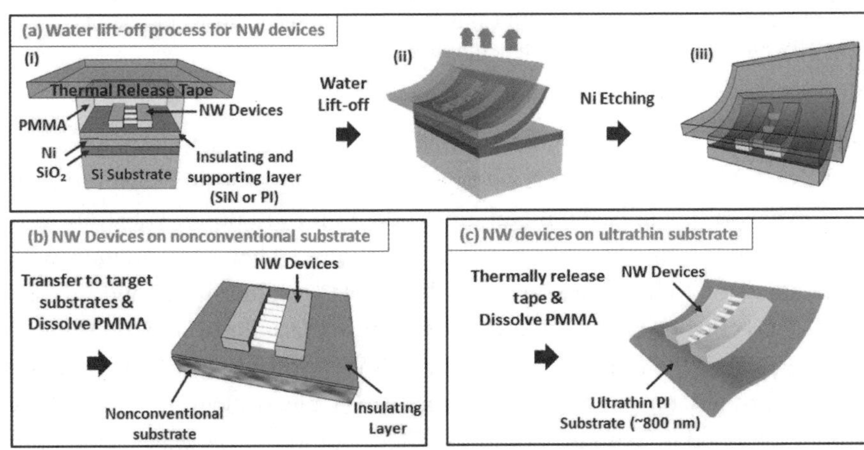

Figure 16.   Schematics of the water assisted transfer printing technology. (a) Schematics of the entire process, (i) fabricate NW array devices on Ni/SiO₂/Si substrates and envelope them by PDMS, (ii) the device is peeled off by water, (iii) etch the Ni layer to obtain the device, (b) the transferred NW devices on a non-conventional substrate, (c) the transferred NW devices on an ultrathin PI substrate. Reprinted with permission.[141]

SiO_2/Si is used as the substrate and a sacrificing Ni layer is deposited on the surface. The NW arrayed electronic devices are fabricated on the composite substrate $Ni/SiO_2/Si$, which are then buried in flexible plastics. Later, water will penetrate into the Ni/SiO_2 interface and separates the two layers. After the Ni layer is etched from the plastics, NW electronics such as diodes and FETs exhibited similar performances with those fabricated on the rigid SiO_2/Si wafer.[141] Similarly, electronic devices can also be fabricated on GaAs substrate with an AlAs sacrificing layer on the surface, which is subsequently etched by HF or HCl solution to release the device layers on the top.[142] Furthermore, another moderate peeling off technology adopts the thermal expansion strain of the top Ni layer for the fully peel-off and for reuse of the substrate.[49] As a result, the transfer printing technology is promising for the utilization in flexible and stretchable electronics, especially when the substrate is not thermal or chemical stable.

3.3.3. Direct Patterning on Plastics

As mentioned previously, the low-cost and lightweight features are the most important ones for flexible macroelectronics, and the corresponding electronic performance should be enhanced as high as possible under that premise. However, if the lithography process (either by photo- or electron-beam) is obligatory for defining electrodes, the resultant expense would be high due to the relatively complex procedures and costly fabrication instruments. In this context, researchers developed technologies directly imprinting patterns on flexible substrates by microprinting of rubber stamp with the resolution of micrometer[143] and nanoimprint lithography (NIL) by rigid stamp with the resolution in submicrometer range.[144] Taking NIL as an example, it generates patterns by the compression of a rigid inorganic stamp onto the flexible polymer for the subsequent metal electrode deposition, as typically illustrated in Fig. 17. In this process, LOR lift-off resist and SU-8 resist can be used for the molding pattern and imprinted by the predefined features on the SiO_2 stamps. Then, the typical metal electrode deposition and lift-off process can be used for the metallization. This method has a high resolution of hundreds nanometer (e.g., 200 nm in Fig. 17), enabling the fabrication of high performance flexible

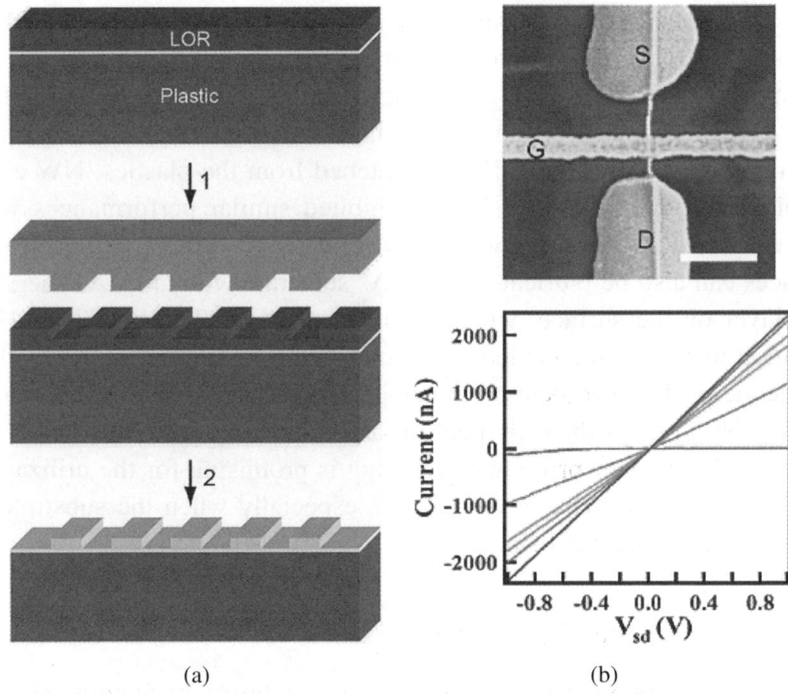

Figure 17.   Schematics of the nanoimprint lithography (NIL) for electrode definition and the subsequent electronic device fabrication. (a) Schematics of NIL process, (1) LOR resist (red) is patterned by the Si/SiO$_2$ stamp (green), (2) metal electrode definition by RIE of LOR, metal deposition and lift-off, (b) SEM image of a Si NWFET by crossing an imprinted metal gate, (c) I_{DS}–V_{GS} curves of the Si NWFET. Reprinted with permission.[145]

electronics. Notably, the inorganic rigid stamps can be reused after the resist removal, and hence this is easily scalable at the low cost.

The NIL defined electrodes on flexible substrates can be used for NW arrayed device fabrication using the technologies mentioned above such as the fluid assisted alignment of NWs as shown in Fig. 17. Specifically, the 20 nm thick p-type Si NWs are grown by the CVD method and are dispersed in ethanol, which are then aligned and fabricated as NW-based electronic devices. The hole mobility of ~200 cm^2/Vs is achieved, which outperforms all the amorphous or nanocrystalline Si TFTs and the organic FETs. Importantly, the NWs can also be integrated into functional circuits such as inverters, paving the way for flexible processors. It should be noted that in order to be compatible with the conventional low-cost roll-to-roll printing technology, rigid stamps configured in the format of roll should be explored in combination with the roll-printing alignment of NWs,[138] which would show the prospect for the NW-based all-organic flexible circuits by the facile printing technology.[10]

3.4. *Summary of the NW-based Flexible Circuit Performance*

Based on the above-mentioned nanomaterials and nanotechnologies, several kinds of functional circuits on flexible substrates such as high frequency FETs, and logic gates etc. have been fabricated with the superior performance. For example, GaAs NWs, obtained by the wet chemical etching, can be transferred onto the flexible substrate and different types of NOR and NAND circuits are fabricated as shown in Fig. 18. The typical

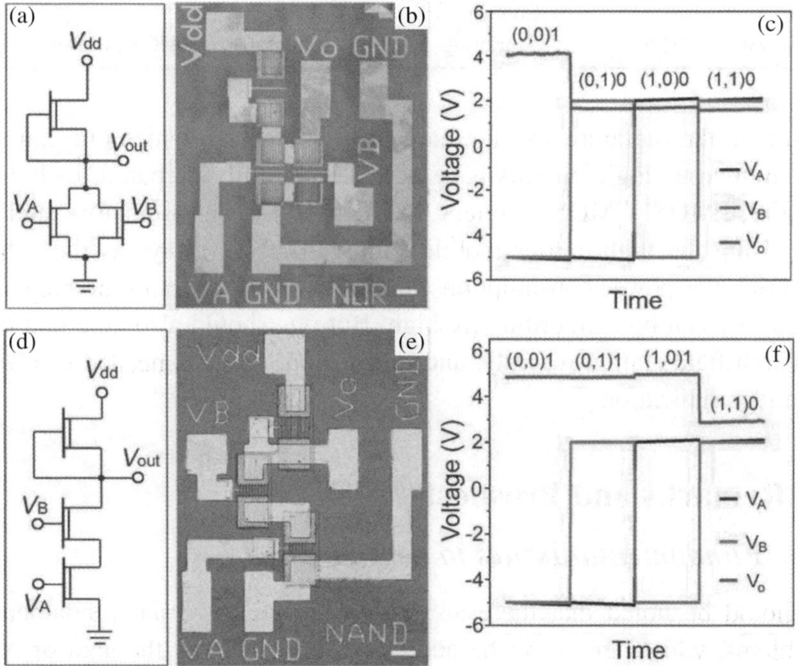

Figure 18. Circuit schematic, images and output–input relationship of GaAs NW arrays based logic circuits. (a)–(c) are NOR and (d)–(e) are NAND. Scale bars are 100 μm. Reprinted with permission.[146]

Table 1. Summary of NW arrays based logic circuits on flexible substrates.

Material	Preparation	Alignment	Fabrication	Circuits	Refs.
Ge/Si NW	VLS	Contact printing	Photolithography	50 MHz PMOS inverter	134
Si strips	Wet etching	*In situ*	Transfer printing	12 GHz TFTs	63
Si ribbon	Wet etching	*In situ*	Transfer printing	515 MHz TFTs	147
Si ribbon	RIE	*In situ*	Transfer printing	2.6 MHz ring oscillator	148
Si NW	VLS	*In situ*	Directly grown	Logic gates	118
Si NW	VLS	Fluid assisted	NIL	PMOS inverter	144
p-Si NW, n-CdS NR	VLS	Fluid assisted	Photolithography	CMOS inverter	149
InAs NW	VLS	Contact printing	Photolithgraphy	1.8 GHz FET	23
GaAs MW	Wet etching	*In situ*	Transfer printing	Logic gates	146
SnO_2 NW	VLS	Contact printing	Photolithography	NMOS inverter	27

results in the literature are summarized in Table 1, in which one can see that numerous logic circuits have been successfully prepared, including NMOS, PMOS, CMOS inverters, NOR, NAND, ring oscillators, and so on. Using the high carrier mobility inorganic NW arrays as the active material, the power consumption can be relatively low and the response frequency can be extraordinarily high. But one should also note that this research field started recently, and much more effort is needed to invest for industrialization.

4. Remarks and Prospects

4.1. *Fundamental Issues to be Addressed*

It should be noted that there are still many controversial fundamental problems, which are yet to be addressed, especially in the area of NW growth mechanisms, integration methods and performance optimization, etc. For example, NWs are more frequently grown on single crystalline substrates using the molecular or chemical beam epitaxy methods (MBE, CBE).[79,150,151] In such methods, the NW orientation and crystal structure are uniform and highly dependent on the mother substrates. However, the associated cost would be relatively high due to the expensive crystalline wafers and the complex growth systems. Recently, researchers explored the growth methods for NWs on cheap substrates such as glass and SiO_2, etc.[90,152,153] Although morphologically good NWs can be obtained with similar VLS/VSS mechanisms, the NW orientation and crystal structure are not that uniform because no confinement exists in this non-epitaxial growth to govern the NWs. This way, as the crystal phase and orientation would significantly influence the corresponding NW device perfor- mances, the preparation–structure–property relationships need more detailed study for the large scale, uniform NW preparation and applica- tions. In this regard, researchers are currently dealing with all these issues, considering the competition between VLS and VSS mechanisms, the influence of the catalyst supersaturation,[81,84] catalyst phase, structure,[112] precursor concentration, ratio,[154] and growth temperature, etc.[137]

In the meanwhile, special attention should be paid to the surface/inter- face problem in the NW device structure, which would influence the electronic properties of the active channel materials greatly either by forming oxide shells or by adsorbing/desorbing oxygen molecules, espe- cially since NWs have an extraordinarily high surface-to-volume ratio. For example, the surface oxide layer of GaAs would provide electron trap states depleting the thin NWs to become p-type in the conductivity, yield- ing medium thick NWs to ambipolar but affect little on the thick n-type NWs.[155] At the same time, the absorbed oxygen would induce similar electron trap states making the PbSe NWs ambipolar, while desorbing oxygen would return them back to the n-type conductivity.[156] On the other hand, the interface of NWs with organic substrates or dielectrics would also induce a great change in their corresponding electrical properties. For instance, as detailed in Fig. 19, the interface of SnO_2 and PI has abundant trap states yielding the I–V curves with large hysteresis, which can be significantly reduced by etching the PI layer and forming a suspended structure.[157] Therefore, great care is indeed needed to invest on the NW surface/interface states manipulating the corresponding surface/interface influence for better electronic performance. Also, various kinds of

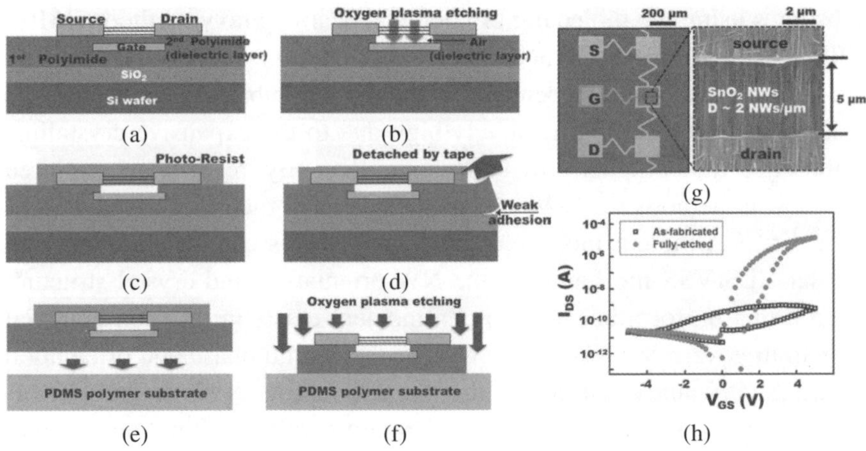

Figure 19.   Schematics of the SnO$_2$ NW arrayed device on stretchable PDMS substrates. (a) Configuration of the backgated FET, (b) suspending the NW arrays by O$_2$ plasma etching, (c) device coated with the photoresist, (d) detaching the device form the Si/SiO$_2$ substrate, (e) placing the device onto PDMS, (f) photoresist etching by the O$_2$ plasma, (g) microscope image and SEM image of the device, (h) IV curves of the flexible FETs with and without the PI layer, showing the effect upon the removal of NW/PI interface states in order to enhance the current density and decrease the hysteresis. Reprinted with permission.[157]

intentionally designed heterostructured core/shell NWs can exhibit much higher device performances than the homogeneous NWs due to the unique design of the band-structures involved.[158–160]

Moreover, the piezoelectric effect of NWs should also be paid attention as a result of the deformation by the flexible substrates. The piezoelectric effect is significant for most nanomaterials as compared with their bulk counterparts, which would influence the output current of Si NWFET greatly as typically shown in Fig. 20.[161] Also, the surface oxide layer would inevitably affect the conductivity and piezoresistance. In this case, if the adopted channel material is piezoelectric, the carrier mobility and concentration would be altered by the strain induced by the deformation of the flexible substrate, which can be exemplified by the ZnO,[162] GaN,[163] and Si NWs,[164] while the channel resistance varies little for B NWs under the strain due to its minimal piezoelectricity.[61] Another related

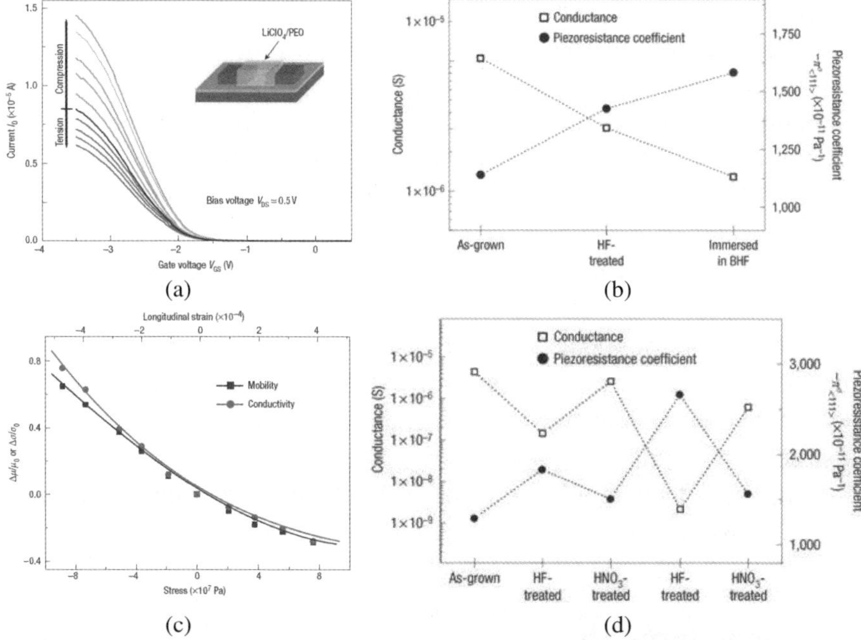

Figure 20.   Electrical properties of Si NWFETs under strain. (a) Transfer characteristics of the Si NWFET under different tensile and compressive strains, the inset gives the device configuration, (b) changes of conductance along with the hole mobility, (c) change of conductance against the piezoresistance coefficient with the tailored surface states. Reprinted with permission.[161]

fundamental aspect is the mechanical properties of NWs, which can also be varied significantly and differently from their bulk counterparts. For example, the long NWs are brittle but the short ones are ductile,[165] while metal NWs can yield plastic deformation.[166] In this regard, there are still many issues to be addressed before semiconductor NWs can be successfully integrated into flexible electronics for enhanced and stable device performances.

4.2. *Future Technologies Prospect*

In summary, there are several key conclusions which can be drawn on flexible electronic technologies here. One would prefer growing high performance NW arrays directly on prepatterned electrodes by a moderate plastic-compatible environment, even though the process condition might be a bit rigorous. In parallel, the roll-printing assembly of NWs grown randomly on a cheap non-crystalline substrate would enable the all-printing flexible electronics including printing the dielectrics and electrodes.[10,45,167,168] Furthermore, the transfer printing of devices fabricated on rigid wafer onto flexible substrates would provide the candidate for flexible processors where high electronic performance is needed. It is noted that flexible and stretchable ionic conductors are well finding their own pathway in the replacement of electrode materials for flexible electronics. However, for certain applications where both high frequency and high voltage operations are demanded, the conventional conductors such as structured metals and graphene are still preferred.[169] As a result, various promising technologies have been developed and competed against each other for the large-scale, low-cost flexible electronics.

Acknowledgments

This work was supported by the General Research Fund of the Research Grants Council of Hong Kong SAR, China (Grant No. CityU 101111), the Early Career Scheme of the Research Grants Council of Hong Kong SAR, China (Grant No. CityU 139413), the National Natural Science Foundation of China (Grant No. 51202205), the Guangdong National Science Foundation (Grant No. S2012010010725), and the Science Technology and Innovation Committee of Shenzhen Municipality (Grant No. JCYJ20120618140624228).

References

1. M. C. LeMieux and Z. N. Bao, *Nat. Nanotechnol.,* **3**, 585 (2008).
2. X. Liu, Y. Z. Long, L. Liao, X. F. Duan and Z. Y. Fan, *ACS Nano,* **6**, 1888 (2012).
3. X. F. Duan, *IEEE T. Electron Dev.,* **55**, 3056 (2008).
4. D. H. Kim and J. A. Rogers, *Adv. Mater.,* **20**, 4887 (2008).
5. B. P. Timko, T. Cohen-Karni, G. H. Yu, Q. Qing, B. Z. Tian and C. M. Lieber, *Nano Lett.,* **9**, 914 (2009).

6. N. Lu and D.-H. Kim, *Soft Robotics,* **1**, 53 (2014).

7. F. Garnier, R. Hajlaoui, A. Yassar and P. Srivastava, *Science,* 265, 1684 (1994).

8. A. R. Brown, A. Pomp, C. M. Hart and D. M. Deleeuw, *Science,* **270**, 972 (1995).

9. G. H. Gelinck, T. C. T. Geuns and D. M. de Leeuw, *Appl. Phys. Lett.,* **77**, 1487 (2000).

10. M. Jung, J. Kim, J. Noh, N. Lim, C. Lim, G. Lee, J. Kim, H. Kang, K. Jung, A. D. Leonard, J. M. Tour and G. Cho, *IEEE T. Electron Dev.,* 57, 571 (2010).

11. G. Gruner, *J. Mater. Chem.,* **16**, 3533 (2006).

12. Y. Z. Long, M. Yu, B. Sun, C. Z. Gu and Z. Y. Fan, *Chem. Soc. Rev.,* **41**, 4560 (2012).

13. K. Takei and T. Takahashi, J. C. Ho, H. Ko, A. G. Gillies, P. W. Leu, R. S. Fearing and A. Javey, *Nat. Mater.,* **9**, 821 (2010).

14. J. Lewis, *Mater. Today,* 9, **38** (2006).

15. S. Wagner and S. Bauer, *MRS Bull.,* **37**, 207 (2012).

16. A. Manekkathodi, M. Y. Lu, C. W. Wang and L. J. Chen, *Adv. Mater.,* 22, 4059 (2010).

17. W. A. MacDonald, M. K. Looney, D. MacKerron, R. Eveson, R. Adam, K. Hashimoto, K. Rakos, *J. Soc. Inf. Display,* **15**, 1075 (2007).

18. A. Laskarakis, S. Logothetidis, S. Kassavetis and E. Papaioannou, *Thin Solid Films,* **516**, 1443 (2008).

19. V. Zardetto, T. M. Brown, A. Reale, A. Di Carlo, *J. Polym. Sci., Part B: Polym. Phys.,* **49**, 638 (2011).

20. S. W. Hwang, G. Park, H. Cheng, J. K. Song, S. K. Kang, L. Yin, J. H. Kim, F. G. Omenetto, Y. G. Huang, K. M. Lee and J. A. Rogers, *Adv. Mater.,* **26**, 1992 (2014).

21. A. L. Salas-Villasenor, I. Mejia, J. Hovarth, H. N. Alshareef, D. K. Cha, R. Ramirez-Bon, B. E. Gnade and M. A. Quevedo-Lopez, *Electrochem. Solid-State Lett.,* **13**, H313 (2010).

22. Q. Cao, M. G. Xia, M. Shim, J. A. Rogers, *Adv. Funct. Mater.,* 16, 2355 (2006).

23. T. Takahashi, K. Takei, E. Adabi, Z. Y. Fan, A. M. Niknejad A. Javey, *ACS Nano,* **4**, 5855 (2010).

24. L. A. Majewski, M. Grell, S. D. Ogier and J. Veres, *Org. Electron.,* 4, 27 (2003).

25. A. Maliakal, H. Katz, P. M. Cotts, S. Subramoney and P. Mirau, *J. Am. Chem. Soc.,* **127**, 14655 (2005).

26. J. Veres, S. Ogier, G. Lloyd and D. de Leeuw, *Chem. Mater.,* **16**, 4543 (2004).

27. G. Shin, M. Y. Bae, H. J. Lee, S. K. Hong, C. H. Yoon, G. Zi, J. A. Rogers and J. S. Ha, *ACS Nano,* **5**, 10009 (2011).

28. J. A. Rogers, T. Someya and Y. G. Huang, *Science,* **327**, 1603 (2010).

29. D. H. Kim, J. L. Xiao, J. Z. Song, Y. G. Huang and J. A. Rogers, *Adv. Mater.,* **22**, 2108 (2010).

30. X. L. Hu, P. Krull, B. de Graff, K. Dowling, J. A. Rogers and W. J. Arora, *Adv. Mater.,* **23**, 2933 (2011).

31. J. A. Fan, W. H. Yeo, Y. W. Su, Y. Hattori, W. Lee, S. Y. Jung, Y. H. Zhang, Z. J. Liu, H. Y. Cheng, L. Falgout, M. Bajema, T. Coleman, D. Gregoire, R. J. Larsen, Y. G. Huang and J. A. Rogers, *Nat. Commun.,* **5**, 3266 (2014).

32. Y. S. Chen, Y. F. Xu, K. Zhao, X. J. Wan, J. C. Deng, W. B. Yan, *Nano Res.,* **3**, 714 (2010).

33. T. Sekitani, Y. Noguchi, K. Hata, T. Fukushima, T. Aida and T. Someya, *Science,* **321**, 1468 (2008).

34. L. B. Hu, H. S. Kim, J. Y. Lee, P. Peumans and Y. Cui, *ACS Nano,* **4**, 2955 (2010).

35. S. K. Lee, B. J. Kim, H. Jang, S. C. Yoon, C. Lee, B. H. Hong, J. A. Rogers, J. H. Cho and J. H. Ahn, *Nano Lett.,* **11**, 4642 (2011).

36. S. De, T. M. Higgins, P. E. Lyons, E. M. Doherty, P. N. Nirmalraj, W. J. Blau, J. J. Boland and J. N. Coleman, *ACS Nano,* **3**, 1767 (2009).

37. J. Y. Lee, S. T. Connor, Y. Cui and P. Peumans, *Nano Lett.,* **8**, 689 (2008).

38. M. S. Lee, K. Lee, S. Y. Kim, H. Lee, J. Park, K. H. Choi, H. K. Kim, D. G. Kim, D. Y. Lee, S. Nam and J. U. Park, *Nano Lett.,* **13**, 2814 (2013).

39. W. Gaynor, G. F. Burkhard, M. D. McGehee and P. Peumans, *Adv. Mater.,* **23**, 2905 (2011).

40. M. G. Kang, H. J. Park, S. H. Ahn and L. J. Guo, *Sol. Energ. Mat. Sol. C.,* **94**, 1179 (2010).

41. A. R. Rathmell and B. J. Wiley, *Adv. Mater.,* **23**, 4798 (2011).

42. I. Park, S. H. Ko, H. Pan, C. P. Grigoropoulos, A. P. Pisano, J. M. J. Frechet, E. S. Lee and J. H. Jeong, *Adv. Mater.,* **20**, 489 (2008).

43. B. Y. Ahn, E. B. Duoss, M. J. Motala, X. Y. Guo, S. I. Park, Y. J. Xiong, J. Yoon, R. G. Nuzzo, J. A. Rogers and J. A. Lewis, *Science,* **323**, 1590 (2009).

44. T. Yokota, T. Sekitani, T. Tokuhara, N. Take, U. Zschieschang, H. Klauk, K. Takimiya, T. C. Huang, M. Takamiya, T. Sakurai and T. Someya, *IEEE T. Electron Dev.,* **59**, 3434 (2012).

45. H. Sirringhaus, T. Kawase, R. H. Friend, T. Shimoda, M. Inbasekaran, W. Wu and E. P. Woo, *Science,* **290**, 2123 (2000).

46. K. Myny, E. van Veenendaal, G. H. Gelinck, J. Genoe, W. Dehaene and P. Heremans, *IEEE J. Solid-State Circuits,* **47**, 284 (2012).

47. T. B. Singh and N. S. Sariciftci, *Annu. Rev. Mater. Res.,* 36, 199 (2006).

48. H. E. Tu and Y. Xu, *Appl. Phys. Lett.,* **101**, 052106 (2012).

49. Y. J. Zhai, L. Mathew, R. Rao, D. W. Xu and S. K. Banerjee, *Nano Lett.,* **12**, 5609 (2012).

50. K. Nomura, A. Takagi, T. Kamiya, H. Ohta, M. Hirano and H. Hosono, *Jpn. J. Appl. Phys.,* **45**, 4303 (2006).

51. D. K. Kim, Y. M. Lai, B. T. Diroll, C. B. Murray and C. R. Kagan, *Nat. Commun.,* **3**, 1216 (2012).

52. C. Wang, J. C. Chien, K. Takei, T. Takahashi, J. Nah, A. M. Niknejad and A. Javey, *Nano Lett.,* **12**, 1527 (2012).

53. Q. Cao, H. S. Kim, N. Pimparkar, J. P. Kulkarni, C. J. Wang, M. Shim, K. Roy, M. A. Alam and J. A. Rogers, *Nature,* **454**, 495 (2008).

54. S. J. Kang, C. Kocabas, T. Ozel, M. Shim, N. Pimparkar, M. A. Alam, S. V. Rotkin and J. A. Rogers, *Nat. Nanotechnol.,* **2**, 230 (2007).

55. J. A. del Alamo, *Nature,* **479**, 317 (2011).

56. A. C. Ford, J. C. Ho, Y. L. Chueh, Y. C. Tseng, Z. Y. Fan, J. Guo, J. Bokor and A. Javey, *Nano Lett.,* **9**, 360 (2009).

57. J. Nah, H. Fang, C. Wang, K. Takei, M. H. Lee, E. Plis, S. Krishna and A. Javey, *Nano Lett.,* **12**, 3592 (2012).

58. X. F. Duan, *MRS Bull.,* **32**, 134 (2007).

59. Z. X. Yang, N. Han, F. Y. Wang, H. Y. Cheung, X. L. Shi, S. Yip, T. Hung, M. H. Lee, C. Y. Wong and J. C. Ho, *Nanoscale,* **5**, 9671 (2013).

60. Z. X. Yang, F. Y. Wang, N. Han, H. Lin, H. Y. Cheung, M. Fang, S. Yip, T. F. Hung, C. Y. Wong and J. C. Ho, *ACS Appl. Mater. Interfaces,* **5**, 10946 (2013).

61. J. F. Tian, J. M. Cai, C. Hui, C. D. Zhang, L. H. Bao, M. Gao, C. M. Shen and H. J. Gao, *Appl. Phys. Lett.,* **93**, 122105 (2008).

62. F. Xu, W. Lu and Y. Zhu, *ACS Nano,* **5**, 672 (2011).

63. L. Sun, G. X. Qin, J. H. Seo, G. K. Celler, W. D. Zhou and Z. Q. Ma, *Small,* **19**, 2553 (2010).

64. Y. G. Sun and J. A. Rogers, *Adv. Mater.,* **18**, 1897 (2007).

65. K. Q. Peng, A. J. Lu, R. Q. Zhang and S. T. Lee, *Adv. Funct. Mater.,* **18**, 3026 (2008).

66. F. Y. Wang, Q. D. Yang, G. Xu, N. Y. Lei, Y. K. Tsang, N. B. Wong and J. C. Ho, *Nanoscale,* **3**, 3269 (2011).

67. J. M. Weisse, C. H. Lee, D. R. Kim and X. L. Zheng, *Nano Lett.,* **12**, 3339 (2012).

68. H. Lin, H. Y. Cheung, F. Xiu, F. Y. Wang, S. P. Yip, N. Han, T. F. Hung, J. Zhou, J. C. Ho and C. Y. Wong, *J. Mater. Chem. A,* **1**, 9942 (2013).

69. Y. G. Sun, D. Y. Khang, F. Hua, K. Hurley, R. G. Nuzzo and J. A. Rogers, *Adv. Funct. Mater.,* **15**, **30** (2005).

70. Y. G. Sun and J. A. Rogers, *Nano Lett.,* **4**, 1953 (2004).

71. E. Menard, K. J. Lee, D. Y. Khang, R. G. Nuzzo and J. A. Rogers, *Appl. Phys. Lett.,* **84**, 5398 (2004).

72. M. Lee, Y. Jeon, T. Moon and S. Kim, *ACS Nano,* **5**, 2629 (2011).

73. A. J. Baca, M. A. Meitl, H. C. Ko, S. Mack, H. S. Kim, J. Dong, P. M. Ferreira and J. A. Rogers, *Adv. Funct. Mater.,* **17**, 3051 (2007).

74. S. Naureen, R. Sanatinia, N. Shahid and S. Anand, *Nano Lett.,* 11, 4805 (2011).

75. R. S. Wagner and W. C. Ellis, *Appl. Phys. Lett.,* **4**, 89 (1964).

76. D. R. Lide, *CRC Handbook of Chemistry and Physics* Boca Raton, CRC Press (2010).

77. A. I. Persson, M. W. Larsson, S. Stenstrom, B. J. Ohlsson, L. Samuelson and L. R. Wallenberg, *Nat. Mater.,* **3**, 677 (2004).

78. Y. W. Wang, V. Schmidt, S. Senz and U. Gosele, *Nat. Nanotechnol.,* **1**, 186 (2006).

79. S. Kodambaka, J. Tersoff, M. C. Reuter and F. M. Ross, *Science,* **316**, 729 (2007).

80. J. L. Lensch-Falk, E. R. Hemesath, D. E. Perea and L. J. Lauhon, *J. Mater. Chem.,* **19**, 849 (2009).

81. N. Han, F. Wang, J. J. Hou, S. Yip, H. Lin, M. Fang, F. Xiu, X. Shi, T. Hung and J. C. Ho, *Cryst. Growth Des.,* **12**, 6243 (2012).

82. N. Han, J. J. Hou, F. Y. Wang, S. Yip, Y. T. Yen, Z. X. Yang, G. F. Dong, T. Hung, Y. L. Chueh and J. C. Ho, *ACS Nano,* **7**, 9138 (2013).

83. C. O'Regan, S. Biswas, C. O'Kelly, S. J. Jung, J. J. Boland, N. Petkov and J. D. Holmes, *Chem. Mater.,* **25**, 3096 (2013).

84. Z. Zhang, Z. Y. Lu, P. P. Chen, H. Y. Xu, Y. N. Guo, Z. M. Liao, S. X. Shi, W. Lu and J. Zou, *Appl. Phys. Lett.,* **103**, 073109 (2013).

85. M. Bar-Sadan, J. Barthel, H. Shtrikman and L. Houben, *Nano Lett.*, **12**, 2352 (2012).

86. S. Breuer, C. Pfüller, T. Flissikowski, O. Brandt, H. T. Grahn, L. Geelhaar and H. Riechert, *Nano Lett.*, **11**, 1276 (2011).

87. P. Krogstrup, R. Popovitz-Biro, E. Johnson, M. H. Madsen, J. Nygård and H. Shtrikman, *Nano Lett.*, **10**, 4475 (2010).

88. P. Hu, F. Yuan, L. Bai, J. Li and Y. Chen, *J. Phys. Chem. C*, **111**, 194 (2007).

89. D. Rudolph, S. Hertenberger, S. Bolte, W. Paosangthong, D. Spirkoska, M. Doblinger, M. Bichler, J. J. Finley, G. Abstreiter and G. Koblmuller, *Nano Lett.*, **11**, 3848 (2011).

90. N. Han, F. Y. Wang, A. T. Hui, J. J. Hou, G. C. Shan, F. Xiu, T. F. Hung and J. C. Ho, *Nanotechnology*, **22**, 285607 (2011).

91. Y. Qin, R. S. Yang and Z. L. Wang, *J. Phys. Chem. C*, **112**, 18734 (2008).

92. X. Yu, H. Wang, Y. Liu, X. Zhou, B. J. Li, L. Xin, Y. Zhou and H. Shen, *J. Mater. Chem. A*, **1**, 2110 (2013).

93. C. W. Cheng, B. Liu, H. Y. Yang, W. W. Zhou, L. Sun, R. Chen, S. F. Yu, J. X. Zhang, H. Gong, H. D. Sun and H. J. Fan, *ACS Nano*, **3**, 3069 (2009).

94. H. Y. Tuan, D. C. Lee, T. Hanrath and B. A. Korgel, *Chem. Mater.*, **17**, 5705 (2005).

95. H. Yu and W. E. Buhro, *Adv. Mater.*, **15**, 416 (2003).

96. J. D. Holmes, K. P. Johnston, R. C. Doty and B. A. Korgel, *Science*, **287**, 1471 (2000).

97. D. D. Fanfair and B. A. Korgel, *Cryst. Growth Des.*, **5**, 1971 (2005).

98. H. Y. Tuan, D. C. Lee, T. Hanrath and B. A. Korgel, *Nano Lett.*, **5**, 681 (2005).

99. F. M. Davidson, R. Wiacek and B. A. Korgel, *Chem. Mater.*, **17**, 230 (2005).

100. J. Zhou, C. Jin, J. H. Seol, X. Li and L. Shi, *Appl. Phys. Lett.*, **87**, 133109 (2005).

101. B. Yoo, F. Xiao, K. N. Bozhilov, J. Herman, M. A. Ryan and N. V. Myung, *Adv. Mater.*, **19**, 296 (2007).

102. W. Wu, S. Bai, M. Yuan, Y. Qin, Z. L. Wang and T. Jing, *ACS Nano*, **6**, 6231 (2012).

103. H. Chen, N. Wang, J. Di, Y. Zhao, Y. Song and L. Jiang, *Langmuir*, **26**, 11291 (2010).

104. D. Li and Y. N. Xia, *Adv. Mater.*, **16**, 1151 (2004).

105. A. J. Baca, J. H. Ahn, Y. G. Sun, M. A. Meitl, E. Menard, H. S. Kim, W. M. Choi, D. H. Kim, Y. Huang and J. A. Rogers, *Angew. Chem. Int. Ed.*, **47**, 5524 (2008).

106. M. S. Gudiksen, L. J. Lauhon, J. Wang, D. C. Smith and C. M. Lieber, *Nature,* **415**, 617 (2002).

107. Y. N. Xia, P. D. Yang, Y. G. Sun, Y. Y. Wu, B. Mayers, B. Gates, Y. D. Yin, F. Kim and Y. Q. Yan, *Adv. Mater.,* **15**, 353 (2003).

108. Y. Y. Wu and P. D. Yang, *J. Am. Chem. Soc.,* **123**, 3165 (2001).

109. G. Bronstrup, C. Leiterer, N. Jahr, C. Gutsche, A. Lysov, I. Regolin, W. Prost, F. J. Tegude, W. Fritzsche and S. Christiansen, *Nanotechnology,* **22**, 385201 (2011).

110. N. Han, J. J. Hou, F. Wang, S. Yip, H. Lin, M. Fang, F. Xiu, X. Shi, T. Hung and J. C. Ho, *Nanoscale Res. Lett.,* **7**, 1 (2012).

111. J. Johansson, K. Dick, P. Caroff, M. Messing, J. Bolinsson, K. Deppert and L. Samuelson, *J. Phys. Chem. C,* **114**, 3837 (2010).

112. N. Han, A. T. Hui, F. Wang, J. J. Hou, F. Xiu, T. F. Hung and J. C. Ho, *Appl. Phys. Lett.,* **99**, 083114 (2011).

113. R. E. Algra, V. Vonk, D. Wermeille, W. J. Szweryn, M. A. Verheijen, W. J. P. van Enckevort, A. A. C. Bode, W. L. Noorduin, E. Tancini, A. E. F. de Jong, E. P. A. M. Bakkers and E. Vlieg, *Nano Lett.,* **11**, 44 (2011).

114. J. J. Hou, F. Wang, N. Han, F. Xiu, S. Yip, M. Fang, H. Lin, T. F. Hung and J. C. Ho, *ACS Nano,* **6**, 9320 (2012).

115. N. Han, F. Wang, Z. Yang, S. Yip, G. Dong, H. Lin, M. Fang, T. Hung and J. Ho, *Nanoscale Res. Lett.,* **9**, 347 (2014).

116. S. A. Fortuna, J. G. Wen, I. S. Chun and X. L. Li, *Nano Lett.,* **8**, 4421 (2008).

117. A. Pevzner, Y. Engel, R. Elnathan, T. Ducobni, M. Ben-Ishai, K. Reddy, N. Shpaisman, A. Tsukernik, M. Oksman and F. Patolsky, *Nano Lett.,* **10**, 1202 (2010).

118. D. R. Kim, C. H. Lee and X. L. Zheng, *Nano Lett.,* **10**, 1050 (2010).

119. D. Whang, S. Jin, Y. Wu and C. M. Lieber, *Nano Lett.,* **3**, 1255 (2003).

120. D. W. Wang, Y. L. Chang, Z. Liu and H. J. Dai, *J. Am. Chem. Soc.,* **127**, 11871 (2005).

121. G. H. Yu, A. Y. Cao, C. M. Lieber, *Nat. Nanotechnol.,* **2**, 372 (2007).

122. F. Kim, S. Kwan, J. Akana and P. D. Yang, *J. Am. Chem. Soc.,* **123**, 4360 (2001).

123. F. Xu, J. W. Durham, B. J. Wiley and Y. Zhu, *ACS Nano,* **5**, 1556 (2011).

124. Y. Huang, X. F. Duan, Y. Cui, L. J. Lauhon, K. H. Kim and C. M. Lieber, *Science,* **294**, 1313 (2001).

125. Y. Huang, X. F. Duan, Q. Q. Wei and C. M. Lieber, *Science,* **291**, 630 (2001).

126. W. Lu and C. M. Lieber, *Nat. Mater.,* **6**, 841 (2007).

127. D. Wang, R. Tu, L. Zhangand H. Dai, *Angew. Chem. Int. Ed.,* **44**, 2925 (2005).

128. X. F. Duan, Y. Huang and C. M. Lieber, *Nano Lett.,* **2**, 487 (2002).

129. C. H. Lee, D. R. Kim and X. L. Zheng, *Nano Lett.,* **10**, 5116 (2010).

130. E. M. Freer, O. Grachev, X. F. Duan, S. Martin and D. P. Stumbo, *Nat. Nanotechnol.,* **5**, 525 (2010).

131. T. Lee, W. J. Choi, K. J. Moon, J. H. Choi, J. P. Kar, S. N. Das, Y. S. Kim, H. K. Baik and J. M. Myoung, *Nano Lett.,* **10**, 1016 (2010).

132. Z. Y. Fan, J. C. Ho, Z. A. Jacobson, R. Yerushalmi, R. L. Alley, H. Razavi and A. Javey, *Nano Lett.,* **8**, 20 (2008).

133. Z. Y. Fan, J. C. Ho, T. Takahashi, R. Yerushalmi, K. Takei, A. C. Ford, Y. L. Chueh and A. Javey, *Adv. Mater.,* **21**, 3730 (2009).

134. A. Javey, S. Nam, R. S. Friedman, H. Yan and C. M. Lieber, *Nano Lett.,* **7**, 773 (2007).

135. S. Nam, X. C. Jiang, Q. H. Xiong, D. Ham and C. M. Lieber, *Proc. Natl. Acad. Sci. USA*, **106**, 21035 (2009).

136. A. C. Ford, J. C. Ho, Z. Y. Fan, O. Ergen, V. Altoe, S. Aloni, H. Razavi, A. Javey, *Nano Res.,* **1**, 32 (2008).

137. J. J. Hou, N. Han, F. Wang, F. Xiu, S. Yip, A. T. Hui, T. Hung and J. C. Ho, *ACS Nano,* **6**, 3624 (2012).

138. R. Yerushalmi, Z. A. Jacobson, J. C. Ho, Z. Fan and A. Javey, *Appl. Phys. Lett.,* **91**, 203104 (2007).

139. C. H. Lee, D. R. Kim and X. L. Zheng, *Proc. Natl. Acad. Sci. USA*, **107**, 9950 (2010).

140. A. Carlson, A. M. Bowen, Y. G. Huang, R. G. Nuzzo and J. A. Rogers, *Adv. Mater.,* **24**, 5284 (2012).

141. C. H. Lee, D. R. Kim and X. L. Zheng, *Nano Lett.,* **11**, 3435 (2011).

142. C. W. Cheng, K. T. Shiu, N. Li, S. J. Han, L. Shi and D. K. Sadana, *Nat. Commun.,* **4**, 1577 (2013).

143. J. A. Rogers, Z. Bao, K. Baldwin, A. Dodabalapur, B. Crone, V. R. Raju, V. Kuck, H. Katz, K. Amundson, J. Ewing and P. Drzaic, *Proc. Natl. Acad. Sci. USA,* **98**, 4835 (2001).

144. M. C. Mcalpine, R. S. Friedman and C. M. Lieber, *Proc. IEEE,* **93**, 1357 (2005).

145. M. C. McAlpine, R. S. Friedman and D. M. Lieber, *Nano Lett.,* **3**, 443 (2003).

146. Y. G. Sun, H. S. Kim, E. Menard, S. Kim, I. Adesida and J. A. Rogers, *Small,* **2**, 1330 (2006).

147. J. H. Ahn, H. S. Kim, K. J. Lee, Z. T. Zhu, E. Menard, R. G. Nuzzo, J. A. Rogers, *IEEE Electron. Dev. Lett.,* **27**, 460 (2006).

148. D. H. Kim, J. H. Ahn, H. S. Kim, K. J. Lee, T. H. Kim, C. J. Yu, R. G. Nuzzo, J. A. Rogers, *IEEE Electron. Dev. Lett.,* **29**, 73 (2008).

149. X. F. Duan, C. M. Niu, V. Sahi, J. Chen, J. W. Parce, S. Empedocles J. L. Goldman, *Nature,* **425**, 274 (2003).

150. M. T. Borgstrom, G. Immink, B. Ketelaars, R. Algra and E. P. A. M. Bakkers, *Nat. Nanotechnol.,* **2**, 541 (2007).

151. M. Piccin, G. Bais, V. Grillo, F. Jabeen, S. De Franceschi, E. Carlino, M. Lazzarino, F. Romanato, L. Businaro and S. Rubini, *Physica E,* **37**, 134 (2007).

152. V. Dhaka, T. Haggren, H. Jussila, H. Jiang, E. Kauppinen, T. Huhtio, M. Sopanen and H. Lipsanen, *Nano Lett.,* **12**, 1912 (2012).

153. A. T. Hui, F. Wang, N. Han, S. P. Yip, F. Xiu, J. J. Hou, Y. T. Yen, T. F. Hung, Y. L. Chueh and J. C. Ho, *J. Mater. Chem.,* **22**, 10704 (2012).

154. H. J. Joyce, J. Wong-Leung, Q. Gao, H. H. Tan and C. Jagadish, *Nano Lett.,* **10**, 908 (2010).

155. N. Han, F. Wang, J. J. Hou, F. Xiu, S. Yip, A. T. Hui, T. Hung and J. C. Ho, *ACS Nano,* **6**, 4428 (2012).

156. D. K. Kim, Y. M. Lai, T. R. Vemulkar and C. R. Kagan, *ACS Nano,* **5**, 10074 (2011).

157. G. Shin, C. H. Yoon, M. Y. Bae, Y. C. Kim, S. K. Hong, J. A. Rogers and J. S. Ha, *Small,* **7**, 1181 (2011).

158. W. Lu, P. Xie and C. M. Lieber, *IEEE T. Electron Dev.,* **55**, 2859 (2008).

159. H. Y. Li, O. Wunnicke, M. T. Borgstrom, W. G. G. Immink, M. H. M. van Weert, M. A. Verheijen and E. P. A. M. Bakkers, *Nano Lett.,* **7**, 1144 (2007).

160. F. Z. Li, L. B. Luo, Q. D. Yang, D. Wu, C. Xie, B. Nie, J. S. Jie, C. Y. Wu, L. Wang and S. H. Yu, *Adv. Energy Mater.,* **3**, 579 (2013).

161. R. R. He and P. D. Yang, *Nat. Nanotechnol.,* **1**, 42 (2006).

162. S. S. Kwon, W. K. Hong, G. Jo, J. Maeng, T. W. Kim, S. Song and T. Lee, *Adv. Mater.,* **20**, 4557 (2008).

163. R. M. Yu, L. Dong, C. F. Pan, S. M. Niu, H. F. Liu, W. Liu, S. Chua, D. Z. Chi and Z. L. Wang, *Adv. Mater.,* **24**, 3532 (2012).

164. Y. M. Niquet, C. Delerue and C. Krzeminski, *Nano Lett.,* **12**, 3545 (2012).

165. Z. X. Wu, Y. W. Zhang, M. H. Jhon, H. J. Gao and D. J. Srolovitz, *Nano Lett.,* **12**, 910 (2012).

166. P. E. Marszalek, W. J. Greenleaf, H. B. Li, A. F. Oberhauser and J. M. Fernandez, *Proc. Natl. Acad. Sci. USA,* **97**, 6282 (2000).

167. T. Makela, S. Jussila, H. Kosonen, T. G. Backlund, H. G. O. Sandberg and H. Stubb, *Synth. Met.,* **153**, 285 (2005).

168. K. Jain, M. Klosner, M. Zemel and S. Raghunandan, *Proc. IEEE,* **93**, 1500 (2005).

169. C. Keplinger, J. Y. Sun, C. C. Foo, P. Rothemund, G. M. Whitesides and Z. G. Suo, *Science,* **341**, 984 (2013).

CHAPTER 5

GRAPHENE DEVICES FOR HIGH-FREQUENCY ELECTRONICS AND THz TECHNOLOGY

Guangcun Shan[*,†,§], Ruguang Ma[*],
Xinghai Zhao[*] and Wei Huang[‡]

*College of Science and Engineering,
City University of Hong Kong Tat Chee Avenue,
Kowloon 1006, Hong Kong SAR
†Max Planck Institute for Chemical Physics of Solids,
Dresden 01187, Germany
§gshan2-c@my.cityu.edu.hk; guangcunshan@mail.sim.ac.cn
‡Nanjing Tech University, South Puzhu Road,
Nanjing 211816, P.R. China

Recent years have witnessed many exciting breakthroughs in graphene as a promising material in electronics and optoelectronics. The novel physical properties of graphene afford multiple functions of signal emitting, transmitting, modulating, and detection to be realized in one material. This chapter provides an introduction to physical properties of graphene, and then graphene high-frequency transistors are discussed and are compared to silicon and III–V transistors. The latest progresses and prospective on the circuit implementations of graphene field-effect transistors (FETs) are looked into including mixers, frequency multipliers, and inverters. These recent pioneering developments open up a route towards the integration of graphene in hybrid silicon circuits to embrace the use of monolithic electronic integrated circuits to maximize system functionality, improve service flexibility, and simplify operations.

1. Introduction

Two-dimensional (2D) graphene as a monolayer of carbon atoms tightly packed into a 2D honeycomb lattice has received a great deal of attention due to its unique and interesting electrical, transport, optical and mechanical properties.[1–10] Electrons propagating through the 2D structure of graphene have a linear relation between energy and momentum, and thus behave as massless Dirac fermions.[1–6] Consequently, graphene exhibits electronic properties of a 2D gas of charged particles governed by the relativistic Dirac equation, rather than the non-relativistic Schrödinger equation with an effective mass,[1–5] with carriers mimicking particles with zero mass and an effective 'speed of light' of around 10^6 ms^{-1}. Graphene also exhibits a variety of transport phenomena that are characteristic of 2D Dirac fermions, such as specific integer and fractional quantum Hall effects,[1–4] a 'minimum' conductivity of ~$4e^2$/h even when the carrier concentration tends to zero, and Shubnikovde–Haas oscillations with a π phase shift due to Berry's phase.[1] In combination with the near-ballistic transport property at room temperature, very high mobilities (μ) of up to 10^6 cm^{2}V^{-1}s^{-1} are observed in suspended samples, rendering graphene a potential material for nanoelectronics.[7,8] Graphene also shows excellent optical properties and magnetic properties.[11–23] The linear dispersion of the Dirac electrons makes broadband applications possible. Saturable absorption is also observed as a consequence of Pauli blocking,[14,15] and non-equilibrium carriers lead to hot luminescence.[16–21,24] Notably, one hybrid graphene-BiTeI sandwich structure has recently been developed to be a non-trivial topological insulator with one Dirac-cone topological surface state,[23] making it viable for room-temperature applications in spintronic devices. Moreover, given recent progress in graphene-based terahertz (THz) and infrared (IR) emitters and detectors,[23–33] graphene may offer some interesting solutions for future THz technologies, which promise a myriad of applications including imaging, spectroscopy, and communications.[24,34–42] More importantly, graphene has the important advantage of flexibility and mechanical strength, which ensures that graphene-based devices will probably dominate flexible applications. These properties make it an ideal material candidate for novel electronic and optoelectronic devices.

To review recent progress in utilizing graphene to realize the graphene nanodevices, we begin in Section 2 with a discussion of the basic properties of graphene. Section 3 discusses recent experiment demonstration of several prototypical graphene devices for high-frequency application from megahertz (MHz) to THz range. Finally, Section 4 concludes by briefly outlining some prospective issues in the graphene devices for high-frequency electronics and THz technology.

2. Basic Physical Properties

2.1. *Electronic Structure of Graphene*

The electronic structure of single-layer graphene (SLG) or monolayer graphene can be described using a tight-binding Hamiltonian.[3,5,6] Because the bonding and anti-bonding σ-bands are well separated in energy (>10 eV at the Brillouin zone center Γ), they can be neglected in semi-empirical calculations, retaining only the two remaining π-bands as shown in Fig. 1. The electronic wavefunctions from different atoms on the hexagonal lattice overlap. And the p_z electrons, which form the π-bonds, can generally be treated independently from the other valence electrons. Within this π-band approximation it is easy to describe the electronic spectrum of the total Hamiltonian and to obtain the dispersion relations $E_{\pm}(k_x, k_y)$ restricted to first-nearest-neighbour interactions only:

$$E_{\pm}(k_x, k_y) = \pm\gamma_0 \sqrt{1 + 4\cos\frac{\sqrt{3}k_x a}{2}\cos\frac{\sqrt{3}k_y a}{2} + \cos^2\frac{\sqrt{3}k_y a}{2}}, \quad (1)$$

where $a = \sqrt{3}a_{cc}$ (with $a_{cc} = 1.42\ \text{Å}$ being the carbon–carbon bond length) and γ_0 is the transfer integral between the nearest-neighbor π-orbitals (typical values for γ_0 are 2.9–3.1 eV). The $\mathbf{k} = (k_x, k_y)$ vectors in the first Brillouin zone constitute the ensemble of available electronic momenta.

With one p_z electron per atom in the $\pi - \pi^*$ model, the (−) band (negative energy branch) in Eq. (1) is fully occupied, whereas the (+) branch is totally empty. These occupied and unoccupied bands touch at the **K** points. Note that the Fermi level E_F is the zero-energy reference, and the Fermi surface is defined by **K** and **K′**. Moreover, by expanding Eq. (1) at **K** (**K′**), the linear π- and π^*-bands of the low-energy band structures for Dirac fermions are given by:

$$E_{\pm}(\kappa) = \pm\hbar\nu_F, \tag{2}$$

where $\kappa = \mathbf{k} - \mathbf{K}$ and ν_F is the electronic group velocity, which is given by $\nu_F = \sqrt{3}\gamma_0 a / (2\hbar) \approx 10^6$ ms^{-1}. The linear dispersion given by Eq. (2) is the solution to the following effective Hamiltonian located at the two inequivalent Brillouin zone corners **K** and **K′**:

$$\mathbf{H} = \pm\hbar\nu_F \left| \kappa \right|, \tag{3}$$

where $\kappa = -i\nabla$ and $\boldsymbol{\sigma}$ are the pseudo-spin Pauli matrices operating on the A–B sublattices of graphene.[3,5]

2.2. *Optical Transitions and Photoconductivity of Graphene*

The linear dispersion of the Dirac electrons implies that for any excitation there will always be an electron–hole pair in resonance. To understand the electron–hole dynamics, the kinetic equation for the electron and hole distribution functions, $f_e(p)$ and $f_h(p)$, p being the momentum counted from the Dirac point can be solved.[12] And it is found that if the relaxation times are shorter than the pulse duration, then during the pulse the electrons reach a stationary state, and collisions put electrons and holes into thermal equilibrium at an effective temperature.[19,21] As a result, the electron and hole densities, total energy density and a reduction of photon absorption per layer, due to Pauli blocking, by a factor of $\Delta A/A = [1 - f_e(p)][1 - f_h(p)] - 1$, can be determined by the populations. Assuming efficient carrier–carrier relaxation (both intraband and interband) and efficient cooling of the graphene phonons, the main bottleneck is energy transfer from electrons to phonons.

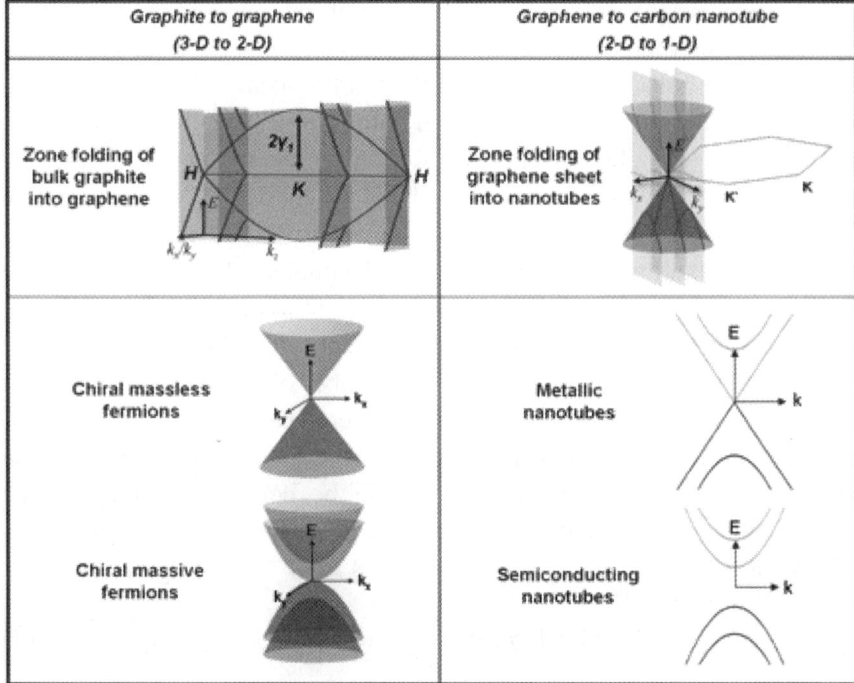

Figure 1.   Application of zone folding to obtain the electronic states of few-layer graphene (FLG) from graphite compared with the generation of the electronic states of carbon nanotubes by zone folding of graphene. The left column shows the generation of 2D chiral massless and massive fermions in FLG from zone folding the three-dimensional (3D) Brillouin zone of bulk graphite. The upper panel displays the band structure of bulk graphite in two spatial dimensions and the zone-folding scheme that generates planes cutting at specific values of kz satisfying Eq. (1). The lower panel presents the resulting fundamental building blocks of the electronic structure of FLG: the massless and the massive components. For comparison, the right column displays the standard procedure for generating 1D metallic and semiconducting carbon nanotubes from zone folding of the 2D Brillouin zone of graphene. The upper panel is a schematic representation of the 2D electronic structure of SLG and the corresponding zone-folding scheme that generates states satisfying the periodic boundary conditions for nanotubes. The lower panel presents the resulting states: metallic and semiconducting nanotubes. Figure reprinted with permission from Ref. 20. (Copyright © 2014 National Academy of Sciences.)

For linear dispersions near the Dirac point, pair–carrier collisions cannot lead to interband relaxation, thereby conserving the total number of electrons and holes separately.[12,19] Interband relaxation by phonon emission can occur only if the electron and hole energies are close to the Dirac point (within the phonon energy). Radiative recombination of the hot electron–hole population has also been suggested.[14–17] For graphite flakes, the dispersion is quadratic and pair–carrier collisions can lead to interband relaxation. Thus, in principle, decoupled SLG can provide the highest saturable absorption for a given amount of material. Interestingly, the optical image contrast can be used to identify graphene on top of a Si/SiO_2 substrate.[12,13] This scales with the number of layers and is the result of interference, with SiO_2 acting as a spacer. The contrast can be maximized by adjusting the spacer thickness or the light wavelength.[11–13] The transmittance of a freestanding SLG can be derived by applying the Fresnel equations in the thin-film limit for a material with a fixed universal optical conductance $G_0 = e^2/(4\hbar) \approx 6.08 \times 10^{-5}\ \Omega^{-1}$, to give:

$$T = (1 + 0.5\pi\alpha)^{-2} \approx 1 - \pi\alpha \approx 97.7\%, \tag{4}$$

where $\alpha = e^2/(4\pi\varepsilon_0\hbar c) = G_0/(\pi\varepsilon_0 c) \approx 1/137$ is the fine-structure constant.[13] Graphene only reflects <0.1% of the incident light in the visible region, rising to ~2% for 10 layers.[12,13] In a FLG sample, each sheet can be seen as a 2D electron gas, with little perturbation from the adjacent layers, making it optically equivalent to a superposition of almost non-interacting SLG. Thus, the optical absorption of graphene layers could be taken to be proportional to the number of layers, each absorbing $A \approx \pi\alpha \approx 2.3\%$ over the visible spectrum. The absorption spectrum of SLG is quite flat from 300 to 2,500 nm with a peak in the ultraviolet region (~270 nm), due to the exciton-shifted van Hove singularity in the graphene density of states. The dependence of optical conductivity with Fermi level indicates that the graphene optical conductivity can be modified by controlling the Fermi level, i.e., carrier concentration. For linear dispersions near the Dirac point, pair–carrier collisions cannot lead to interband relaxation, thereby conserving the total number of electrons and holes separately.[13] Interband relaxation by phonon emission can occur only if the electron and hole energies are close to the Dirac point (within the phonon energy). Radiative

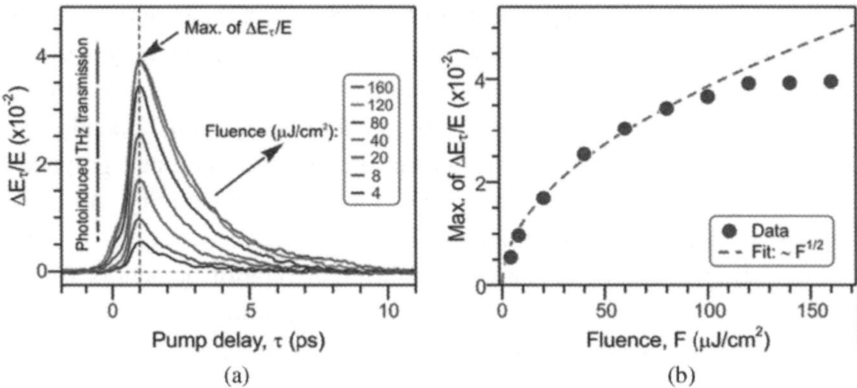

Figure 2.   Transient THz transmission response of photoexcited graphene layer: (a) temporal evolution of the change in the maxima of the transmitted THz waveforms normalized by the THz probe field, i.e., $\Delta E_\tau(t)/E(t)$, for different applied fluences of optical pump pulse. (b) Maxima of $\Delta E_\tau(t)/E(t)$ from part a as a function of applied fluence, which saturates at higher fluences. The dashed line varies as the square root of the fluence, as a guide to the eye. Figure adapted with permission from Ref. 22. (© 2013 ACS.)

recombination of the hot electron–hole population has also been suggested.[14–17]

In the far-IR and THz spectral region, the intraband response in graphene becomes pronounced, leading to the possibility of strong extinction in SLG, [12–14] as well as of plasmon excitation through appropriate coupling. [18,19,21–23] As shown in Fig. 2, Heinz *et al.*[22] have measured the THz frequency-dependent sheet conductivity and its transient response following femtosecond optical excitation for SLG samples grown by chemical vapor deposition (CVD). The optical excitation of the graphene is implemented by a femtosecond pulse and probing of the THz response using a time-domain spectroscopy approach. By recording the transmitted THz electric field at different fixed delay times between the pump pulse and THz probe pulse, the differential waveform results show that rather than a simple change in the amplitude of the transmitted field with pump excitation, the change in THz waveform indicates the role of the finite carrier scattering time in graphene samples. The physical origin of the observed effect can be understood by considering the Drude conductivity of graphene with a real conductivity of $\sigma = D/\pi\Gamma$ in the low-frequency limit.

Thus, photoinduced changes in both the Drude weight D and scattering rate Γ can affect the conductivity. Under excitation, D may increase through a rise in the carrier concentration, while Γ may increase through a rise in the effective temperature of the system. As a result, depending on the initial Drude weight, the scattering rate of the unpumped graphene, and the pump fluence, the ratio of D/Γ and the conductivity can either increase or decrease after photoexcitation.

Interestingly, graphene can exhibit different type of behaviors for the transient response, depending on the detailed conditions. When the initial doping level and scattering time are both relatively high, the main effect is to be that of change in the scattering rate, rather than the change in the Drude weight, that is, a response resembling more that of a metal. This is the case observed in Heinz's measurements.[22] Figure 2(a) shows the increase in the THz waveform induced by the pump beam, measured at the maximum of the time domain waveform, as a function of the delay time. It should be noted that the dominant change is an increase in the carrier scattering rate, rather than an increase in the Drude weight. This explains the observed negative THz photoconductivity response. At relatively high pump fluence, the increase of the Drude weight will become more and more important. This will give rise to an increase in the graphene conductivity, which would compensate the decrease in the conductivity caused by an increased electron scattering rate. The balance of these factors provides a natural explanation for the apparent saturation behavior in the photoresponse with increasing pump fluence (Fig. 2(b)). In addition to this factor related to the amplitude of the response, there is a modest increase in the decay time.[22] The reduced sensitivity of the response of the conductivity to the electronic temperature at high fluences can explain this trend, since the initial rise in the THz response will then be diminished and the apparent decay time lengthened.

2.3. *Luminescent Properties*

Graphene could be made luminescent by inducing a bandgap, following two main routes. One is by cutting it into ribbons and quantum dots, the other is by chemical or physical treatments, to reduce the connectivity of the π-electron network. Although graphene nanoribbons have been produced with varying bandgaps,[4,7,10] no photoluminescence has been reported from them thus far. However, bulk graphene oxide dispersions and solids do show a broad photoluminescence. [4,24,43] The combination of photoluminescent and conductive layers could be used in sandwich light-emitting diodes covering the IR, visible, and blue spectral ranges.[23–26]

Even some groups have ascribed photoluminescence in graphene oxide to bandgap emission from electron-confined sp^2 islands.[40] Whatever the origin, fluorescent organic compounds are of importance to the development of low-cost optoelectronic devices. Although widely used for bio-labeling and bio-imaging, the nanotoxicity and potential environmental hazard of luminescent quantum dots limit widespread use and *in vivo* applications. Fluorescent bio-compatible carbon-based nanomaterials may be a more suitable alternative. Fluorescent species in IR and near-infrared ranges (NIR) are useful for biological applications, because cells and tissues show little auto-fluorescence in this region.

Wang *et al.* have reported a gate-controlled, tunable gap up to 250 meV in bilayer graphene.[18] This may make new photonic devices possible for far-infrared light generation, amplification and detection. Broadband nonlinear photoluminescence is also possible following non-equilibrium excitation of untreated graphene layers, as recently reported by several groups.[17–21] Emission occurs throughout the visible spectrum, for energies both higher and lower than the exciting one, in contrast with conventional photoluminescence processes.[11,17–21] This broadband nonlinear photoluminescence is thought to result from radiative recombination of a distribution of hot electrons and holes, generated by rapid scattering between photoexcited carriers after the optical excitation.[17–20] In addition, electro-luminescence was also reported in pristine graphene.[11]

3. Roadmap of Graphene Electronics

Despite the fact that graphene will make it into high-performance integrated logic circuits as an excellent planar channel material within the next decade because of the absence of a bandgap, yet, many other graphene electronics for flexible electronics and radio-frequency (RF) applications are being developed, using the available material in terms of quality.

3.1. *Flexible Graphene Electronics*

Transparent conductive coatings are widely used in electronic consumer products such as touch screen displays, electronic paper (e-paper) and organic light-emitting diodes (OLEDs) and require a low sheet resistance with high transmittance (of over 90%) depending on the specific application.

Graphene meets the electrical and optical requirements (sheet resistance reaching 30 V per square of 2D area in highly doped samples) and an excellent transmittance of 97.7% per layer,[4,13,14] though the indium tin oxide (ITO) still demonstrates slightly better characteristics.

However, considering the quality improvement of graphene every year (already making the difference in performance marginal), while ITO would become more expensive and ITO deposition is already expensive, graphene does have an opportunity to get a good fraction of the market. Graphene also has excellent mechanical flexibility and chemical durability — very important characteristics for flexible electronic devices,[4,10] where ITO usually fails. The advantage of graphene electrodes in touch panels is that graphene's endurance far exceeds that of any other available candidate at the moment. Moreover, the fracture strain of graphene is ten times higher than that of ITO, which makes it successful to be applied to bendable and rollable devices. The requirements of electrical properties (for example, sheet resistance) for each electrode type differ from application to application. Depending on the production methods, various grades of transparent conductive coating could be made from graphene. As a result, electrodes for touch screens, which require an expensive CVD method of production, tolerate a relatively high sheet resistance (50–300 V per square) for a transmittance of ~90%.

Rollable e-paper is a very appealing electronic product, which requires a bending radius of 5–10 mm. It is easily achievable by a graphene electrode. In addition, graphene's uniform absorption across the visible spectrum is beneficial for color e-papers.[12,13] However, the contact resistance between the graphene electrode and the metal line of the driving circuitry still remains a problem. A working prototype is expected by 2016, but the manufacturing cost needs to decrease before it will appear on the market.

OLED devices have become an attractive technology and the first (non-graphene) products have been in the market since 2013. Besides the primary crucial parameter of the sheet resistance (below 30V per square), other crucial parameters for OLED devices are the work function and the electrode's surface roughness, which effectively govern the performance. The tunability of graphene's work function could improve the efficiency, and its atomically flat surface would help avoid electrical shorts and leakage currents. It is worthwhile highlighting that advanced flexible or foldable OLED devices could be introduced soon after 2016 once device integration issues (such as conformal deposition of graphene on 3D structures and contact resistance between graphene and the source/drain) are resolved.

Liquid-phase exfoliation produces such graphene coatings for mass production without the use of expensive vacuum technology in the low-cost sector way. Although the resistance of these films is on the high side, they still perform well enough for smart windows, solar cells, and some touch screen applications. Graphene has the important advantage of flexibility and mechanical strength, which ensures that graphene-based devices will probably dominate flexible applications.

3.2. *High-Frequency Electronics: From Transistors to Circuits*

The fundamental building blocks of digital electronics are logic gates which must be capable of cascading such that more complex logic functions can be realized. Although many graphene devices have been demonstrated as discrete devices, recently there have been several high-frequency circuit demonstrations, including high frequency transistors, multipliers, inverters and mixers.

3.2.1. High-Frequency Transistors

Graphene has been intensively studied for high-frequency transistor applications.[8] However, it has to compete against more mature technologies such as compound semiconductors (III–V materials). Thus, graphene will probably be used only when even III–V materials fail to satisfy the stringent requirements of high-frequency transistors. Projections show that III–V materials will no longer be able to sustain the required cutoff frequency $f_T = 850$ GHz (the top frequency for current modulation) and maximum oscillation frequency $f_{max} = 51.2$ THz (the top frequency for power modulation) after 2020s because device requirements will become more stringent. A recent progress report on graphene[44] presented a value of f_T as high as 300 GHz, with the possibility of extending it up to 1.5 THz at a channel length of about 100 nm.[24]. On the other hand, f_{max} has only reached 30 GHz in traditional graphene structures thus far, which is far from the 330 GHz in Si-based high-frequency transistor performance, according to the 2011 International Technology Roadmap for Semiconductors (ITRS). Thus, the principal remaining issue is the low value of f_{max} for graphene transistors, which trails f_T by an order of magnitude in a comparable conventional device. There are two ways to improve f_{max}: by lowering the gate resistance or the source–drain conductance at pinch-off.[24,45] The former approach could be done using well-established semiconductor processes. The latter will require current saturation in the high-frequency graphene transistor, which will probably involve finding a new dielectric layer with properties similar to those of boron nitride [46] and compatible with modern semiconductor technology. An f_{max} of 58 GHz has been reported using graphene on top of an exfoliated hexagonal boron nitride film.[47–49]

3.2.2. Multipliers

Frequency multipliers are one of the important components to realize frequency conversion in the RF communication system or even future sub-THz communication systems. Conventional frequency multipliers are either field-effect tansistor (FET) or diode-based — the former offers good conversion efficiency (30%) but no gain while the latter offers gain but a lower efficiency (~15%). Ambipolar transport properties of graphene have been used to demonstrate full-wave signal rectification as well as frequency doubling.[50] By correctly biasing an ambipolar graphene field-effect transistor in common-source configuration shown in Fig. 3, a sinusoidal voltage applied to the transistor gate is rectified at the drain electrode. By using a common source configuration, a sinusoidal voltage applied to the backgate was rectified at the drain electrode; a 10-kHz signal was demonstrated to double to 20 kHz with good spectral purity. Therefore, by using

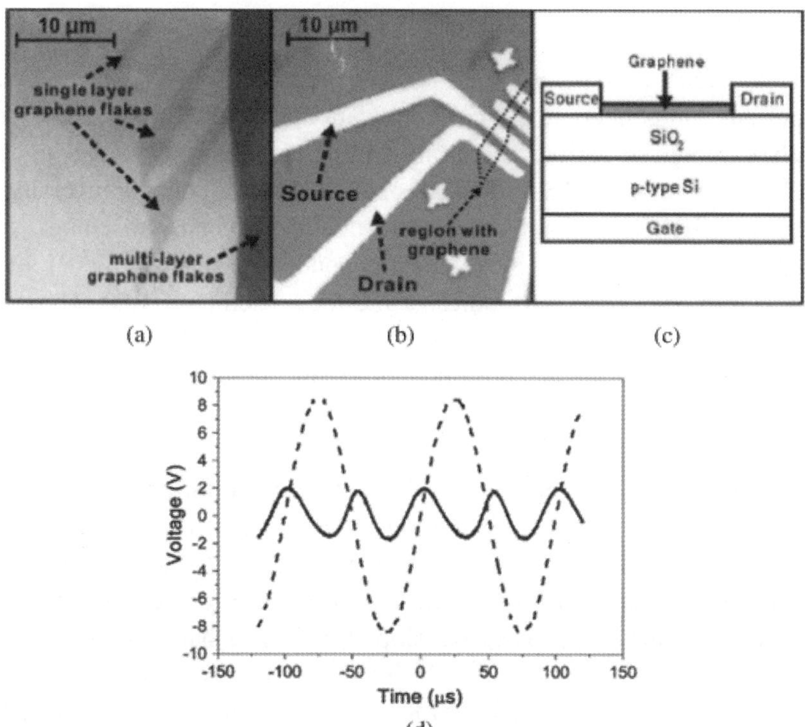

Figure 3.   Structure of the fabricated Gra-FET. (a) Optical micrograph showing SLG flakes. (b) Optical micrograph of the structure of the final device. The device is back gated through the p-type Si wafer. (c) Schematic of the vertical structure of the device. (d) Measured input (dotted curve) and (solid curve) output of the frequency-multiplier circuit when the graphene device is biased near the minimum conduction point. For clarity of display, the output signal is multiplied by 50, and both signals are ac coupled. The low amplitude of the measured output signal is due to the poor on/off current ratio of the fabricated device, and it can be improved by using a Gra-FET with better transfer characteristics. Reproduced with permission from Ref. 50. (Copyright © 2009, IEEE.)

this concept, frequency multiplication of a 10 kHz input signal depicted in Fig. 3(d) has been experimentally demonstrated.[50] Although this frequency is much less than the 1 THz multipliers possible using Si diodes, graphene multiplier devices point to the possibility of using graphene for the application in THz technology. This high efficiency, combined with the high electron mobility of graphene, makes graphene-based frequency multipliers a very promising option for signal generation at ultrahigh frequencies even up to sub-THz region.

3.2.3. Mixers

Mixers, which are typically used for frequency conversion are one of the important components in RF communication systems. It has been extensively explored in various research institutes including United Kingdom RAL Laboratories, NASA JPL Laboratory, United States University of Virginia, VIRGINIA Diodes Inc. (VDI), City University of Hong Kong etc. In a SHM the mixing mechanism is conducted between the RF or IF signals and one of the harmonics of the local oscillator (LO). Thus, the nonlinear mixer device (diode, metal semiconductor FET, etc.) performs both mixing and frequency multiplication. The output component of interest is $f_{IF} = f_{RF} - f_{LO}$, where f_{RF} is the input RF frequency and f_{LO} is the input LO frequency.

Various traditional microwave mixers working at millimeter wave frequency have been successfully demonstrated by some researchers.[51–55] In 1997, Jeffrey L. Hesler *et al.* have developed 585 GHz harmonic mixer,[51] where DSB conversion loss reached 7.3 dB noise temperature 2380 K. And then Thomas *et al.* have developed 330 GHz harmonic mixer[52] to achieve 5.7 dB DSB (Double Side Band, DSB) conversion loss, where noise temperature is less than 930 K in 2005. Compared to the SIS HEB mixer, graphene-based mixer outperforms with some distinctive features such as working at room temperature, low cost, low conversion loss, moderate noise temperature, etc. Graphene mixers would be very economical at millimeter wave frequencies since low-frequency low-cost microwave sources (as LO) can be adopted as higher-order harmonics will perform.

CVD graphene mixer was used to demonstrate a mixer upto 10 MHz.[56] An ideal graphene FET shows symmetric transfer characteristics that is infinitely differentiable; then the drain current can be described as:

$$I_q = c_0 + c_2 \left(V_g - V_{g\min} \right)^2 + c_4 \left(V_g - V_{g\min} \right)^4 + \cdots, \tag{5}$$

where c_0, c_2, and c_4 are constants. From this model, ideal graphene FETs biased at $V_{g\min}$ should show only the difference of sum frequencies and other even-order terms; thus, odd-order intermodulations can be significantly reduced compared to conventional unipolar mixers. Conventional mixers need more complicated circuits to achieve good performance while graphene FETs need to display symmetric characteristics but can use much simpler circuits. The implemented circuit using a graphene FET is shown in Fig. 4(a). When two RF signals at 10 MHz and 10.5 MHz are applied to the gate, the graphene FET mixes them to generate $(f_{RF} + f_{LO})$ and $(f_{RF} - f_{LO})$ at the output. The power at these frequencies is more than 10 dB, higher than the power at the input/fundamental frequencies. In addition, the power at the odd-order frequencies like $2f_{LO} - f_{RF}$ and $2f_{RF} - f_{LO}$ are significantly suppressed. The frequency limit can be calculated from $f_T = g_m/(2\pi C)$. In the studied devices, there was significant asymmetry in the transfer curves, perhaps because of contact doping or adsorbates. Since the device operation was within ± 1 V of V_{gmin}, the asymmetry in I–V characteristic curve didn't cause a noticeable degradation of the mixing operation.

The mixer used for the above demonstration was limited in frequency more by parasitics rather than graphene FET performance. An integrated mixer was demonstrated in Ref. 57 that could operate up to 10 GHz. The circuit was fabricated using graphene grown on SiC, and the footprint, including contact pads, was less than 1 mm^2. The circuit implementation is depicted in Fig. 4(b). The graphene FET is modulated both by f_{RF} and f_{LO}, and the drain current contains the sum and difference of frequencies. The inductors complement the graphene FET — L1 resonates out the

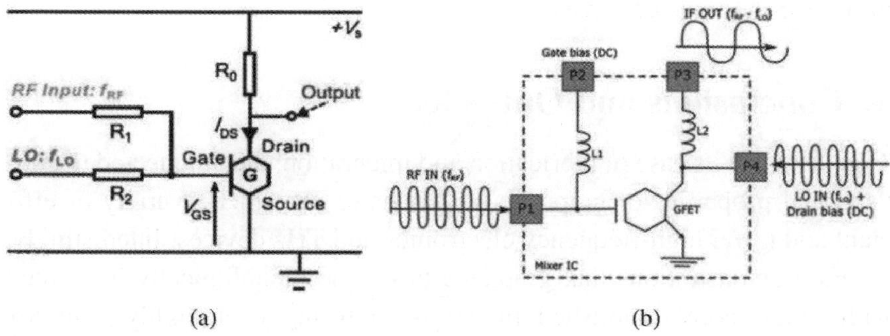

Figure 4. Circuit implementation for GFET RF mixer (a) discrete CVD-grown GFET showed operation up to 10 MHz (Reproduced with permission from Ref. 56. (Copyright © IEEE.) (b) Integrated epitaxial–graphene mixer operated at frequencies up to 10 GHz (Adapted from Ref. 57).

parasitic capacitances from the input RF pads and the gate of the graphene FET while L2 provides an input match to the LO signal and also acts as a low-pass filter between the drain and output port P3. The graphene was two to three layers thick, and meanwhile the Si-face of SiC was used. The $I_d \sim V_{gs}$ curve was linear and did not show a V_{gmin} — perhaps because of the strong n-type doping from the substrate. The output characteristics showed a linear I_d–V_{ds} for V_{ds} up to 1.6 V. This resulted in a g_m that increased with rising V_{ds}. At $V_{ds} \sim 1.6$ V, a current density of 2 mA mm^{-1} and g_m of 80 μS/μm was achieved. R_c was measured to be 600 Ω·μm. With the measured g_m and C_G, f_T is calculated to be 9 GHz for L $\sim$ 550 nm and $V_{ds} = 1.6$ V. The drain current can be described as $I_d \sim A(B + g_m V_g) V_{ds}$ where A and B are constants. Since the output signal is proportional to I_d, the power of f_{IF} is proportional to the product of g_m and g_d. Current saturation was not achieved, and most devices showed a triode-like behavior. It was found that gm slightly decreases with V_g; this was also evident in the conversion loss of the mixer as a function of V_g. This strong correlation showed that the graphene mixer performance was determined by the graphene FET rather than by parasitics. The conversion loss of the graphene mixer showed good thermal stability, when the temperature was increased from 300 K to 400 K. This was attributed to the degenerate doping level in the graphene FETs and a temperature-independent scattering mechanism associated with optical phonons at high biases. Note that conventional semiconductor mixers need an additional feedback circuitry to minimize thermal sensitivity.

4. Conclusions and Outlooks

Together with its ease of fabrication and integration, the unique and unconventional properties of graphene enable us to construct a variety of efficient and novel high-frequency electronics and THz devices. Interestingly, several demonstrations that graphene has to be monolithically integrated with other more established materials to form novel highly complex devices open up real application prospects in low cost photodetectors, nanolasers and transmitter modules by releasing the need for temperature control, isolators, and external modulators. Moreover, implementation of graphene-based photonic devices is especially promising for applications in future THz technology. In the future, large-scale integrated circuits on one single silicon chip will require capabilities of controlling and manipulating the direction, phase, polarization, and amplitude of signal waves or beams. The graphene-based configurable RF devices can be easily integrated with other THz devices to increase functionality or to realize single chip RF system. Hopefully, recent breakthroughs in the direct deposition of graphene on silicon can open up a route towards the integration of graphene in hybrid silicon integrated circuits (hMICs). And such advanced large-scale hMICs represent a significant technology innovation that simplifies communication system circuit design, reduces space and power consumption, and improves reliability. This chapter has provided a deep understanding and useful guideline for graphene-based nanodevices for future RF electronic circuits and THz technology.

Acknowledgment

This work is supported by General Research Fund (GRF) Project Nos. 9041909 and CityU119212 from RGC, Hong Kong. Moreover, G.C. would gratefully acknowledge the Max-Planck research fellowship support to carry out research work on graphene nanostructures as a postdoc scientist. We thank Dr. Qiyuan He from UCLA, and Prof. Xi Chen from Columbia University for their valuable comments and fruitful discussions.

References

1. K. S. Novoselov, A. K. Geim, D. Jiang, Y. Zhang, S. V. Dubonos, I. V. Grigorieva and A. A. S. Firsov, Electric field effect in atomically thin carbon films. *Science*, **306**, 666 (2004).
2. Y. Zhang, Y.-W. Tan, H. L. Stormer and P. Kim, Experimental observation of quantum Hall effect and Berry's phase in grapheme. *Nature*, **438**, 201 (2005).
3. P. R. Wallace, The band theory of graphite. *Phys. Rev.,* **71**, 622–634 (1947).
4. A. K. Geim and K. S. Novoselov, The rise of grapheme. *Nat. Mater.,* **6**, 183–191 (2007).

5. J. C. Charlier, P. C. Eklund, J. Zhu and A. C. Ferrari, Electron and phonon properties of graphene: Their relationship with carbon nanotubes. *Top. Appl. Phys.*, **111**, 673–709 (2008).

6. M. Liu, X.B. Yin, E. Ulin-Avila, B.S. Geng, T. Zentgraf, L. Ju, F. Wang and X. Zhang, A graphene-based broadband optical modulator. *Nature*, **474**, 64–67 (2011).

7. M. Y. Han, B. Ozyilmaz, Y. Zhang, and P. Kim, Energy band-gap engineering of graphene nanoribbons. *Phys. Rev. Lett.*, **98**, 206805 (2007).

8. Y.-M. Lin, C. Dimitrakopoulos, K. A. Jenkins, D. B. Farmer, H. Y. Chiu, A. Grill and P. Avouris, 100-GHz transistors from wafer-scale epitaxial grapheme. *Science*, **327**, 662 (2010).

9. X. Du, I. Skachko, F. Duerr, A. Luican, and E.Y. Andrei, Fractional quantum Hall effect and insulating phase of Dirac electrons in grapheme. *Nature*, **462**, 192 (2009).

10. K. S. Novoselov, V. I. Fal'ko, L. Colombo, P. R. Gellert, M. G. Schwab and K. Kim, A roadmap for grapheme. *Nature*, **490**, 192–200 (2012).

11. F. Bonaccorso, Z. Sun, T. Hasan and A.C. Ferrari, Graphene photonics and optoelectronics. *Nat. Photonics*, **4**, 611–622 (2010).

12. M. I. Katsnelson, K. S. Novoselov and A. K. Geim, Chiral tunnelling and the Klein paradox in grapheme. *Nat. Phys.*, **2**, 620–625 (2006).

13. C. Casiraghi, A. Hartschuh, E. Lidorikis, H. Qian, H. Harutyunyan, T. Gokus, K. S. Novoselov and A. C. Ferrari, Rayleigh imaging of graphene and graphene layers. *Nano Lett.*, **7**, 2711–2717 (2007).

14. P. Blake, E. W. Hill, A. H. Castro Neto, K. S. Novoselov, D. Jiang, R. Yang, T. J. Booth and A. K. Geim, Making graphene visible. *Appl. Phys. Lett.*, **91**, 063124 (2007).

15. Z. Sun, T. Hasan, F. Torrisi, D. Popa, G. Privitera, F. Wang, F. Bonaccorso, D. M. Basko and A. C. Ferrari, Graphene mode-locked ultrafast laser. *ACS Nano*, **4**, 803–810 (2010).

16. T. Hasan, Z. Sun, F. Wang, F. Bonaccorso, P. H. Tan, A. G. Rozhin and A. C. Ferrari, Nanotube–polymer composites for ultrafast photonics. *Adv. Mater.*, **21**, 3874–3899 (2009).

17. R. R. Nair, P. Blake, A. N. Grigorenko, K. S. Novoselov, T. J. Booth, T. Stauber, N. M. Peres and A. K. Geim, Fine structure constant defines transparency of graphene. *Science*, **320**, 1308–1308 (2008).

18. F. Wang, Y. Zhang, C. Tian, C. Girit, A. Zettl, M. Crommie and Y. R. Shen, Gate-variable optical transitions in graphene. *Science*, **320**, 206–209 (2008).

19. R. J. Stoehr, R. Kolesov, J. Pflaum and J. Wrachtrup, Fluorescence of laser created electron–hole plasma in graphene. *Phys. Rev. B*, **82**, 121408(R) (2010).

20. K. F. Mak, M. Y. Sfeir, J. A. Misewich and T. F. Heinz, The evolution of electronic structure in few-layer graphene revealed by optical spectroscopy. *Proc. Natl. Acad. Sci. USA*, **107** (34) 14999–15004 (2010).

21. C. H. Liu, K. F. Mak, J. Shan, and T. F. Heinz, Ultrafast photoluminescence from graphene. *Phys. Rev. Lett.*, **105**, 127404 (2010).

22. G. Jnawali, Y. Rao, H. Yan and T. F. Heinz, Observation of a transient decrease in terahertz conductivity of single-layer graphene induced by ultrafast optical excitation. *Nano Lett.*, **13**, 524–530 (2013).

23. S. Wu, G. C. Shan, B. Yan, Prediction of Nearly Room-Temperature Quantum Anamalous Hall insulator based on Honeycomd Materials, *Phys. Rev. Lett.* **113**, 256401 (2014).

24. G. C. Shan, M. J. Hu and C. H. Shek, eds., *Graphene Science Handbook* (Vol 3: Electrical and Optical Properties), Chapter 45, in press, CRC Press, Taylor & Francis Group.

25. I. Maeng, S. Lim, S J. Chae, Y. H. Lee, H. Choi and J.-H. Son, Gate-controlled nonlinear conductivity of Dirac Fermion in graphene field-effect transistors measured by terahertz time-domain spectroscopy. *Nano Lett.*, **12**, 551 (2012).

26. X. H. Zhao, G. C. Shan and W. Huang, Steady-state property and dynamics in graphene-nanoribbon-array lasers. *Front. Phys.*, **7**(5), 527–532 (2012).

27. Y. Zhou, X. L. Xu, H. Fan, Z. Ren, J. Bai and L. Wang, Tunable magneto-plasmons for efficient terahertz modulator and isolator by gated monolayer grapheme. *Phys. Chem. Chem. Phys.*, **15**, 5084–5090 (2013).

28. C. H. Liu, Y. C. Chang, T. B. Norris and Z. Zhong, Graphene photodetectors with ultra-broadband and high responsivity at room temperature, *Nat. Nanotechnol.*, **9**, 273–278 (2014).

29. T. J. Echtermeyer, L. Britnell, P. K. Jasnos, A. Lombardo, R. V. Gorbachev, A. N. Grigorenko, A. K. Geim, A. C. Ferrari and K. S. Novoselov, Strong plasmonic enhancement of photovoltage in grapheme. *Nat. Commun.*, **2**, 458 (2011).

30. S. Thongrattanasiri, F. H. L. Koppens and F. J. Garcia de Abajo, Complete optical absorption in periodically patterned grapheme. *Phys. Rev. Lett.*, **108**, 047401 (2012).

31. M. Furchi, A. Urich, A. Pospischil, G. Lilley, K. Unterrainer, H. Detz, P. Klang, A. M. Andrews, W. Schrenk, G. Strasser and T. Mueller, Microcavity-integrated graphene photodetector. *Nano Lett.*, **12** (6), 2773–2777 (2012).

32. M. Engel, M. Steiner, A. Lombardo, A. C. Ferrari, H. V. Löhneysen, P. Avouris and R. Krupke, Light–matter interaction in a microcavity-controlled graphene transistor. *Nat. Commun.*, **3**, 906 (2012).

33. F. Xia, T. Mueller, Y. M. Lin, A. Valdes-Garcia and P. Avouris, Ultrafast graphene photodetector. *Nat. Nanotechnol.*, **4**, 839–843 (2009).

34. E. Lee, K. Balasubramanian, R. T. Weitz, M. Burghard and K. Kern, Contact and edge effects in graphene devices. *Nat. Nanotechnol.*, **3**, 486–490 (2008).

35. A. George Paul, J. Strait, J. Dawlaty, S. Shivaraman, M. Chandrashekhar, F. Rana and M. G. Spencer, Ultrafast optical-pump terahertz-probe spectroscopy of the carrier relaxation and recombination dynamics in epitaxial grapheme. *Nano Lett.*, **8**, 4248–4251 (2008).

36. H. Ito, T. Furuta, S. Kodama and T. Ishibashi, InP/InGaAs uni-travelling-carrier photodiodes with a 340 GHz bandwidth. *Electron. Lett.*, **36**, 1809–1810 (2000).

37. G. C. Shan, X. Zhao, H, Zhu and C.-H. Shek, Critical components in 140 GHz communication systems. *IEEE 2012 International Workshop on Microwave and Millimeter Wave Circuits and System Technology*, **1**, 1–4 (2012).

38. G. C. Shan, X. Zhao, C.-H. Shek and H. Zhu, Sub-THz wireless communication technology at G-band. *2013 IEEE International Workshop on Electromagnetics*, **1**, 57–60 (2013).

39. V. Ryzhii, *Intersubband Infrared Photodetectors* (World Scientific, Singapore, 2003).

40. Z. Y. Tan, Z. Chen, Y. J. Han, R. Zhang, H. Li, X. G. Guo and J. C. Cao, Experimental realization of wireless transmission based on terahertz quantumcascade laser. *Acta Phys. Sin.*, **61**, 098701 (2012).

41. V. Ryzhii, M. Ryzhii, V. Mitin and T. Otsuji, Terahertz and infrared photodetection using p-i-n multiple-graphene-layer structures. *J. Appl. Phys.*, **107**, 054512 (2010).

42. J. Zheng, L. Wang, R. Quhe, Q. H. Liu, H. Li, D. P. Yu, W. N. Mei, J. J. Shi, Z. X. Gao and J. Lu, Sub-10 nm gate length graphene transistors: Operating at terahertz frequencies with current saturation. *Sci. Rep.*, **3**, 1314 (2013).

43. T. Gokus, R. R. Nair, A. Bonetti, M. Böhmler, A. Lombardo, K. S. Novoselov, A. K. Geim, A. C. Ferrari and A. Hartschuh, Making graphene luminescent by oxygen plasma treatment. *ACS Nano*, **3**, 3963–3968 (2009).

44. L. Liao, Y. C. Lin, M. Bao, R.Cheng, J. Bai, Y. Liu, Y. Qu, K. L. Wang, Y. Huang and X. Duan, High-speed graphene transistors with a self-aligned nanowire gate. *Nature*, **467**, 305–308 (2010).

45. S. J. Han, K. A. Jenkins, A. Valdes Garcia, A. D. Franklin, A. A. Bol and W. Haensch, High-frequency graphene voltage amplifier. *Nano Lett.*, **11**, 3690–3693 (2011).

46. I. Meric, C. R. Dean, A. F. Young, N. Baklitskaya, N. J. Tremblay, C. Nuckolls, P. Kim and K. L. Shepard, Channel length scaling in graphene field-effect transistors studied with pulsed current-voltage measurements. *Nano Lett.*, **11**, 1093–1097 (2011).

47. C. R. Dean, A. F. Young, I. Meric, C. Lee, L. Wang, S. Sorgenfrei, K. Watanabe, T. Taniguchi, P. Kim, K. L. Shepard and J. Hone, Boron nitride substrates for high-quality graphene electronics. *Nat. Nanotechnol.*, **5**, 722–726 (2010).

48. I. Meric, C. R. Dean, S. J. Han, L. Wang, K. A. Jenkins, J. Hone and K. L. Shepard, High-frequency performance of graphene field effect transistors with saturating IV-characteristics (*IEEE Electron Devices Society*, 2011).

49. M. Xu, T. Liang, M. Shi and H. Z. Chen, Graphene-like two-dimensional materials. *Chem. Rev.*, **113** (5), 3766–3798 (2013).

50. H. Wang, D. Nezich, J. Kong and T. Palacios, Graphene frequency multipliers. *IEEE Electr. Device L.*, **30**, 547–549 (2009).

51. J. L. Hesler, W. R. Hall, T. W. Crowe, R. M. Weikle, B. S. Deaver, Jr., R. F. Bradley and S. K. Pan, Fixed-tuned sub-millimeter wavelength waveguide mixers using planar Schottky barrier diodes. *IEEE T Microw. Theory*, **45**(5), 653–658, (1997).

52. B. Thomas, A. Maestrini and G. Beaudin, A low-noise fixed-tuned 300–360-GHz sub-harmonic mixer using planar Schottky diodes. *IEEE Microw. Wirel. Co.*, **15**(12), 865–867 (2005).

53. Q. Xue, K. M. Shum and C. H. Chan, Low conversion-loss fourth subharmonic mixers incorporating CMRC for millimeter-wave applications. *IEEE T Microw. Theory*, **51**, 1449–1454 (2003).

54. X. Zhao, G. C. Shan, Y. Zheng, Y. Du, Y. Chen, J. Liu, C. Wang, J. Bao and Y. Gao, MEMS rectangular waveguide filter at 140GHz. *J. Infrared Millim. W.*, **32**(2), 165–169 (2013).
55. M. Liang, X. Tang and T. Wu, Design and analysis of the eighth sub-harmonic mixers at W-band. In *Microwave, Antenna, Propagation and EMC Technologies for Wireless Communications, 2007 International Symposium on*, pp. 389–391 (2007).
56. H. Wang, A. Hsu, J. Wu, J. Kong and T. Palacios, Graphene-based ambipolar RF mixers. *IEEE Electr. Device L.*, **31**, 906–908 (2010).
57. Y. M. Lin, A. Valdes-Garcia, S. J. Han, D. B Farmer, I. Meric, Y. Sun, Y. Wu, C. Dimitrakopoulos, A. Grill, P. Avouris and K. A. Jenkins, Wafer-scale graphene integrated circuit. *Science*, **332**, 1294–1297 (2011).

CHAPTER 6

DESIGN OF NANOSTRUCTURES FOR FLEXIBLE ENERGY CONVERSION AND STORAGE

Zhuoran Wang,* Di Chen[†,§] and Guozhen Shen*[,‡]

*State Key Laboratory for Superlattices and Microstructures,
Institute of Semiconductors,
Chinese Academy of Sciences, Beijing 100083, P.R. China
†School of Mathematics and Physics,
University of Science and Technology Beijing,
Beijing 100083, P.R. China
‡gzshen@semi.ac.cn
§dichen09@gmail.com

Flexible energy conversion and storage applications have attracted great attention for decades as they can, to a certain extent, contribute to releasing the pressure caused by fossil energy shortage. Designing nanostructures for flexible energy conversion and storage is a promising way due to their unique physical and chemical properties for future research. In this chapter, we provide a comprehensive review of the state-of-the-art research activities related to nanostructures based flexible energy conversion and storage. In the energy conversion part, we highlight the significance of nanostructures-based flexible photovoltaic (PV) applications as well as the brief discussions on the possibilities to make other forms of energy conversion applications flexible through nanostructure design. In the energy storage part we specially give emphasis to flexible lithium-ion batteries (LIBs) and supercapacitors

189

for the great potential they have shown for practical use. All these applications are believed to be developed towards high performance and low cost direction due to the booming perspective of nanotechnology.

1. Introduction

Flexibility is a general feature of the next generation electronic applications. During the past decades, researchers have tried many endeavors in order to make every application flexible. For example, the intensively investigated inorganic thin-film technologies have realized the aim of fabricating flexible and bendable electronic devices.[1–3] Also, the booming research of organic electronics including organic photovoltaic (PV) and Organic Light Emitting Diodes (OLED) provides another clue to perfectly fulfil the flexibility of devices.[4,5] However, traditional bulk thin-film fabrication processes like sputtering, atomic layer deposition, thermal evaporation, and chemical vapor deposition (CVD) always require a high-cost and super clean environment. Otherwise they will cause huge damage to the microstructure of the active material, resulting in sacrifice of the device performances. What is worse, the compact thin-film layer usually cannot bear severe bending or folding and hence it limits the flexibility of the devices. For organic electronic applications, even though thin-film can be formed using simple techniques like printing or spin-coating, sealing becomes a vital problem since organic materials are relatively unstable under ambient condition. Therefore, opportunities as well as challenges still exist in the field of flexible device fabrication.

Nanotechnology has made many previous fantasies come true in recent years because of the unique features of materials shown in nanoscale. It is well recognized that different morphologies usually have different characteristics. For instance, nanostructures like nanoparticles have unique surface effect and small size effect, which are mainly due to the significantly enlarged specific surface area. These advantages have been largely utilized in applications like photocatalysis,[6] PVs,[7] etc. One-dimensional (1D) nanostructures including nanorods, nanowires and nanotubes are generally accepted to have superior charge transport and collection properties, which make it promising to develop high quality electronic devices such as nanowire field-effect transistors (FET),[8,9] photodetectors,[10] and also energy storage applications like lithium-ion batteries (LIBs).[11] Three-dimensional (3D) structures assembled by zero-dimensional (0D) or 1D fundamental nanostructure are also promising for various application areas since they may make a good combination of all the superiorities of its sub-units.[12,13] All the advantages of nanotechnology together with the increasingly developed structure design methods guarantee the prosperous future for nanodevices. More importantly, it can help to fabricate the perfectly flexible devices which overcome such flaws caused by the traditional thin-film depositing process as identified above. Using the high quality nanomaterials obtained, thin-film flexible devices can be made using facile techniques like printing or coating, because of which scale-up and industrialization could be conveniently realized.[14] Besides, there are also many ways to synthesize nanomaterials directly on substrates including low temperature hydrothermal method,[15] solvothermal method,[16] electrochemical method,[17] chemical or physical deposition methods[18,19] and so on, through which almost any morphology required can be successfully obtained. So long as we choose the suitable deposition methods for the specific substrates and nanomaterials, flexibility of devices can be realized easily.

We are facing increasingly serious problems of fossil energy shortage and environmental pollution. As a result, finding effective routes for energy conversion and storage is urgent today. Nanotechnology opens great opportunities for making high quality energy conversion applications like solar cells,[20] as well as high performance electrochemical energy storage applications including LIBs[21] and supercapacitors.[22] Compared to the traditional energy applications, flexible energy devices are easy to be integrated with other consumer goods and portable to make them more attractive to the public for practical use. Nano-based flexible energy conversion and storage will surely bring new concepts and may even lead to a revolution for the next generation energy field.

The significance of nanostructure based energy conversion and storage have been highly valued already in research field as several important related review articles have been published.[23–26] However, till today none of them has provided a systematic overview specifically focusing on nanostructure based flexible energy conversion and storage. To fill in this gap, in this article we will make a detailed review of the state-of-the-art research progresses on different nanostructures based flexible energy conversion and storage, covering the basic materials design techniques and the advanced device applications. The second part of the article describes the general methods for nanostructures design ranging from 0D to 3D hierarchical architectures. Following this, we introduce the research progresses related to designing nanostructures for flexible energy conversion, and in this part we mainly focus on nano-based flexible solar cell since it is the most important and widely investigated flexible energy conversion application. In the next part, we show the examples of flexible electricity energy storage applications such as LIBs and supercapacitors. Finally, we conclude this article with an overview of the stimulating progresses up-to-date and an outlook for future development of nano-based flexible energy conversion and storage.

2. Strategies for Nanostructure Design

Materials are fundamental to the final performance of devices. High quality materials may have the potential to make a breakthrough against the current performance limitation of certain applications. Nanostructure design method, or morphology controlling in such a small scale, is very important for developing high performance, low-cost, new generation applications and great changes may happen to the traditional technologies as well. The concept of nano-design means to fabricate nanostructures with controllable size, dimension and structure, and hence we can test and find the optimized route to achieve the highest device performance. For flexible energy applications, nanostructure design method is even more important since nanomaterials have been proven to play an increasingly predominant role when applying in energy conversion and storage field, and they can also help to solve many key problems related to flexible applications. Therefore, in order to develop the new generation flexible energy applications, a thorough understanding of the different nanostructure design strategies and a wise choice of fabricating energy devices are important. Research has shown that different vnanostructures may have their preferred synthesizing techniques.

2.1. *Synthesis of Nanoparticles*

In order to obtain high quality 0D nanostructures, the most important thing is to control the growth rate of materials since it is very easy for nanoparticles to aggregate and form bulk materials. Therefore, there are various synthesis methods to slow down the reaction process and thereby to obtain particles in nanoscale.

Colloidal suspension (Sol) method is a popular and versatile method to synthesize inorganic nanomaterials. A general scheme for preparing mono-disperse nanostructures requires a single, temporally short nucleation event followed by slower growth on the existing nuclei. Nanocrystal (NC) size increases with increased reaction time as more material is added to its surfaces and also with increased temperature as the rate of addition of material to the existing nuclei increases. When the NC sample reaches the desired size, further growth is arrested by quickly cooling the solution. Figure 1 shows a transmission electron microscope (TEM) image of the spherical CdSe nanoparticles which were obtained by firstly heating a chosen solvent and a cadmium precursor to about 250–360°C under Ar flow and forming an optically clear solution, then injecting a selenium solution quickly into the reaction system.[27] The temperature of the reaction mixture decreased to 200–300°C by the injection and maintained at this temperature for the growth of the NCs. By controlling the synthesis conditions, the size of NCs can be varied in a wide range.

Sol-gel method is another effective method to obtain nanoparticles. In a typical sol-gel process, precursors like inorganic metal salts or metal organic compounds such as metal alkoxides are added into aqueous solution to begin the hydrolysis and polymerization reactions, which then lead to forming a sol. Further, complete polymerization and loss of solvent will lead to the transition from liquid sol into a solid gel phase. TiO_2 nanoparticles can be obtained through this method with controllable morphology (Fig. 2).[28]

Hydrothermal method is also widely used in synthesizing 0D nanostructures. Since this method is relatively low-cost and easy to operate, intensive investigations have been carried out. For instance, ZnO nanoparticles can be simply obtained through the hydrothermal route by immersing zinc source into slightly alkaline solutions under specific reaction

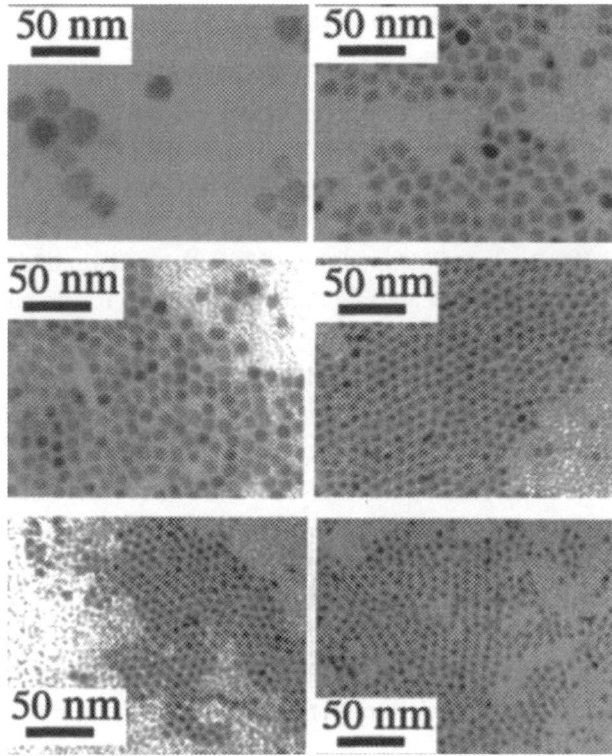

Figure 1.   Transmission electron micrographs of CdSe nanoparticles with varied sized synthesized by sol method. Reprinted with permission from Ref. 27. (Copyright: 2001 American Chemical Society.)

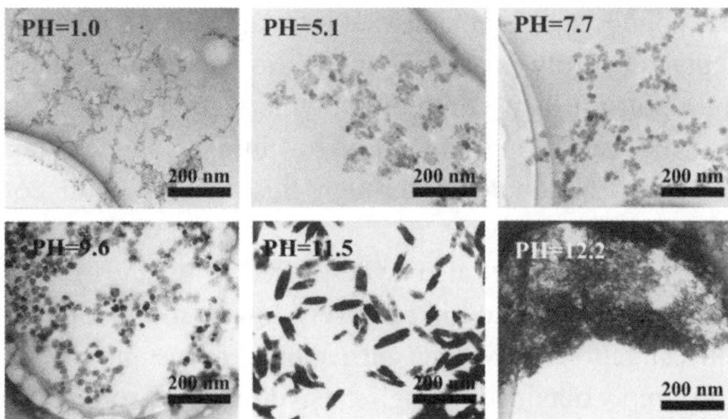

Figure 2.   TEM images of uniform anatase TiO$_2$ nanoparticles with different pH value. Reprinted with permission from Ref. 28. (Copyright: 2003 Elsevier.)

(a)   (b)

Figure 3.   Scanning electron microscope (SEM) images of Nanoparticles synthesized by hydrothermal method. (a) α-Fe_2O_3 nanoparticles. Reprinted with permission from Ref. 30. (Copyright: 2010 WILEY-VCH.) (b) SnO_2 nanoparticles. Reprinted with permission from Ref. 31. (Copyright: 2009 WILEY-VCH.)

condition.[29] Single crystalline α-Fe_2O_3 nanoparticles can be obtained by a facile hydrothermal method with the aid of F^- anions (Fig. 3(a)).[30] Single crystalline SnO_2 nanoparticles can also be obtained by hydrolysis of $SnCl_4$ in acid solution assisted with PVP, which can be observed in Fig. 3(b).[31] Nanoparticle research is quite a thermal topic nowadays since it could have unique properties due to the dramatically enlarged specific surface area when the size is getting small enough. Finding a suitable and common route to obtain nanoparticles with different particle sizes is essential for device applications.

2.2. *Synthesis of 1D Nanostructures*

Due to the superior charge transport properties of 1D nanostructures like nanowires, nanorods, and nanotubes, intensive research was carried out in this field. So as to obtain the large aspect ratio, controlling the material growth in a confined domain or along a specific direction is essential. Until now 1D nanostructures have been obtained according to a whole variety of methods covering almost all chemical and physical routes for materials synthesis, among them CVD and hydrothermal/solvothermal methods are the most popular in research field currently since they can be used to effectively induce the growth towards a certain direction.

CVD is the most effective way to obtain nanowires with high degree of crystallinity. And it is also an important route to synthesize ultralong and

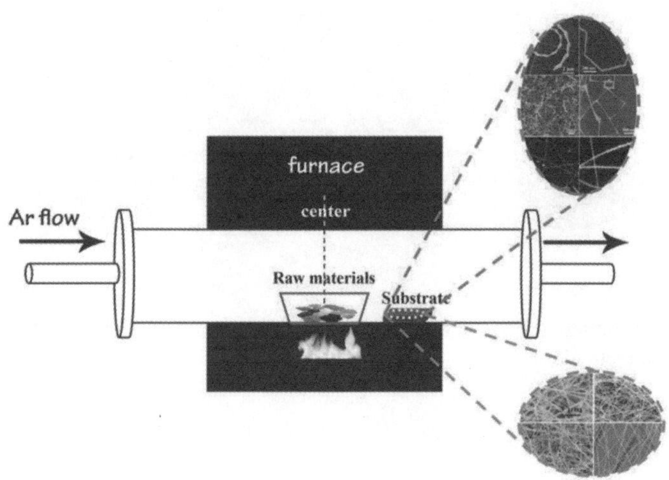

Figure 4. Scheme for a typical CVD method.

thin 1D nanostructure. A typical CVD reaction scheme is shown in Fig. 4. Generally, the source powder is heated at hundreds of or around 1000°C under a flow of carrier gas (usually Ar flow) and the nanomaterials are deposited on proper substrates located downstream or upstream to the source powder.[8] Usually, metal droplets such as gold particles thin-film or colloids are introduced first as catalysts. These catalysts capture the source vapor and form liquid alloy droplets, and then supersaturation and nucleation happen at the liquid–solid interface, leading to axial crystal growth. Owing to the catalyzed mechanism, the nanowires have diameters equal to the droplets, and only grow at sites activated by metal droplets. Besides, self-catalysis and vapor solid method with higher temperature are introduced as alternative routes to obtain nanowires. Using CVD methods, high quality In_2O_3 nanowires of different nanostructure obtained by Shen *et al.* are shown in Fig. 5.[32–34] In a typical process, a simple laser-ablation CVD method was introduced to obtain 1D nano-In_2O_3. Specifically, cleaned Si/SiO_2 wafer was used as the collecting substrate coated with a thin layer of gold nanoparticles as the catalysts for nanowire growth. Pure Ar was flown through the system at a certain rate to prevent the system from oxidation. A pulsed neodymium-doped, yttrium aluminum garnet (Nd:YAG) laser was used to provide power to stimulate the

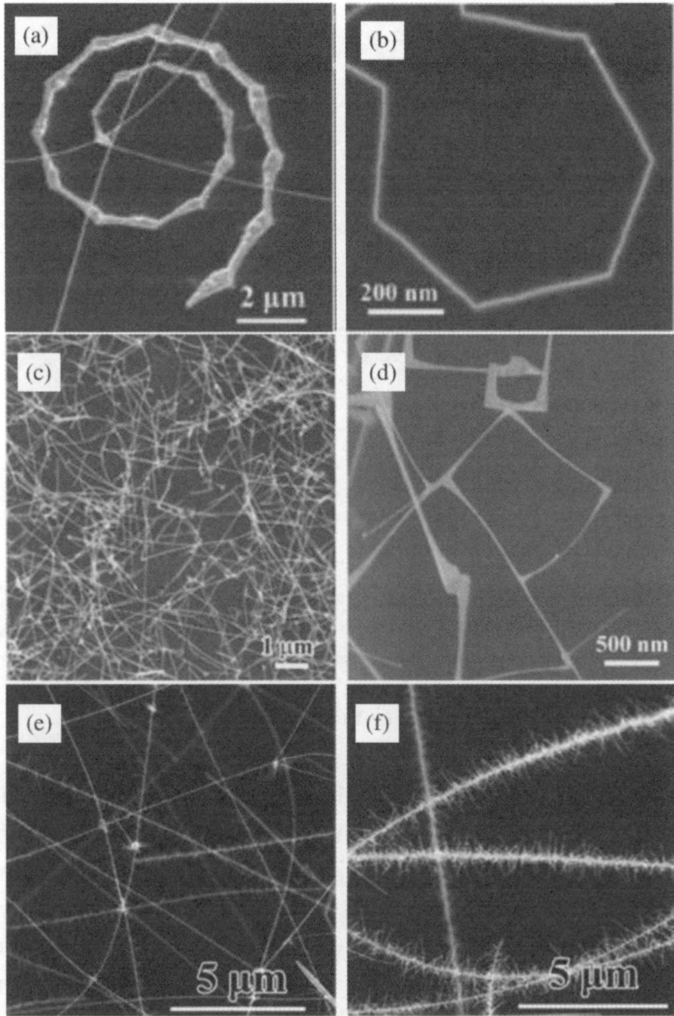

Figure 5. (a), (b) SEM images of In_2O_3 nanospirals made of kinked nanowires. Reprinted with permission from Ref. 33. (Copyright: 2011 American Chemical Society.) (c), (d) In_2O_3 nanowires thin-film. Reprinted with permission from Ref. 34. (Copyright: 2011 WILEY-VCH.) In_2O_3 ultrathin nanowires obtained after reaction for (e) 25 minutes and (f) 45 minutes. Reprinted with permission from Ref. 32. (Copyright: 2011 American Chemical Society.)

Figure 6.   (a) SEM images of a) Zn_2GeO_4 and (b) In_2GeO_4 nanowires synthesized by CVD method. Reprinted with permission from Ref. 35. (Copyright: 2012 The Optical Society.) (c) $ZnSnO_2$ nanowires. Reprinted with permission from Ref. 36. (Copyright: The Royal Society of Chemistry 2011.) (d) ZnO nanowires. Reprinted with permission from Ref. 37. (Copyright: 2009 IOP Publishing Ltd.)

reaction and different morphologies can be obtained on substrates finally. Many other kinds of 1D nanostructures have also been successfully synthesized according to a similar CVD method. Zn_2GeO_4 and $In_2Ge_2O_7$ nanowires were synthesized by our fellow researchers using GeO_2, Zn, carbon powders (for Zn_2GeO_4 shown in Fig. 6(a)), or GeO_2, In_2O_3, carbon powders (for $In_2Ge_2O_7$ shown in Fig. 6(b)) as raw materials through a simple CVD method.[35] Besides, Zn_2SnO_4 (Fig. 6(c)),[36] ZnO (Fig. 6(d)),[37] CdSe, etc., can be conveniently synthesized according to the same route. Because the 1D morphology diversity can be realized easily with CVD method, it has become a popular way in developing high quality materials for high standard electronic and optical applications and numerous exciting progresses have already been achieved. However, there are also some flaws existing in this method. High energy consumption and the relatively low production rate may limit the potential of scale-up application. Many efforts on solving such problems are ongoing.

Figure 7.   (a) SEM images of ZnO nanowires and (b) ZnO nanorods synthesized by hydrothermal method. Reprinted with permission from Ref. 38. (Copyright: The Royal Society of Chemistry 2011.) (c) Zn-doped SnO_2 nanorods. Reprinted with permission from Ref. 39. (Copyright: 2012 IOP Publishing Ltd.) (d) In_2O_3 nanobelts. Reprinted with permission from Ref. 40. (Copyright: The Royal Society of Chemistry 2011.)

Hydrothermal/solvothermal method is also widely used for synthesizing nanowires, nanorods, and nanotubes because of the relatively low cost and the simple operating process. Besides, by controlling the basic factors like reaction time, temperature, pressure and the amount of added raw materials we may easily find out the formation mechanism of nanostructures and hence use the optimized conditions to get the one we need. Figures 7(a) and 7(b) show ZnO nanowires and nanorods synthesized by a simple hydrothermal method. In general, zinc-oleate complex and different carbonates (Li_2CO_3, Na_2CO_3, K_2CO_3, Ru_2CO_3, and $(NH_4)_2CO_3$) were dissolved in ethanol and water. The reaction was performed under relatively low temperature (120°C) for 12 hour. And different kinds of carbonates were introduced to control the morphology.[38] Zn-doped SnO_2 was also obtained through a similar hydrothermal method. Specifically, $SnCl_4$, $ZnCl_2$ were added into NaOH water and ethanol solution and mixed until uniform. Reaction was carried out at 200°C for a whole day and cooled down at room temperature. SEM image of the synthesized needle-like SnO_2 nanorods is displayed in Fig. 7(c).[39] In Fig. 7(d), porous In_2O_3 nanobelts can be seen and this structure was achieved using $InCl_3$ and KOH as raw materials to react in PEG-600 solution.[40] There were also many other types of 1D nanomaterials synthesized through hydrothermal method like TiO_2,[41] WO_3,[42] ZnS[43] and so on. Due to great superiorities in controllable synthesizing 1D nanostructures with high yielding rate, hydrothermal method will be quite promising in future research field. However, comparing with CVD method, the degree of crystallinity of hydrothermal product is usually poor which may limit the device performance.

Besides, there are also many different methods to obtain 1D nanostructures such as template-introduced sol-gel method, electrospinning and so on. To get high quality 1D nanostructure for advanced device application, we need to develop an efficient nano design strategy which is low-cost, energy saving and environmental friendly.

2.3. *Synthesis of 3D Nanostructures and Nano Arrays on Substrates*

3D nanostructures which consist of the basic sub-units like nanoparticles or nanowires can be synthesized through a number of methods. To obtain this structure, designing a route to induce sub-units to assemble and then form hierarchy architecture is essential. Nanorod assembled hollow fiber, nanorod film, nanoflowers, nanospike, urchin-like nanostructure, straw-bundles, and so on were obtained in a number of ways.[12,44–49] More specific, in our group, TiO_2 nanorods assembled hollow fibers were synthesized through a typical microwave assisted hydrothermal method using carbon fibers as sacrificing templates (Fig. 8(a)). TiO_2 nanorods thin-film, $Ni(OH)_2$ nanoflowers, SnO_2 nanospike, and urchin-like $NiCo_2O_4$ nanostructures were all achieved according to a simple and effective hydrothermal self-assembling technique shown in Figs. 8(b)–8(e) respectively. In contrast, 3D nanostructures of Zn_2SiO_4 straw-bundles were synthesized via CVD methods. By controlling the temperature for experiment, the morphology of the straw-bundles can be changed from nanorods, to straw bundles, and then to spheres.

Figure 8.   SEM images of 3D nanostructures. (a) TiO_2 nanorods assembled hollow fibre. Reprinted with permission from Ref. 12. (Copyright: 2011 American Chemical Society.) (b) Nanorod-assembled Co_3O_4 hexapods. Reprinted with permission from Ref. 49. (Copyright: The Royal Society of Chemistry 2011.) (c) $Ni(OH)_2$ nanoflower. Reprinted with permission from Ref. 41. (Copyright: The Royal Society of Chemistry 2012.) (d) SnO_2 nanospike. Reprinted with permission from Ref. 46. (Copyright: The Royal Society of Chemistry 2011.) (e) Urchin-like $NiCo_2O_4$ nanostructures. Reprinted with permission from Ref. 47. (Copyright: The Royal Society of Chemistry 2012.) (f) Zn_2SiO_4 straw-bundles. Reprinted with permission from Ref. 48. (Copyright: The Royal Society of Chemistry 2012.)

In order to conveniently use the obtained nanostructures for device applications, we prefer directly designing nanostructure arrays onto specific substrates since there would be better electronic contact formed. Nanomaterials can be well deposited onto suitable substrates, so choosing substrates which match certain kind of nanomaterials will benefit the

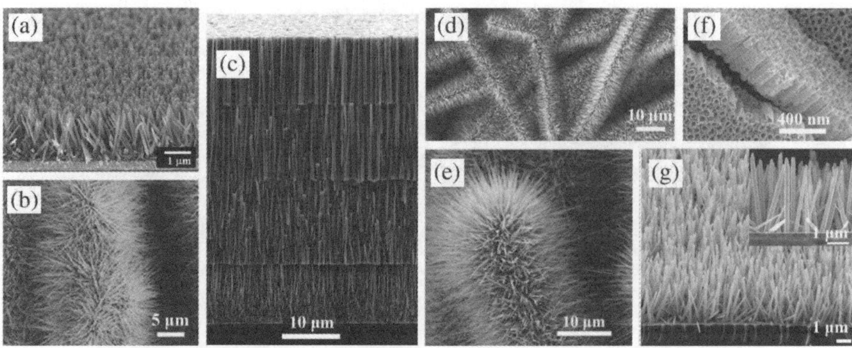

Figure 9.   SEM images of growing nanoarrays directly on substrates by (a–c) hydrothermal method, (d,e) CVD method and (f,g) electrochemical methods. (a) TiO_2 nanorod arrays on fluorine tin oxide (FTO) substrate. Reprinted with permission from Ref. 15. (Copyright: 2009 American Chemical Society.) (b) $ZnCo_2O_4$ nanowires on carbon fibres. Reprinted with permission from Ref. 11. (Copyright: 2012 American Chemical Society.) (c) ZnO nanowire arrays on FTO substrates. Reprinted with permission from Ref. 51. (Copyright: 2011 American Chemical Society.) (d) ZnO nanowires on carbon fibers, Reprinted with permission from Ref. 52. (Copyright: 2008 American Institute of Physics.) (e) WO_3 nanowires on carbon fibers. Reprinted with permission from Ref. 53. (Copyright: 2010 WILEY-VCH.) (f) TiO_2 nanotube arrays obtained on Ti foil by electrochemical anodization. Reprinted with permission from Ref. 54. (Copyright: 2006 American Chemical Society.) (g) ZnO nanowires on indium tin oxide (ITO) substrates according to electrochemical deposition. Reprinted with permission from Ref. 55. (Copyright: 2010 Elsevier.)

material depositing process.[50] Also, special treatment like "seeds" pre-deposition or slight-corrosion of the substrates is sometimes helpful.[12] Generally, hydrothermal method and CVD are both intensively used to obtain nanoarrays on substrates. Highly ordered TiO_2 nanorods arrays were successfully synthesized by Liu with controllable length and density shown in Fig. 9(a), by introducing a simple hydrothermal method under acid condition.[15] Carbon fiber is another type of substrate that has been widely used due to its high flexibility and conductivity. $ZnCo_2O_4$ nanowires were uniformly deposited on the surface of carbon fiber cloth using one-step reaction in aqueous solution, which can be seen in Fig. 9(b).[11] Figure 9(c) displays the ZnO nanorods arrays synthesized according to a novel layer by layer hydrothermal method and finally achieved a length of 40 μm for each nanorod grown on FTO substrates.[51] The enhanced thickness of film eventually increased the performance of device. CVD is another popular route to obtain nanostructure arrays. According to Unalan *et al.*'s work, zinc oxide nanowires can also be perfectly formed on the surface of carbon fibers in a typical carbothermal reduction of ZnO powder at a high temperature (Fig. 9(d)).[52] WO_3 nanowire arrays have also been deposited onto carbon fiber cloth according to a high temperature, catalyst-free, physical evaporation deposition process, which is vividly shown in Fig. 9(e).[53] Besides, there are also many other techniques to obtain nanostructure arrays. For example, TiO_2 nanotube arrays and ZnO nanowire arrays have both been successfully produced according to electrochemical methods on Ti foil and ITO substrate shown in Figs. 9(f) and 9(g) respectively.[54,55] TiO_2 was formed according to electrochemical anodization of titanium foil performed in solution containing F^- while ZnO nanowire arrays were achieved by electrochemically depositing nanostructures onto ITO substrate within the solution containing Zn^{2+}.

Except for all these techniques introduced above, there are still many novel and promising methods under development. Finding a suitable route towards better physical and chemical properties of nanomaterials will be beneficial for further device applications. Synthesizing nanomaterials with high specific surface area, good charge transport and collecting ability, high stability as well as special optical and electronic characteristics is the first step for the whole nano-energy applications. As for the field of flexible energy devices where many traditional nano design processes are no longer adaptable since there are many limitations for device fabrications to realize the flexibility, optimizing old routes and developing new strategies are solutions to meet these demands. Therefore, in order to realize new generation flexible energy applications, except for high quality nanomaterials, novel concepts of device structure optimization and new fabricating processes are urgently needed.

3. Design of Nanostructures for Flexible Energy Conversion

Energy exists in natural world as different kinds of forms including light, heat, water, wind, mineral, etc. The law of energy conservation offers the possibility to convert energy from one form to the others. Electric energy, doubtless, is one of the greatest inventions that has significantly promoted the fast development of the human society because it is the most efficient, clean and quiet secondary energy form we can use directly. Therefore, to effectively convert all those primary energy resources into electricity has been the general aim of human beings for a long time.

Fossil fuel and minerals are the main energy resources that most people on earth rely on in both past and present. However, serious shortage of coal and crude oil compels us to find alternative solutions to live a sustainable life. Researchers are turning to develop clean and renewable energy resources and trying to use them for electric power generation instead of burning coal, oil, and natural gas. For instance, solar energy is one of the most attractive alternative resources for its abundant distribution on the earth among all renewable energy forms, and hence great efforts have been taken to utilize solar energy to generate power.

The highly developed nanotechnology further promotes the energy conversion field. On one hand, nanotechnology has promoted the development of conventional energy conversion applications towards a highly efficient and environment friendly direction; on the other hand, it is devoting to novel applications like flexible and portable ones. Unlike large-scale power plant, flexible energy conversion devices like flexible and portable PV cells or modules can make people's life enter a whole different era. Flexible energy conversion components can be integrated conveniently with clothes or other daily consumer goods, which will equip everyone with a mini power plant and energy shortage pressure may be reduced to a large extent. In this chapter, we will make an up-to-date summary of research work about using nanostructure design methods to develop high performance flexible energy conversion devices.

3.1. *Design of Nanostructures for Flexible Photon-to-Electron Conversion*

Since the first practical silicon based PV cell was born in 1954 at Bell labs in USA, this great idea that using sunlight directly to generate electricity had attracted significant interests globally. Until now, it is generally accepted that there are three generations of solar cells: bulk silicon solar cells as the first generation, thin-film solar cells including compound and amorphous silicon solar cells as the second generation. New generation or third generation solar cells are proposed to make a breakthrough of the previous performance limitations, to which nanotechnology is making great contributions at present. Combining different PV technologies with various nano-fabrication techniques has already begun to show the potential in promoting the area of flexible photon-to-electron energy conversion.

3.1.1. Flexible Dye-Sensitized Solar Cells (DSSCs)

Speaking of nanostructured solar cells, an important example is the highly successful dye-sensitized solar cell (DSSC), which traditionally uses a network of sintered titanium oxide nanoparticles as a transport medium for photo-generated electrons. In 1991 the concept of using dye-sensitized colloidal nanostructured films was first brought up with an initial overall efficiency of 7.1–7.9% by Gratzel's group[7] and till now this record has been renewed to over 12%.[56] General operation scheme of a typical DSSC is shown in Fig. 10. The photoanode, made of a mesoporous dye-sensitized semiconductor, receives electrons from the photo-excited dye which is thereby oxidized, and which in turn oxidizes the mediator, a redox species dissolved in the electrolyte. The mediator is regenerated at the cathode by receiving the electrons circulated through the external circuit.[20] Nanostructure has played a vital role in DSSCs, which can strongly affect the global efficiency of a certain electrochemical system. TiO_2 nanoparticles, the most commonly used anode materials, act as skeletons to support dye molecules to form the 3D networks and hence largely improve the light harvest efficiency. The comparison made between a singlecrystal anatase cut in the (101) plane and nanocrystalline anatase film-based DSSC are shown in Figs. 10(a) and 10(b), respectively. The huge difference of their quantum efficiency vividly indicates the importance of the nanostructured porous thin-film. Therefore, designing suitable nanostructures for DSSC is definitely an effective means to further enhance the performance, so that can make it more suitable for practical applications.

Since DSSCs function through a solution based electrochemical process by using only nanomaterials to transport charge carriers instead of built-in electric field induced electron-hole pairs separation. This design is more tolerant of defects and impurities in the solar cell materials than conventional planar p–n junction cells. Therefore, the fabrication environment

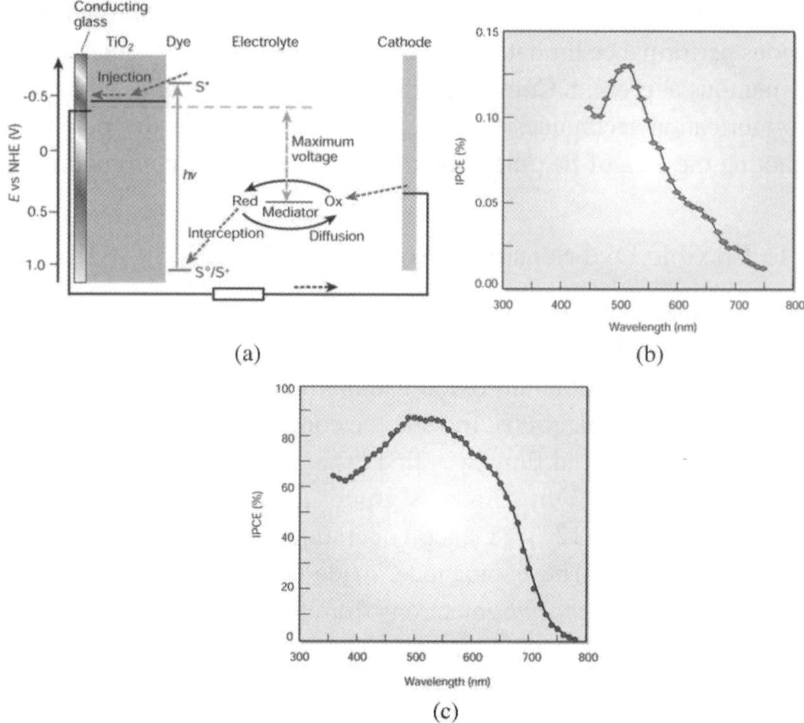

Figure 10.   (a) Schematic of operation of the dye-sensitized electrochemical PV cell. (b)The incident photon-to-current conversion efficiency of a singlecrystal anatase cut in the (101) plane. (c) nanocrystalline anatase film based DSSC. Reprinted with permission from Ref. 20. (Copyright: 2001 Macmillan Magazines Ltd.)

and techniques are relatively simple. Doctor blading, screen printing and coating methods are widely used to form thin-film for DSSCs, which offers possibilities for fabricating low-cost flexible DSSC. In fact, there are many reports and investigation focusing on this area. Because of the fast emerging printing electronics, flexible DSSC is becoming an increasingly competitive member among the whole PV family.

3.1.1.1. *Printing TiO₂ Nanoparticles Paste on Transparent Substrates for Flexible DSSC Anodes*

TiO₂, the whitest materials in the natural world, has been widely used as photoanode materials for DSSC due to its suitable physical and chemical properties. To obtain flexible DSSC, firstly we need to fabricate the flexible photoanode which plays a predominant role in the whole cell. Printing or coating paste made from TiO_2 nanoparticles onto charge collectors is a very popular way. By using these methods, the flexibility of solar cells is determined by the flexible substrates applied. Traditionally, the just coated thin-film on transparent conductive oxide (TCO) glass will need a high temperature calcination process, usually 450–500°C for 30–60 minutes to increase the degree of crystallinity, remove the organic binder and form good electric contact for the photoanode. However, when considering the transparent flexible substrates which are usually TCO coated polymer materials like ITO/PET (Polyethylene terephthalate) or ITO/PEN, most of these substrates cannot tolerate temperatures higher than 150°C and hence the poor crystallinity becomes a tough problem which limits the performance of flexible solar cells. But still, researchers have tried several alternative ways to solve such problems. Instead of using organic binders like ethylcellulose and terpineol, different kinds of less viscous, liquid-like TiO_2 colloid solutions were intensively investigated for decades and progresses have been made. Hence, in this way the flexibility of device can be realized. But another problem is that being free of organic binder may lead to bad contact without high temperature sintering. During the testing, bending or rolling, active materials may easily be detached, leading to the efficiency lose of devices. Therefore, some process including compression,[57] microwave radiation with certain frequency,[58] UV–ozone treatment,[59] hydrothermal treatment[60] and so on have been used to replace heating process and have already proved to be effective.

Zhang *et al.* reported a method for low temperature fabrication of titania photoelectrodes by hydrothermal crystallization at the solid/gas interface.[60] In their work, they introduce an aqueous or ethanolic mixed paste of nanocrystalline TiO_2 powder and titanium sources such as $TiCl_4$, $TiOSO_4$, Ti alkeoxides instead of using organic binders and then "steam-cook" the coated plastic substrate in the water–gas phase in an autoclave at 100°C. The added titanium salts are hydrolyzed and crystallized into either rutile or anatase TiO_2, which acts as a "glue" to form connection between thin-film and plastic substrates. From the morphology and crystallinity changes after the hydrothermal solid/gas treatment we can notice the importance of this process on the performance of a flexible photoanode. After 100°C steam treatment, the microstructure of this film becomes quite similar to the ones prepared by high temperature sintering methods (Figs. 11(a)–11(c))[60] and peaks for rutile and anatase can be observed in XRD patterns (inset of Fig. 11(c)). This work introduced a very simple method to obtain flexible photoanodes with an efficiency of 4.2% achieved on glass substrate. Then in later report, instead of using $TiCl_4$ and $TiOSO_4$ as "glue", the paste was prepared using Ti-tetraisopropoxide (TTIP) since the as-printed thin-film is

Figure 11.   (a) SEM photographs of the TiO_2 films resulted from the mixed paste as the raw film. (b) that after 12 hour hydrothermal treatment. (c) the same film at lower magnification. Inset shows the XRD patterns where a, b, c, d indicate the film dried in oven at 100°C, or hydrothermal treated by $Ti(OiPr)_4$, $TiOSO_4$ and $TiCl_4$ respectively. Reprinted with permission from Ref. 60. (Copyright: 2003 Wiley-VCH.)

more stable against solvents such as water and ethanol even when it was not subjected to the hydrothermal treatment.[59] Except for heating, a novel UV–ozone method was introduced then. The main efforts of using this treatment was concluded to remove the organic contaminant introduced by TTIP, which hence provided an effective route to replace high temperature heating as well as the potential to produce large scale flexible photoanodes. Additionally, pre-heating the nanoparticles before making paste has also been proved to have a significant contribution to efficiency because it can help to remove the absorbed water in the air previously. With pre-heating treatment, using TTIP as "glue" and UV–ozone treatment, the rigid FTO substrates based solar cells can achieve an efficiency of 4.89% and 3.27% by using flexible ITO/PET as substrates. Except for heating, chemical treatment is another effective option. The general aim for this approach is to develop a paste with high viscosity which can be coated onto substrates to form good contact and also will not significantly affect performance for the less binder introduced. What is more, since the paste with appropriate chemical treatment is of high viscosity, cracks can be easier to avoid for devices. In an early report, acidic TiO_2 colloid solution was found to be very viscous when ammonia solution was added (Fig. 12).[61] Also, some other organic chemicals like PVP has been found effective to realize this aim.[62]

All these are methods aimed at avoiding high temperature directly on the substrates. It is the first and the most important step for fabricating a planar flexible DSSC photoanode. Even though they have been proved

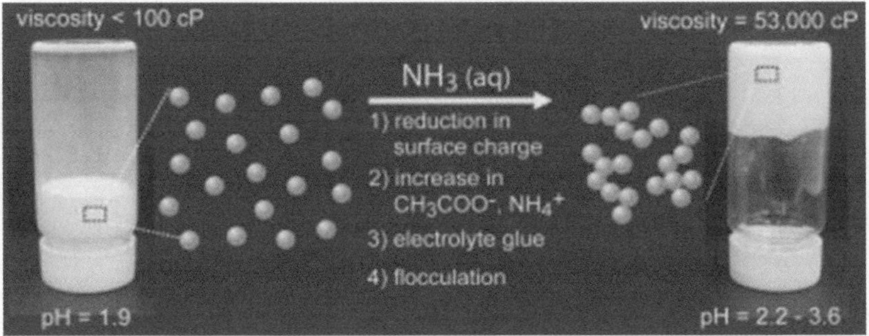

Figure 12   Effect of ammonia solution on the viscosity of acidic TiO_2 colloid solution. Photo images of colloid solution before (left) and after adding certain amount of ammonia (right). Reprinted with permission from Ref. 61. (Copyright: 2003 WILEY-VCH.)

effective, the organic contaminants can hardly be totally removed and the crystallization may still remain poor. Focusing on this issue, unconventional heating methods were introduced. Uchida *et al.* have brought up an alternative method by using microwave to provide power so as to enhance the quality of the as-prepared thin-film.[58] This process operated well due to the different absorption co-efficiency of microwave with certain frequency. Therefore, finding a suitable microwave source which can be absorbed well by nanomaterials-based thin-film and also be transparent to the substrate is essential. In this work, temperature changes of different nanomaterials were thoroughly investigated in Fig. 13. From these figures we can easily notice that there are great enhancements of temperature with increased irradiation time for both SnO_2 and In_2O_3. In contrast, TiO_2 was found to exhibit moderate coupling to microwaves while there was no relation between SiO_2 and microwave. When considering the relatively high

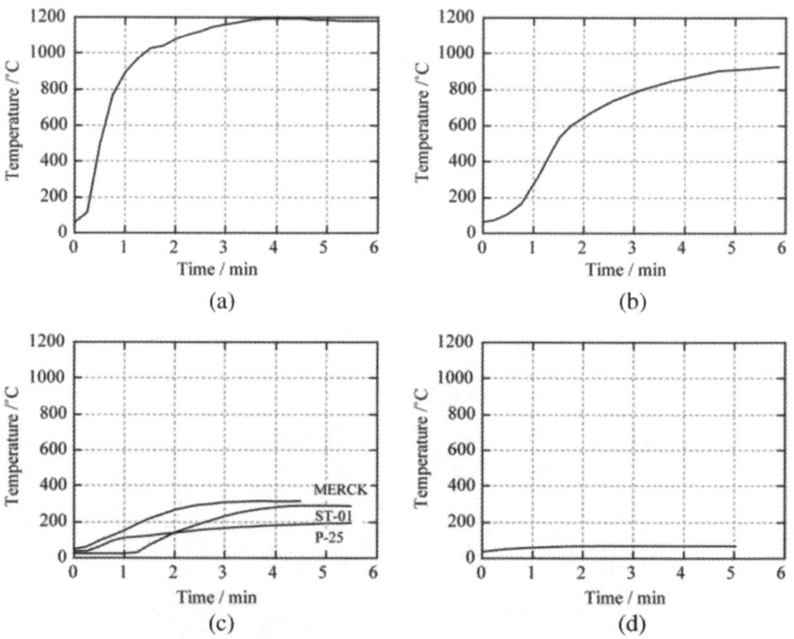

Figure 13.   Temperature as a function of time during 28 GHz, 1 kW microwave irradiation of (a) SnO_2. (b) In_2O_3. (c) TiO_2. (d) SiO_2. Reprinted with permission from Ref. 58. (Copyright: 2004 Elsevier.)

electrical conductivity of SnO_2 and In_2O_3 which are non-stoichiometric semiconductors, the good match with microwave may be due to their higher concentration of conduction electrons. However, increasing temperature of a material caused by microwave–material interactions are mostly attributed to induction losses from electric conduction. TiO_2, a stoichiometric semiconductor, showed moderate coupling to microwaves because of its low electrical conductivity and low magnetic induction loss. The absorbed energy P (Wm^{-3}), material–microwave interaction is explained with the generalized energy loss equation as follows:

$$P = 2\pi f \varepsilon_0 \varepsilon_r \tan \delta E^2 V_s \Theta, \tag{1}$$

where f, ε are the frequency, dielectric constant, $\tan \delta$ stands for the dielectric and magnetic loss factor value. E, Vs are the electric field values inside the sample, volume, and shape factor, respectively. An efficiency of 5.51% has been achieved under 0.7 KW microwave radiation for the rigid substrate based DSSC, and 2.16% was obtained when applying to the flexible ones. In addition, the frequency of microwave is also important, which is shown in Fig. 14. The best efficiency was achieved using 1KW, 28 GHz microwave radiation for flexible photoanodes. The importance of

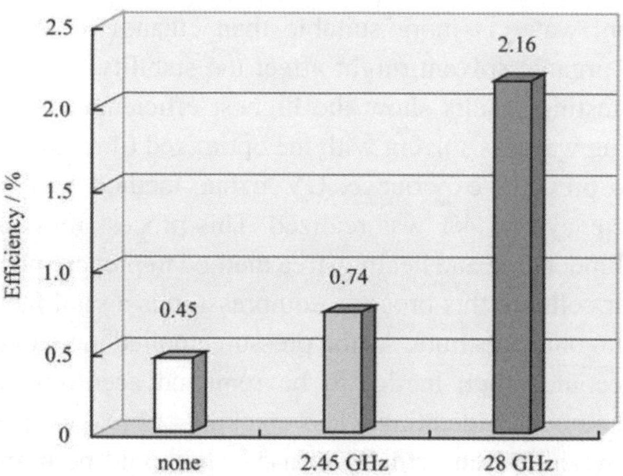

Figure 14. Comparison of efficiencies in different microwave irradiation at 1 kW for five minutes for PET–ITO electrodes. Reprinted with permission from Ref. 58. (Copyright: 2004 Elsevier.)

this research progress is that it provided a new concept of finding new energy sources to stimulate reaction, including electromagnetic wave that can be selectively absorbed by certain materials. Even though the efficiency here was still low, if we could find the optimal electromagnetic wave that can perfectly heat the nano-thin-film while being totally transparent to flexible substrates, flexible flat photoanodes will be successfully prepared without significant reduction of performance.

Later, a much higher solar cell efficiency of 7.4% based on PEN flexible substrate was achieved using a press method without any heat treatment at all.[57] Here, unlike the techniques introduced before, only TiO_2 powder and ethanol were used to make the paste without any other organic chemicals. With the optimized TiO_2 particles with different diameters, investigations were carried out to understand the effect of different treatment. Interestingly, there is a decrease of efficiency after applying heat treatment compared to compression only. Also from the fact observed after the heat treatment, the conducting plastic film was warped more than before. Hence it can be concluded that heat treatment might decrease the adhesive force between the TiO_2 nanoparticles and the conducting plastic substrate. Therefore, in such a situation without any organic chemicals as binders, heat treatment is unnecessary. Further, with only compression method, comparison was made between water and ethanol as solution. In this situation, water is more suitable than ethanol to form the paste, because the organic solvent might affect the stability of the plastic substrate. The testing results show the highest efficiency of over 7% was obtained using water as solvent with the optimized film thickness. Finally, according to previous experience, UV–ozone method was used and the highest efficiency of 7.4% was realized. This process introduced to us a completely binder-free and heating-free method to prepare photoanode for flexible solar cells. In this process, compression is a vital factor to attach TiO_2 thin-film onto substrate. As the pressure applied increased, the anode materials become much harder to be removed according to ISO/DIS 15184 measurement, indicating a less cracking or breaking possibility for the flexible device when actually in use.[67] It should be noticed that the applied compression melt the TiO_2 nanoparticle surface and formed the welded particles, leading to a good mechanical adhesion strength. This also gives us a clue to realize good flexibility of devices. Also, only water used as solvent further opens great opportunities to a simple and low cost fabrication process towards high efficient flexible PV industry, which also has the potential to promote the development of printable electronics.

Additionally, before making TiO_2 nanoparticles into slurry, special treatments to the nanostructures, to a certain extent, will help to enhance the performance. For example, the high temperature pre-annealing will help to remove the absorbed water and increase the degree of crystallization as mentioned above. Also, an interesting ball milling process was widely applied.[63–65] The effect of pre-ball milling of TiO_2 nanoparticles was proved by Weerasinghe *et al.*[63] Milling of P-25 TiO_2 powder was found to increase the thin-film quality and its mechanical stability, as well as enlarge the surface area. It is obviously shown in Fig. 15 that thin-film prepared by milled particles will lead to fewer cracks (Figs. 15(b) and 15(d)) than the non-milled particles (Figs. 15(a)) and 15(c)), which is proved to be caused by the de-aggregation of TiO_2 powers during milling process. De-aggregation leads to much uniform diameter distributions and

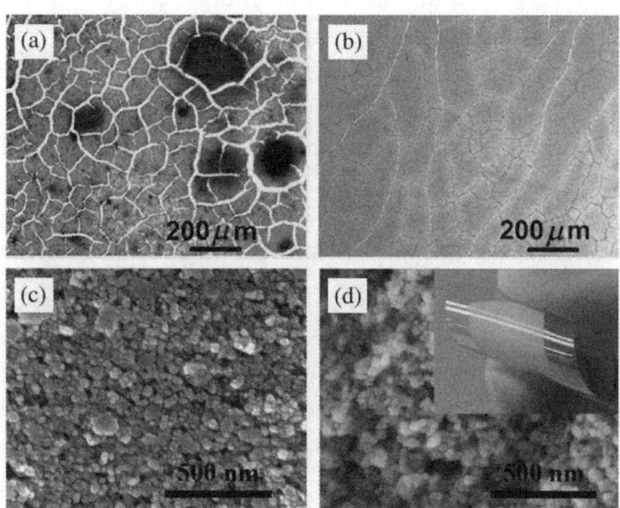

Figure 15. Optical microscopic images of TiO_2 films prepared by using (a) non-milled an (b) milled slurries and SEM images of TiO_2 films prepared by using (c) non-milled, and (d) milled samples (milled time of both samples is 20 hour). Reprinted with permission from Ref. 63. (Copyright: 2004 Elsevier.)

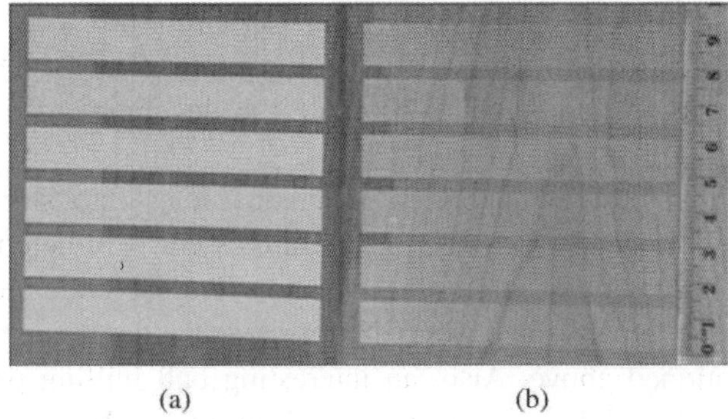

Figure 16.   (a) Appearance of the TiO$_2$ films on an ITO-coated polyethylene naphthalate (PEN) substrate as-printed and (b) after cold isostatic pressing when the films were laid on a wooden bench. Reprinted with permission from Ref. 64. (Copyright: 2011 John Wiley & Sons, Ltd.)

hence compact thin-film with fewer cracks can be formed. Fewer cracks formed on electrodes are very important for high quality flexible optoelectronic devices. Meanwhile, this approach can significantly enhance the dye-loading ability so that the overall energy conversion efficiency is improved. The optimized efficiency was 4.2% with commercial P-25 TiO$_2$ power. Further, improved work has been carried out by the same group with an improved process based on ball milling.[64] After printing the thin-film, compression was applied to enhance the quality of photoanodes. An apparent change could be noticed after pressing. And the photoanode became much more transparent which increased the performance of photoanode as well as its aesthetic value (Fig. 16). The high efficiency of 6.3% and 7.4% were obtained under illumination of 100 and 15 mW/cm^2, respectively, which are the highest among all the flexible DSSC prepared by using P-25 commercial nanoparticles.

Till here, all progresses introduced above in this section are related to finding an alternative way of preparing printable titanium dioxide paste with simple and low temperate process. And all these exciting progresses will open new vistas for printable PVs. When using plastic substrates, it has the possibility to absorb light from both sides, which is a great advantage for the printed flexible DSSCs compared to other technologies.[66] High transparency will also offer the potential to be used in special applications like windows of houses and cars with no needs to occupy extra room (Fig. 17(a)). Also, the property of mechanical flexibility allows a large-area module to be set on a round surface which is exposed to diffuse light with multiple incident angles. Figure 17(b) demonstrates an example of a large belt-shaped cell made on ITO-PEN films, in which the photocurrent is collected by metal films at the edges of the cell. The high voltage of above 0.7 V can be maintained under exposure to indoor illumination. If only one side illumination can be acceptable, metal foil like Ti and stainless steel can be introduced as substrates and hence the process is the same as using rigid TOC glass since there are fewer limitations for such substrates except for sacrificing the transparency (Fig. 17(c)).[68,69] Therefore, in order to get the transparent flexible DSSC, the efforts on low temperature and moderate photoanodes fabrication processes are necessary.

Figure 17. (a) A flexible film-type DSSC made on a belt-shaped plastic ITO-PEN. (b) A film-type bifacial module of a DSSC comprised of a series connection of six unit cells. Reprinted with permission from Ref. 66. (Copyright: 2011 American Chemical Society.) (c) Ti foil based dye-sensitized photoande. Reprinted with permission from Ref. 68. (Copyright: 2011 John Wiley & Sons, Ltd.) (d) Stainless steel as substrate for flexible DSSC. Reprinted with permission from Ref. 65. (Copyright: 2005 Elsevier.)

3.1.1.2. *Alternative Methods for Designing TiO_2 Nanostructures on Flat Flexible Substrates as Photoanodes*

Even though the highest efficiency for DSSC was achieved using nanoparticles printed photoanode,[56] different TiO_2 nanostructures have been widely investigated for applications in DSSC and stimulating progresses has been achieved on rigid TCO like 1D [16,50,70] and hierarchy architectures.[71–73] As for flexible photoanodes fabrication, directly growing nanostructures on flexible conductive substrates is a convenient and promising strategy.

Ti metal foil is of good flexibility and can be both the raw materials and substrates for TiO_2 thin-film deposition. Using a piece of Ti foil, TiO_2 nanotube arrays can be obtained according to electrochemical anodization, which is a very popular method for research.[17,54,74–79] The general anodization procedure of TiO_2 foil has been briefly described in the former chapter, and there are various changes taking place to meet the more optimized device performance. The first effort on dye-sensitized TiO_2 tubes arrays based solar cells was carried out by Macak *et al.*[80] However, only very low efficiency of 0.036% was achieved at that time, because the tube geometry and other experimental factors are far from being optimized. Later, much better performance was obtained for pure TiO_2 nanotubes according to a protective, mild growth process, an efficiency of 5.2% under AM 1.5 illumination was tested by the same group.[81] This work demonstrates vividly how the performance could be enhanced due to the optimized nanostructures. Before electrochemical anodization, a positive photoresist coating was applied on Ti foil in order to protect the initial layer during the anodization procedure. An observed feature of the non-protected layers is that without precautions, the upper tube end exhibits some disordered morphologies in the form of "nanograss" (Fig. 18(a)) or nucleated or bundled nanotubes (Fig. 18(b)) after anodization. This is a consequence of the permanent thinning of the tube walls caused by the etching electrolyte. Alternatively, if layers are formed under this protective condition, after anodization, remnants of an initiation layer are still preserved which may lead to an efficient use of the nanotube layer (Fig. 18(c)). With the completed top open morphology, the highest efficiency up to now for pure TiO_2 nanotubes was obtained (Fig. 18(d)) and it was concluded to be due to the optimization of charge transport and light

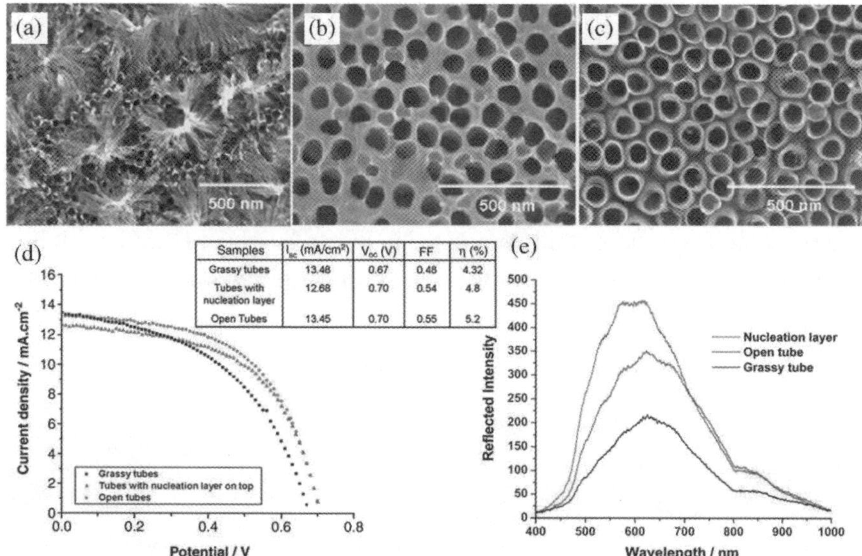

Figure 18. SEM images top view of TiO_2 nanotubes with different tube tops including (a) grassy tubes, (b) open tubes with nucleation layer on the top, and (c) completely open TiO_2 tubes obtained using a photoresist coating. (d) *J–V* characteristics under AM 1.5 solar light illuminations for 16 μm long TiO_2 nanotubes with different tube top morphologies, inset: PV performance parameters of DSSC based on TiO_2 nanotubes. (e) Reflectance spectra of three different nanotube layers. Reprinted with permission from Ref. 81. (Copyright: 2010 Elsevier.)

absorption for such nanostructures. However, it seems this nanostructure is facing great limitation of performance enhancement since the electrochemical strategies have been thoroughly investigated by enormous endeavors already, and it is now hard to make a significant breakthrough except for introducing new concepts. Hierarchy structures are then introduced to meet this challenge. In fact, since nanoparticles introduced composites[73] and surface decoration[82] have been proven to have superiorities in increasing dye-loading or reducing recombination, many efforts have been taken to combine such 0D structure with 1D morphologies so as to make good use of the advantages of different sub-units.[77,83] Stimulatingly, significant increase of performance has been made by many research groups via this strategy. A high global energy conversion efficiency of nearly 7% was achieved by Lin *et al.*[78]

It is the first report to introduce the concept that using synthesized TiO_2 nanotube thin-film as an underlayer for providing a better connection between TiO_2 nanoparticles and Ti substrate, which can be explained using the scheme shown in Fig. 19(a). According to this explanation, electron transport ability may be highly improved. Significant change also happened when using nanoparticles for surface decoration. $TiCl_4$ post-treatment has been successfully promoted in research field of DSSC, which is considered as an effective route to reducing the recombination as well

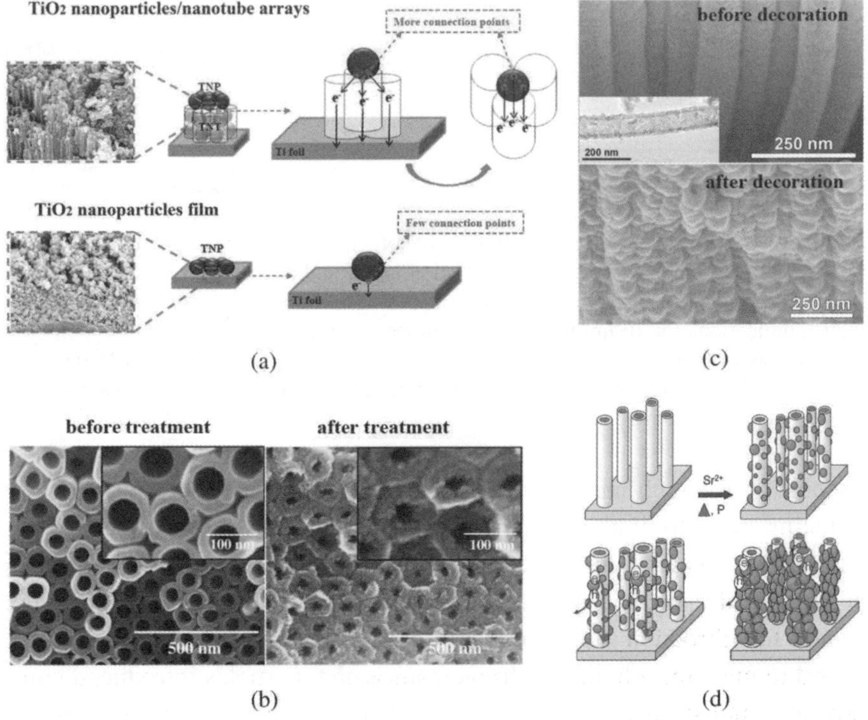

Figure 19 TiO_2 nanotubes based hierarchy structures for flexible dye solar cells. (a) Scheme of the enhanced contact formed using TiO_2 underlayer structures. Reprinted with permission from Ref. 78. (Copyright: 2011 Elsevier.) (b) SEM images of top view TiO_2 nanotubes before (left) and after (right) $TiCl_4$ treatment. Reprinted with permission from Ref. 84. (Copyright: 2011 Elsevier.) (c) SEM images of cross section view of TiO_2 nanotubes before and after SrTiO3 decoration and (d) mechanism of the formation of $SrTiO_3$ and blocking layer formed by too much decoration. Reprinted with permission from Ref. 87. (Copyright: 2009 American Chemical Society.)

as roughening the surface. In the work reported by Schmuki's group, the TiO_2 nanoparticles deposited on the surface of TiO_2 nanotubes by hydrolysis of $TiCl_4$ eventually doubled the initial efficiency from 1.9% to 3.8% under optimized condition, which is caused by the apparently roughened surface of nanotubes (Fig. 19(b)).[84] Besides, hetero-decoration was also regarded effective to enhance the electrochemical performance of TiO_2 nanotubes.[85–87] Zhang *et al.* has introduced a novel method by using a simple hydrothermal technique to directly decorate $SrTiO_3$ onto nanotubes' surface to make a good match of bandgap structure, leading to the improved charge transfer characteristics.[87] Fig. 19(c) shows the effect of the $Sr(OH)_2$ treatment on the pure TiO_2 nanotubes, the roughened surface indicates the successful hetero-decoration. And Fig. 19(d) demonstrates the concluded function of $SrTiO_3$, appropriate amount of decoration will help to promote the photoelectrochemical activity while long time treatment may lead to form a blocking layer that may negatively affect the performance.

Except for direct anodization of the metal substrate, thin-film can also be formed using electrophoretic deposition method. Unlike other electrochemical methods, electrophoretic deposition processes are usually carried out within TiO_2 nano-powder dispersed solution under applied DC bias. An efficiency of 6.5% was achieved by Chen *et al.* by using this method to deposit commercial TiO_2 nanoparticles onto titanium foil.[88] Figure 20(a) shows the optical photo images of the prepared flexible photoanode and the inset displays its SEM images. However, to obtain such an acceptable performance, compression and high temperature of 350°C were needed, which may limit the opportunities for developing flexible transparent photoanodes.

Facing this problem, this technique was then simplified and extended to plastic substrate. Without post-annealing, thin-film deposited via electrophoretic method successfully got an efficiency of 5.25% for DSSC application.[89] The anode materials in this work were the lab-made TiO_2 nanoparticles which were proved to have better performance than the commercial ones, and it has been reconfirmed by investigating the effect of using varied particle size comparing with P25 particles to perform the electrophoretic deposition process in another report.[90] When using the electrophoretic method, an overall conversion efficiency of 6% was

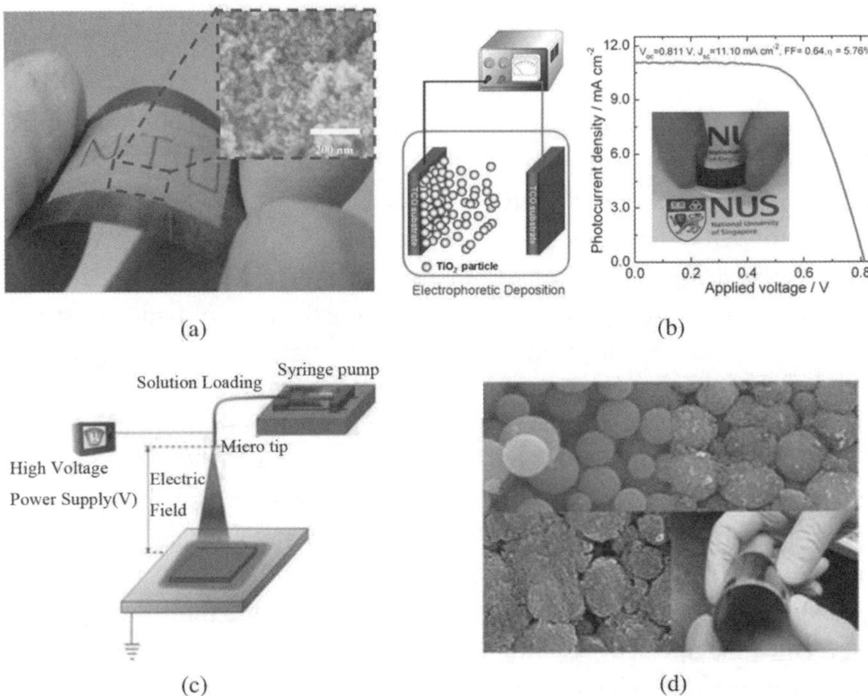

(a)

(b)

(c)

(d)

Figure 20. Electrodepositing TiO_2 nanoparticles on flexible substrates (a) Photo image of the prepared Ti foil based flexible photoanode using electrophoretic deposition method and inset shows the SEM image of the deposited photoanode. Reprinted with permission from Ref. 88. (Copyright: 2011 Elsevier.) (b) Scheme of electrophoretic deposition and the *I–V* curve of the full-plastic solar cells made from P25 TiO_2. Inset shows the appearance of flexible dye-sensitized photoanode on ITO/PEN. Reprinted with permission from Ref. 91. (Copyright: 2012 American Chemical Society.) (c) Scheme of the electrospray process and (d) SEM images of the sprayed TiO_2 morphology with specific treatment and photo image of a fabricated flexible DSSC. Reprinted with permission from Ref. 92. (Copyright: 2012 American Chemical Society.)

obtained for 19 nm self-made nanoparticles, which was much higher than that of commercial Degussa P25 TiO_2 (5.2%). Under optimized condition in later report by Yin *et al.* an enhanced efficiency of 5.76% was obtained for the full-plastic DSSC using electrophoretic method to deposit Degussa P25 on ITO/PEN as photoanodes (Fig. 20(b)).[91] There are many other convenient physical deposition methods as well. Electrospray process was developed recently by Lee *et al.* and an efficiency of 5.57% was achieved after compression and low-temperature chemical treatment.[92] Figures 20(c) and 20(d) vividly show the general process of deposition and the obtained micromorphology, respectively. Besides, laser direct-write technique,[93] atomic layer deposition,[94] sputtering method[95], and so on were, though not so widely, but proved quite useful in developing flexible dye solar cell photoanodes. Overall, all these methods seem to be effective alternative ways to fabricate flexible nanoparticles based photoanodes. Provided we could design high quality and suitable nanoparticles, it would be possible to make a significant breakthrough in the field of flexible DSSCs.

Since plastic substrate cannot bear high temperature annealing as well as strong acid or alkaline during synthesis procedure, transferring or transplanting is a very promising strategy intensively investigated. To conveniently transfer active materials on substrate, the first thing is to obtain high quality, large-scale and self-standing thin-film materials.[44,71,73,96,97] The most popular and widely investigated kind of nanostructures that can be transferred to substrates for DSSC applications are TiO_2 nanotube arrays synthesized by electrochemical anodization. The idea of detaching this nanostructure from Ti foil was first brought up by Park *et al.*[96] As is mentioned in the work, the traditional anodization method was first repeated to obtain the highly ordered nanotubes. Then a simple hydrochloric acid treatment process was applied and the TiO_2 thin film could be separated from the substrate automatically. And this thin-film consisted of TiO_2 nanotubes was transferred onto FTO glass and achieved a very high overall energy conversion efficiency of 7.6%. Further, based on 1D nanostructure and the transplanting technique, together with appropriate structure design by introducing nanoparticles (Fig. 21(a)), a much higher performance has been realized (8.8%).[73] However, this approach contains a serious flaw. For such a piece of TiO_2 thin-film, each nanotube is aligned very intensively as can be seen from the SEM images in Fig. 21(a). As a result, if the nanotube thin-film is used to fabricate flexible solar cell, the device cannot suffer from bending or stretching since there will seldom be inner space for the compact anode materials to release pressure. Therefore, finding a suitable transplanting nanostructure is essential. In our recent work, a template introduced novel cloth-shape thin-film consisting of enormous nanorods were synthesized with wonderful transferability.[12]

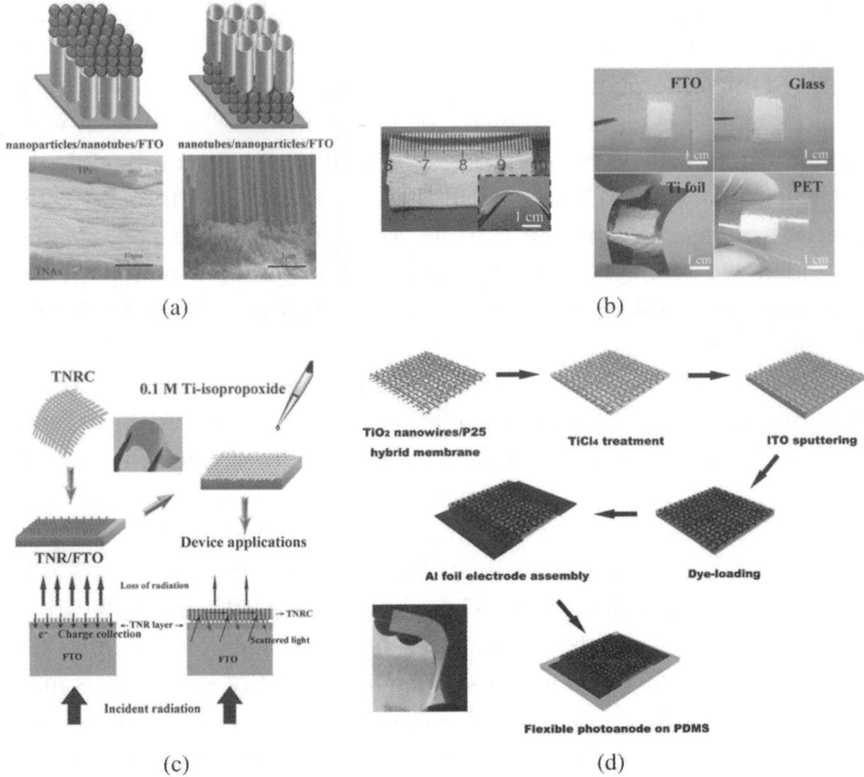

Figure 21.   Transfer method to fabricate DSSCs (a) Scheme and SEM images of devices structure constructed using nanorods and nanotubes together. Reprinted with permission from Ref. 73. (Copyright: 2011 American Chemical Society.) (b) Transferring cloth-shape nanostructured thin-film onto different substrates. Reprinted with permission from Ref. 12. (Copyright: 2011 American Chemical Society.) (c) Enhanced performance by using the concept of device structure optimization. Multilayer device fabrication process (left) and the mechanism of enhanced performance (right). Reprinted with permission from Ref. 13. (Copyright: The Royal Society of Chemistry 2012.) (d) Inverted method to transfer the already dye-sensitized TiO_2 nanowires/nanoparticles thin-film. Inset shows the free-standing hybrid thin-film. Reprinted with permission from Ref. 98. (Copyright: 2011 Elsevier Ltd.)

More importantly, as can be seen from the left of Fig. 21(b), a large piece of free-standing TiO_2 cloth can be conveniently bended to form a perfect arch without any damage to the original structure. The good flexibility is due to the preserved texture structure containing enough interspace and holes to release pressure under mechanical forces. The right of Fig. 21(b) demonstrates the possibilities of using it for flexible applications since theoretically it can be transferred onto any substrate conveniently. This interesting nanostructure was also investigated for DSSC application but just over 2% global efficiency was achieved. Further, with the concept of structure optimization, directly deposited TiO_2 nanorod arrays were used as the underlayer to enhance connection between anode materials and the conductive contact (Fig. 21(c)).[13] Meanwhile, by adjusting the layers of transplanted thin-films, the initial efficiency has been doubled. Even though all these works have been done by using rigid FTO substrates, great potentials do exist when considering flexible applications as the free-standing nanostructure films can be easily transferred onto any substrate using this simple method. Relating to this approach, another urgent issue is to find out an effective way to finish the transplanting process, that is, to find a suitable "glue" to form good attachment between active materials and plastic electrodes. The solution to such problem can be referred back to the low-temperature crystallizing techniques as mentioned before, because all those pastes in such processes can act as glue to adhere TiO_2 nano-thin-film on substrates. Interestingly, flexible DSSC photoanodes could be fabricated by an inverted method to transfer the already dye-sensitized free-standing TiO_2 nanowires/nanoparticles thin-film onto flexible substrates like polydimethylsiloxane (PDMS) (Fig. 21(d)). Conductive layer (ITO) was formed directly on materials before the transfer process.[98] This approach has offered a perfect alternative method to transfer thin-film for optoelectronic devices to avoid the problem of using glue. Advantages of using the large-scale, free-standing nanostructure thin-film for device application are obvious. For instance, the anode materials are already of high crystallization, fabrication procedure is easy to operate, so the final performance of devices will be largely determined by the quality of the thin-films themselves, namely, wakening other interference factors for performance. Also, the flexibility requirement should be satisfied, which means the free-standing thin-film materials should be carefully tested before device fabrication to see if their mechanical properties can match well with the flexible substrates. Since the transferring methods for fabricating solar cells is emerging very fast recently, we believe it could be, together with printing electronics, promising for developing the next generation flexible DSSCs.

3.1.1.3. *Flat Flexible Photoanodes Based on ZnO Nanostructures*

Admittedly, the research of DSSCs is dominated by TiO_2 since it is the most widely investigated material in this field. The DSSC is a complicated system and nearly all other components like dyes and electrolyte currently match TiO_2 better than other kinds of wide-bandgap materials. It may make us easily ignore the potential of other materials. But still, we have reasons to believe there would be some other more suitable candidates when applying for DSSC photoanodes. ZnO is considered as quite a promising material for DSSC applications and many stimulating progresses have been achieved.[99–105] Especially for the application of ZnO nanowire arrays, the highest efficiency of 7% was achieved by using interesting techniques to grow ZnO arrays with length of 40 μm on substrate by Xu *et al.*[106] The relatively simple material depositing methods led to high efficiency of the fabricated devices, making zinc oxide very promising for flexible PV applications. Similarly, ZnO nanoparticles can also be made into paste by printing or coating process to fabricate DSSCs.[65] However, the performance of flexible porous nanoparticulate ZnO based DSSCs can hardly compare to that of TiO_2 DSSC since TiO_2 photoanodes have functioned within a relatively optimized electrochemical system. Then it was the intensive investigations relating to growing 1D ZnO nanostructure arrays on flexible substrates through a low temperature and mild condition that started to reveal the superiorities of ZnO solar cells. Generally, ZnO nanowire/nanorod arrays can be obtained by a two-step method: firstly, a "seeds" layer was deposited onto substrate by spin-coating, and then nanowire arrays can be formed through a low temperature (around 120–150°C) hydrothermal method. Therefore, it can be easily deposited onto flexible substrate like paper,[107] metal[108] or plastics.[109]

The advantage of using ZnO nanowire arrays as photoanode materials is apparent when comparing to the compact or porous thin-film consisting of nanoparticles. As shown in Fig. 22(a), the bending stress in a mesoporous network has to be released through network deformation, leading to cracks or peeling, while orderly aligned nanowires can efficiently release the bending stress in the film through the gap between nanowires.[109] To test the mechanical bendability of the flexible nanowire

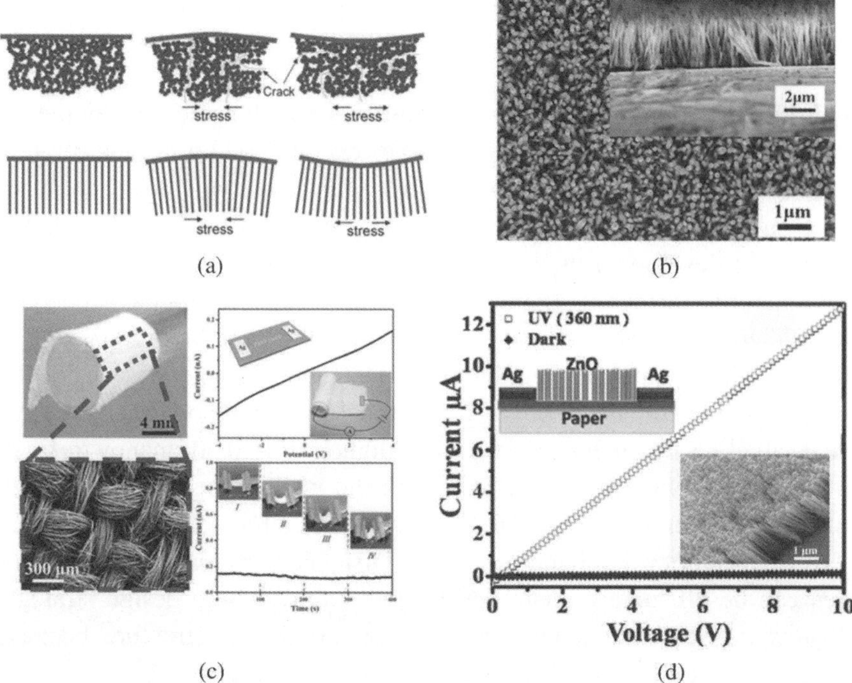

Figure 22.   ZnO nanostructures for flexible photoanodes of DSSC. (a) Schematic shows of the bending stress release in nanocrystalline films of nanowire array and mesoporous network. Reprinted with permission from Ref. 109. (Copyright: 2008 American Institute of Physics.) (b) SEM image of ZnO nanowire arrays grown on ITO/PEN. Reprinted with permission from Ref. 110. (Copyright: 2012 Elsevier.) (c) Digital photograph and SEM image of flexible ZnO nanoparticles assembled cloth and its application for bendable photodetectors. Reprinted with permission from Ref. 115. (Copyright: The Royal Society of Chemistry 2012.) (d) UV detector performance of ZnO nanorod arrays on paper and the schematic and SEM images of the device are shown as insets. Reprinted with permission from Ref. 107. (Copyright: 2010 Wiley-VCH.)

substrates, the film was manually bended to an extreme bending radius of 2 mm for several cycles and was further bended 2,000 cycles with bending radius of 5 mm using a bending machine. After bending, there are no visible cracks or signs of peeling off in the film, while after several cycles of bending test with a bending radius of 10 mm, visible cracks and peeling off could be clearly seen in the porous ZnO film. The results confirmed the superiority of 1D nanoarrays when used for flexible devices.

Figure 22(b) shows a typical example of growing ZnO nanowire arrays on conductive plastic substrate.[110] Obviously, carefully adjusting the width of the gaps is important to avoid cracks. But the conventional hydrothermal processes are always full of unpredictable developments, which makes it hard to find the suitable reaction condition to accurately control the distance between nanowires. Herein we would like to introduce an interesting work done by Kim *et al.* who brought up a novel concept of developing the periodically aligned single crystalline vertical ZnO nanorods arrays.[111] Using this structure design technique, stress can be efficiently released and hence good bendability can be preserved. The highly ordered nanorod bundles with designed distance were achieved using a simple zinc metal pillar arrays transferring method. Then after the weak alkaline treatment, periodically aligned zinc oxide nanowires can eventually be obtained. The DSSC with the well-aligned ZnO NR array released the bending strain efficiently, and therefore maintained device performance after 500 bending cycles. Distance between individual bundles can be effectively controlled by this nanostructure design strategy. However, a well-designed distance is necessary for this structure, because as we know the enlarged distance can reduce the possibility of cracks while the dye-loading ability may be negatively affected (lower density of ZnO). Solar cell efficiency can be further improved by means of using compact layer to enhance the contact and charge collection,[112] introducing hierarchy construction such as adding nanoparticles to the system,[109,110] or using other materials like TiO_2 and SnO_2 to make a composite anode structure.[113,114] Except for 1D nanostructure, many other interesting zinc oxide morphologies or architectures have been obtained and used for various electronic applications. It may also give great opportunities to have a thorough understanding of the influence caused by different ZnO nanostructures on the overall performance of flexible PV devices in future research. For example, we have synthesized a ZnO-nanoparticle-assembled cloth with a simple template removing method and used it for flexible photodetectors and recyclable photocatalysts (Fig. 22(e)).[115] Just as discussed before, this type of flexible self-standing nanostructured thin-film also contains great potential in applying for flexible DSSCs as it can be easily attached onto plastic substrates via a simple transfer method.[12,13] After 20, 60, 100 cycles of severe bending, the electrical property has been perfectly preserved, proving the excellent flexibily of this material. Also, in another work, the possibility of growing ZnO nanorods arrays on paper substrates for optoelectronic applications has been confirmed (Fig. 22(f)),[107] indicating a great potential to successfully develop novel applications like paper-shaped DSSCs, provided more highly conductive and flexible paper-like substrates were invented in the future.

Except for ZnO and TiO_2, other n-type materials like SnO_2,[116–118] WO_3,[119,120] or p-type materials such as NiO[121,122] and $CuAlO_2$[123–125] have been investigated in terms of their potentials in applying for photoanodes of DSSCs. However, there is still a long way to go in order to use these materials for practical PV applications. Since their efficiency still remains quite low compared to TiO_2 based devices, reports about nanomaterials based flexible DSSC apart from TiO_2 and ZnO are very few. Mostly they are playing important roles in effectively assisting the mainstream materials like TiO_2 and ZnO by surface decoration or making composite in enhancing the optoelectronic performance,[77,83,126] which can be considered as another promising direction to promote the development of current flexible DSSC research.

3.1.1.4. *Design of Nanostructures for Counter-Electrodes of DSSC*

Counter electrode is another very important part within the system of DSSCs. The coated thin-films on its surface function as the catalysts to promote the electron delivering and accepting between electrolyte and the back contact. The quality of counter electrode materials and structures will, to a large extent, affect the final performance of a whole device. Traditionally, platinum has been widely used as active materials for counter electrode in electrochemical applications because of its good catalytic property and high conductivity. In a typical counter electrode fabrication process, thermal decomposition and sputtering are intensively used to form high quality thin-film on rigid TCO glass.[7,127] There will be problems when extending to the flexible electrochemical PV field, such unbearable high temperature (400–500°C) for decomposition of certain chemicals (chloroplatinic acid) and relatively expensive process with poor transparency caused by sputtering. Many efforts have been taken to solve such problems and offer alternative solutions. For example, electrodeposition can help to realize the cost effective aim of depositing Pt onto flexible substrate instead of sputtering and thermal chemical methods.[128,129] Figure 23(a) illustrates an example of electrodepositing platinum clusters onto ITO/PET to form highly transparent counter electrode.[128] Also, thermal treatment can be replaced by alternative chemical methods to reduce chloroplatinic acid.[130]

However, no matter how well we can optimize the depositing process, the expensive platinum we used leads to an unavoidable high cost of fabricating a whole solar cell. Instead, carbon or graphene nanostructures were introduced and tested for its potential in photoelectrochemical cell applications. Surprisingly, it showed competitive performance in many reports and was even superior in specific related areas such as inorganic/organic hybrid DSSCs[131] and flexible DSSC. Similar to the traditional photoanode preparation, the cathodes can also be produced using the carbon nanostructure based paste coating process.[132] Chen *et al.* has reported a heating-free method to prepare activated carbon paste assisted with terpineol and SnO_2 nanoparticles. High efficiency of 6.46% was achieved by depositing this paste onto flexible graphite sheet as a pure carbon counter electrode, which can be compared to that of Pt/FTO cathode based DSSC (6.37%). As a result, this counter electrode structure may open new strategies towards high performance and totally flexible photoelectrochemical applications. Besides, there are some other ways to deposit carbon materials onto flexible substrate. For instance, laser technique can also be used to fulfil this aim.[133] Pulsed laser deposition (PLD) route was used to prepare graphite counter electrodes. PLD is a promising low-cost method for the preparation of thin-films on a wide range of substrates including plastics. Figure 23(b) shows the optical image of the laser deposited graphite film, from which we can notice, the transparency of the substrate was still preserved and the color indicates the uniform distribution of graphite. Also, an interesting concept of using the dispersed plasma modified freestanding flexible single-wall carbon nanotube (SWNT) films (bulk paper) to fabricate the counter electrode was introduced by Roy *et al.*[134] This material could be instrumental in developing Pt-free DSSCs, especially in cases where flexibility is important. Even though in this work, solar cells were made from rigid substrate, it can use the same process to conveniently deposit carbon materials on plastic substrates since

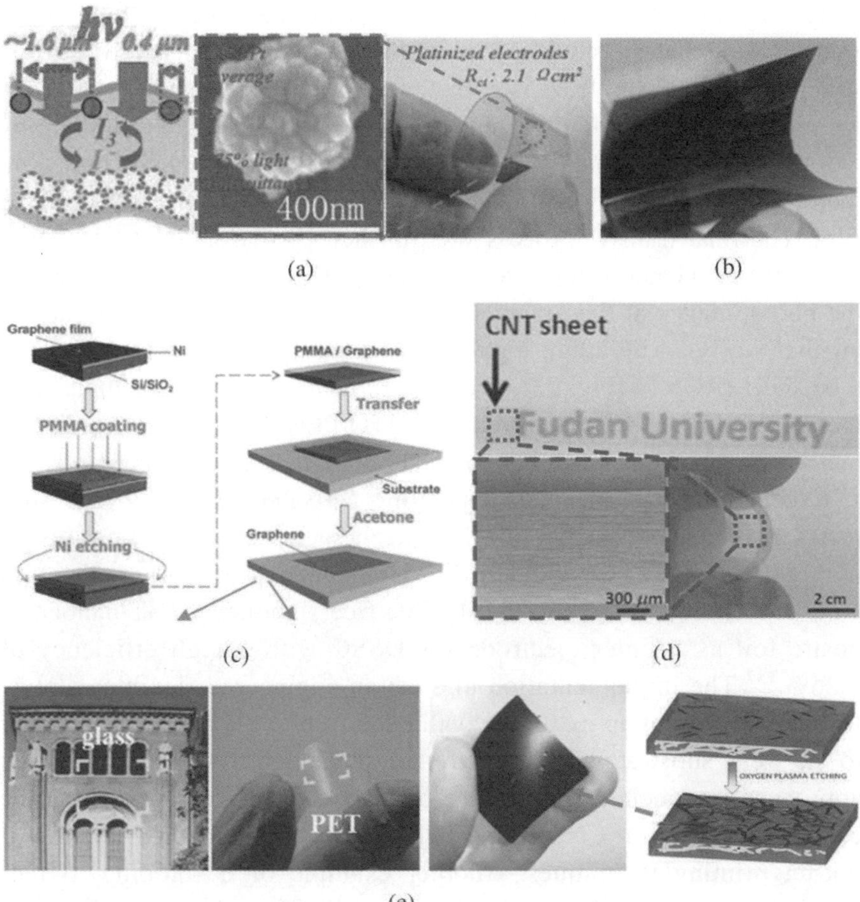

Figure 23. Design nanostructure for counter electrodes of flexible PV applications. (a) Electrodepositing platinum clusters onto ITO/PET to form highly transparent counter electrode. Reprinted with permission from Ref. 128. (Copyright: 2012 American Chemical Society.) (b) The optical image of the laser deposited graphite film. Reprinted with permission from Ref. 133. (Copyright: 2010 American Institute of Physics.) (c) Schematic of the CVD graphene transfer process onto transparent substrates. Photographs show highly transparent graphene films transferred onto glass (left) and PET (right). Reprinted with permission from Ref. 135. (Copyright: 2010 American Chemical Society.) (d) Optical image of a CNT sheet with thickness of 20 nm on a marked paper (upside) and A flexible CNT sheet on ITO/PEN (left); SEM image of a CNT sheet (right). Reprinted with permission from Ref. 136. (Copyright: 2011 Wiley-VCH.) (e) Real picture (left) of a CNT-based monolithic counter electrodes and a schematic representation (right) of etching treatment effect. Reprinted with permission from Ref. 138. (Copyright: 2011 American Chemical Society.)

there were no heat or other extra treatments. Furthermore, the large scale, free-standing materials hold the possibility to be directly used for flexible solar cells with a simple transfer method. For instance, a facile process to form the transferrable transparent graphene thin-film was brought up by De Arco *et al.*[135] including a CVD method to grow graphene thin-film, then a polymer coating and metal etching step to form the transferrable film. The final transfer process was finished by dissolving the polymer with acetone. General steps are presented in Fig. 23(c) which also shows the photo images of the graphene coated glass (left) and PET (right). In another work, an ultrathin, highly transparent and aligned carbon nanotube sheet (ACNT) (Fig. 23(d)) can be formed using a very simple spun method.[136] It can be transferred onto ITO/PEN substrates by a simple method as can be seen from the optical photograph in this figure.

Except for transferring or depositing activated materials onto subjects, free-standing carbon thin-film itself seems to be quite promising in directly using for flexible DSSC. Malara and his co-workers reported a novel technique to obtain the flexible free-standing ACNT/nanocomposite foil as counter electrode for DSSC with a high efficiency of 7.26%.[137] The implementation of such an engineered flexible catalyst foil may represent an extremely valid solution to the problems relating to the weak substrate adhesion strength attributed to the coating-based approaches (especially in the case of plastic substrates) as well as the extreme volatility of CNTs deposited by conventional spray-coating or ink-jet printing techniques. Another example of developing a free-standing composite plate and using it for DSSC cathodes is shown in Fig. 23(e), in which the carbon based composite bulk materials were formed by mixing carbon nanotubes together with polymers, following a double-side etching treatment to further improve the conductivity and catalytic activity.[138]

Using carbon materials as counter electrodes for DSSC has already proved to be quite efficient and can be compared to that made of Pt. In some situations, composites made by mixing platinum particles together with carbon materials showed even better performance.[139,140] Nanostructured conductive polymer was also introduced to the area of flexible counter electrode for DSSC. In the article published by Peng *et al.* conductive polyaniline–polylactic acid composite nanofibers was formed by electrospun method and applied for DSSC cathodes.[141] The photoelectric conversion efficiency of the DSSCs based on such rigid and flexible polymer counter electrodes achieves 5.3% and 3.1% under 1 sun illumination (AM 1.5), respectively, which is close to that of sputtered Pt-based DSSCs. Therefore, it may lead to a prospective development for new generation flexible cathode for DSSCs.

Still, new strategies and alternative materials are intensively being investigated, the new record of efficiency caused by invention of novel counter electrodes nanostructure are being updated almost every minute. Even though this field is predominantly platinum and carbon based, we should have the courage to explore other materials with new nanostructure since some of them begin to show stimulating progresses now.[142,143] A high quality counter electrode for flexible DSSCs must be with high electrochemically catalytic activity, high conductivity, good transparency and of course the superior flexibility and bendability. Designing nanostructure towards that goal is an important direction to developing high quality flexible electrochemical energy conversion devices.

3.1.1.5. *Novel Constructions of Flexible DSSCs*

DSSCs can be considered a great invention in this century not just because it can help to solve severe problems in traditional PVs, it also helps to boost the imagination and innovation in the research field. Since these electrochemical solar cells can be fabricated by simply sandwiching the basic components together, it offers the possibility to form different kinds of device structure by depositing nanostructured active materials onto novel substrates or introducing new concept of assembling the whole cell. Especially, novelty exists more often when designing flexible DSSC.

Wire-shaped or fiber based DSSCs are developing really fast recently and enormous progresses focusing on this area have been reported. Using wire-shaped substrates instead of flat ones is the general strategy to obtain the 1D device macrostructure. High flexibility is obviously a very significant advantage for fiber-shaped devices, which made them easy to be integrated into daily products like clothes and especially promising for military applications. The wire-like solar cells have the possibility to harvest sunlight illumination from any direction no matter whether the substrates are transparent or not. As a result, metals like copper, titanium and stainless steel wires were widely applied as substrates for DSSC research since they have good temperature-resistance and excellent flexibility. In an early, interesting work reported by Zou and co-workers,[144] using the traditional paste coating process TiO_2 thin-film was deposited on the surface of stainless steel fiber, which was then twisted together with a platinum wire functioning as the counter electrode. The general structure is vividly displayed in Fig. 24, through which the general construction can be observed and an energy conversion efficiency of 2% was eventually

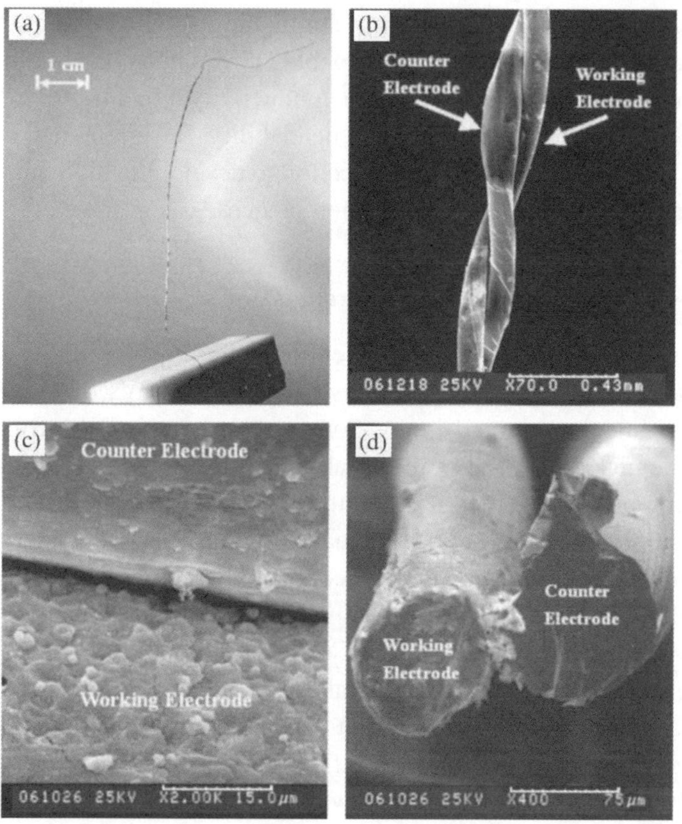

Figure 24.   (a) Optical photo of a twisted WSF-DSSC (Uncut; Radius:~ 0.2 mm); (b) and (c) SEM photo of a WSF-DSSC (top view); (d) SEM photo of a WSF-DSSC (sectional view); Reprinted with permission from Ref. 144. (Copyright: 2008 WILEY-VCH.)

achieved by the optimized device structure. Later, there were efforts taken to replace the liquid electrolytes with the solid ones such as CuI, which effectively helped to solve the sealing problem of the liquid ones since its performance remained stable after 500 hour.[145] However, even though solid state will be one of the main directions for the development of future electronics, the current performance is too poor to be practically used. This field still has quite a long way to go. Finding a facile technique to encapsulate the conventional liquid-state fiber-shaped solar cells must be even more urgent for current research. Lv and co-workers in the same group brought up a novel sealing method to keep the liquid electrolyte based DSSC stable.[146] In a typical sealing step, a glass capillary was introduced to keep electrolyte and electrodes inside, two ends of which were later sealed by wax. This process was frequently referred and further developed by other researchers focusing on wire-shaped solar cells.[147] Further investigation on the promising device structure with appropriate optimization was then carried out, the efficiency of TiO_2 nanoparticle colloids based wire-like dye DSSC alone reached 7.02% and the maximum power output was enhanced by a factor of two and five, respectively, when the fiber was in conjunction with a diffuse reflector or a microlight concentrating groove.[148]

Apart from dip-coating, nanostructures especially 1D constructions grown directly on surface of fibers are considered to form better electric contact and hence lead to a better electron transport property. Titanium wires were intensively used as anode electrode since it specially matches well with TiO_2. Consequently, it is easy to bring up the idea of using anodized Ti wire for fiber-like DSSC photoanode.[149] In Lv *et al.*'s work, the photoelectric conversion efficiency of a completely flexible fiber-shaped DSSC based on TiO_2 nanotube arrays approaches 7% under standard solar simulator. Highly ordered nanorods can also be deposited around conductive fibers. It is well known that ZnO can be easily deposited on TCO substrates so that it is easy for researchers to think of the idea of using fiber-shape TCO substrates to make flexible solar cells. Using ITO coated optical fiber, Weintraub *et al.* has successfully fabricated wire-like DSSC based on ZnO nanorod arrays/optical fiber (Fig. 25(a)) and the device construction is shown in Fig. 25(b),[150] which made a good combination of PV effect and optical waveguide and proposed to open a new route to

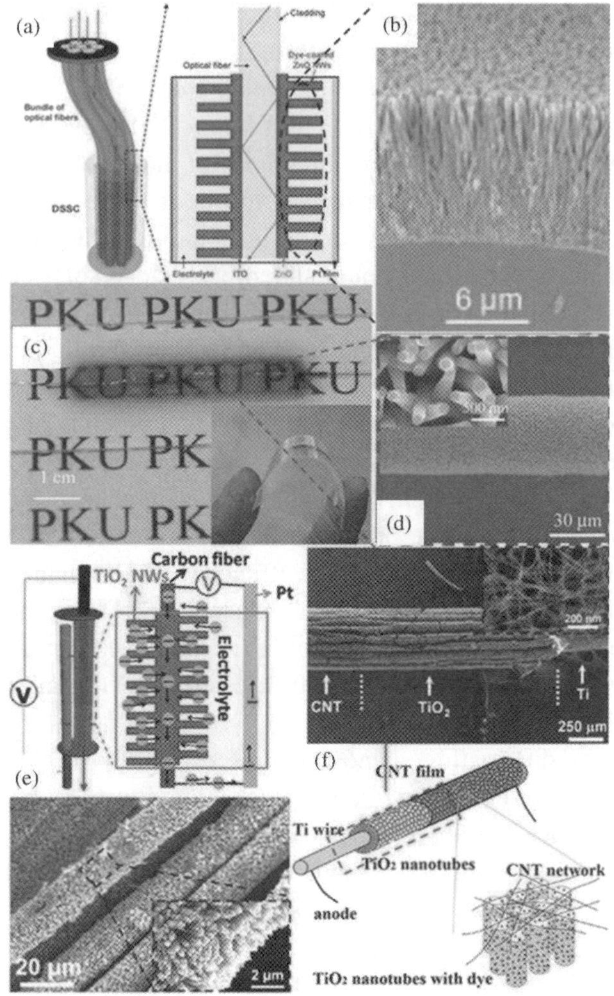

Figure 25.   1D nanostructures grown on metal wires for fiber-shape DSSCs. (a) Structure of the wire-like DSSC based on ZnO nanorod arrays/optical fiber and (b) SEM image of ZnO nanowire arrays on ITO/optical fiber. Reprinted with permission from Ref. 150. (Copyright: 2009 Wiley-VCH.) (c) Digital photograph of ZnO nanowires based wire-shape DSSC and (d) the SEM image of the photoanode. Reprinted with permission from Ref. 108. (Copyright: 2012 Wiley-VCH.) (e) The construction and SEM image of TiO₂ nanorod arrays/ carbon fiber based DSSC. Reprinted with permission from Ref. 147. (Copyright: 2012 American Chemical Society.) (f) The structure of singlewire dye solar cell and its microstructure. Reprinted with permission from Ref. 152. (Copyright: 2011 American Chemical Society.)

develop fiber solar cells for special applications. ZnO has been synthesized on the surface of stainless steel as well. The PV device construction under digital camera and the microstructure of photoanode are illustrated in Figs. 25(c) and 25(d), respectively.[108] In our previous work mentioned in the former chapter, highly ordered TiO_2 nanorods were deposited onto carbon fibers,[12] which may hold the potential to develop the low-cost and highly efficient fiber-shape PV devices. In fact, this idea has been successfully realized later by Guo *et al.* via using the similar TiO_2/carbon fiber structure as photoanode (Fig. 25(e)).[147] A high efficiency of 1.28% was obtained for the first time using carbon fibers as substrates for photoanodes. The concept was even extended to make a full-carbon based fiber DSSC with an efficiency of 2.94% instead of using platinum as counter electrode.[151] The idea of using carbon fiber as both anode and cathode help to significantly enhance the flexibility of devices as well as largely reduce the cost. In fact, carbon materials can be obtained with various nanostructures and hence it can help to promote performance enhancement on one hand and the relatively light weight makes carbon fibers promising in developing portable, bendable and textile optoelectronic applications on the other. Nowadays, the two-wire twisted solar cell structures are developing very well for the advantages including low fabrication cost, easy to handle, etc. However, problems do exist at the mean time. For instance, failing to thoroughly cover the anode substrate will lead to the direct contact between two electrodes, resulting short-circuit accident may cause disasters. Encapsulation is another problem when using two-wire system, which may, to some extent, limit the possibilities of weaving into textiles in the future. Therefore, singlewire DSSC is urgently demanded and research focusing in this direction is highly valued. Figure 25(f) shows an example of how to integrate anode (TiO_2 nanotubes/Ti wire) and cathode (carbon nanotubes) on a single metal wire.[152] A singlewire structure also promises better device stability and flexibility, which is important for integration in textile and fabric electronics.

Fiber-structure PV devices attract great attention today, for which the most important reason is probably the potential to integrate into large-scale flexible textiles by traditional weaving process. Hence it offers possibilities to fulfill the early fairy tales that we can wear clothes woven by solar cell wires. To fulfill that aim, directly using substrates with woven structure can be an alternative choice. Metal meshes have been widely used as substrates for DSSCs since they have great superiorities compared to metal foils like less weight and much high transparency. These advantages certainly can help to provide possibilities to harvest sunlight from both sides which seems to be impossible for the traditional metal foils. Except for the wire-shaped cells, Zou's research group also devoted to mesh-like cells in relatively early time.[153] Even though the efficiency was quite low (1.49%), it offered a new direction towards TCO-free fully transparent DSSCs. From Fig. 26(a) we can vividly notice that the ultralight and highly transparent photoanodes can be obtained based on stainless steel mesh. Similar to the fiber DSSC, TiO_2 nanotubes were obtained

Figure 26. Metal meshes based photoanodes for flexible DSSC. (a) Optical photo of the as-prepared electrode (upper) and the net substrate (lower) and the corresponding SEM images (right). Reprinted with permission from Ref. 153. (Copyright: 2007 American Institute of Physics.) (b) Pictures of TiO_2 nanotube arrays on Ti mesh (50-mesh) fabricated by anodization. Reprinted with permission from Ref. 79. (Copyright: 2009 American Chemical Society.) (c) Optical photographs of the flexible TiO2/Ti working electrode. Reprinted with permission from Ref. 154. (Copyright: 2012 IOP Publishing Ltd.) (d) Structure diagram and the computational formula of light transmittance of the Ti mesh. Reprinted with permission from Ref. 155. (Copyright: 2012 Elsevier.)

on Ti mesh by Liu *et al.* shown in Fig. 26(b) which shows the macroscopic and microscopic images of the photoanodes clearly.[79] Recently, a much higher efficiency of 5.3% was achieved by optimization of the nanostructure of photoanodes and 11 μm TiO_2 nanotubes were proved to have the best performance.[154] Unique advantages of titanium mesh includes, as the photograph demonstrated in Fig. 26(c), availability of front illumination, high surface area, and low sheet resistance, which made it a competitive candidate in future flexible DSSC applications. Figure 26(d) shows the structure diagram of the Ti mesh and the computational formula of light transmittance of the Ti mesh.[155] Where TB is the diagonal line breadth, TL is the diagonal line length of the Ti mesh and W is the width of the Ti mesh infarction. As a result, we can choose the specifically designed Ti mesh to make a better optimization of conductivity, power generation density and more importantly, transmittance of certain devices. In this publication, an effective surface treatment performed by NaOH and HF was brought up and much better performance could be obtained for pt/Ti mesh counter electrodes. Besides, many stimulating progresses have been made by designing various nanostructure on titanium and stainless steel mesh as photoanodes.[156–159] All these efforts have proved the possibilities to make high-quality DSSC by using metal mesh electrode.

Comparing with traditional liquid state solar cells, quasi-solid and all-solid state electrolyte can be more conveniently applied for fabricating flexible devices for its relative stability.[157,160] Figure 27 shows an interesting scheme of using the electrolyte sheet to prepare dye solar cells with

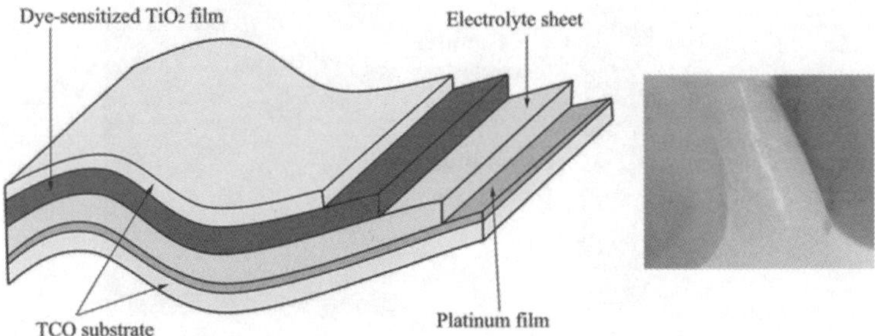

Figure 27.   Scheme of using the electrolyte sheet to prepare dye solar cells and the inset shows the photograph of prepared electrolyte sheet. Reprinted with permission from Ref. 160. (Copyright: 2011 Elsevier Ltd.)

an efficiency of 2.25% and optical digital image is depicted in its inset,[160] which indicates a promising route towards facile solid state devices fabrication. The next generation flexible DSSCs must be full of novelty, highly efficient as well as solid state, which is the general aim for current researchers to fulfil.

3.1.2. Quantum Dots Based Flexible PV Devices

DSSC is considered as the most important success of nano-based PV application. Except for that, quantum dot solar cell is a newly emerging area using quantum confinement to fabricate devices with optimized light harvesting ability. Instead of using organic or rare-metal complex sensitizer, quantum dots are used to synthesize wide-bandgap inorganic nanostructure (TiO_2, ZnO, etc.,) to prepare solar cells. Because of its relatively low efficiency for the rigid applications, reports on flexible quantum dot sensitized solar cells are limited, but still some exciting progress were made in this area. For example, CdS/CdSe quantum dot sensitized solar cells have also been made into fiber-like twisted construction and showed significant PV effect by Meng's group,[161] which is shown in Fig. 28. Mesh based and flat flexible quantum dot sensitized solar cells have also been developed and acceptable performance has been achieved.[162–164] Comparing with other widely used artificial sensitizer, superiorities including low-cost, controllable bandgap width make quantum dots competitive candidates in light capturing for solar cells.

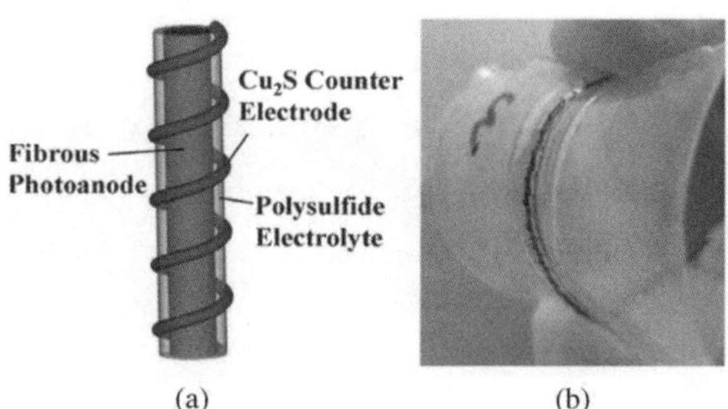

Figure 28.   Configuration (a) and photo (b) of a fibrous QD-SSC. Reprinted with permission from Ref. 161. (Copyright: 2010 IOP Publishing Ltd.)

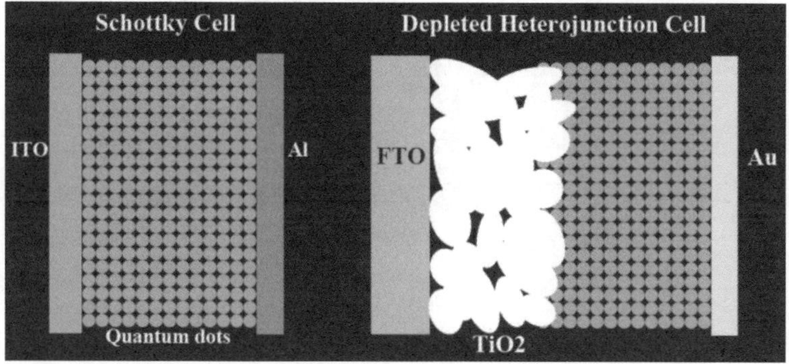

Figure 29.   Architectures of quantum dots based Schottky cell and depleted heterojunction cell.

Apart from electrochemical cells, quantum dots based schottky cells and depleted heterojunction cells do attract great attention. The general construction of the two types of solar cells are shown in Fig. 29. Highest efficiency of over 5% was achieved with these constructions by Sargent's group.[165] However, there are still many fundamental and practical problems existing in this area, and hence most of the reports focus merely on fabricating devices on rigid substrates. Quantum dots are usually CdS, CdSe, PbS, etc., which consists of abundant elements on the earth. It opens vistas for fabricating low-cost PV devices and then modules in the next step. The goal of lowering module costs is being pursued by constructing solar cells on flexible substrates and employing semiconductors and conductors that can be applied for the liquid phase at low temperatures. Flexible, lightweight substrates offer additional advantages such as lowering the solar module weight, which reduces installation costs. The solution-processed quantum dot solar cells can be made through various techniques such as spray painting, reel-to-reel printing and ink-jet printing (Fig. 30), indicating its significant potential in new generation flexible PV applications.[166] Quantum dot solar cells, with great possibility to perfectly match the solar spectra through bandgap optimization, may bring a great revolution to the thin-film PV field, especially for flexible solar cells. Therefore, flexible quantum dot solar cell is believed to become a very attractive research direction.

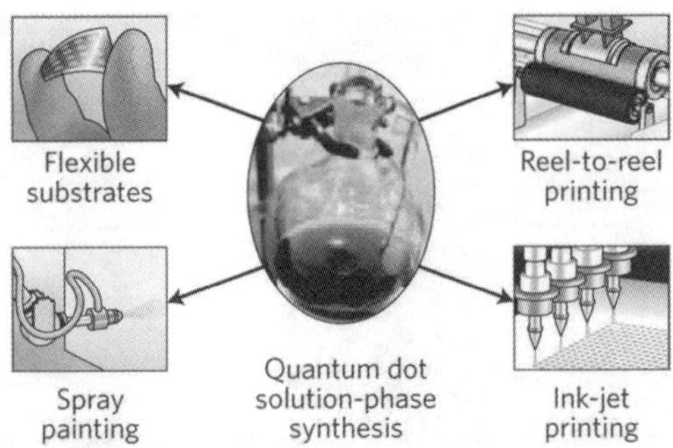

Figure 30.   After synthesis, the colloidal quantum dots are deposited onto a fexible substrate at low temperature. Processing techniques for achieving this include spin-coating, spray-coating, reel-to-reel printing and ink-jet printing. Reprinted with permission from Ref. 166. (Copyright: 2012 Macmillan Publishers Limited.)

3.1.3. Other Flexible PV Technologies

Since the first generation solar cells fulfil the initial fantasy of using sunlight directly for power generation, people begin to think about the possibilities to make this technology more flexible and portable, which was made possible by the emerging of thin-film photovoltaics. When the active materials become a very thin layer (micrometer scale) deposited onto flexible substrates like polymer or metal foils, devices usually can be bent without serious cracks. For example, $Cu(In,Ga)Se_2$ (CIGS) and CdTe thin-film solar cells with efficiency of more than 18% and 13% have been achieved on flexible substrates by Tiwari's group.[1] Fabrication techniques were either thermal evaporation or sputtering according to a layer-by-layer microscale thin-film deposition method. The roll-to-roll method was also introduced to fabricate large-scale flexible solar cells where the active materials and buffer layers were deposited through co-evaporation and chemical bath deposition.[167] The general configuration is shown in Fig. 31(a). The use of roll-to-roll equipment results in a 100 m long and 0.2 m wide solar cell band, as seen in Fig. 31(b). The general roll-to-roll process to fabricate CIGS solar cells is displayed in Fig. 31(c), through which flexible solar cells and other large-scale flexible electronic applications can be easily and efficiently realized. Printing is another effective means for fabricating flexible solar cells as mentioned in the former sections, which will surely simplify the thin-film fabrication process and lower its cost if CdTe or CIGS can be printable.

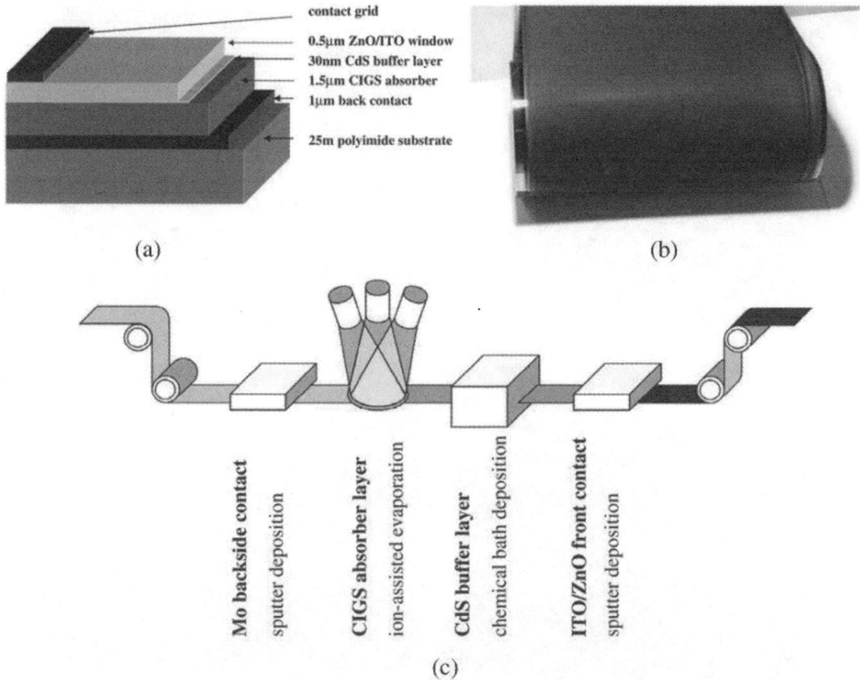

Figure 31. Roll-to-roll methods to fabricate large-scale thin-film solar cells. (a) General configuration of CIGS solar cells. (b) Picture of a large-scale solar band. (c) Schematic drawing of a continuous roll-to-roll fabrication process for the fabrication of flexible thin-film CIGS solar cells. Reprinted with permission from Ref. 167. (Copyright: 2005 Elsevier.)

In fact, a few progresses focusing on this area have already been reported.[168–172] Panthani and co-workers have successfully synthesized CIGS crystal "ink" with controllable particle sizes for printable PVs.[173] The In/Ga ratio in the CIGS nanocrystals could be controlled by varying the In/Ga reactant ratio in the reaction, and the optical properties of the $CuInS_2$ and CIGS nanocrystals correspond to those of the respective bulk materials. Figures 32(a)–32(d) depicts the TEM images of CIGS with varied In/Ga ratio, indicating the high quality and the dispersed nanocrystals can be successfully obtained with solution based chemical method. The photo images of CIGS ink are shown in Figs. 32(e) and 32(f), which vividly described how "ink" can be deposited onto substrate for solar cell applications. Although the devices fabricated through this method exhibited relatively

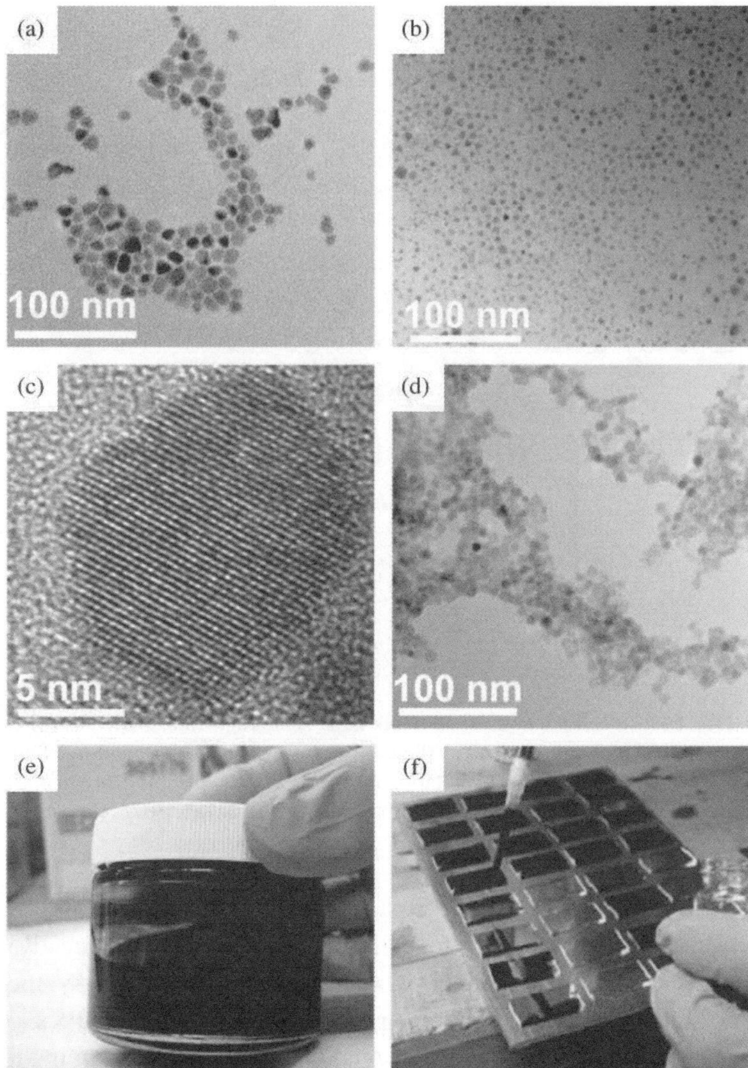

Figure 32.   TEM images of CuIn$_x$Ga$_{1-x}$Se$_2$ nanocrystals with (a) $x = 0.79$, (b) 0.56, (c) 0.21, and (d) 0. (e) Photograph of a CuInSe$_2$ nanocrystal dispersion and (f) the deposition of thin-films on an array of glass substrates. Reprinted with permission from Ref. 173. (Copyright: 2008 American Chemical Society.)

low performance, the idea of nanocrystal "ink" for printing electronics is believed to open a new direction towards future flexible thin-film PVs. Nanostructure can be perfectly designed with a low-cost, solution-based method as photo active materials, and it can be further dispersed into suitable chemicals to prepare "ink". But still, there are many obstacles that make it hard to develop this process. For instance, the solution of the "ink" is hardly to be entirely removed after deposition. Poor contact will be formed between semiconductor and electrodes with printing deposition. Quality of the thin-film is relatively low and cracks and holes are easily formed. In order to effectively solve these identified problems, high quality photo active nanostructures are needed to be synthesized and a convenient route to perform the deposition process is urgently required.

Except for all the inorganic thin-film technologies, organic PV is another competitive candidate in flexible photon-to-electricity conversion field. Similar to DSSCs, organic photovoltaic (OPV) cells have advantages of rich sources, low cost, and easily controlled molecular structures of materials. Moreover, since organic materials are flexible themselves, they can be continuously produced through low-cost thin-film-forming processes such as spin-coating, ink-jet printing, and roll-to-roll etc., and are expected to develop new bendable and foldable PV devices that can bring great convenience in use. In recent years, OPV cells have attracted great academic interests, and also have drawn enormous investments from the industrial communities to realize practical applications. The general configuration of organic solar cells is shown in Fig. 33(a), for which a high efficiency of over 8% has been achieved.[174]

Later, concerning the serious problem of instability for the conventional configuration, inverted polymer solar cells were developed, in which air-stable high-work-function metals (typically, Ag or Au) are used as the anode and ITO is used as the cathode (Fig. 33(b)).[175] In this novel type device, n-type metal oxides such as titanium oxide (TiO_x), zinc oxide (ZnO), cesium carbonate (Cs_2CO_3) and others are used to modify the cathode electrode, leading to the more efficient (as high as $8.4\%^4$) and stable PV performance. Among all the inorganic n-type metal oxide, ZnO nanostructures are the most widely used.[176–178] Traditionally, very thin layer (nanoscale) of ZnO thin-film is formed by spin coating of ZnO precursor solution and then annealing.

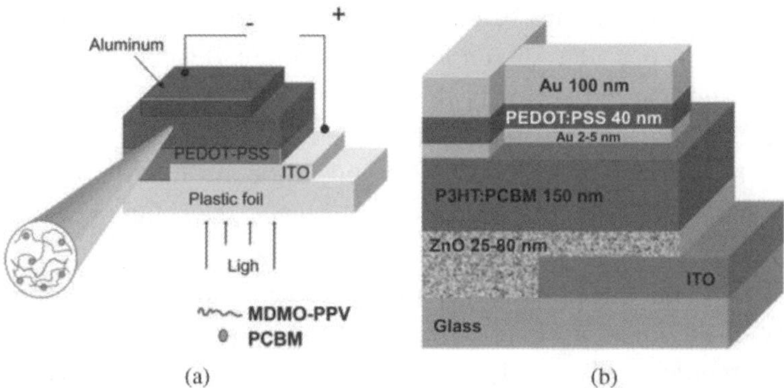

Figure 33   (a) Bulk heterojunction configuration in the organic solar cell. Reproduced with permission from Ref. 174. (Copyright: 2007 American Chemical Society.) (b) An example of inverted organic solar cells. Reproduced with permission from Ref. 175. (Copyright: 2012 Elsevier.)

Besides, different morphologies of nano-ZnO have also been investigated for its application in inorganic/organic hybrid solar cells. For example, ZnO nanoarrays were used for the hybrid structure solar cells and significant optoelectronic response were obtained.[178,179] Liu *et al.* brought up the idea of using ZnO nanorods for the inorganic/organic solar cells to enhance its performance.[180] SEM images of ZnO nanorods are shown in Figs. 34(a) and 34(b). Figure 34(c) clearly displays the general construction of the solar cells. ZnO nanorods can be obtained through a low temperature hydrothermal method, which offers quite a promising future to promote the development of new generation flexible solar cell. ZnO can be further decorated by other materials in order to achieve better performance.[178] CdS was used for this function reported by Wu *et al.* The CdS modification at the optimum condition can dramatically improve all PV parameters and increase the power conversion efficiency of the hybrid solar cells by over 100% due to the cascaded band structure of the heterojunction. Figures 34(d) and 34(e) demonstrate the SEM images of ZnO/CdS nanofibers while Fig. 34(f) shows the general device structure. All works in this area guarantee the significant potential of the inorganic/organic electronic devices which can make a good combination of the advantages of both inorganic and organic materials. To fulfil that goal, carefully designed nanostructures and the whole configuration of devices

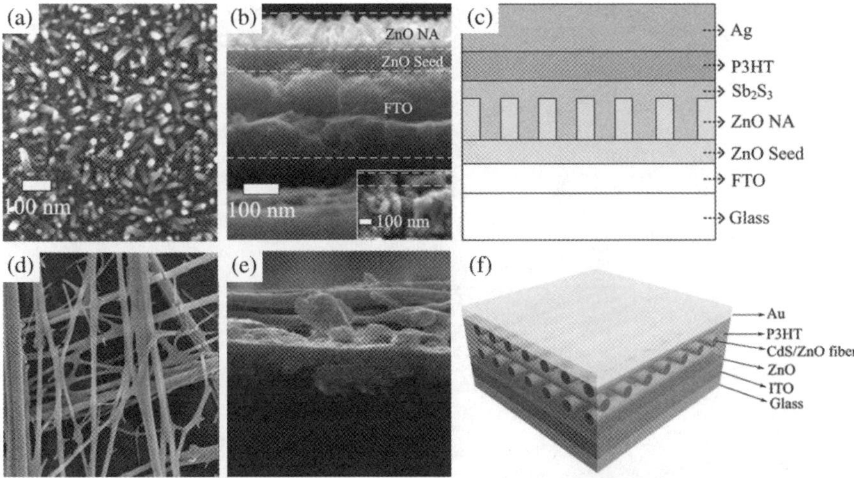

Figure 34. ZnO nanostructures for hybrid solar cells. (a), (b) SEM images of ZnO nanowire arrays on FTO substrates and (c) the general structure of the whole device. Reprinted with permission from Ref. 180. (Copyright: 2012 American Institute of Physics.) (d), (e) SEM images of top view and cross section of ZnO decorated with CdS respectively. (f) Construction of this type of hybrid solar cells. Reprinted with permission from Ref. 178. (Copyright: 2012 Elsevier.)

are important. Comparing to the traditional inorganic or organic solar cells, this new emerging member of photovoltaic family exhibits relatively low performance so that it limits the possibility for extending to flexible PV applications. But still, we believe that with enormous endeavors taken by generations later, hybrid solar cells will become a competitive member in the solar cell field and also play an important role in flexible optoelectronic areas.

As the main active material of OPV, polymer itself is flexible and bendable, which makes the study of flexible organic based solar cells focus on how to find suitable conductive substrates. Similar to the requirements of electrodes materials for DSSCs, high conductivity and transparency are needed when selecting electrodes for OPV. Except for the widely used TCO that requires high cost and rigid process to be fabricated and then deposited onto flexible substrates, metal nanowires have attracted great interests in this field. For instance, silver nanowires have been dispersed onto flexible substrates like plastics through a convenient route and hence

Figure 35.   (a) Ag NW ink in ethanol solvent with concentration of 2.7 mg/mL. (b) Meyer rod coating setup for scalable Ag NW coating on plastic substrate. The PET plastic substrate is put on a flat glass plate and a Meyer rod is pulled over the ink and substrate, which leaves a uniform layer of Ag NW ink with thicknesses ranging from 4 um to 60 um. (c) Finished Ag NW film coating on PET substrate. The Ag NW coating looks uniform over the entire substrate shown in the figure. (d) A SEM image of Ag NW coating shown in panel c. The sheet resistance is 50Ω/o. Reprinted with permission from Ref. 181. (Copyright: 2010 American Chemical Society.)

formed an ultrathin and highly conductive film. The whole process is illustrated in Fig. 35 reported by Cui's group.[181] Increasing the densities of silver nanowires can help to increase the conductivity of thin-film while leading to the decrease of transparency. Therefore, finding the optimal condition is vital to achieve high quality flexible electrodes. Well-designed flexible electrodes must have good mechanical properties, which means that the thin-films are needed to be bendable as well as stretchable. Lee and co-workers have reported a facile transfer method to prepare silver electrode.[182] In their work, highly stretchable and highly conductive metal conductors were prepared by solution filtration and transfer of very long Ag NWs and formation of NW percolation networks on a highly stretchable polymer substrate. The outstanding stretchability and bendability were tested and shown in Fig. 36, where a LED circuit was introduced to

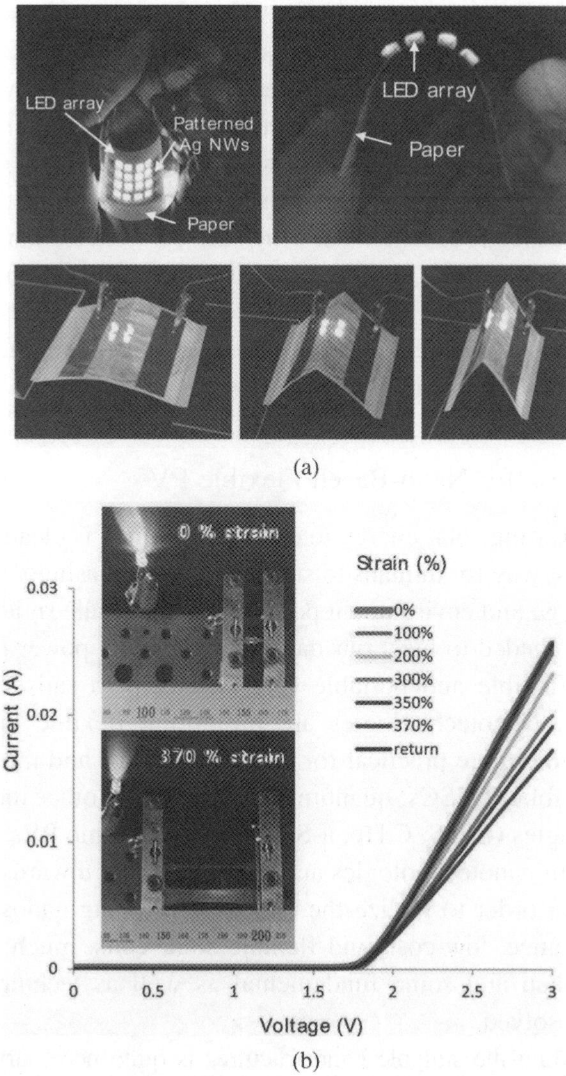

Figure 36. Applications of (a) Images of a flexible paper display LEDs arrays with Ag nanowire electrodes (top pictures) and images of a flexible paper display under various bending radii. (b) Current–voltage measurement of the LED integrated stretchable electrode using the prepared Ag conductor on a highly elastic substrate at various tensile strains (over 370%) and after recovery from its stretching stage (black line). Insets show the digital images of the LED integrated circuits operating at 0 and 370% strain. Reprinted with permission from Ref. 182. (Copyright: 2012 Wiley-VCH.)

help confirm the high quality of such silver electrode and meanwhile, the potential for its future application was also demonstrated. Besides, other highly conductive metal nanostructures like Cu and Au as well as carbon based materials have also been applied for fabricating transparent, and flexible electrodes for energy conversion field.[183,184] Overall, nanotechnologies have offered various strategies to get highly conductive, transparent, and stable electrode materials, which are believed to change the future organic photovoltaics by entirely replacing the current dominant ITO or FTO. Moreover, the fast emerging printable nanoelectrodes may lead a great revolution to the whole flexible energy conversion field, which we are looking forward to witnessing.

3.1.4. Outlook for Nano-Based Flexible PV

Directly converting solar energy to electricity, which is clean and quiet, is surely the best way for humans to survive from increasingly serious fossil energy shortage and environment pollution. Large-scale rigid solar panels are certainly needed to meet our daily demands like power plants. In the mean time, flexible and portable PVs will help to satisfy our special requirements. Nanotechnologies are promoting flexible optoelectronic areas to become more practical for real applications and more approachable to the public. DSSCs, quantum dot solar cells, other inorganic thin-film technologies (CIGS, CdTe, a-Si, etc.,) and organic PVs have already benefited from nanotechnologies and are developing towards more promising ways. In order to realize the aim of developing nanostructures for high-performance, low-cost, and flexible solar cells, much more efforts should be taken and some fundamental as well as technical problems needed to be solved.

Firstly, finding the suitable nanostructures is quite necessary for specific solar cells. Different nanostructures have different superiorities and may be better fitted into different conditions. Having a better understanding of the physical and chemical nature of nanostructures ranging from 0D to 3D and making thorough investigations into the role played by each nanostructure in certain types of solar cells will, to a large extent, help researchers find more easily the optimal condition to make the best devices. Secondly, after obtaining the proper nanomaterials, using a low-cost and room temperature process to form thin-film is vital for future flexible PVs. Development of advanced nano-printing techniques such as ink-jet printing, screen printing and spray printing may help to settle such issues. Further, concerning different situations of application, careful selection of device configuration is important. Unlike planar flexible solar cells that can be suitable to integrate into daily commercial products or roll up to be conveniently carried around, wire or fiber shaped devices have attracted great attention for their novel construction and the possibilities to be woven as texture or even "cloth" to wear. Currently, this idea is still a fantasy that can be only partly realized in lab conditions. Enormous endeavors are needed to be taken to fulfil that. Provided imagination could be further more developed in this field, more interesting and useful device constructions would be brought up. Finally, more effective sealing methods must be tried and applied. Nowadays, encapsulation is a very serious problem for nano-flexible solar cells. Leakage of electrolyte for liquid electrochemical solar cells and instability of organic based PV devices when exposed to air are proposed to be solved with low-cost effective means in the future.

Materials are the very core of all devices. Designing high quality nanostructure for active materials will lead to a booming development of new generation flexible photovoltaics. Based on all the exciting progresses made today, we are hopeful that the wide use of flexible solar cells will make our life different in future.

3.2. *Design of Nanostructures for Other Forms of Flexible Energy Conversion*

3.2.1. Nano-Based Flexible TE Devices

Thermal power plant is the most traditional means of heat-to-power energy conversion that has benefited human beings for centuries. Since we are facing an increasingly serious fossil energy shortage, people have to turn part of their attention towards renewable energy resources like solar energy. Together with the emergence of solar cells which can directly change sunlight into electricity, solar towers and solar furnaces can make use of the sunlight radiation as the form of heat to generate power as the power plants do. But all these technologies require large-scale spaces with abundant sunlight illumination, and in most cases such locations are always quite far from city areas. Therefore, these limitations lead to relatively high cost to generate and transport electricity. Then researchers begin to think: whether it would be possible to develop the small-scale thermal-to-electricity devices like photovoltaic panels that can be accessible to the public? The research of TE devices seems to be able to answer this question.

The TE effects arise because charge carriers in metals and semiconductors are free to move much like gas molecules, while carrying electrons as well as heat. When a temperature gradient is applied to a material, the mobile charge carriers at the hot end tend to diffuse to the cold end. The build-up of charge carriers results in a net charge (negative for electrons, e^-, positive for holes, h^+) at the cold end, producing an electrostatic potential (voltage). Equilibrium is thus reached between the chemical potential for diffusion and the electrostatic repulsion due to the build-up of charge. This property, known as the Seebeck effect, is the basis of thermoelectric power generation.[185] General construction of a TE device is shown in Fig. 37 which consists of many basic units (inset of Fig. 37). TE generators have already reliably supplied power in remote terrestrial locations. And the new generation TE devices are proposed to not only act as a small standalone power plant, but also harness waste heat and provide efficient electricity through co-generation.

The performance of a TE material can be estimated by the figure of merit (ZT): $ZT = \sigma S^2 T/\kappa$, where σ is the electrical conductivity, κ is the thermal conductivity, S is the Seebeck coefficient, and T is the average temperature of the hot and cold sides of the devices $((T_{hot} + T_{cold})/2)$. However, up-to-date, most TE devices are based on rigid yet complicated design, which, together with the high manufacture/installation cost and limited scalability, set a severe barrier for large-scale deployment.[186] Nanomaterials do provide a promising direction in developing high-quality flexible TE applications. For example, TE active nanomaterials can be coated onto the flexible substrates to fabricate the devices. Liang *et al.* first brought up an idea to fabricate flexible fiber-shaped device for high-performance thermoelectric energy harvesting.[187] The microstructure of the

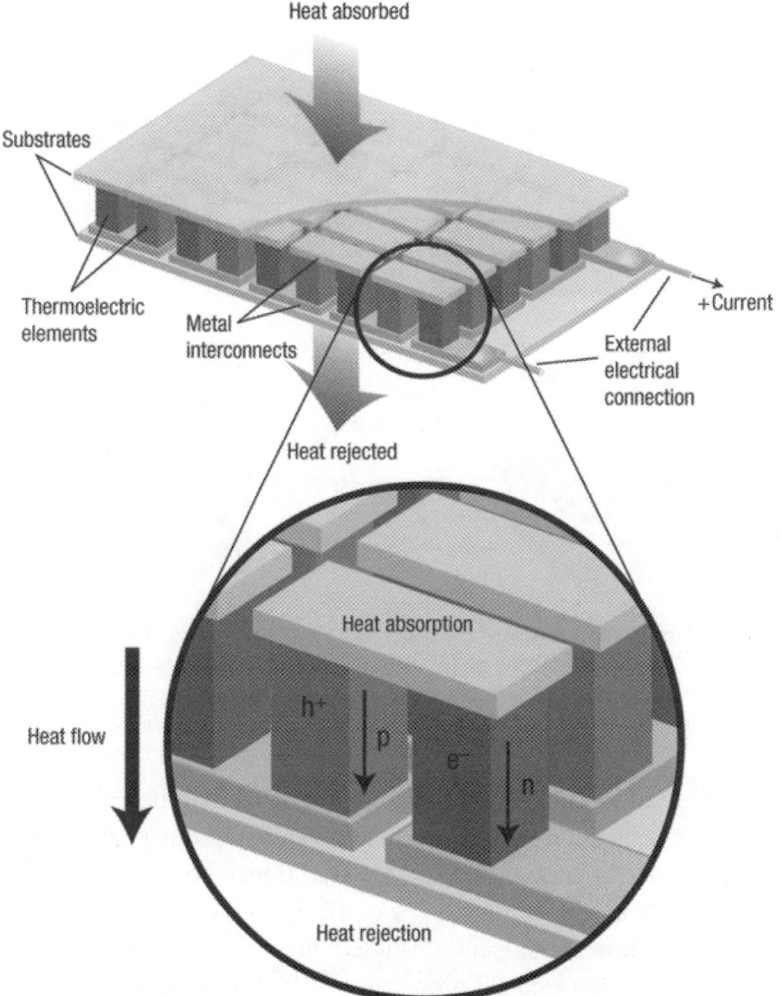

Figure 37.   General configuration of TE power generator. Reprinted with permission from Ref. 185. (Copyright: 2008 Nature Publishing Group.)

PbTe nanocrystal coated glass fiber is shown in Fig. 38(a), and the photograph of bendability testing is illustrated in Figs. 38(b)–38(d), depicting the device configuration for energy conversion application and the generated voltaic through TE effect using such devices, respectively. It shows

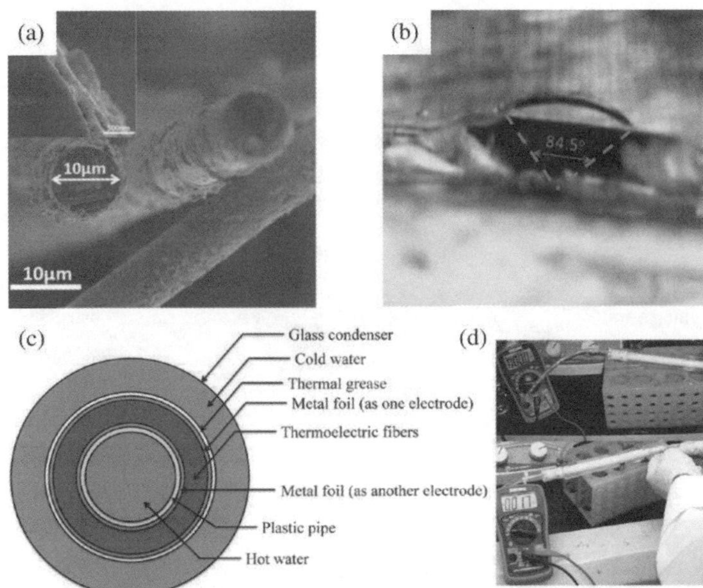

Figure 38.   Fiber shaped flexible TE devices. (a) Morphology of PbTe nanocrystal coated glass fiber. (b) Bendability testing of TE fiber. (c), (d) Demonstration of using flexible TE fibers for harvesting energy from industrial pipes. Reprinted with permission from Ref. 187. (Copyright: 2012 American Chemical Society.)

the flexibility of the fibers and the curvature is 84.5°. The voltage is relatively stable, indicating the great potential for using this type of structure to supply power in special conditions. Traditional TE materials are basically low bandgap semiconductors, e.g., Bi_2Te_3, PbTe, Sb_2Te_3, and Co_4Sb_3 etc.[188,189] Besides, Si and carbon-based nanomaterials do attract great attention for the enhanced TE properties.[190–192] Investigations into flexible TE devices are quite rare, but still the significant potential it holds makes us believe that this novel application will help to improve the present electric energy conversion area.

TE effect is proposed to be an important member to supply power directly together with solar cells, wind turbine, biomass, and other forms of renewable energy applications in the future. TE devices play another role which is even more promising, that is, to utilize the waste heat to generate electricity. For example, when PV panels are operating under sun illumination, a significant part of solar energy goes waste as the form

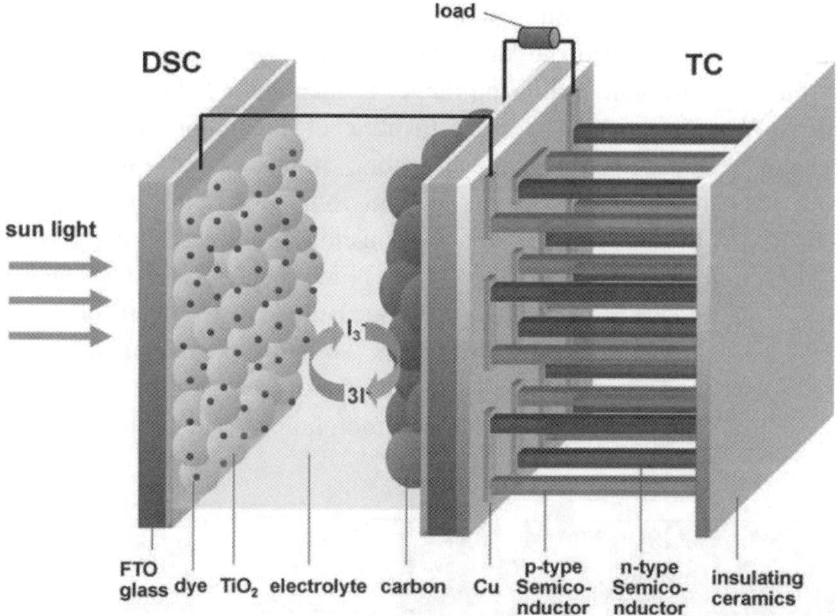

Figure 39.   Schematic structure of a PV and TE hybrid device. Reprinted with permission from Ref. 193. (Copyright: 2010 Elsevier.)

of heat rather than photons absorbed by solar cells to produce electricity, and also the extra heat is the nightmare of solar cells for their efficiency could be negatively affected by this. Using TE devices can effectively solve the problem and increase the whole efficiency of devices. Meng's group has reported a novel hybrid structure that combined a TE device with DSSCs modules (Fig. 39), which shows a significant overall efficiency increase of 10%.[193]

3.2.2. Design of Nanostructures for Flexible Mechanical-to-Electrical Energy Conversion

Ever since the wide range applications of portable electronics like laptop and cell phones, demands for small-scale, flexible power sources are increasing rapidly. PV cells and TE applications that use environmental energy (photon and heat) to generate power can be used to fulfil the aim. Besides, mechanical energy is another abundant energy resource that is always used in industrial-scale to supply electricity based on

254 *Z. Wang, D. Chen & G. Shen*

electromagnetic induction. Lately, since nanoelectronics is developing rapidly, the power needed to drive a micro/nano-system is rather small, in the range of micro- to milli-Watt range. To meet these technological challenges, Wang's group at Georgia Institute of Technology proposed the concept of Piezoelectric Nanogenerators, aiming at harvesting energy from the environment to power the micro/nano-systems based sensor network.[194] In their initial work, ZnO nanowire arrays were grown on α-Al$_2$O$_3$ substrate, whose SEM and TEM images are displayed in Figs. 40(a) and 40(b) respectively. The measurements were performed by AFM using a Si tip coated with Pt film, which has a cone angle of 70°. The tip scanned over the top of the ZnO nanowires, and the tip's height was adjusted according to the surface morphology and local contacting

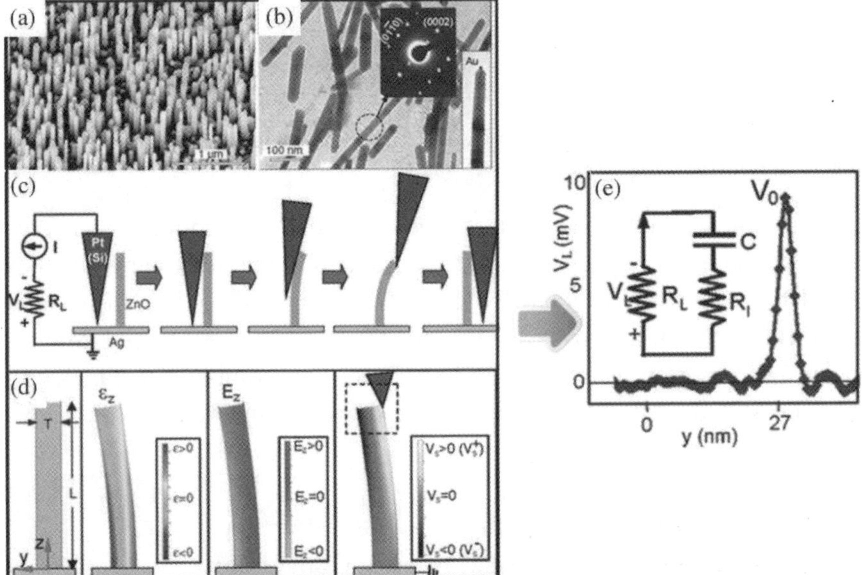

Figure 40. ZnO nanowire arrays based piezoelectric nanogenerator. (a), (b) SEM and TEM images of ZnO nanowires grown on –Al$_2$O$_3$ substrate, respectively. (c) Experimental setup and procedures for generating electricity by deforming a PZ NW with a conductive AFM tip. (d) Stimulation of single ZnO nanorods and the scheme of distribution of strain (ε), electrical field (E), and voltage generated (V_s). (e) Line profile of the voltage output signal when the AFM tip scans across a vertical NW. Reprinted with permission from Ref. 194. (Copyright: 2006 the American Association for the Advancement of Science.)

force (Fig. 40(c)). The distribution of strain, electrical field and proposed generated voltage around a single ZnO nanowire have been simulated and illustrated vividly in Fig. 40(d), revealing the principle of how the nanowire arrays can utilize piezoelectric effect to operate as a nanogenerator. Figure 40(e) shows the testing result by using AFM tip to scan across a vertical nanowire, which indicates a clear strain response. Using a simplified calculation, an efficiency of 17–30% has been received for a cycle of the resonance. It was a relatively high efficiency for energy conversion so that proves the value to continue intensive research on such a novel nanostructure based energy conversion topic.

Later, ultrasonic waves were applied to the nanogenerator and significant response was observed, which is shown in Fig. 41(a).[195] Using a novel zigzag top electrode, ZnO nanowire arrays based nanogenerator was successfully fabricated and sealed. The testing results demonstrate the possibilities for the nanogenerators to be used in practical conditions. Nanowire nanogenerators built on hard substrates were demonstrated for harvesting local mechanical energy produced by high-frequency ultrasonic waves in this work. To harvest the energy from vibration or disturbance originating from footsteps, heartbeats, ambient noise, and air flow, it is important to explore innovative technologies that work at low frequencies and that are based on flexible soft materials. For instance, fiber-shaped nanogenerator has also been developed within the same group.[196] As can be seen from Fig. 41(b), with high flexibility, the Kevlar fiber was orderly deposited with vertical ZnO nanowires around its surface. With a pair of entangled fiber of this type, when applying an external pulling force at a low motor speed, apparent voltage response were observed, which indicates that we are able to generate electricity by collecting the energy of ordinary movements in lab condition. However, it was still theoretically feasible since the output power signal was too tiny to drive any commercial electronic applications. Then in 2010, Xu and co-workers solved that problem by fabricating the integrated large-scale nanogenerators.[197] Figure 41 shows the flexible multiple lateral-nanowire-array integrated nanogenerator consisting of many rows of nanowire-arrays (700 in total). Many sub-units of small nanogenerators formed the high-output flexible one, generating a peak voltage of 1.26 V at a low strain of 0.19%, which is potentially sufficient to charge a standard battery. Further, the

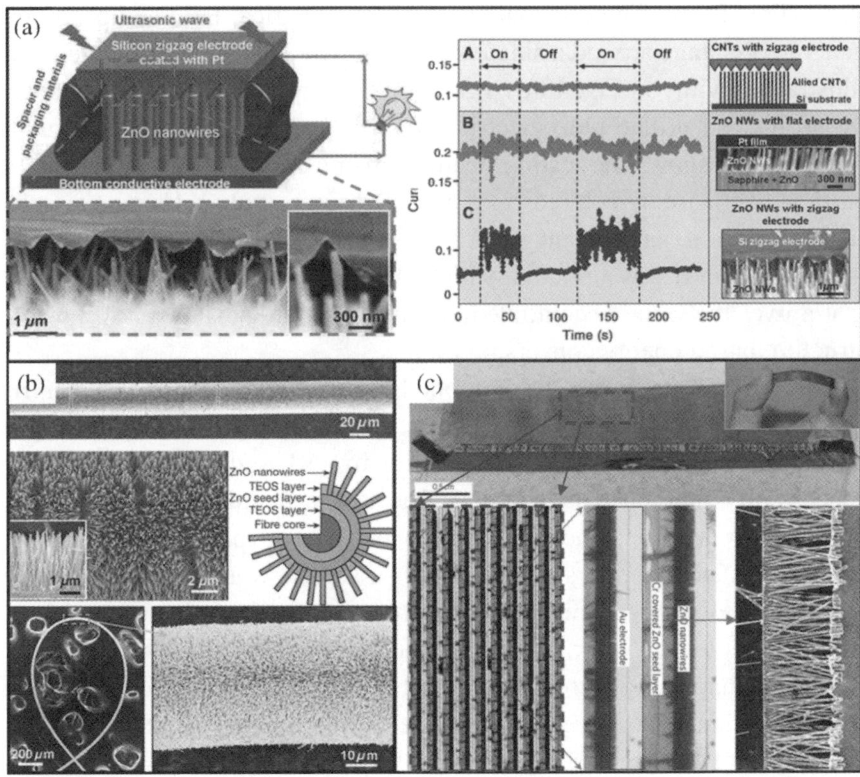

Figure 41. Advanced piezoelectric mechanical-to-electrical energy conversion devices. (a) ZnO nanowire arrays based nanogenerator driven by ultrasonic waves, under the stimulation of ultrasonic wave, the fabricated device starts to output electric signal on the right side. Reprinted with permission from Ref. 195. (Copyright: 2007 the American Association for the Advancement of Science.) (b) SEM images and scheme configuration of ZnO nanowires grown on flexible microfiber for nanogenerator. Reprinted with permission from Ref. 196. (Copyright: 2008 Nature Publishing Group.) (c) Flexible high-output multiple lateral-nanowire-array integrated nanogenerator. Reprinted with permission from Ref. 197. (Copyright: 2010 Macmillan Publishers Limited.)

prepared devices were used to supply power for a nanowire pH sensor and a nanowire UV sensor in their work, thus demonstrating a self-powered system composed entirely of nanowires.

Well-designed nanowires have been playing vital role in flexible mechanical-to-electrical energy conversion. In future, flexible nanogenerator seems to be promising to supply power for small-scale applications in special conditions. Still, there are enormous doubts cast on this novel concept, and when the nanogenerator will stop being just a theory that is only realized in labs, has not been clearly determined. The problems like instability, high-cost, immature fabrication process limit the development of this technology currently. But we still think this fascinating invention may make our lives different in the future due to unremitting efforts paid by generations.

3.3. *Conclusion and Outlook for Flexible Nano-Energy Conversion*

Energy conversion is a wide-range topic since the energy on the earth exists in many different kinds of forms which can convert from one to another. However, research on flexible energy conversion applications is relatively limited. Since the significance of flexible devices is highly valued, persistent efforts should be taken in this field. Doubtless, nanotechnology offers great possibilities to open new directions to fabricate flexible devices via facile and low-cost methods such as printing and dipcoating on the one hand, and on the other, high performance flexible energy conversion applications can be realized due to the superiority of nanostructured materials to the conventional bulk ones. Additionally, devices can also be made into micro/nanoscale based on nanostructures, which makes it more portable and easy to be integrated. As a result, we provide a general summary of different flexible energy conversion applications based on various nanostructures, which we hope could give a better understanding of current technologies in flexible nano-energy field in terms of advantages and flaws and hence promote new thoughts for future development.

Majority of the progress in flexible nano-energy conversion focus on PV cells in this chapter. DSSC is the dominant type of nano-based flexible energy conversion device in research field. Printable electronics have been making this technology increasingly attractive and booming. Designing nanostructure for flexible photoanodes, counter electrodes and different flexible device configurations including planar devices, fibershaped, waveguide enhanced and texture based construction have provided plenty of space for imagination to be fulfilled. Other types of solar cell technologies mentioned above such as quantum dots PVs, compound thin-film technologies (CIGS, CdTe, etc.,) and organic solar cells have been fully or partly realized flexibility for applications using nanostructure design methods. All these guarantee a promising prospective for new generation PV applications.

Besides, other forms of flexible nano-energy conversion areas such as TE and piezoelectric nanogenerator are also introduced. They have attracted great attention for their novelty and significant potential. The flexible TE device can be either used as power generator by itself or integrated with other energy conversion applications (PV cells) to play an enhancing role by absorbing the extra waste energy. Nanowire based piezoelectric effect offers the possibilities to make portable, small-scale and flexible nanogenerators to utilize the mechanical energy produced by ordinary movements like walking, typing or a blowing wind. Overall, nanomaterials have great potential to make fantasy come true. Generally, flexible energy conversion device can be realized through either flexible substrates or flexible active thin-film materials. It is believed more effective nanotechnologies will be brought up in the near future focusing on those two directions so as to offer new strategies toward flexible devices to more efficiently convert energy from one form into another.

4. *Design of Nanostructures for Flexible Energy Storage*

Energy storage strategies are the effective solutions when facing the intermittent energy conversion and power supply process especially for most of the renewable energy resources (solar, wind, tide, etc.). Since energy conversion applications can be made into flexible and portable ones, it is necessary to develop bendable energy storage units to match the power generators. As roll-up and portable electronic field has experienced dramatic development, the need for flexible energy storage seems to be an immediate requirement. Nanomaterials, doubtless, not only help to enhance the general performance of energy storage applications, but also to make the devices more flexible and portable. In this section, we will give a brief review of two main energy storage technologies: LIB and supercapacitors, which are the most widely investigated areas utilizing nanostructures to fulfil the flexibility of high quality devices.

4.1. *Design of Nanostructures for Flexible LIBs*

LIB is considered as one of the most promising applications nowadays for the enormous potential advantages including long cycle life, low self-discharge, high operating voltage, wide temperature window, and no "memory effect".[198] Indeed, LIBs have been widely used currently in our daily products for their provided capacity to deliver the stored chemical energy as the form of electricity with very high energy conversion efficiency and no gaseous exhausts. A typical lithium-ion device consists of a negative electrode or anode (usually carbon materials), a non-aqueous liquid electrolyte, and a positive electrode or cathode (usually formed from layered $LiCoO_2$) as shown in Fig. 42.[26] During the charging process, lithium-ions are de-intercalated from the layered $LiCoO_2$ intercalation host, transported through the electrolyte and then intercalated between the layers of anode materials. Discharging process reverses this process. Electricity is formed in external circuit during the Li^+ ion intercalation/de-intercalation process.

There are significant advantages for introducing nanostructures into LIBs. First, fast electron transport can be led due to the increased rate of lithium-ion insertion and removal. Because the reduced dimension of nanomaterials allows shorter transport distances for lithium-ion within the active materials.

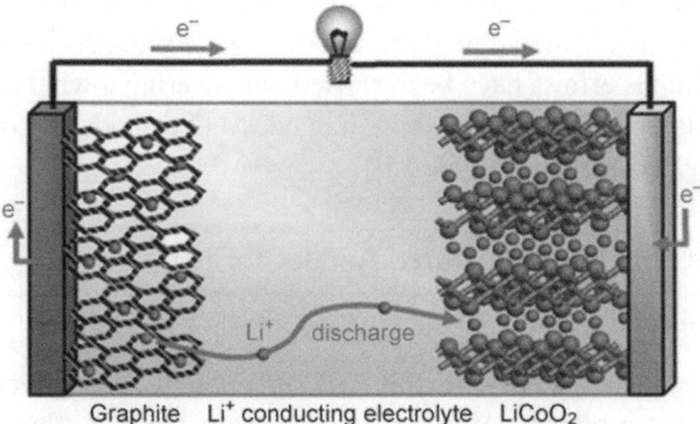

Figure 42. A typical operating schematic diagram of a LIB. Reprinted with permission from Ref. 26. (Copyright: 2008 Wiley-VCH.)

The diffusion time constant (t) is hence reduced, leading to a better energy storage performance, as can be seen from the following formula:

$$t = \frac{L^2}{2D},\qquad(2)$$

where L refers to the diffusion length and D is the diffusion constant. Second, the relatively high specific surface area of nanomaterials can lead to thorough insertion/extraction process, indicating an enhanced reversible specific capacity for nano-based LIB. By tuning the size of active materials within nanoscale, the chemical potentials for lithium- ions and electrons may be modified and hence the resulting electrode potential could be changed. Finally, certain nanostructures like porous thin-films, core/shell or hollow nanospheres will, to a certain extent, help to release the strain and tolerate volume change during the intercalation/de-intercalation process, which will prevent the electrodes from being destroyed during application and hence increase the life time of batteries.[25]

Since flexible LIB is urgently needed, many exciting progresses have been made already. Assisted with nanotechnologies, highly efficient flexible rechargeable battery has been fabricated. Unlike solar cells, transparency is not a necessity for LIB, which means there are more choices of flexible substrates or sealing techniques available. However, there are still problems existing in this topic such as high-cost and low-performance compared to the coin-cells or other rigid substrates based batteries. Stability and safety are also bothering issues which need to be further considered when being applied for practical use. Facing such tough problems, enormous efforts have been carried out covering a wide range from designing the bending construction of anodes and cathodes, to optimizing assembling process for flexible LIB.

4.1.1. Flexible Nanostructured Anodes for LIBs

4.1.1.1. *Flexible Carbon Anodes*

In order to realize the fully flexible LIB, attention has been given to every component of a battery. As the widely investigated part, flexible anode materials have attracted great attention. Among all active anode materials, carbonaceous materials, especially graphite, are the most popularly used for rechargeable LIBs. They can avoid the problem of Li dendrite formation by reversible intercalation of Li into carbon host lattice, which guarantees the good cyclability and long life span for the batteries. Unfortunately, the theoretical capacity for graphite is relatively low because of the specific reaction mechanism (the most Li-enriched compound only has LiC_6). To solve this problem, designing nanostructure for carbon materials is vital to obtain high-performance carbon materials. 1D carbon nanostructures especially nanotubes (CNT) have been widely investigated for their superior performance as anode materials for LIBs,[199–203] through which the specific capacity can be obtained more than 460 mAh/g and reach up to 1,116 mAh/g. Graphene, a novel 2D aromatic monolayer of honeycomb carbon lattice, has attracted great attention as an anode material in rechargeable LIBs in recent years because of its outstanding Li-storage abilities.[204–207] All these exciting progresses offer a great potential for carbon materials to be used for next-generation LIBs.

Traditionally, active materials are made into slurry to coat onto metal current collectors to prepare electrodes, and then sealing into standard coin-cells. The process of making slurry needs introducing organic binder to form a good connection between active materials and the substrates, which will affect the conductivity of the whole electrode and hence lead to a bad performance eventually. What is worse, during the paste making process, the microstructure of nanomaterials are always seriously destroyed for the violent grinding or milling process. It will also cause poor electrochemical properties. Therefore, developing self-standing and binder-free anode materials can be beneficial to the device performance. More importantly, high quality flexible LIB can be more conveniently realized in this way. For example, Liu *et al.* have brought up a novel method to use graphene aerogel to fabricate folded structured graphene paper (Fig. 43(a)).[208] From the photo images we can notice that large-scale, free-standing and flexible anode materials were obtained. It was reported previously that the significant irreversible capacity during the initial charge/discharge cycle is the primary limitation in the performance of graphene paper anodes.[209] The poor reversibility is due to the formation of a solid electrolyte interface (SEI) on the paper surface, which stabilizes the battery during charge and discharge at the expense of battery performance.[210] In the work, the folded structure of graphene paper was supposed to have slightly increased interlayer space and nucleation site, which can control the formation of SEI and lead to the optimized reversibility. The cycling properties under varied changing/discharging rates in Fig. 43(a) demonstrates the great potential for the graphene paper to be used as the flexible anode material for new generation LIB. Since it is generally accepted that the aligned 1D nanostructures have superb electron transport ability so that intensive endeavors have also been taken to develop orderly aligned CNTs for anodes of LIB. The large-scale production of ACNT arrays was first reported by Dai and co-workers,[211] providing an opportunity to develop highly ordered, high-surface-area electrodes with excellent electronic and mechanical properties. Later, Chen *et al.* introduced a novel and facile route to develop free-standing carbon nanotube arrays for application in LIB.[212] The whole fabrication process is illustrated in Fig. 43(b), where the insets show the SEM and optical photo images of the flexible carbon nanotube arrays during fabrication. There are several advantages of such anode materials. First, much higher capacity was observed for this electrode structure, indicating the potential for future high performance LIB. Another significant improvement is that this free-standing ACNT/PEDOT/ PVDF electrode with excellent electronic and mechanical properties does not require a metal substrate, which will significantly lower the weight of flexible batteries. Furthermore, without using copper as charge collector, long-term stability could be guaranteed for avoiding copper dissolution caused by impurities in the electrolyte. Polymer can help to enhance the mechanical strength which is considered as an important factor for flexible devices. Recently, a stretchable, highly porous polymer (PDMS)– carbon nanotube composite electrode with excellent conductivity was reported and used for flexible LIBs (Fig. 43(c)).[213] It is demonstrated that the porous CNT-embedded PDMS nanocomposites are capable of good electrochemical performance with mechanical flexibility, suggesting these novel kind of materials could be outstanding anode candidates for application in flexible LIBs. Similarly, there are many other strategies to synthesize free-standing carbon based battery anodes with large specific area, good electrical conductivity, and electrochemical activity, they altogether offer new thoughts for developing the low-cost, ultralight, and high performance flexible LIBs.

4.1.1.2. *Flexible LIB Anodes Based on Other Nanomaterials*

Except for carbon, nanostructured metal-oxides provide great potential as high performance anode materials for flexible LIBs because of the diversity of their chemical and physical properties which can lead to relatively high reversible capacity for energy storage. Transitional metal oxides (MO_x, where M = Fe, Co, Ni, Cu, Mo, Ni, Cr, Ru) are the most important candidate materials for LIB anodes, which mainly follow the conversion reaction mechanism.[214–216] During the reaction, these oxides are converted

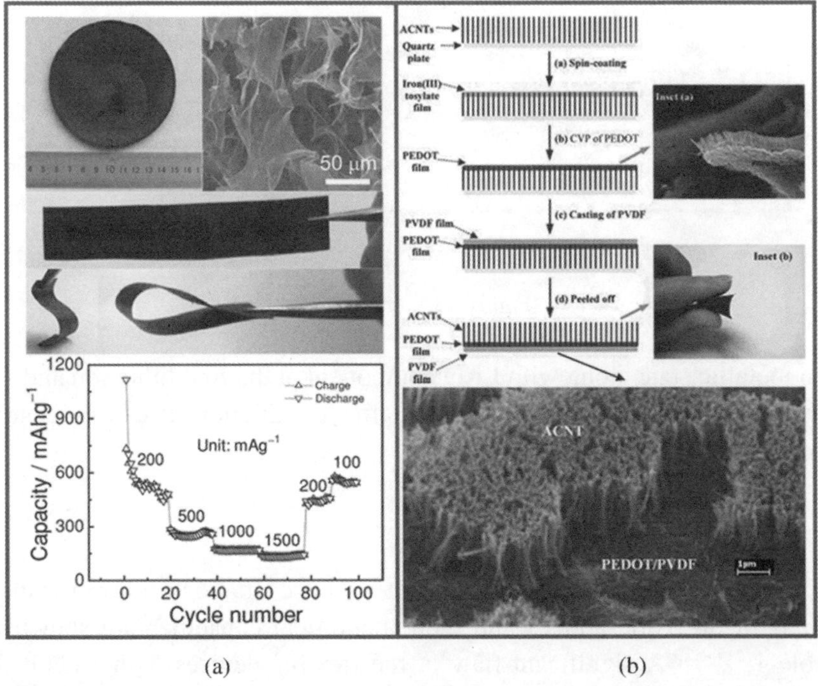

<table>
<tr><td>(a)</td><td>(b)</td></tr>
</table>

Figure 43. Carbon materials based flexible anodes for LIBs. (a) Flexible graphene paper obtained from graphene aerogel as anodes for LIB. Reprinted with permission from Ref. 208. (Copyright: 2012 Wiley-VCH.) (b) Illustration of the procedures for the preparation of a free-Standing and highly conductive ACNT/PEDOT/PVDF membrane electrode for LIB anodes. Reprinted with permission from Ref. 212. (Copyright: 2007 American Chemical Society.) (c) The procedure to fabricate the porous flexible PDMS–CNT nano-composites and its application for LIB. (Insets show the photographs of the flexible anode materials and its application for lighting a LED.) Reprinted with permission from Ref. 213. (Copyright: 2012 Wiley-VCH.)

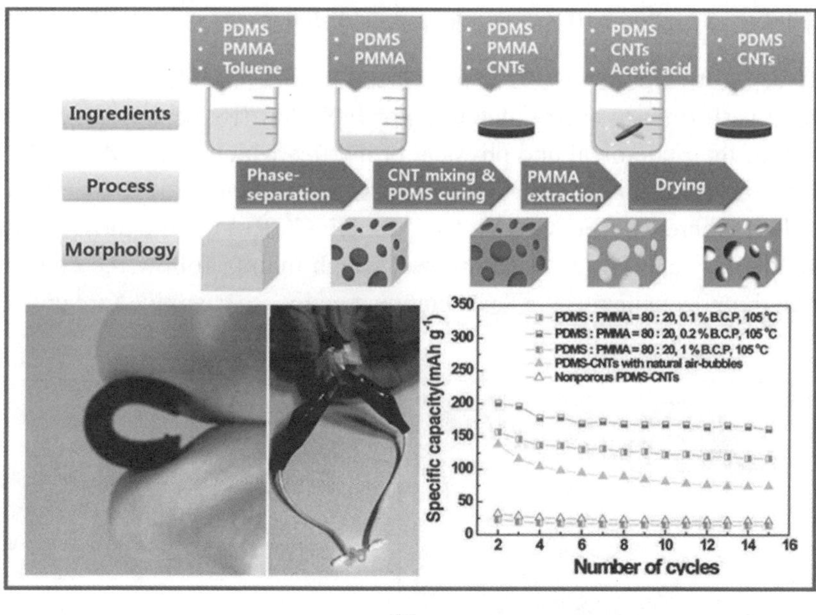

(c)

Figure 43.   (*Continued*)

to a metallic state along with Li_2O component at the first lithiation and are reversibly returned to its initial state after delithiation. It can be clearly explained using formula[198]:

$$M_xO_y + 2yLi^+ + 2ye^- \leftrightarrow xM + yLi_2O, \tag{3}$$

A general list of conversion reaction based nanostructured transitional metal oxide anodes and their theoretical capacities are shown in Table 1.[217–229] A significant flaw of the flexible devices is the relatively poor performance due to the limitation of fabrication process and sealing techniques. Concerning this issue, metal oxides with significantly high theoretical capacities (see from Table 1) compared to carbon materials demonstrate a promising future for the high performance flexible LIB applications. For instance, a novel nanostructure design method using virus as template to assemble Co_3O_4 nanowires for LIB application was first mentioned by Nam *et al.*[218] Negative electrode was prepared by depositing Cu as charge collector on the assembled Co_3O_4 nanowires/

Table 1. Summarization of conversion reaction-based nanostructured transitional metal oxide anodes.

Metal Oxide		Theoretical capacities (mAh/g)
Iron oxides	Fe_2O_3, Fe_3O_4	1,007[217], 926[219]
Cobalt oxides	Co_3O_4, CoO	890[218,220], 715[221]
Manganese oxides	MnO_x	700–1,000[222,223]
Molybdenum oxides	MoO_3, MoO_2	1,111, 830[224]
Copper oxides	CuO, Cu_2O	674[225], 375[226]
Chromium oxides	Cr_2O	1,058[227]
Nickel oxides	NiO	718[228]
Ruthenium oxides	RuO_2	1,130[229]

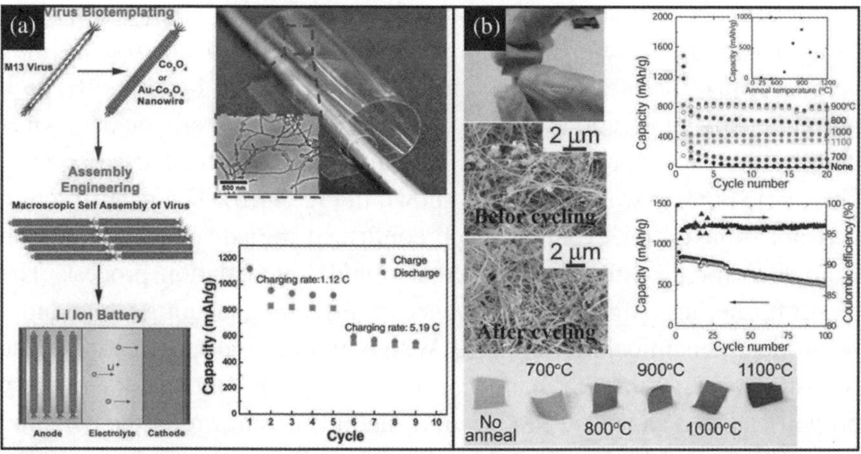

Figure 44. (a) Virus-enabled synthesis and assembly of nanowires for LIB electrodes. Reprinted with permission from Ref. 218. (Copyright: 2006 by the American Association for the Advancement of Science.) (b) Flexible silicon nanowire fabric as a LIB electrode material. Reprinted with permission from Ref. 230. (Copyright: 2011 American Chemical Society.)

polymer layer (shown in photograph from Fig. 44(a)) using E-beam evaporation. The specific capacity can reach up to 750 mAh/g, which was about twice that of current carbon-based negative electrodes. Good rate battery properties can be observed for these type of anodes. They have altogether proved the possibilities for this bio-template method towards high quality, self-standing and flexible anodes for LIB.

Besides, Li can be electrochemically alloyed with a number of metallic and semi-metallic elements in groups IV and V, such as Si, Sn, Ge, Pb, P, As, Sb, and Bi, and also some other metal elements. Silicon is one of the "hot" materials in LIB for its extremely high theoretical specific capacity (4,200 mAh/g).[231–233] However, the volume of silicon anode changes by about 400% during cycling so that Si films and particles tend to pulverize. It hence leads to the poor cycling property. Cui and co-workers brought up an idea to use Si nanowires to solve such problems.[231] Facile strain relaxation in the nanowires allows them to increase in diameter and length without breaking. Additional, electron can be transported more efficiently through the 1D structure, leading to a better performance than other constructions. Silicon nanowires can be further developed for new generation practical energy storage applications especially for the flexible ones. Due to its superb Li storage capacity, designing suitable nanostructure may help to realize high-performance, roll-up and portable batteries for high power density applications. A standalone anode made of silicon nanowires without the need for additional conductive fillers (activated carbon) or polymeric binders was reported as shown in Fig. 44(b). The nanostructure was not ruined after cycling, which confirmed the advantages of silicon nanowires for the lithium-ion intercalation/de-intercalation process. The report further investigated the influence of different annealing temperature on its electrochemical performance. With optimized thermal treatment, the free-standing flexible Si nanowire fabric exhibits outstanding battery properties (high capacity, good cycling stability), confirming the potential for application in high performance flexible LIB.

4.1.1.3. *Carbon Introduced Composites for Flexible LIB Anodes*

In fact, the significance of combining carbon with other materials for flexible anodes are even more valued compared to the single phase material. This approach is proposed to make good use of both the high capacity of metal oxides or semi-metallic materials and high conductivity of carbon to the maximum benefits of performance.[234–238] Since the electrochemical performance could be improved, this approach can also be beneficial to the

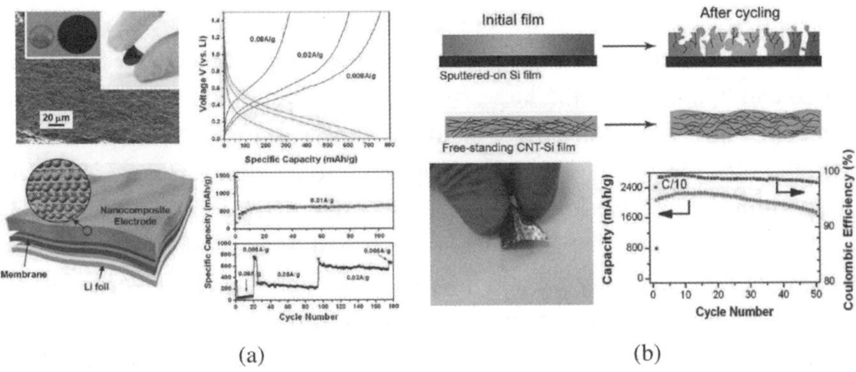

Figure 45. Carbon based composite flexible anodes. (a) SnO_2-graphene nanocomposites for flexible anodes of lithium-ion battery. Reprinted with permission from Ref. 207. (Copyright: 2010 American Chemical Society.) (b) Illustration of enhanced cycling properties of CNT-Si films as flexible anodes for lithium-ion battery. Reprinted with permission from Ref. 245. (Copyright: 2010 American Chemical Society.)

development of flexible anodes for better-performed LIB. Wang *et al.* demonstrated a ternary self-assembly approach using graphene as fundamental building blocks to construct ordered metal oxide-graphene nanocomposites.[207] The self-assembly method can also be used to fabricate free-standing, flexible metal oxide-graphene nanocomposite films for electrodes of LIBs (Fig. 45(a)). The electrochemical testing results indicated that the SnO_2-graphene nanocomposite films can achieve near theoretical specific energy density (780 mAh/g) without significant charge/discharge degradation. The enhanced performance is due to the improved conductivity of the electrode materials by introducing graphene and good electric contact between SnO_2 and conductive graphene during the charge/discharge process. Additionally, as well known that due to the unique microstructure, carbon based materials can be made with good mechanical strength (diamond) as well as strechability (carbon fibers). With the support of carbon materials, good mechanical properties were formed, which hence makes the self-standing thin-film promising for developing flexible batteries. There are also many progresses using carbon materials and silicon to fabricate electrochemical energy storage devices with enhanced performance.[239–244] Flexible free-standing silicon–carbon composites are even more attractive in terms of the enhanced conductivity and cycling properties.[245] Less damage will be caused due to the "ripple up" of CNT-Si films which can help to relax the large strain during lithium ion cycling as illustrated in Fig. 45(b). Such free-standing films successfully integrated the current collector and anode active material into a single sheet of film. The bifunctional films have low sheet resistance due to the infiltrated CNT network and high energy capacity due to the use of Si as anode material. High energy storage capacity (2,000 mAh/g) and good cycling performance which were much superior to pure sputter-on Si film with similar thickness have been demonstrated. All these contribute to the possibilities to use this hybrid flexible anode material for high performance bendable LIBs.

Apart from carbon nanotubes or graphene, in recent research, carbon texture has aroused great interest in flexible LIBs. Because of the advantages of low cost, good flexibility and high conductivity, and stability, carbon texture/cloth is one of the best choices for flexible LIB batteries. In our group, $ZnCo_2O_4$ nanowires were successfully deposited onto carbon cloth to form the unique hierarchical 3D composite architecture.[11]

Figure 46 vividly demonstrates the process of materials grown and the corresponding FESEM images. When applying for LIBs, the novel flexible and self-standing electrodes delivered very high specific capacity (~1,530 mAh/g), excellent cycling stability and extremely high coulombic efficiency. Luo *et al.* have also reported a similar construction of anode materials based on TiO_2 at α-Fe_2O_3 core/shell nanowire arrays on carbon texture and used it for LIBs.[246] The carbon texture materials have been proved to have significant influence on the performance of lithium batteries, because the unique 3D constructions can lead to better contact with the anode materials, large interfacial area for lithium insertion/extraction, and reduced ion diffusion pathways. What is more, the large-scale, free-standing, highly conductive and flexible carbon texture with excellent physical and chemical stability can perfectly meet the requirements of developing scale-up and flexible energy storage devices so as to bring convenience to our future.

4.1.2. Flexible Nanostructured Cathodes for LIBs

Cathode is another very important constituent part for LIBs. And most of the currently commercialized cathode materials are $LiCoO_2$ and $LiFePO_4$.

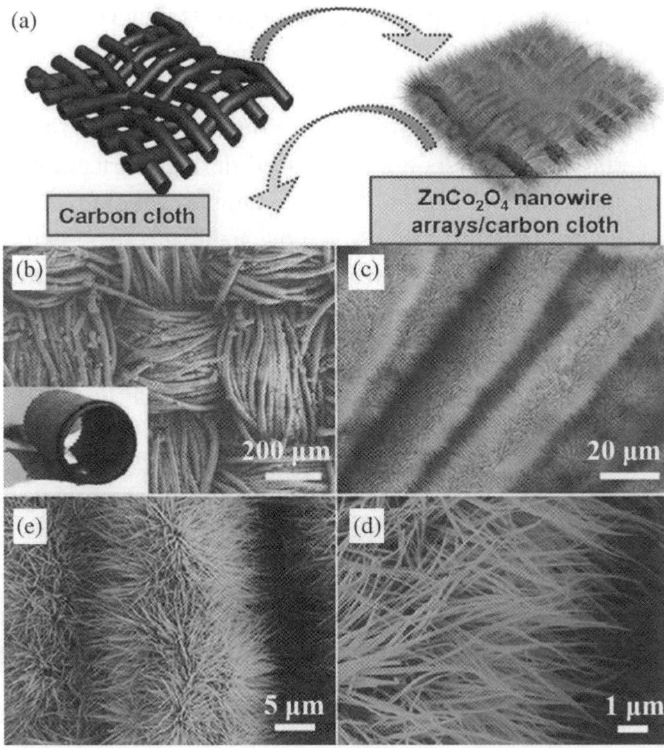

Figure 46. Hierarchical 3D $ZnCo_2O_4$ nanowire arrays on carbon cloth anodes for flexible LIBs. Reprinted with permission from Ref. 11. (Copyright: 2012 American Chemical Society.)

Besides, other materials like $LiMn_2O_4$, $Li(Ni_{1/2}Mn_{1/2})O_2$, and V_2O_5 have attracted great attention in research field which are promising as alternative battery materials. However, due to the relatively complex composition of cathode materials, much less reports are related to flexible nanostructured cathodes for LIBs in research field compared with that of anode materials. And most of the commercial cathodes are prepared through a ball-milling, slurry making and printing or coating process currently. But still, some exciting progresses using the flexible, free-standing cathodes have been reported and attracted great interest, providing possibilities to develop novel types of flexible LIBs so that it can benefit our future to a certain extent.

Complex ternary or multiple compound cathode materials like $LiFePO_4$, $LiCoO_2$, and $LiMn_2O_4$ are more likely to be obtained as the form of nano-particles or bulk materials, which makes it hard to form the self-standing -film with suitable mechanical strength for flexible LIBs. To fulfil that aim, carbon materials or polymers are introduced during fabrication process as skeleton to support the thin-film.[247] For example, $LiFePO_4$ nanoparticles have been obtained by Kim *et al.* using a mechanical activation method that comprised magnetic stirring, rotary evaporation, and ball-milling.[248] The highly flexible and stretchable thin-film cathode was prepared by coating polypyrrole on $LiFePO_4$ through oxidative chemical polymerization of pyrrole, which was then casted onto Al/carbon film to apply for LIB application, and relatively reasonable performance was achieved for such a highly flexible cathode. However, polymer may play a negative role in such process that can affect the performance due to the poor electric conductivity. Instead of introducing polymers, direct synthesis of cathode materials together with carbon can help to effectively solve this problem. Lu and co-workers developed a direct-growth method to make high performance flexible cathodes.[249] Using $LiMn_2O_4$ based flexible cathodes as an example, the $LiMn_2O_4$/CNT composites were firstly synthesized, and then the vacuum filtration of the composites created free-standing cathodes that are binder-free and flexible. There are some unique advantages for the free-standing cathode comparing to the conventional coated one: first, these cathodes consist of networks of CNTs and $LiMn_2O_4$ nanocrystals. The CNT networks provide a flexible, conductive and porous scaffold, facilitating ion and electron transport, and meanwhile the small size of the $LiMn_2O_4$ nanocrystals shortens lithium-ion diffusion length within the crystals. Furthermore, as-formed $LiMn_2O_4$ nanocrystals are threaded through the CNTs, forming intimate and robust interfaces between the nanocrystals and the conductive CNT network. All these features are essential to ensuring good electrode performance and hence this construction can be beneficial to promoting the development of high quality flexible cathodes for LIBs.

As mentioned in the former section, nanowires have superiorities in electron transport and collection, which makes this structure more suitable for electronic devices. Since nanowires can be fabricated into fabric with good mechanical strength and bendability for anode preparation, this interesting nanostructured architecture is also promising in cathode applications. For example, V_2O_5 have been intensively reported in this field since it can be easily obtained with 1D morphology as a newly emerging cathode material.[250–256] Free-standing V_2O_5 electrode for flexible LIBs was obtained by Seng *et al.*[257] In their work, V_2O_5 nanowires were synthesized first via hydrothermal method described by Zhai *et al.* and then the dispersed nanowires were filtered using PVDF.[258] After a simple peeling-off process, the free standing membrane with excellent flexibility was displayed in inset of Fig. 47(a). The corresponding electrochemical testing results were shown in Fig. 47(b), indicating good cycling

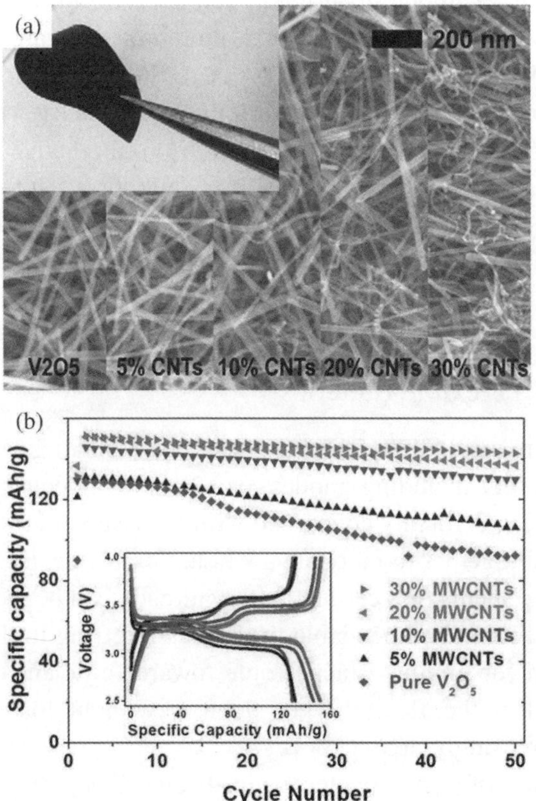

Figure 47. Free-standing nanowire V_2O_5 electrode. (a) SEM images of V_2O_5 nanowires composites with different proportion of CNTs and inset shows the photograph of flexible V_2O_5 cathode. (b) LIB cycling properties testing based on the V_2O_5/carbon flexible thin film. Reprinted with permission from Ref. 257. (Copyright: 2010 Elsevier.)

properties for the flexible nanowire-based cathode material. Additionally, using carbon nanotubes to make a composite can help to further enhance its stability as well as capacity.

To increase the mechanical strength of as-prepared flexible thin-film cathodes, polymer was sometimes introduced during fabrication process.[259] But the overall performance affected by the organic molecules, as identified previously, should be well considered and optimized. Besides, other nanowire materials with high capacity like sulphur based flexible cathodes have been used for novel type LIBs with great potential value for practical applications.[260] However, research on such flexible, large-scale and free-standing cathode materials is relatively insufficient currently in terms of real applications. In this field, new routes toward the traditional materials ($LiFePO_4$, $LiCoO_2$, etc.,) based flexible cathode should be highly developed and novel nanostructures especially 1D architecture should be introduced. Meanwhile, research on alternative nanomaterials for flexible cathode deserves much more attention. High quality standalone and flexible cathode is essential to the whole device and the breakthrough of this component will definitely lead a revolution to the whole field of new generation flexible energy storage.

4.1.3. Toward Flexible Batteries

Nowadays there are a few researches focusing on making each component flexible separately including anodes, cathodes and electrolyte. However, rare reports are specifically concerned with the issue of how to effectively assemble them together to fabricate a whole flexible device. Indeed, since nearly all components of a certain battery already can be well designed to realize free-standing and bendable properties referring to the former sections, the need for finding a facile route towards efficiently sealing them together to fulfil the flexibility of devices without losing capacity is becoming increasingly urgent.

The emerging of printable electronics brings a new thought to flexible nano-energy research as discussed before, including the field of flexible LIBs as well. If all the electrodes, separators, electrolytes and the charge collectors can be designed as paints and applied sequentially to build a complete battery on any arbitrary surface, it would have significant impact

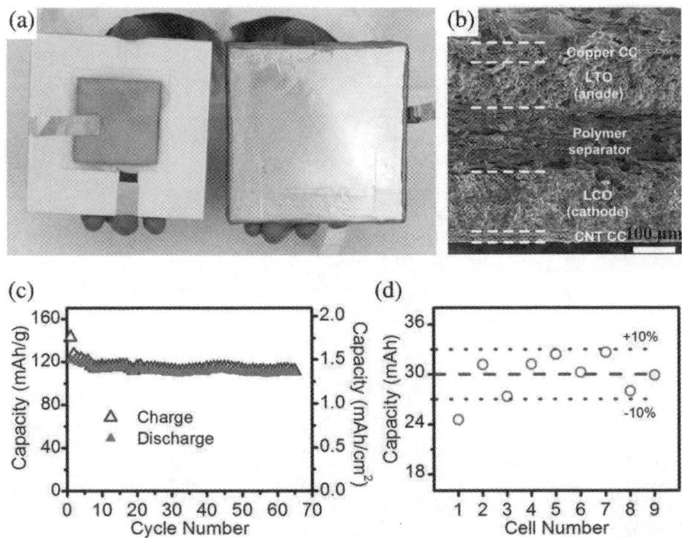

Figure 48. (a) (Left) Glazed ceramic tile with spray painted Li-ion cell (area~535cm^2, capacity~30 mAh) shown before packaging. (Right) Similar cell packaged with laminated PE-Al-PET sheets after electrolyte addition and heat sealing inside glove box. (b) Cross-sectional SEM micrograph of a spray painted full cell showing its multilayered structure. (c) Specific capacity versus cycle numbers for the spray painted full cell. (d) Capacities of eight out of nine cells fall within 10% of the targeted capacity of 30 mAh. Reprinted with permission from Ref. 261.

on the design, implementation and integration of energy storage devices. Singh *et al.* from Rice University have applied the spray printing method for fabricating LIBs on glazed ceramic tile (Fig. 48(a)) and the structure of a device is shown in Fig. 48(b).[261] Good cycling performance can be obtained using this novel fabrication process (Fig. 48(c)). Moreover, capacities of eight out of nine cells fall within 10% of the targeted capacity of 30 mAh (Fig. 48(d)), suggesting good process control over a complex device even with manual spray painting.

This spray printing method can be more convenient to prepare batteries onto transparent and flexible substrates, and can also be useful to make scaled-up and integrated devices, which has been vividly confirmed by the examples shown in Fig. 49. The potential of printable electronics is infinite when used for energy conversion and storage. As a result, much attention should be paid in this field to further develop more suitable routes

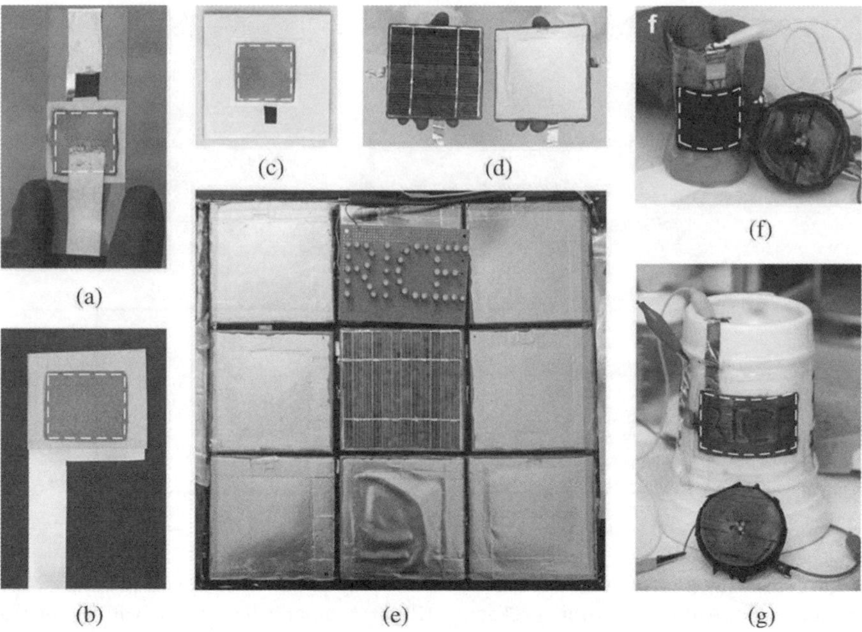

Figure 49. Printing batteries onto different substrates and integration. Examples of printed batteries on (a) glass slide, (b) stainless steel sheet, (c) glazed ceramic tile. (d) (Left) A packaged and charged tile cell and (right) a similar tile cell charged with a PV panel mounted on the tile. (e) Fully charged battery of 9 parallely connected powering 40 red LEDs. (f) A flexible spray-painted Li-ion cell fabricated on a PET transparency sheet, powering LEDs. (g) Spray painted Li-ion cell fabricated on the curved surface of a ceramicmug, powering LEDs. The top electrode (LTO/Cu) was sprayed through a stencilmask to spell 'RICE'. The cell area in a, b, c, f, and g has been highlighted by dashed line for clarity. Reprinted with permission from Ref. 261.

towards preparing printable paste to fabricate each component of a single device. With the highly developed technology of printable batteries in the near future, we can imagine that we may carry around roll-up batteries to supply power wherever we want. We may drive a car with batteries painted on the surface, well-designed battery paintings may be hung inside our houses as part of an electric grid to supply power as well as function as a fancy indoor decoration.

The fast emerging solid-state electrolytes significantly promote the development of flexible electrochemical applications, because the liquid leaking caused sealing problems can be easily avoided. Thanks to polymer engineering and the research on inorganic solid Li-ion conductors, seldom but still some exciting processes have been reported relating to all-solid-state flexible LIBs,[262] meanwhile, unremitting efforts are being taken world-wide to solve all fundamental and practical problems that exist in this field. Based on polymer electrolyte, Amaratunga and co-workers developed well-sealed, all-solid, flexible zinc–carbon batteries with superb performance,[263] which also guarantees possibilities for further extending to rechargeable batteries like LIBs.[264] Flexible batteries could be assembled layer by layer with a sandwich construction using polymer electrolyte (Fig. 50(a)). The cathode was made by depositing manganese oxide/SWCNT mixture onto carbon fiber mat which is also a very good flexible charge collector to support active materials. Strained measurement was performed in Fig. 50(d), indicating the high stability of these type of devices even under severe bending. On the other hand, inorganic solid Li-ion conductors, except for polymeric lithium-ion conductors, are of great interest as potential solid electrolytes in lithium batteries such as perovskite-type lithium lanthanum titanates, NASICON-type, LiSICON- and Thio-LiSICON-type Li-ion conductors, as well as garnet-type Li-ion conducting oxides.[265] A successful example of using a lithium phosphorus oxynitride electrolyte (LiPON) to realize the perfect flexibility shown in Fig. 50(e) was reported by Koo *et al.*[266] They thought of an interesting and simple method to directly transfer LIB from mica to flexible and highly transparent PDMS film. The whole battery was entirely sealed inside PDMS, which guarantees a wonderful mechanical strength and outstanding stability. In fact, PDMS is a very useful polymer material in electronic industry. Using it for lamination of energy devices including solar cells and batteries will definitely help to enhance their possibilities for practical applications. Further, a LED lighting system has been sealed by similar technique based on a thin-film OLED with this flexible battery structure (Fig. 50(h)), where the device novelty has been increased and the practical value has been highly confirmed as well.

Even though the development of all-solid-state devices have attached great importance for the convenience of encapsulation, problems like high cost, instability and reduced performance altogether limit the current application. There is still quite a long way to go so as to commercialize

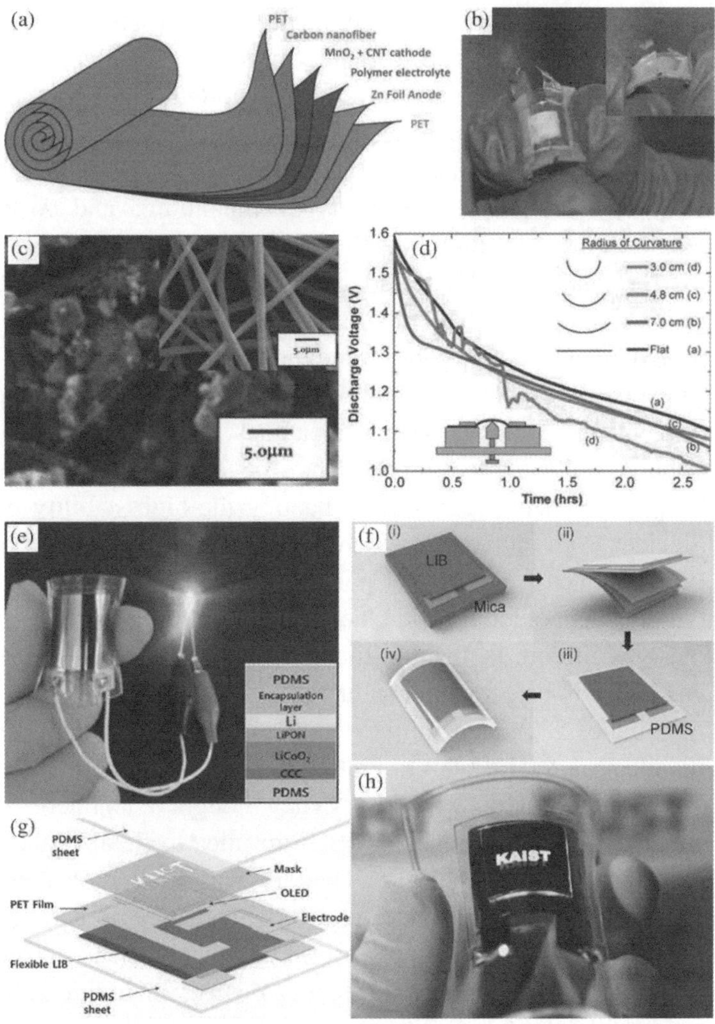

Figure 50. Solid state flexible batteries. (a) Structure of the zinc–carbon battery studied in this work and (b) optical photograph of the zinc–carbon flexible battery. (c) SEM image of manganese oxide + SWCNT mixture pasted onto the carbon fiber mat while the inset shows the morphology of electrospun carbon fiber. (d) Discharge characteristics showing performance of the solid-state battery at different bending strains. Reprinted with permission from Ref. 263. (Copyright: 2010 American Chemical Society.) (e) Photograph of a bendable LIB turning on a blue LED in bent condition. The inset shows the stacked layers in the flexible LIB. (f) Schematic illustration of the process for fabricating flexible LIBs. (g) Schematic diagram of an all-flexible LED system. (h) Picture of an all-in-one flexible LED system integrated with a bendable LIB. Reprinted with permission from Ref. 266. (Copyright: 2012 American Chemical Society.)

the all-solid-state batteries. Since liquid electrolyte has been highly developed for energy storage application with much higher performance, suitable route for effectively sealing all components for the flexible applications is more urgently needed now. Using nanomaterials deposited carbon texture as the anode material, large-scale flexible thin-film battery has been made as shown in Fig. 51(a).[11] Plastic shell was used to carefully

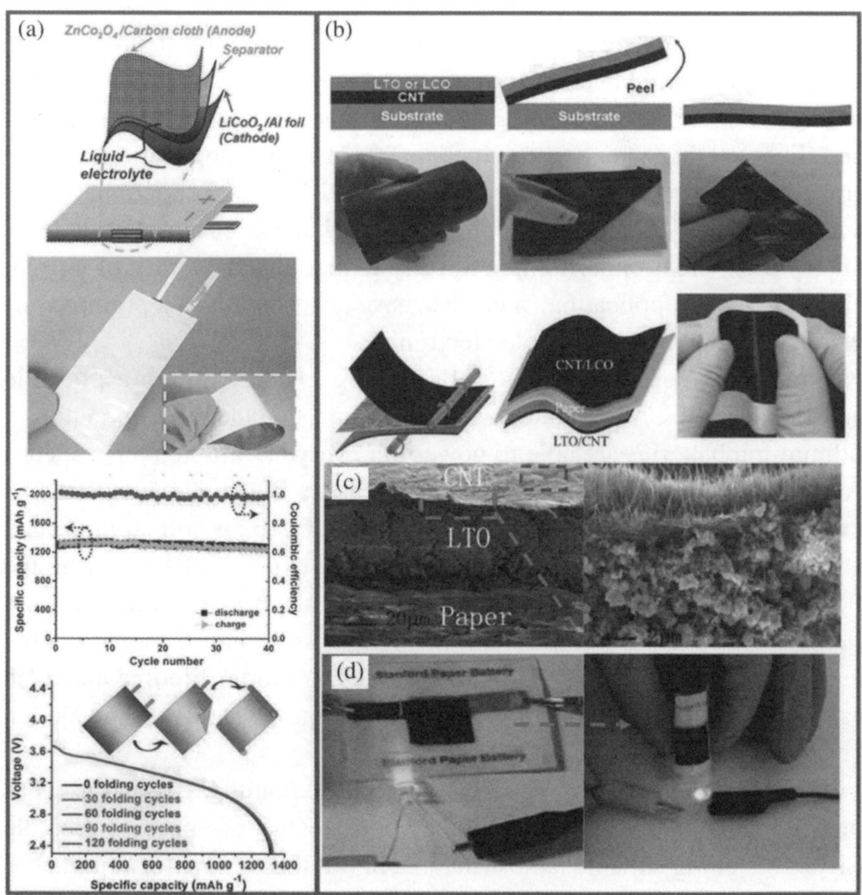

Figure 51. Flexible liquid electrolyte based flexible LIB. (a) ZnCo$_2$O$_4$ nanowire arrays/ carbon Cloth based flexible thin-film LIB. Reprinted with permission from Ref. 11. (Copyright: 2012 American Chemical Society.) (b) Fabrication process of Li-ion batteries. (c) SEM images of construction of the Li-ion paper batteries. (d) Using the paper battery to drive a small LED. Reprinted with permission from Ref. 267. (Copyright: 2010 American Chemical Society.)

encapsulate the devices to prevent liquid electrolyte from leaking outside during application. Wonderful cycling performance and good bendability have been proved for the novel structure of liquid electrolyte based flexible LIBs. The convenient strategy to fabricate high-performance, large-scale and flexible LIBs will surely contribute to the industrialization of thin-film flexible lithium-ion battery in the near future. Novelty can be increased when different substrates are introduced. Cui and co-workers in Stanford University have come up with an interesting idea of using Xerox paper as both substrate and separator (Fig. 51(a)).[267] Its microstructure and examples of application have been depicted in Figs. 51(c) and 51(d), respectively. Such rechargeable energy storage devices are thin, flexible, and lightweight, which are excellent for various applications where embedded power devices are needed, such as RFID tags, functional packaging, and new disposable applications. All these encapsulating techniques offer great opportunities to fully realize the flexibility of current energy storage application, and also have the potential to promote the whole research area of flexible electronics.

Future device applications will develop toward the more portable and flexible ones. With the enormous advantages indentified previously, lithium-ion batteries are able to power our everyday life in many aspects. As a result, flexible LIBs deserve great attention. But still, many problems exist in this topic. For instance, the high-cost of LIBs is still an important issue that needs to be considered, let alone the expense of fabricating the flexible ones. The regular performance cannot always be guaranteed for a flexible application especially during bending or folding. Last but not the least, instability of the flexible LIB will lead to shorter lifetime and even cause safety problems. Such problems can be quite serious and needed to be solved before actual commercialization.

Flexible, free-standing and large-scale anodes, cathodes, different types of electrolyte and encapsulation methods are the essences of flexible LIBs. Designing suitable nanostructures for all these constituent parts has been proved to have significantly positive effect on their performance, which hence opens vistas for fabricating the high-quality, flexible LIBs in the near future by solving the existing problems. The booming nanotechnologies have already changed and will continue promoting the energy storage field as a whole.

4.2. *Design of Nanostructures for Flexible Supercapacitors*

Supercapacitors, or electrochemical capacitors, have emerged since the first patent brought up by Becker in 1957.[268] During more than half century of rapid development, supercapacitors, together with rechargeable batteries, have been attached great importance for the promising future in powering new generation applications. Supercapacitors can be fully charged or discharged in seconds, indicating a much higher power output with a lower energy density compared to rechargeable batteries, which is clearly illustrated in Fig. 52.[269] With recent advances in understanding charge storage mechanisms and the development of advanced nanostructured materials,

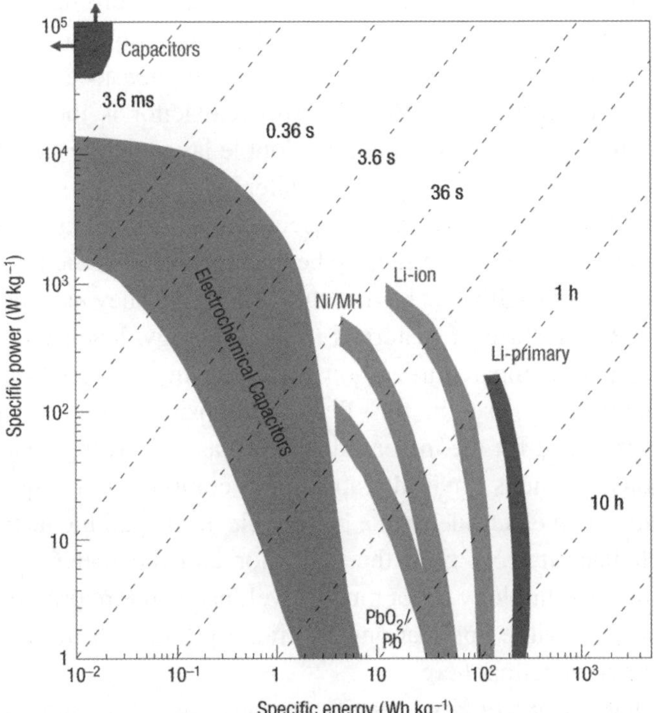

Figure 52. Specific power against specific energy, also called a Ragone plot, for various electrical energy storage devices. Reprinted with permission from Ref. 269. (Copyright: 2008 Macmillan Publishers Limited.)

supercapacitors are playing an increasingly important role in complementing or replacing batteries in the energy storage field, such as for uninterruptible power supplies (back-up supplies used to protect against power disruption) and load-leveling. A very important potential application for the electrochemical capacitors is the electric vehicle which is promising to be a part of national grid system. However, due to space constraints in the vehicle hood, finding extra room for supercapacitors and other necessary components is challenging. Less-weight, ultrathin, and flexible supercapacitors can help to provide effective solutions to such issues, and great amounts of endeavors have been taken in this field.

Generally accepted, supercapacitors store energy using two mechanisms: ion adsorption (electrochemical double layer capacitors) and fast surface redox reactions (pseudo-capacitors). Double layered electrochemical capacitor can electrostatically store the charge according to reversible adsorption/desorption of ions within the electrolyte onto active materials with good electrochemical stability as well as highly accessible specific surface area. Carbon materials based supercapacitor is the typical type device that follows the electrochemical double layer charge storage mechanism, which we will discuss in detail later. Metal oxides such as RuO_2, Fe_3O_4, and MnO_2,[270,271] as well as electronically conducting polymers (polypyrrole, polyaniline, etc.,), have been extensively studied for decades for the pseudo-capacitive behavior they showed. They use fast, redox reactions at the surface of materials to store energy, leading to a higher specific capacity compared to carbon while causing the instability for the redox reaction they use. In this field, composite materials have been widely investigated for the increased stability as well as high energy storage capacity. Besides, hybrid capacitors, combining a capacitive or pseudo-capacitive electrode with a battery electrode, are the latest kind of EC, which benefit from both the capacitor and the battery properties. Since this new technology is not much developed and rare progresses have been made in flexible applications, we mainly focus on the former two types of supercapacitors here.

Due to the extremely large specific surface area, designing nanostructure for supercapacitors is an effective route to enhance its performance. For flexible energy storage applications, nanostructure is especially important because it has the potential to perfectly realize the device flexibility through various fabricating methods without too much performance decay. These nano-design methods also offer opportunities to develop more practical, low-cost and novel constructions of supercapacitors. In this section, we are going to discuss recent progresses about how nanostructured materials have been used for flexible supercapacitors from different aspects, and hence take an outlook for the future development of new generation energy storage applications.

4.2.1. Nanocarbon Materials for Flexible Supercapacitors

As mentioned above, high performance supercapacitors require high physical and chemical stability as well as large specific area for the active materials. Well-designed carbonaceous nanomaterials can meet this demand perfectly. Carbon nanotubes have aroused interests in research field of energy storage applications for so many years as mentioned in the former sections and intensive progresses have been made as well for flexible supercapacitors. Either flexible carbon nanotubes based free-standing electrodes or coating them onto flexible substrates can be conveniently obtained. For example, Niu *et al.*, have brought up a novel method to prepare large-scale, free-standing SWNT films for flexible supercapacitors.[272] Many pieces of separated free-standing carbon nanotube thin-films were connected layer by layer and end-to-end so as to form the large-scale free-standing construction. Using the plastic separator, the devices can be rolled up and sealed conveniently for practical applications. A power density of 197 kW/kg, an energy density of 43.7 Wh/kg and a capacitance of 35 F/g were achieved for this device, which were much higher than the active carbon materials and post-treated single-walled nanotubes. This approach has introduced a novel fabrication process to assemble the supercapacitor in roll-up design using the directly grown SWCNT films as both the electrode and current collector, simplifying the architecture and reducing the weight of supercapacitors, providing new thoughts for further development of large-scale flexible energy devices. Few-walled carbon nanotube has also been prepared using the organic acid assist method.[273] After a simple suction filtration of the dispersed solution, free-standing and flexible thin-film has been obtained with interwoven nanostructure as can be seen from Figs. 53(a) and 53(b).

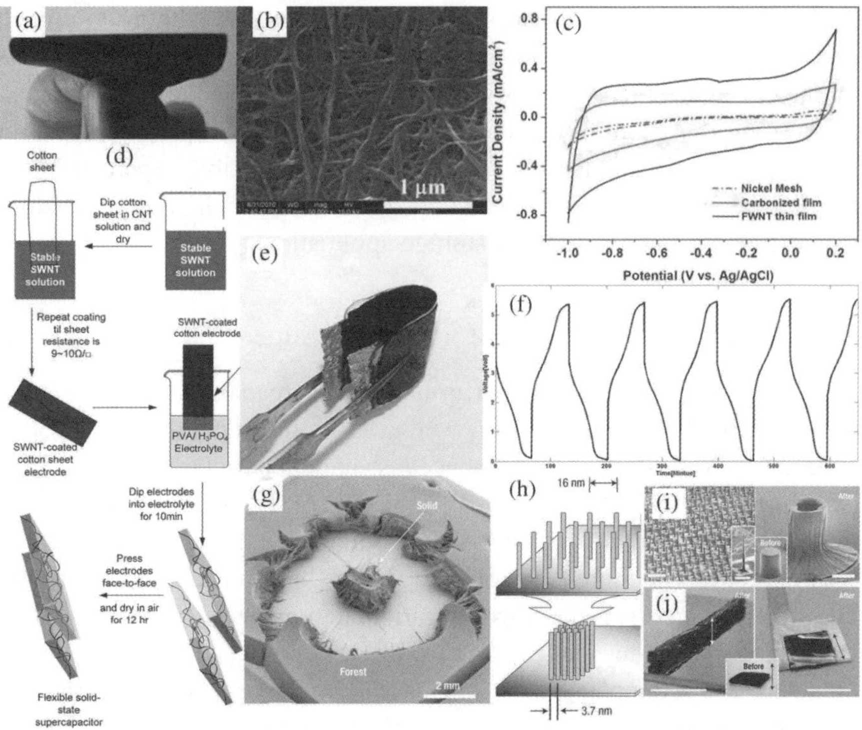

Figure 53. Carbon nanotubes based flexible supercapacitors. (a) Prepared EBA-f-FWNT flexible thin-film and (b) its SEM image. (c) CV testing of the FENT film and Ni mesh. Reprinted with permission from Ref. 273. (Copyright: 2011 American Chemical Society.) (d) Fabrication process of SWNT-based, flexible, and solid-state supercapacitor. (e) Photograph of the bendable supercapacitor. (f) Galvanostat charging-discharging curve at 2 mA. Reprinted with permission from Ref. 274. (Copyright: 2012 American Institute of Physics.) (g) SEM image of SWNT-forest structural collapse from a single drop of liquid. (h) Schematic diagram of the collapse of the aligned low-density as-grown forest (above) to the highly densely packed SWNT solid (below). (I) SEM image of an array of lithographically designed solid needles and microvolcano. Scale bar, 100 μm and 250 μm, respectively. (J) SWNT solid engineered into a rigid 'bar' (left) and flexible SWNT solid engineered into a flattened sheet adhered to a copper sheet (right). Scale bar, 1 cm. Reprinted with permission from Ref. 22. (Copyright: 2006 Nature Publishing Group.)

The composite film exhibits high values of tensile stress and elongation at break approaching 80 MPa, perfect conductivity of over 29,400 S/m and the superior capacitive behaviors nearing 133 F/g for the thin-films (Fig. 53(c)), indicating the great potential for high-performance flexible capacitors with good mechanical strength.

Printing is another very effective route to deposit nanomaterials onto substrates for nano-energy applications. Gruner and co-workers have thoroughly investigated the printable thin-film supercapacitor using SWNTs[14] The work has already covered the typical range of different electrolytes based conventional supercapacitor devices. In combination with the established field of printed electronics, the concept of printed power offers a new platform of all kinds of lightweight devices. We have reasons to be faithful that a completed flexible system including electronic circuit, energy converter and storage component could be built just via printing in the future. Alternatively, for the lightweight and high physical attachment ability of CNT, it can also be conveniently deposited onto flexible porous substrates via a simple dip-coating method to fabricate the whole devices (Fig. 53(d)).[274] Using solid state polymer as electrolyte, flexible devices can be easily encapsulated (Fig. 53(e)) and high specific capacitor and energy density with the value of 115.83 F/g and 48.86 Wh/kg, respectively have been measured in Fig. 53(f). This approach can help to develop nanostructure based materials onto almost any substrate, so it has opened new vistas for promoting the field of low-cost, flexible electronics. Besides, nanostructure design methods can help to increase the performance of supercapacitors such as tuning the morphologies of carbon nanotubes or controlling the loading density of CNTs onto a certain substrate. Researchers in Japan provided a really simple but promising nano-shape engineering method to increase the density of loaded CNTs and hence increased the energy storage capacity per volume.[22] Figures 53(g) and 53(h) displayed the SEM image and scheme respectively to illustrate how to use the zipping effect of liquids to draw tubes together. By controlling the fabrication process, it is possible to fabricate a wide range of solids in numerous shapes and structures (Figs. 53(I) and 53(J)). All these interesting morphologies encouraged us to think about different physical routes toward large specific surface area nanomaterials for high performance electrochemical applications.

Except for the 1D nanotubes, 2D nanostructure carbon materials like graphene have shown great potential for electrochemical applications. Ultrathin layer of graphene (nanoscale) was transferred onto transparent substrate like PET using a facile technique described by Chen *et al.* to be applied for transparent supercapaciors.[275] The importance of transparent energy storage devices is significant. Currently, for a portable electronic device, the display component can be made transparent, integrated

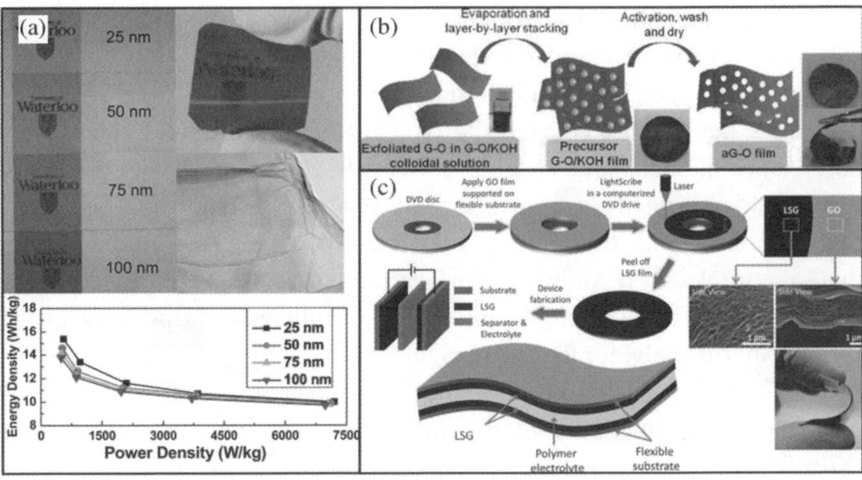

Figure 54.   Graphene based flexible supercapacitors. (a) Thin layer of graphene on PET and the electrochemical capacitor testing. Reprinted with permission from Ref. 275. (Copyright: 2010 American Institute of Physics.) (b) Process to prepare highly conductive and porous activated rGO films for high-power supercapacitors. Reprinted with permission from Ref. 276. (Copyright: 2012 American Chemical Society.) (c) Laser scribing of high-performance and flexible graphene-based electrochemical capacitors. Reprinted with permission from Ref. 279. (Copyright: 2012 the American Association for the Advancement of Science.)

controlling circuit has already been realized transparent, however, since the active materials of power supply equipment like batteries or supercapacitors are always of dark color, fully transparent device is still a great challenge. From Fig. 54(a), we can observe that with varied thickness of graphene thin-film, the transparency has been changed.[276] Surprisingly, with the thinnest layer of graphene, the corresponding supercapacitor exhibits highest performance (specific capacitor of 135 F/g under a current density of 0.75A/g), indicating a great possibility for fully integrating into printable, wearable and transparent electronics. To increase the capacity of a certain supercapacitor, using nano-design method to obtain materials with high BET is the most efficient means. Figure 54(b) clearly demonstrates a way to fulfil this aim. Specifically, at first, an aqueous colloidal suspension was prepared by addition of KOH solution dropwise into graphite oxide (G-O) colloidal suspension; such a colloid contained graphene oxide platelets suspended in the water with KOH dissolved and was heated until form the "ink paste". Films composed of stacked and over-lapped G-O platelets decorated with KOH were obtained through brief vacuum filtration. Next, an activation step was carried out. The final free-standing porous activate reduced G-O films were obtained after washing and drying. Impressively, these flexible carbon thin-films possess a very high specific surface area of 2,400 m^2/g with a high in-plane electrical conductivity of 5,880 S/m, resulting in the highest reported power delivery of 500 kW/kg while maintaining the excellent specific capacitance and energy density of 120 F/g and 26 Wh/kg, respectively. This nanomaterial may have the potential to power future transportation for its significant output ability. Similar nanostructure design method can be referred in order to prepare high quality materials with large specific surface area, which renders a promising direction to develop future field of power sources. Currently, graphene-based materials derived from G-O as just mentioned can be manufactured on the ton scale at low-cost, making them potentially approachable materials for charge storage devices.[277] However, the specific capacitance of such materials (just around 130 F/g) falls far below the theoretical value of 550 F/g calculated for the singlelayer graphene,[278] which is concluded to attribute to the restacking of graphene sheets during its processing as a consequence of the strong sheet-to-sheet van der Waals interactions. And hence the resulting reduction in the specific surface area of graphene accounts for the overall low capacitance. In terms of this problem, Kaner and co-workers presented a novel strategy for the production of graphene-based supercapacitors through a simple all-solid-state approach that avoids the restacking of graphene sheets and hence induces a large specific surface area up to 1,520 m^2/g as can be seen in Fig. 54(c).[279] Interestingly, a standard Light Scribe DVD optical drive was introduced to do the direct laser reduction of graphite oxide films. A very high specific capacitance of 265 F/g was achieved for this flexible electrode as expected. Further, a fully flexible all-solid device was fabri-cated whose construction was clearly illustrated in Fig. 54(c). This idea exhibits a new route to fabricate high capacitance flexible supercapacitors through a very easy way using conventional daily equipment, indicating a great potential to commercialize flexible supercapacitors in the near future.

Additionally, there are post-treatment methods to tailor the properties of such thin-film, flexible, carbon-based devices in order to meet the optimized performance. For instance, compression has been applied to understand the energy storage mechanism as well as find the optimized condition for the best performance.[280,281] All the experimental data reveal that certain pressure applied onto such kind of flexible capacitors can help to significantly improve their capacitive performance. An important factor that determines the minimum amount of pressure applied is the ratio of solvated ion size to the pore size of the carbon electrodes. These results pave the way for further enhancing performance of flexible carbon-based supercapacitors after encapsulation. Since the exciting progresses have been made, carbon materials based electrochemical capacitors are leading a prosperous future in energy storage especially for the high power output applications. Also, it provides the possibility to utilize the advantages of carbon materials by making composites with different materials to maximize the benefits of electrochemical capacitor performance.

4.2.2. Nanostructured Metal Oxides-Based Flexible Supercapacitors

Even though, carbon materials played a dominant role in energy storage field, instead of electrostatical energy storage, metal oxides/hydroxides store charges with surface faradaic (redox) reactions, which enable higher energy density compared to carbon. Despite these efforts, practical energy storage applications still require higher specific capacitance. One solution is to design nanometer-scale electrode materials with very large surface areas and structural stability. High quality materials are urgently needed especially in the field of flexible energy applications, which can provide great opportunities for utilizing metal oxide nanostructure to fabricate flexible supercapacitors. MnO_2, NiO, $Ni(OH)_2$, CoO_x, etc. have been widely investigated as electrode materials for supercapacitors since they have shown superb electrochemical properties.[282–287] For example, ultrathin MnO_2 can be coated around flexible metal charge collector (Ni foam probably) and be used for flexible supercapacitor application as a core-shell nano-network.[288] And a capacitance of 214 F/g can be obtained under a very large current density of 20 A/g, confirming the advantages of the nanometal oxides as the active materials for supercapacitors. In our group, related research has also been carried out. Ternary metal oxides like $NiCo_2O_4$ have been proved to have significant ability to store energy reported by Wang *et al.*[47] This work took a deep insight into the morphology evolution of the urchin-like $NiCo_2O_4$ and influence of different nanostructure on supercapacitor performance. High specific capacitances of 1,650 F/g and 1348, F/g were tested at current densities of 1 A/g and 15 A/g, respectively. The superior energy storage performance leads to a great potential for these type of materials being used for real applications. Moreover, a concept of designing nanostructure to optimize the device performance has also been proved to be effective, and hence guarantees the future of nano-energy applications.

Besides, by combining unique properties of individual constituents, improved performance has been demonstrated in such a composite electrode. Liu *et al.* have brought up an interesting approach to constructing a novel hybrid metal oxide core/shell nanowire arrays (Fig. 55).[289] The Co_3O_4 at MnO_2 hybrid nanowire arrays exhibit a high capacitance of around four- to ten-fold increase in areal capacitance with respect to pristine Co_3O_4

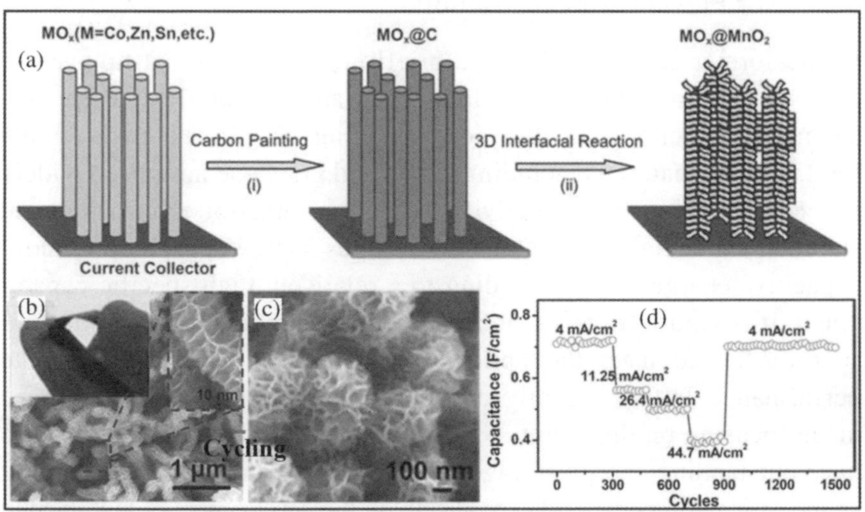

Figure 55.   Hybrid metal oxide core/shell nanowire arrays for flexible supercapacitors. (a) Scheme of fabricating the hybrid nanowire array constructions. (b) SEM image of Co_3O_4 at MnO_2 core/shell nanowires. (c) SEM images of the same materials after 5,000 electrochemical testing cycles. (d) Cycling stability of the hybrid array at progressively varied current density. Reprinted with permission from Ref. 289. (Copyright: 2011 Wiley-VCH.)

array, depending on current rates (480 F/g at 2.67 A/g) with good cycle performance and remarkable rate capability. The significance of this report are the highest performance obtained for hybrid metal oxide systems in the absence of a conducting matrix up-to-date, and also, a novel idea to design 3D nanowire array constructions of different hybrid electrodes. This work uses $Co_3O_4@MnO_2$, ZnO at MnO_2, $SnO_2@MnO_2$ as examples to show that the technique can be further developed for other architecture designing in nanoscale. As can be seen in their later work, MnO_2–NiO composite nanotube arrays have been obtained through an additional treatment of ZnO@ MnO_2 obtained based on the method shown in Fig. 55(a).[290] Since the enhancement of performance by combining metal oxide nanostructures in supercapacitor applications have been confirmed by various publications.[291–293] Endeavors are needed to be taken persistently in order to develop high-performance flexible supercapacitors.

4.2.3. Design of Nanostructured Carbon-based Composites for Flexible Supercapacitors

In the former sections, we have briefly discussed the advantages of carbon materials and metal oxides when applied for flexible electrochemical capacitors and the corresponding progresses made in the field. Carbon materials including CNTs and graphene have been widely proved to have high conductivity as well as superior electrochemical stability, and metal oxides that store charges mainly depend on pseudocapacitive charge storage, leading to a relatively high specific capacitance. It is natural to think about making a combination of both their superiorities to meet the optimized electrochemical energy storage performance. In fact, recently, enormous research progresses have been made focusing on this point. Nanostructured composites are proved to have outstanding energy storage ability, and they are currently playing an increasingly important role specifically in the development of high-quality flexible supercapacitors.

Carbon materials can be made into composites for flexible energy applications by two main ways. The first is to use free-standing thin-film directly for electrochemical applications or deposit carbon nanostructure/ metal oxides mixture onto substrates. A variety of progress have been reported relating to this approach. Yuan *et al.* have introduced facile one-step hydrothermal method for large-scale production of well-designed flexible and free-standing Co_3O_4 /rGO/carbon nanotubes (CNTs) hybrid paper as an electrode for electrochemical capacitors.[294] Similarly, V_2O_5 nanowires have been mixed together with carbon nanotubes to prepare the free-standing, flexible, composite paper for a binder-free supercapacitor electrode,[295] and free-standing TiC/carbon film have been developed from the precursor of electrospun titanium carbide (TiC) nanobelts with satisfied electrochemical performance.[296] As mentioned before, these binder-free, standalone, flexible materials can be used directly for electrochemical application with enormous advantages, guaranteeing the promising direction of developing energy storage application in the near future. Currently, the most popular process to prepare electrode for either flexible supercapacitors or batteries is to coat slurry onto flexible substrates like stainless foil or Ni foam.[297–300] However, in order to easily integrate the device into transparent electronic systems, a transparent supercapacitor is necessary, and a carefully designed composite electrode can also be helpful to meet this requirement. Zhou and co-workers have fabricated flexible and transparent electrodes by dispersing In_2O_3 nanowires and carbon nanotubes onto PET substrates.[301] The fabricated supercapacitor with high flexibility and transparency can be observed in Fig. 56(a). With almost the same concept, ZnO nanorod arrays were deposited onto PET substrate with graphene layer, leading to a highly transparent and flexible supercapacitor with good electrochemical energy storage capacity.[302]

Another route to use composite for flexible supercapacitors is to deposit nanostructured metal oxides onto highly conductive carbon based substrates. The most convenient way is to use the highly conductive carbon fiber paper or cloth directly as substrate to support the active materials for energy storage.[303] Intensive research progresses have been made using this concept. MnO_2 was synthesized using commercially available carbon cloth as substrate to achieve a capacitance of 425 F/g with aqueous electrolyte.[305] Later, facing the problem of instability of liquid phase device, Yuan *et al.* deposited carbon nanoparticles/MnO_2 nanorods using a very easy technique onto carbon fabric for solid state supercapacitors.[306] The all solid electrochemical device using (PVA)/H_3PO_4 as electrolyte guaranteed wonderful stability and displayed the practical application value

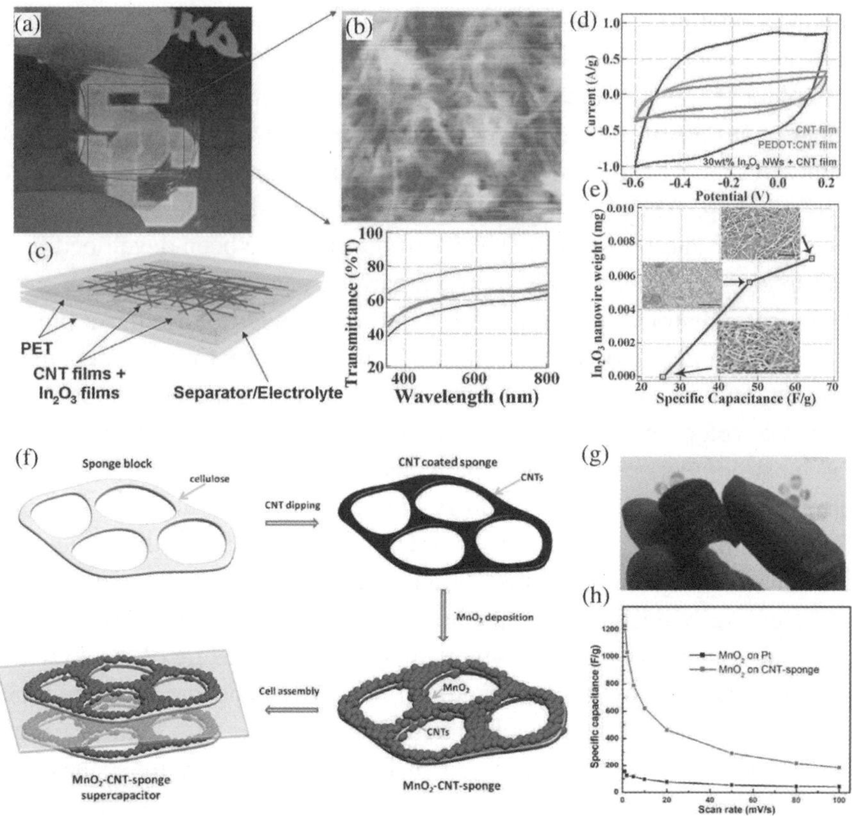

Figure 56. Carbon-based hybrid flexible electrodes for flexible supercapacitors. (a) Photograph of a flexible and transparent supercapacitor fabricated using CNT films. (b) AFM image of entangled CNT networks sitting on a transparent PET substrate. (c) Schematic of a flexible and transparent supercapacitor and transmittance spectra of three different electrochemical capacitors and a single CNT film. (d) and (e) Electrochemical testing of the as-prepared supercapacitor. Reprinted with permission from Ref. 301. (Copyright: 2009 American Institute of Physics.) (f) Fabrication process of MnO_2-CNT-sponge supercapacitors. (g) Optical photograph of a flexible MnO_2-CNT-sponge supercapacitor. (h) Specific capacitance comparison between MnO_2-Pt and MnO_2-CNT-sponge. Reprinted with permission from Ref. 304. (Copyright: 2011 American Chemical Society.)

since it can power a LED device for as long as 20 minutes. Similar work was also reported using either carbon cloth or carbon paper coated with different nanomaterials to fabricate flexible supercapacitors.[307,308] The advantages of using the carbon texture are apparent. The interwoven microstructure can help to support large amount of materials so that the loading density of active materials was increased. Also, the highly conductive carbon fibers with perfect physical and chemical stability provided efficient channels for charge collection and transfer, leading to the superb electric properties. More importantly, the perfect flexibility of carbon texture can effectively benefit the highly flexible applications. But when considering the cost, those commercial carbon textiles require special treatment of raw materials that will sometimes exceed our budget. Besides, the relatively smooth surface of carbon microfiber may cause a poor attachment with active materials. Therefore, alternative routes have been introduced to make the regular textiles like paper or cloth highly conductive through very simple techniques. Figure 56(f) demonstrates an example of dipping a highly porous sponge into a CNT ink suspension to form a highly conductive textile.[304] MnO_2 nanoparticles were then attached onto the sponge network through galvanostatic electrochemical deposition to prepare the flexible electrode (Fig. 56(g)). A specific capacitance of 1,230 F/g (based on the mass of MnO_2) can be reached, which is very close to the theoretical specific capacitance of MnO_2 (1,370 F/g),[309] demonstrating great superiority compared to Pt substrate based device (Fig. 56(h)). This "dip and dry" approach has been widely developed and reported frequently in recent years for high performance supercapacitors.[310–315] In this approach, the "conductive ink" for dip-coating can be prepared using carbon nanotubes and graphene. By simply adjusting the dipping times, conductivity of the textiles can be perfectly controlled. This facile approach provides a concept that we can store energy and supply power conveniently with these daily consumer goods like clothes, sponge and porous paper. There are also other methods to pre-prepare the conductive texture for electrochemical applications. Bao *et al.* introduced a facile heat involved treatment to convert cotton T-shirt into activated carbon textiles and then MnO_2 was electrochemically deposited for energy storage applications.[316] Changing daily textile products into conductive substrates for energy storage must be a great breakthrough as it is leading the field of flexible energy storage to develop towards a practical, low-cost and high performance direction.

Except for carbon/metal oxides composites, combination can be made using carbon materials and conductive polymers.[317–319] Large-scale thin-film can either be formed by using the mixture of composites or orderly deposition of polymers onto carbon materials. Hybrid nanomaterials based electrodes have displayed superior electrochemical performance due to a perfect combination of all the advantages of each single component. Thin flexible devices can be even more convenient to fabricate since using composites can be an effective route that helps to form the free-standing thin-film. Overall, for the great advantages shown above, the carbon based composite materials have greater potential to be utilized in future flexible energy storage applications. Designing nanostructures for the carbon composites based electrode will, on one hand, help to develop high performance flexible supercapacitors, and on the other hand can improve the current technologies towards the direction of scale-up and industrialization.

4.2.4. Novelty of Flexible Supercapacitors

Flexible energy devices have been made into many different constructions based on the basic operating mechanism. In the former sections, solar cells and batteries have been made into wire-shape, microtype, cloth, and paper-like constructions to realize the flexibility of devices. For the relatively simple constructions, supercapacitors do have the potential to be made into novel constructions for special applications. To fulfil this aim, making good use of our imagination together with considering the urgent requirements in energy storage field can be essential. During device fabrication, carefully designing nanostructure of active materials as well as the configuration of whole devices is vital.

Cloth-shape or textile based construction have attracted enormous attentions especially in the field of electrochemical capacitors. Just as mentioned before, the textiles can be used to support active materials such as CNTs to directly perform as electrodes, which can be further used to realize the fantasy that we can wear clothes that supply power to the portable products including cell phones and laptops. Another way to develop the flexible and wearable energy devices is to firstly fabricate the highperformance fiber-based supercapacitors and then integrate it into largescale and textile devices. Wang's group has abundant experience in making single-fiber flexible energy devices.[147,150,196] As can be seen from Fig. 57(a), they have also presented a prototype of a high-efficiency

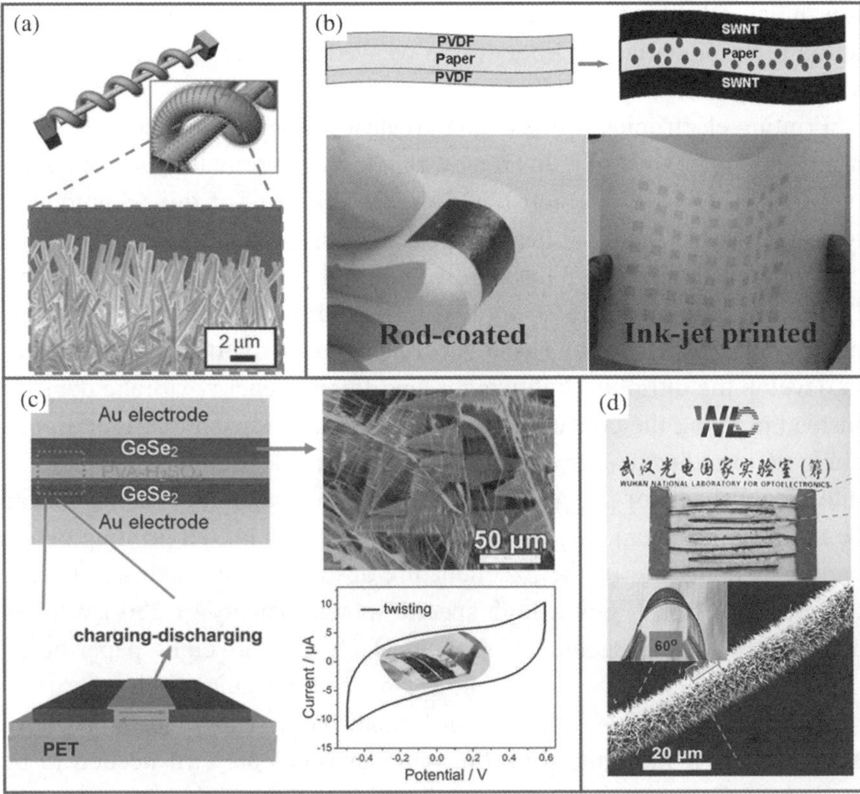

Figure 57. Novel constructions of supercapacitors. (a) Wire shape ZnO nanowires based supercapacitor. Reprinted with permission from Ref. 320. (Copyright: 2011 Wiley-VCH.) (b) Printing supercapacitors on paper. Reprinted with permission from Ref. 322. (Copyright: 2010 American Institute of Physics.) (c) In-plane all solid supercapacitor. Reprinted with permission from Ref. 323. (Copyright: 2012 Wiley-VCH.) (d) Flexible planar-integrated all-solid-state fiber supercapacitors. Reprinted with permission from Ref. 324. (Copyright: 2012 Wiley-VCH.)

fiber-based electrochemical microsupercapacitor (MSC) using ZnO nanowires as electrodes.[320] The specific capacitance can be further improved significantly by coating MnO_2 and using PVA/H_3PO_4 solid state electrolyte. Similar twisted fiber-shaped supercapacitor has also been reported by Fu *et al.* who directly used commercial pen ink dip-coating for electrodes preparation.[321] The well-designed encapsulation method together with the outstanding energy storage performance indicate a promising future for this all-solid, nanostructured, fiber-shaped MSC and a great potential for fibers to be weaved together as flexible energy storage textiles.

Printing electronics is one of the greatest inventions in recent decades. Many examples shown in this review ranging from printable photovoltaics to LIBs have vividly demonstrated the superiorities of this technique in making flexible devices. In fact, pintable supercapacitors can be more conveniently realized for the relatively simple symmetric constructions. Different flexible substrates applied in printable electronics increase the novelty of devices. Paper based supercapacitors provide the possibilities to develop the ultrathin, lightweight, and flexible energy storage devices. Instead of using the conventional multilayered sandwich configuration,[14] Hu *et al.* introduced a novel method to integrate electrodes and separators into single sheets of paper (Fig. 57(b)).[322] This paper supercapacitor can be fabricated conveniently by Meyer rod coating or ink-jet printing onto a paper substrate due to the excellent ink absorption of paper. A specific capacitance of 33 F/g at a high specific power density of 250 kW/kg is achieved with an organic electrolyte. Such a lightweight paper-based supercapacitor could be used to power paper electronics such as transistors or displays. Even though the paper-based supercapacitors have already been made quite thin, the two electrodes are still needed to be assembled face-to-face in a vertical direction and the added separator will inevitably increase the whole thickness of devices. What is worse, an accidently broken separator may lead to short-circuit, and prove disastrous to an electrochemical device. In-plane construction, in which both the electrodes are in the same plane, has perfectly solved those identified problems, leading to a new direction towards the thin-film flexible energy storage technologies. Stimulating progresses have been made in our group relating to the novel construction of flexible in-plane supercapacitors. Figure 57(c) illustrates an all-solid-state, in-plane supercapacitor using the layered structured nano-$GeSe_2$ as active materials.[323] PET was used as substrate so that guaranteed the high flexibility of this device. Even under seriously twisted condition, the flexible supercapacitor still preserves a good energy storage property as a double layered electrochemical supercapacitor. Additionally, the practical application value of this novel supercapacitor has been proved through using it directly to power the nanomaterials based photodetectors. However, flaws do exist in this construction. The enlarged charge separation distance and decreased effective electrode surface area compared to the conventional face-to-face design will lead to relatively low specific capacitance and energy density. And this planar structure also requires more horizontal space since both electrodes are located within the same plane, limiting the opportunities for promoting further development of this construction. Facing such problems, later, carbon microfibers have been introduced in designing the in-plane supercapacitors. Shen and co-workers also deposited $ZnCo_2O_4$ nanowires around carbon fibers and integrated many fiber-shaped supercapacitors in to the same plane supported by PET substrate.[324] This interesting structure can effectively avoid such identified problems of in-plane capacitors by forming 3D constructions in microscale. The supercapacitors with excellent performance were fabricated via a low-cost and facile method and their corresponding enhanced distributed-capacitance effect was proved in this work. This strategy enables highly flexible new structured supercapacitor with maximum functionality and minimized size, thus making it possible to be readily applied in flexible and portable electronic devices.

Device performance is the vital factor considered in most cases. However, sometimes, in order to fit in special applications, the performance like capacitance, power density can be properly sacrificed in order to fulfil the overall requirements. Nowadays, micro or nano electronics are developing dramatically, but the advances of micro/nanopower sources cannot keep pace with such developments, which limits the practical application for a whole microelectronic system. Therefore, developing the micro/nano-energy storage devices is extremely important. Since the nanoelectronic devices do not need too much power to operate, the microbatteries or capacitors only need to supply minor amount of power to satisfy the demands, which means the energy capacitance is not the mean factor to be concerned. The real challenge in this field is how to build an energy storage component into micro/nano-scale. Herein we show an example made by Wang *et al.* to provide some clue for the mini-construction of energy devices.[325] Figure 58(a) illustrates the fabrication procedure for the MSC. Polyaniline (PANI) nanorod arrays were used as the active material and acceptable performance was achieved for this device (588 F/cm^3).

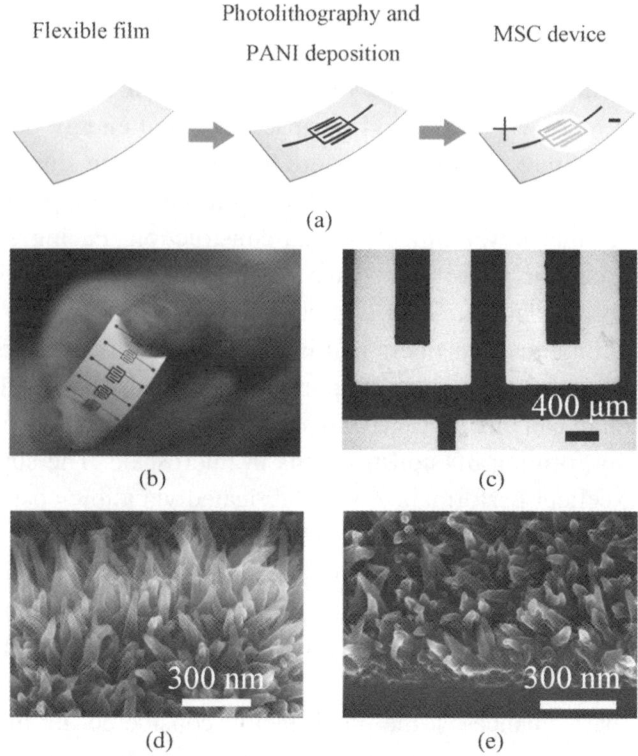

Figure 58.   An all-solid-state flexible MSC on a Chip. (a) A schematic image of fabrication process of PANI nanowire arrays based on MSC flexible film; (b) the optical picture of flexible MSC unit arrays on PET film; (c) optical microscopy of MSC 400; (d, e) SEM image of PANI nanowire arrays obtained from different views. Reprinted with permission from Ref. 325. (Copyright: 2011 Wiley-VCH.)

These energy storage sub-units can be integrated in series or in parallel to supply required current or voltage, revealing a potential for practical application. Another advantage of the pattern based supercapacitors is the convenience in integrating it into electronic systems, as it can be produced at the bottom of a chip, or be a part of integrated circuit board. However, this approach is still far from real application, many fundamental and technical problems need to be solved by future generations.

Supercapacitors are promising energy devices that bridge the gap between batteries and conventional capacitors since they are able to provide higher energy density than conventional capacitors and much higher power density than batteries. Recent development of supercapacitors has made this technology more superior in energy storage field that can compensate or even replace batteries in some cases. Flexibility can be more conveniently realized for supercapacitors because of the relatively simple device construction compared to batteries and other electrochemical devices. In order to get high quality supercapacitors, a careful design of nanostructures for the active materials is essential. And to lower the cost to make the device more accessible to the public, new electrodes fabrication strategies like printing or dip-coating need to be developed. Further, for special and novel applications, imagination have been boosted to bring up a variety of fascinating device configurations such as wire-shape, in-plane and MSC, leading to a booming future for flexible electrochemical energy storage to benefit our real life.

4.3. *Conclusion and Outlook for Flexible Nano-Energy Storage*

In this chapter we have taken a brief review of the recent progresses made for LIBs and supercapacitors, mainly focusing on the flexible devices. A completed LIB consists of an anode, a cathode, an electrolyte and a separator, of which the relatively complex construction limits the possibilities for flexible applications. Using nanostructured materials for both anodes and cathodes together with high performance solid-state electrolyte and well-designed device encapsulation will help to develop flexible LIBs. Breakthroughs are still needed in several aspects such as:

1. Large-scale, free-standing, high-capacitance, binder-free, flexible thin-film anode materials with good mechanical strength are required to be obtained through a low-cost and convenient method.
2. Free-standing and flexible thin-film cathodes are even more urgently needed since there are seldom progresses made in this area, and also alternative nanomaterials should be introduced as cathode materials.
3. Solid-state electrolytes are needed to be further improved in terms of performance and stability.

4. Techniques of encapsulation deserve great attention for both liquid and solid-state batteries.
5. New thoughts should be provided for the construction of flexible LIB in order to be fitted into special applications.

Flexible LIB is an increasingly attractive research area with great potential for future electrical energy storage and power supply. A booming prospect is awaiting us provided the identified problems are well solved. Due to the high power density, fast rates of charge-discharge, reliable cycling life and safe operation, electrochemical capacitors are believed to play an important role in powering the new generation transportations together with LIBs. Nanomaterials with various morphologies do promote the development of supercapacitors in all aspects. With the simple device construction, flexibility can be much more conveniently achieved using simple methods for materials deposition and device assembling. The next research of flexible supercapacitors we concluded should focus on two main factors to guarantee a prosperous future — performance and novelty. High performance flexible supercapacitor can be achieved by applying well-designed nanostructures and suitable combination of different kinds of materials, and hence can satisfy the basic requirements of our daily life. Novel configurations like cloth-shape, wire-shape, planar and microscale supercapacitors are urgently needed as well for the increasing demands of special application in military field, dramatic development microelectronics, and the needs to make our lives more colorful and convenient. Flexible supercapacitors, together with new generation LIBs, will probably change our current lifestyle to a large extent in the next few decades.

Certainly, energy storage is a wide concept that is not merely limited to LIBs and supercapacitors as we discussed. Nanostructured materials may also offer great advantages for molecular hydrogen storage by providing high surface area, or encapsulating or trapping hydrogen in microporous media.[326,327] Besides, mechanical, electromagnetic, and thermal energy storage that are being widely used in many areas also contain great value for further research.[328–330] However, most of these forms focus on large-scale energy applications currently, the flexible and portable features are not quite urgently required or even unnecessary in some cases, which explains why there are fewer reports in those areas concerning flexible device fabrications. But still, new directions and novel concepts are believed to be boosted once nanostructure based flexible energy storage applications in such fields are practically realized. Therefore, designing nanostructures to make other energy storage applications flexible and portable should be a direction that deserves certain amount of attention. There is still a long way to go for humans to enjoy a more convenient lifestyle brought by flexible nano-energy storage.

5. Conclusion

Flexible energy applications have attracted attention for quite a long time. Until recently, booming research on nanoscience and nanotechnology has promoted the flexible energy field to develop towards high-performance and low-cost direction. Due to the unique properties of small size and high surface area, nanostructures offer promising perspective for enhancing the overall device performance in both energy conversion and storage field. Also, the dramatic developments of nano-fabrication engineering have significantly promoted the field of flexible electronics, making it possible to fabricate flexible devices via the low-cost, environment friendly, energy-saving routes. Designing nanostructures is an effective way leading to the new generation flexible energy conversion and storage. Enormous stimulating progresses have been achieved in related fields as introduced in the former chapters. Using well-designed nanomaterials for each component of a certain energy device has helped to meet optimized performance. In this field, free-standing and flexible thin-films consist of nano-sub-units playing an increasingly important role in perfectly satisfying the demands of device flexibility. Facile techniques to prepare electrodes such as printing and transferring have made flexible energy devices more easy to be obtained. Various configurations including sandwiched, wire-shaped, cloth-like and in-plane ones have made devices more easy to be encapsulated with a promising future for specifically fitting into different applications. All these endeavors are devoted to making flexible energy conversion and storage applications more accessible to the public.

There are still gaps between the experimental feasibility and practical applications of flexible nano-energy. Many fundamental and technical problems like reduced performance, instability, poor mechanical strength are still needed to be reconsidered thoroughly, which is the first important aspect we should continue working on in order to narrow such gaps. Another very attractive future research direction is integration. Integration here does not only mean to integrate single energy units to fabricate scaled-up modules, but also to make a combination of flexible modules with different functions to build a system using nanotechnology. There are already some very interesting examples of integrating energy conversion and storage into a single unit.[331–333] These multi-functional devices can be used as standalone systems to serve special conditions, demonstrating the possibility of using nanostructure to design a complex, flexible, and portable energy system. Novelties are full in this research area, which requires further exploration of our imagination. By designing nanostructures for flexible energy conversion and storage, we are gradually changing the way to live our lives.

Acknowledgments

This work was supported by the National Natural Science Foundation (Grant Nos. 61377033, 91123008). The authors are indebted for the kind permission from the corresponding publishers and authors to reproduce their materials, especially figures, used in this review.

References

1. A. Chirila, S. Buecheler, F. Pianezzi, P. Bloesch, C. Gretener, A. R. Uhl, C. Fella, L. Kranz, J. Perrenoud, S. Seyrling, R. Verma, S. Nishiwaki, Y. E. Romanyuk, G. Bilger and A. N. Tiwari, *Nat. Mater.*, **10**, 857 (2011).
2. D. Graham-Rowe, *Nat. Photonics*, **1**, 433 (2007).
3. K. T. Park, Z. Guo, H. D. Um, J. Y. Jung, J. M. Yang, S. K. Lim, Y. S. Kim and J. H. Lee, *Opt. Express*, **19**, A41 (2011).
4. T. Yang, M. Wang, C. Duan, X. Hu, L. Huang, J. Peng, F. Huang and X. Gonng, *Energ. Environ. Sci.*, **5**, 8208 (2012).
5. L. S. Zhou, A. Wang, S. C. Wu, J. Sun, S. Park and T. N. Jackson, *Appl. Phys. Lett.*, **88**, 083502 (2006).

6. R. Asahi, T. Morikawa, T. Ohwaki, K. Aoki and Y. Taga, *Science*, **293**, 269 (2001).

7. B. O'Regan and M. Gratzel, *Nature*, **353**, 737 (1991).

8. H. Huang, B. Liang, Z. Liu, X. Wang, D. Chen and G. Z. Shen, *J. Mater. Chem.*, **22**, 13428 (2012).

9. X. P. A. Gao, G. F. Zheng and C. M. Lieber, *Nano Lett.*, **10**, 547 (2010).

10. P. D. Yang, H. Kind, H. Q. Yan, B. Messer and M. Law, *Adv. Mater.*, **14**, 158 (2002).

11. B, Liu, J. Zhang, X. F. Wang, G. Chen, D. Chen, C. W. Zhou and G. Z. Shen, *Nano Lett.*, **12**, 3005 (2012).

12. Z. R. Wang, H. Wang, B. Liu, W. Z. Qiu, J. Zhang, S. H. Ran, H. T. Huang, J. Xu, H. W. Han, D. Chen and G. Z. Shen, *ACS Nano*, **5**, 8412 (2011).

13. Z. R. Wang, S. H. Ran, B. Liu, D. Chen and G. Z. Shen, *Nanoscale*, **4**, 3350 (2012).

14. M. Kaempgen, C. K. Chan, J. Ma, Y. Cui and G. Gruner, *Nano Lett.*, **9**, 1872 (2009).

15. B. Liu and E. S. Aydil. *J. Am. Chem. Soc.*, **131**, 3985 (2009).

16. X. J. Feng, K. Shankar, O. K. Varghese, M. Paulose, T. J. Latempa and C. A. Grimes, *Nano Lett.*, **8**, 3781 (2008).

17. F. W. Zhuge, J. J. Qiu, X. M. Li, X. D. Gao, X. Y. Gan and W. D. Yu, *Adv. Mater.*, **23**, 1330 (2011).

18. S. K. C. Lee, Y. H. Yu, O. Perez, S. Puscas, T. H. Kosel and M. Kun, *Chem. Mater.*, **22**, 77 (2010).

19. J. M. Weisse, C. H. Lee, D. R. Kim and X. L. Zheng, *Nano Lett.*, **12**, 3339 (2012).

20. M. Gratzel, *Nature*, **414**, 338 (2001).

21. J. M. Tarascon and M. Armand, *Nature*, **414**, 359 (2001).

22. D. N. Futaba, K. Hata T. Yamada, T. Hiraoka, Y. Hayamizu, Y. Kakudate, O. Tanaike, H. Hatori, M. Yumura and S. Iijima, *Nat. Mater.*, **5**, 987 (2006).

23. C. Liu, F. Li, L. P. Ma and H. M. Cheng, *Adv. Mater.*, **22**, E28 (2010).

24. J. Baxter, Z. X. Bian, G. Chen, D. Danielson, M. S. Dresselhaus, A. G. Fedorov, T. S. Fisher, C. W. Jones, E. Maqinn, U. Kortshaqen, A. Manthiram, A. Nozik, D. R. Rolison, T. Sands, L. Shi, D. Sholl and Y. Wu, *Energy Environ. Sci.*, **2**, 559 (2009).

25. X. Chen, C. Li, M. Gratzel, R. Kostecki and S. S. Mao, *Chem. Soc. Rev.*, **41**, 7909–7937 (2012).

26. P. G. Bruce, B. Scrosati and J. M. Tarascon, *Angew. Chem. Int. Ed.*, **47**, 2930 (2008).

27. L. Qu, Z. A Peng and X. Peng, *Nano Lett.*, **1**, 333 (2001).

28. T. Sugimoto, X, Zhou and A. Muramatsu, *J. Colloid Interf. Sci.*, **259**, 43 (2003).

29. P. Gao, Y. Chen, Y. Wang, Q. Zhang, X. Li and M. Hu, *Chem. Commun.*, **2762** (2009).

30. B. Lv, Z. Liu, H. Tian, Y. Xu, D. Wu and Y. Sun, *Adv. Funct. Mater.*, **20**, 3987 (2010).

31. X. Han, M. Jin, S. Xie, Q. Kuang, Z. Jiang, Y. Jiang, Z. Xie and L. Zheng, *Angew. Chem. Int. Ed.*, **48**, 9180 (2009).

32. G. Z. Shen, B. Liang, X. F. Wang, H. T. Huang, D. Chen and Z. L. Wang, *ACS Nano*, **5**, 6148 (2011).

33. G. Z. Shen, B. Liang, X. F. Wang, P. C. Chen and C. W. Zhou, *ACS Nano.*, **5**, 2155 (2011).

34. G. Z. Shen, J. Xu, X. F. Wang, H. T. Huang and D. Chen. *Adv. Mater.*, **23**, 771 (2011).

35. Z. Liu, H. T. Huang, B. Liang, X. F. Wang, Z. R. Wang, D. Chen and G. Z. Shen, *Opt. Express*, **20**, 2982 (2012).

36. D. Chen, J. Xu, B. Liang, X. Wang, P. C. Chen, C. Zhou and G. Z. Shen, *J. Mater. Chem.*, **21**, 17236 (2011).

37. L. Yanbo, D. V. Florent, S. Mathieu, Y. Ichiro and D. Jean-Jacques, *Nanotechnology*, **20**, 045501 (2009).

38. J. Xu, Y. G. Zhu, H. T. Huang, Z. Xie, D. Chen and G. Z. Shen, *Cryst. Eng. Comm.*, **13**, 2629 (2011).

39. H. T. Huang, S. Q. Tian, J. Xu, Z. Xie, D. W. Zeng, D. Chen and G. Z. Shen, *Nanotechnology*, **23**, 105502 (2012).

40. Y. Li, J. Xu, J. Chao, D. Chen, S. Ouyang, J. Ye and G. Z. Shen, *J. Mater. Chem.*, **21**, 12852 (2011).

41. T. Kasuga, M. Hiramatsu, A. Hoson, T. Sekino and K. Niihara, *Langmuir*, **14**, 3160 (1998).

42. Z. J. Gu, T. Y. Zhai, B. F. Gao, X. H. Sheng, Y. B. Wang, H. B. Fu, Y. Ma and J. N. Yao, *J. Phys. Chem. B*, **110**, 23829 (2006).

43. G. H. Yue, P. X. Yan, D. Yan, X. Y. Fan, M. X. Wang, D. M. Qu and J. Z. Liu, *Appl. Phys. A Mater.*, **84**, 409 (2006).

44. Y. Liu, H. Wang, Y. C. Wang, H. M. Xu, M. Li and H. Shen, *Chem. Commun.*, **47**, 3790 (2011).

45. S. H. Ran, Y. G. Zhu, H. T. Huang, B. Liang, J. Xu, B. Liu, J. Zhang, Z. Xie, Z. R. Wang, J. H. Ye, D. Chen and G. Z. Shen, *CrystEngComm.*, **14**, 3063 (2012).

46. J. Xu, Y. Li, H. T. Huang, Y. G. Zhu, Z. R. Wang, Z. Xie, X. F. Wang, D. Chen and G. Z. Shen, *J. Mater. Chem.*, **21**, 19086 (2011).

47. Q. F. Wang, B. Liu, X. F. Wang, S. H. Ran, L. M. Wang, D. Chen and G. Z. Shen, *J. Mater. Chem.*, **22**, 21647 (2012).

48. X. F. Wang, H. T. Huang, B. Liu, B. Liang, C. Zhang, Q. Ji, D. Chen and G. Z. Shen, *J. Mater. Chem.*, **22**, 5330 (2012).

49. L. M. Wang, B. Liu, S. H. Ran, H. T. Huang, X. F. Wang, B. Liang, D. Chen and G. Z. Shen, *J. Mater. Chem.*, **22**, 23541 (2012).

50. A. Kumar, A. R. Madaria and C. W. Zhou. *J. Phys. Chem. C.*, **114**, 7787 (2010).

51. C. Xu, J. Wu, U. V. Desai and D. Gao, *J. Am. Chem. Soc.*, **133**, 8122 (2011).

52. H. E. Unalan, D. Wei, K. Suzuki, S. Dalal, P. Hiralal, H. Matsumoto, S. Imaizumi, M. Minagawa, A. Tanioka, A. J. Flewitt, W. I. Milne and G. Amaratunga, *Appl. Phys. Lett.*, **93** (2008).

53. X. H. Zhang, L. Gong, K. Liu, Y. Z. Cao, X. Xiao, W. M. Sun, X. J. Hu, Y. H. Gao, J. Chen, J. Zhou and Z. L. Wang, *Adv. Mater.*, **22**, 5292 (2010).

54. G. K. Mor, K. Shanka, M. Paulose, O. K. Varghese and C. A. Grimes, *Nano Lett.*, **6**, 215 (2006).

55. O. Lupan, V. M. Guérin, I. M. Tiginyanu, V. V. Ursaki, L. Chow, H. Heinrich and T. Pauporte, *J. Photoch. Photobio. A: Chemistry.* **211**, 65 (2010).

56. A. Yella, H. W. Lee, H. N. Tsao, C. Yi, A. K. Chandiran, M. K. Nazeeruddin, E. W. G. Diau, C. Y. The, S. M. Zakeeruddin and G. Michael, *Science*, **334**, 629 (2011).

57. T. Yamaguchi, N. Tobe, D. Matsumoto and H. Arakawa, *Chem. Commun.*, **45**, 4767 (2007).

58. S. Uchida, M. Timiha, H. Takizawa and M. Kawaraya, *J. Photoch. Photobio. A*, **164**, 93 (2004).

59. D. S. Zhang, T. Yoshida, T. Oekermann, K. Furuta and H. Minoura, *Adv. Funct. Mater.*, **16**, 1228 (2006).

60. D. S. Zhang, T. Yoshida and H. Minoura, *Adv. Mater.*, **15**, 814 (2003).

61. N. G. Park, K. M. Kim, M. G. Kang, K. S. Ryu, S. H. Chang and Y. J. Shin, *Adv. Mater.*, **17**, 2349 (2005).

62. C. X. Cheng, J. H. Wu, Y. M. Xiao, Y. Chen, L. Q. Fan and M. L. Huang, *Electrochim. Acta.*, **56**, 7256 (2011).

63. H. C. Weerasinghe, P. M. Sirimanne, G. P. Simon and Y. B. Cheng, *J. Photoch. Photobio.: A*, **206**, 64 (2009).

64. H. C. Weerasinghe, P. M. Sirimanne, G. P. Simon, Y. B. Cheng, *Prog. Photovolt. Res. Appl.*, **20**, 321 (2012).

65. X. Z. Liu, Y. H. Luo, H. Li, Y. Z. Fan, Z. X. Yu, Y. Lin, L. Q. Chen and Q. B. Meng, *Chem. Commun.*, **2847** (2007).

66. T. Miyasaka, *J. Phys. Chem. Lett.*, **2**, 262 (2011).

67. T. Yamaguchi, N. Tobe, D. Matsumoto, T. Nagai and H. Arakawa, *Sol. Energ. Mat. Sol. C*, **94**, 812 (2010).

68. L. Y. Lin, C. P. Lee, K. W. Tsai, M. H. Yeh, C. Y. Chen, R. Vittal C. G. Wu and K. C. Ho, *Prog. Photovoltaics.*, **20**, 181 (2012).

69. M. G. Kang, N. G. Park, K. S. Ryu, S. H. Chang and K. J. Kim. *Sol. Energ. Mat. Sol. C*, **90**, 574 (2006).

70. L. Francis, A. S. Nair, R. Jose, S. Ramakrishna, V. Thavasi and E. Marsano, *Energy*, **36**, 627 (2011).

71. D. V. Bavykin, A. N. Kulak and F. C. Walsh, *Langmuir*, **27**, 5644 (2011).

72. W. Zhang, D. Zhang, T. X. Fan, J. J. Gu, R. Ding, H. Wang, Q. X. Guo and H. Ogawa, *Chem. Mater.*, **21**, 33 (2009).

73. Q. Zheng, H. Kang, J. Yun, J. Lee, J. H. Park and S. Baik, *ACS Nano*, **5**, 5088 (2011).

74. D. Kuang, J. Brillet, P. Chen, M. Takata, S. Uchida, H. Miura, K. Sumioka, S. M. Zakeeruddin and M. Gratzel, *ACS Nano*, **2**, 1113 (2008).

75. K. Nakayama, T. Kubo and Y. Nishikitani, *Appl. Phys. Express*, **11**(3), C23–C26 (2008).

76. D. A. Wang, Y. Liu, C. W. Wang, F. Zhou and W. M. Liu, *ACS Nano*, **3**, 1249 (2009).

77. Y. H. Chen, K. C. Huang, J. G. Chen, R. Vittal and K. C. Ho, *Electrochim Acta*, **56**, 7999 (2011).

78. L. Y. Lin, M. H. Yeh, C. P. Lee, Y. H. Chen, R. Vittal and K. C. Ho, *Electrochim Acta*, **57**, 270 (2011).

79. Z. Y. Liu, V. Subramania and M. Misra, *J. Phys. Chem. C*, **113**, 14028 (2009).

80. J. M. Macák, H. Tsuchiya, A. Ghicov and P. Schmuki, *Electrochem. Commun.*, **7**, 1133 (2005).

81. P. Roy, S. P. Albu and P. Schmuki, *Electrochem. Commun.*, **12**, 949 (2010).

82. P. M. Sommeling, B. C. O'Regan, R. R. Haswell, H. J. P. Smit, N. J. Bakker, J. J. T. Smits, J. M. Kroon and J. A. M. van Roosmalen, *J. Phys. Chem. B*, **110**, 19191 (2006).

83. L. Y. Lin, M. H. Yeh, C. P. Lee, C. Y. Chou, R. Vittal and K. C. Ho, *Electrochim Acta*, **62**, 341 (2012).

84. P. Roy, D. Kim, I. Paramasivam and P. Schmuki, *Electrochem. Commun.*, **11**, 1001 (2009).

85. J. G. Chen, C. Y. Chen, C. G. Wu, C. Y. Lin, Y. H. Lai, C. C. Wang, H. W. Chen, R. Vittal and K. C. Ho, *J. Mater. Chem.*, **20**, 7201 (2010).

86. A. Benoit, I. Paramasivam, Y. C. Nah, P. Roy and P. Schmuki, *Electrochem. Commun.*, **11**, 728 (2009).

87. J. Zhang, J. H. Bang, C. Tang and P. V. Kamat, *ACS Nano*, **4**, 387 (2009).

88. H. W. Chen, K. C. Huang, C. Y. Hsu, C. Y. Lin, J. G. Chen and C. P. Lee, *Electrochim, Acta*, **56**, 7991 (2011).

89. H. W. Chen, C. P. Liang, H. S. Huang, J. G. Chen, R. Vittal, C. Y. Lin, K. C. W. Wu and K. C. Ho, *Chem. Commun.*, **47**, 8346 (2011).

90. Z. S. Xue, W. Zhang, X. Yin, Y. M. Cheng, L. Wang and B. Liu, *RSC Adv.*, **2**, 7074 (2012).

91. X. Yin, Z. S. Xue, L. Wang, Y. M. Cheng and B. Liu, *ACS Appl. Mater. Inter.*, **4**, 1709 (2012).

92. H. Lee, D. Hwang, S. M. Jo, D. Kim, Y. Seo and D. Y. Kim, *ACS Appl. Mater. Inter.*, **4**, 3308 (2012).

93. H. Kim, G. P. Kushto, C. B. Arnold, Z. H. Kafafi and A. Pique, *Appl. Phys. Lett.*, **85**, 464 (2004).

94. J. A. van Delft, D. Garcia-Alonso and W. M. M. Kessels, *Semicond. Sci. Tech.*, **27** (2012).

95. P. Chen, Y. W. Lo, T. L. Chou and J. M. Ting, *J. Electrochem. Soc.*, **158**, H1252 (2011).

96. J. H. Park, T. W. Lee and M. G. Kang, *Chem. Commun.*, 2867 (2008).

97. X. Z. Liu, L. Wang, Z. S. Xue and B. Liu, *RSC Adv.*, **2**, 6393 (2012).

98. B. L. Chen, H. Hu, Q. D. Tai, N. G. Zhang, F. Guo and B. Sebo, *Electrochim. Acta*, **59**, 581 (2012).

99. T. P. Chou, Q. F. Zhang, G. E. Fryxell and G. Z. Cao, *Adv. Mater.*, **19**, 2588 (2007).

100. S. Yodyingyong, Q. F. Zhang, K. Park, C. S. Dandeneau, X. Y. Zhou, D. Triampo and G. Z. Cao, *Appl. Phys. Lett.*, **96**, 073115 (2010).

101. H. M. Cheng and W. F. Hsieh, *Energ. Environ. Sci.*, **3**, 442 (2010).

102. F. Xu and L. T. Sun, *Energ. Environ. Sci.*, **4**, 818 (2011).

103. S. H. Lee, S. H. Han, H. S. Jung, H. Shin and J. Lee, *J. Phys. Chem. C*, **114**, 7185 (2010).

104. S. H. Ko, D. Lee, H. W. Kang, K. H. Nam, J. Y. Yeo, S. J. Hong, C. P. Grigoropoulos and H. J. Sung, *Nano Lett.*, **11**, 666 (2011).

105. Y. Y. Li, X. Dong, C. W. Cheng, X. C. Zho, P. G. Zhang, J. S. Gao and H. Q. Zhang, *Phys. B*, **404**, 4282 (2009).

106. C. K. Xu, J. M. Wu, U. V. Desai and D. Gao, *J. Am. Chem. Soc.*, **133**, 8122 (2011).

107. A. Manekkathodi, M. Y. Lu, C. W. Wang and L. J. Chen, *Adv. Mater.*, **22**, 4059 (2010).

108. W. Wang, Q. Zhao, H. Li, H. W. Wu, D. C. Zou and D. P. Yu, *Adv. Funct. Mater.*, **22**, 2775 (2012).

109. C. Y. Jiang, X. W. Sun, K. W. Tan, G. O. Lo, A. K. K. Kyaw and D. L. Kwong, *Appl. Phys. Lett.*, **92** (2008).

110. Z. Qin, G. Zhang, Q. Liao, Y. Qiu, Y. Huang and Y. Zhang, *Colloids and Surfaces A: Physicochemical and Engineering Aspects*, **402**, 127 (2012).

111. J. J. Kim, K. S. Kim and G. Y. Jung, *J. Mater. Chem.*, **21**, 7730 (2011).

112. F. Y. Hu, Y. J. Xia, Z. S. Guan, X. Yin and T. He, *Electrochim. Acta*, **69**, 97 (2012).

113. W. Chen, H. Zhang, I. M. Hsing and S. Yang, *Electrochem. Commun.*, **11**, 1057 (2009).

114. C. K. Xu, J. M. Wu, U. V. Desai and D. Gao, *Nano Lett.*, **12**, 2420 (2012).

115. B. Liu, Z. R. Wang, Y. Dong, Y. G. Zhu, Y. Gong, S. H. Ran, Z. Liu, J. Xu, Z. Xie, D. Chen and G. Z. Shen, *J. Mater. Chem.*, **22**, 9379 (2012).

116. X. C. Dou, D. Sabba, N. Mathews, L. H. Wong, Y. M. Lam and S. Mhaisalkar, *Chem. Mater.*, **23**, 3938 (2011).

117. K. Wijeratne, J. Akilavasan, M. Thelakkat and J. Bandara, *Electrochim. Acta*, **72**, 192 (2012).

118. M. Liu, J. Y. Yang, S. L. Feng, H. Zhu, J. S. Zhang and G. Li, *Mater Lett.*, **76**, 215 (2012).

119. H. D. Zheng, Y. Tachibana and K. Kalantar-zadeh, *Langmuir*, **26**, 19148 (2010).

120. K. Hara, Z. G. Zhao, Y. Cui, M. Miyauchi, M. Miyashita and S. Mori, *Langmuir*, **27**, 12730 (2011).

121. C. Y. Hsu, W. T. Chen, Y. C. Chen, H. Y. Wei, Y. S. Yen, K. C. Huang, K. C. Ho, C. W. Chu and J. T. Lin, *Electrochim. Acta*, **66**, 210 (2012).

122. S. Uehara, S. Sumikura, E. Suzuki and S. Mori, *Energ. Environ. Sci.*, **3**, 641 (2010).

123. B. Mahrov, G. Boschloo, A. Hagfeldt, L. Dloczik and T. Dittrich, *Appl. Phys. Lett.*, **84**, 5455 (2004).

124. M. Z. Yu, G. Natu, Z. Q. Ji and Y. Y. Wu, *J. Phys. Chem. Lett.*, **3**, 1074 (2012).

125. T. Prakash and S. Ramasamy, *Science of Advanced Materials*, **4**, 29 (2012).

126. Y. F. Wei, L. Ke, J. H. Kong, H. Liu, Z. H. Jiao, X. H. Lu, H. J. Du and X. W. Sun, *Nanotechnology*, **23** (2012).

127. A. Hagfeldt, G. Boschloo, L. Sun, L. Kloo and H. Pettersson, *Chem. Rev.*, **110**, 6595 (2010).

128. N. Q. Fu, X. R. Xiao, X. W. Zhou, J. B. Zhang and Y. Lin, *J. Phys. Chem. C*, **116**, 2850 (2012).

129. D. Fu, P. Huang and U. Bach, *Electrochim. Acta*, **77**, 121–127 (2012).

130. L. L. Chen, W. W. Tan, J. B. Zhang, X. W. Zhou, X. L. Zhang and Y. Lin, *Electrochim Acta*, **55**, 3721 (2010).

131. H. Wang, G. Liu, X. Li, P. Xiang, Z. Ku, Y. Rong, M. Xu, L. Liu, M. Hu, Y. Yang and H. W. Han, *Energ. Environ. Sci.*, **4**, 2025 (2011).

132. J. K. Chen, K. X. Li, Y. H. Luo, X. Z. Guo, D. M. Li, M. H. Deng, S. Q. Huang and Q. B. Meng, *Carbon*, **47**, 2704 (2009).

133. K. P. Acharya, H. Khatri, S. Marsillac, B. Ullrich, P. Anzenbacher and M. Zamkov, *Appl Phys Lett.*, **97** (2010).

134. S. Roy, R. Bajpai, A. K. Jena, P. Kumar, N. Kulshrestha and D. S. Misra, *Energy Environ. Sci.*, **5**, 7001 (2012).

135. L. G. De Arco, Y. Zhang, C. W. Schlenker, K. Ryu, M. E. Thompson and C. W. Zhou, *ACS Nano*, **4**, 2865 (2010).

136. Z. B. Yang, T. Chen, R. X. He, G. Z. Guan, H. P. Li, L. B. Qiu and H. S. Peng, *Adv Mater*, **23**, 5436 (2011).

137. F. Malara, M. Manca, M. Lanza, C. Hubner, E. Piperopoulos and G. Gigli, *Energy Environ. Sci.*, **5**, 8377 (2012).

138. F. Malara, M. Manca, L. De Marco, P. Pareo and G. Gigli, *ACS Appl. Mater. Inter.*, **3**, 3625 (2011).

139. V. Tjoa, J. Chua, S. S. Pramana, J. Wei, S. G. Mhaisalkar and N. Mathews, *ACS Appl. Mater. Inter*, **4**, 3447 (2012).

140. Y. M. Xiao, J. H. Wu, G. T. Yue, J. M. Lin, M. L. Huang and Z. Lan, *Electrochim. Acta*, **56**, 8545 (2011).

141. S. J. Peng, P. N. Zhu, Y. Z. Wu, S. G. Mhaisalkar and S. Ramakrishna, *RSC Adv*, **2** (2012).

142. M. X. Wu, Q. Y. Zhang, J. Q. Xiao, C. Y. Ma, X. Lin, C. Y. Miao, Y. J. He, Y. R. Gao, A. Hagfeldt and T. L. Ma, *J. Mater. Chem.*, **21**, 10761 (2011).

143. J. Z. Chen, B. Li, J. F. Zheng, J. H. Zhao, H. W. Jing and Z. P. Zhu, *Electrochim. Acta*, **56**, 4624 (2011).

144. X. Fan, Z. Z. Chu, F. Z. Wang, C. Zhang, L. Chen, Y. W. Tang and D. C. Zou, *Adv. Mater.*, **20**, 592 (2008).

145. X. Fan, Z. Z. Chu, L. Chen, C. Zhang, F. Z. Wang, Y. W. Tang, J. L. Sun and D. C. Zou, *Appl. Phys. Lett.*, **92** (2008).

146. Z. Lv, Y. Fu, S. Hou, D. Wang, H. Wu, C. Zhang, Z. Chu and D. Zou, *Phys. Chem. Chem. Phys.*, **13**, 10076 (2011).

147. W. X. Guo, C. Xu, X. Wang, S. H. Wang, C. F. Pan, C. J. Lin and Z. L. Wang, *J. Am. Chem. Soc.*, **134**, 4437 (2012).

148. Y. Fu, Z. Lv, S. Hou, H. Wu, D. Wang, C. Zhang, Z. Chu, X. Cai, X. Fan, Z. L. Wang and D. Zou, *Energy Environ. Sci.*, **4**, 3379 (2011).

149. Z. B. Lv, J. F. Yu, H. W. Wu, J. Shang, D. Wang, S. C. Hou, Y. P. Fu, K. Wu and D. C. Zou, *Nanoscale*, **4**, 1248 (2012).

150. B. Weintraub, Y. G. Wei and Z. L. Wang, *Angew. Chem. Int. Ed.*, **48**, 8981 (2009).

151. T. Chen, L. Qiu, Z. Cai, F. Gong, Z. Yang, Z. Wang and H. Peng, *Nano Lett.*, **12**, 2568 (2012).

152. S. Zhang, C. Y. Ji, Z. Q. Bian, R. H. Liu, X. Y. Xia, D. Q. Yun, L. H. Zhang, C. H. Huang and A. Y. Cao, *Nano Lett.*, **11**, 3383 (2011).

153. X. Fan, F. Z. Wang, Z. Z. Chu, L. Chen, C. Zhang and D. C. Zou, *Appl. Phys. Lett.*, **90**, 073501 (2007).

154. W. He, J. J. Qiu, F. W. Zhuge, X. M. Li, J. H. Lee, Y. D. Kim, H. K. Kim and Y. H. Hwang, *Nanotechnology*, **23** (2012).

155. Y. M. Xiao, J. H. Wu, G. T. Yue, J. M. Lin, M. L. Huang, L. Q. Fan and Z. Lan, *J. Power Sources*, **208**, 197 (2012).

156. S. I. Cha, Y. Kim, K. H. Hwang, Y. J. Shin, S. H. Seo and D. Y. Lee, *Energy Environ. Sci.*, **5**, 6071 (2012).

157. X. W. Huang, P. Shen, B. Zhao, X. M. Feng, S. H. Jiang, H. J. Chen, H. Li and S. T. Tan, *Sol. Energy Mater. Sol. C*, **94**, 1005 (2010).

158. V. Vijayakumar, A. Du Pasquier and D. P. Birnie, *Sol. Energy Mater Sol. C*, **95**, 2120 (2011).

159. H. Dai, Y. Zhou, Q. Liu, Z. D. Li, C. X. Bao, T. Yu and Z. G. Zhou, *Nanoscale*, **4**, 5454 (2012).

160. J. K. Chen, H. Lin, X. Li, X. C. Zhao, F. Hao and S. M. Dong, *Electrochim. Acta*, **56**, 6026 (2011).

161. H. Shuqing, Z. Quanxin, H. Xiaoming, G. Xiaozhi, D. Minghui, L. Dongmei, L. Yanhong, S. Qing, T. Taro and M. Qingbo, *Nanotechnology*, **21**, 375201 (2010).

162. Y. R. Smith and V. Subramanian, *J. Phys. Chem. C*, **115**, 8376 (2011).

163. J. Chen, W. Lei, C. Li, Y. Zhang, Y. P. Cui, B. P. Wang and W. Q. Deng, *Phys. Chem. Chem. Phys.*, **13**, 13182 (2011).

164. X. M. Huang, S. Q. Huang, Q. X. Zhang, X. Z. Guo, D. M. Li, Y. H. Luo, Q. Shen, T. Toyoda and Q. B. Meng, *Chem. Commun.*, **47**, 2664 (2011).

165. A. G. Pattantyus-Abraham, I. J. Kramer, A. R. Barkhouse, X. Wang, G. Konstantatos, R. Debnath, L. Levina, I. Raabe, M. K. Nazeeruddin, M. Grätzel and E. H. Sargent, *ACS Nano*, **4**, 3374 (2010).

166. E. H. Sargent, *Nat. Photonics*, **6**, 133 (2012).

167. K. Otte, L. Makhova, A. Braun and I. Konovalov, *Thin Solid Films*, **511**, 613 (2006).

168. M. Kaelin, D. Rudmann and A. N. Tiwari, *Sol. Energy*, **77**, 749 (2004).

169. E. Lee, S. J. Park, J. W. Cho, J. Gwak, M. K. Oh and B. K. Min, *Sol. Energy Mat. Sol. C*, **95**, 2928 (2011).

170. V. K. Kapur, A. Bansal, P. Le and O. I. Asensio, *Thin Solid Films*, **431**, 53 (2003).

171. V. A. Akhavan, B. W. Goodfellow, M. G. Panthani, D. K. Reid, D. J. Hellebusch, T. Adachi and B. A. Korgel, *Energy Environ. Sci.*, **3**, 1600 (2010).

172. Q. Guo, G. M. Ford, H. W. Hillhouse and R. Agrawal, *Nano Lett.*, **9**, 3060 (2009).

173. M. G. Panthani, V. Akhavan, B. Goodfellow, J. P. Schmidtke, L. Dunn, A. Dodabalapur, P. F. Barbara and B. A. Korgel, *J. Am. Chem. Soc.*, **130**, 16770 (2008).

174. S. Gunes, H. Neugebauer and N. S. Sariciftci, *Chem. Rev.*, **107**, 1324 (2007).

175. B. Gholamkhass, N. M. Kiasari and P. Servati, *Org. Electron.*, **13**, 945 (2012).

176. A. Gadisa, Y. C. Liu, E. T. Samulski and R. Lopez, *ACS Appl. Mater. Inter.*, **4**, 3846 (2012).

177. J. Hu, Z. W. Wu, H. X. Wei, T. Song and B. Q. Sun, *Org. Electron.*, **13**, 1171 (2012).

178. S. J. Wu, J. H. Li, S. C. Lo, Q. D. Tai and F. Yan, *Org. Electron.*, **13**, 1569 (2012).

179. H. E. Unalan, P. Hiralal, D. Kuo, B. Parekh, G. Amaratunga and M. Chhowalla, *J. Mater. Chem.*, **18**, 5909 (2008).

180. C. P. Liu, Z. H. Chen, H. E. Wang, S. K. Jha, W. J. Zhang, I. Bello and J. A. Zapien, *Appl. Phys. Lett.*, **100**, 243102 (2012).

181. L. B. Hu, H. S. Kim, J. Y. Lee, P. Peumans and Y. Cui, *ACS Nano*, **4**, 2955 (2010).

182. P. Lee, J. Lee, H. Lee, J. Yeo, S. Hong, K. H. Nam, D. Lee, S. S. Lee and S. H. Ko, *Adv. Mater.*, **24**, 3326 (2012).

183. A. R. Rathmell, M. Nguyen, M. F. Chi and B. J. Wiley, *Nano Lett.*, **12**, 3193 (2012).

184. G. X. Ni, Y. Zheng, S. Bae, C. Y. Tan, O. Kahya, J. Wu, B. H. Hong, K. Yao and B. Ozyilmaz, *ACS Nano*, **6**, 3935 (2012).

185. G. J. Snyder and E. S. Toberer, *Nat. Mater.*, **7**, 105 (2008).

186. P. F. P. Poudeu, J. D'Angelo, H. Kong, A. Downey, J. L. Short, R. Pcionek, T. P. Hogan, C. Uher and M. G. Kanatzidis, *J. Am. Chem. Soc.*, **128**, 14347 (2006).

187. D. Liang, H. Yang, S. W. Finefrock and Y. Wu, *Nano. Lett.*, **12**, 2140 (2012).

188. M. G. Kanatzidis, *Chem. Mater.*, **22**, 648 (2010).

189. A. I. Hochbaum and P. D. Yang, *Chem. Rev*, **110**, 527 (2010).

190. J. Y. Tang, H. T. Wang, D. H. Lee, M. Fardy, Z. Y. Huo, T. P. Russell and P. D. Yang, *Nano Lett.*, **10**, 4279 (2010).

191. J. L. Liu, J. Sun and L. Gao, *Nanoscale*, **3**, 3616 (2011).

192. C. Yu, K. Choi, L. Yin and J. C. Grunlan, *ACS Nano*, **5**, 7885 (2011).

193. X.-Z. Guo, Y.-D. Zhang, D. Qin, Y.-H. Luo, D.-M. Li, Y.-T. Pang and Q.-B. Meng, *J. Power Sources*, **195**, 7684 (2010).

194. Z. L. Wang and J. Song, *Science*, **312**, 242 (2006).

195. X. Wang, J. Song, J. Liu and Z. L. Wang, *Science*, **316**, 102 (2007).

196. Y. Qin, X. Wang and Z. L. Wang, *Nature*, **451**, 809 (2008).

197. S. Xu, Y. Qin, C. Xu, Y. Wei, R. Yang and Z. L. Wang, *Nat. Nano.*, **5**, 366 (2010).

198. L. Ji, Z. Lin, M. Alcoutlabi and X. Zhang, *Energy Environ. Sci.*, **4**, 2682 (2011).

199. B. Gao, C. Bower, J. D. Lorentzen, L. Fleming, A. Kleinhammes, X. P. Tang, L. E. McNeil, Y. Wu and O. Zhou, *Chem. Phys. Lett.*, **327**, 69 (2000).

200. B. J. Landi, M. J. Ganter, C. D. Cress, R. A. DiLeo and R. P. Raffaelle, *Energy Environ. Sci.*, **2**, 638 (2009).

201. S. Klink, E. Ventosa, W. Xia, F. La Mantia, M. Muhler and W. Schuhmann, *Electrochem. Commun.*, **15**, 10 (2012).

202. J. Y. Eom, H. S. Kwon, J. Liu and O. Zhou, *Carbon*, **42**, 2589 (2004).

203. X. X. Wang, J. N. Wang, H. Chang and Y. F. Zhang, *Adv. Funct. Mater.*, **17**, 3613 (2007).

204. G. X. Wang, X. P. Shen, J. Yao and J. Park, *Carbon*, **47**, 2049 (2009).

205. D. Chen, L. H. Tang and J. H. Li, *Chem. Soc. Rev.*, **39**, 3157 (2010).

206. E. Yoo, J. Kim, E. Hosono, H. Zhou, T. Kudo and I. Honma, *Nano Lett.*, **8**, 2277 (2008).

207. D. H. Wang, R. Kou, D. Choi, Z. G. Yang, Z. M. Nie, J. Li, L. V. Saraf, D. H. Hu, J. G. Zhang, G. L. Graff, J. Liu, M. A. Pope and I. A. Aksay, *ACS Nano*, **4**, 1587 (2010).

208. F. Liu, S. Song, D. Xue and H. Zhang, *Adv. Mater.*, **24**, 1089 (2012).

209. C. Wang, D. Li, C. O. Too and G. G. Wallace, *Chem. Mater.*, **21**, 2604 (2009).

210. O. C. Compton, B. Jain, D. A. Dikin, A. Abouimrane, K. Amine and S. T. Nguyen, *ACS Nano*, **5**, 4380 (2011).

211. Y. Yang, S. Huang, H. He, A. W. H. Mau and L. Dai, *J. Am. Chem. Soc.*, **121**, 10832 (1999).

212. J. Chen, Y. Liu, A. I. Minett, C. Lynam, J. Z. Wang and G. G. Wallace, *Chem. Mater.*, **19**, 3595 (2007).

213. H. Lee, J. K. Yoo, J. H. Park, J. H. Kim, K. Kang and Y. S. Jung, *Advanced Energy Materials*, **2**, 976 (2012).

214. C.-M. Park, J.-H. Kim, H. Kim and H.-J. Sohn, *Chem. Soc. Rev.*, **39**, 3115 (2010).

215. J. Cabana, L. Monconduit, D. Larcher and M. R. Palacín, *Adv. Mater.*, **22**, E170 (2010).

216. J. Jiang, J. P. Liu, X. T. Huang, Y. Y. Li, R. M. Ding, X. X. Ji, Y. Y. Hu, Q. B. Chi and Z. H. Zhu, *Cryst. Growth. Des.*, **10**, 70 (2009).

217. D. Larcher, C. Masquelier, D. Bonnin, Y. Chabre, V. Masson, J. B. Leriche and J. M. Tarascon, *J. Electrochem. Soc.*, **150**, A133 (2003).

218. K. T. Nam, D. W. Kim, P. J. Yoo, C. Y. Chiang, N. Meethong, P. T. Hammond, Y. M. Chiang and A. M. Belcher, *Science*, **312**, 885–888 (2006).

219. Z.-M. Cui, L.-Y. Jiang, W.-G. Song and Y.-G. Guo, *Chem. Mater.*, **21**, 1162 (2009).

220. Z.-S. Wu, W. Ren, L. Wen, L. Gao, J. Zhao, Z. Chen, G. Zhou, F. Li and H.-M. Cheng, *ACS Nano*, **4**, 3187 (2010).

221. C. H. Chen, B. J. Hwang, J. S. Do, J. H. Weng, M. Venkateswarlu, M. Y. Cheng, R. Santhanam, K. Ragavendran, J. F. Lee, J. M. Chen and D. G. Liu, *Electrochem. Commun.*, **12**, 496 (2010).

222. K. Zhong, X. Xia, B. Zhang, H. Li, Z. Wang and L. Chen, *J. Power Sources*, **195**, 3300 (2010).

223. M. M. Thackeray, Progress in Solid State Chemistry, **25**, 1 (1997).

224. X. Ji, P. S. Herle, Y. Rho and L. F. Nazar, *Chem. Mater.*, **19**, 374 (2007).

225. S. Grugeon, S. Laruelle, R. Herrera-Urbina, L. Dupont, P. Poizot and J. M. Tarascon, *J. Electrochem. Soc.*, **148**, A285 (2001).

226. Y. H. Lee, I. C. Leu, S. T. Chang, C. L. Liao and K. Z. Fung, *Electrochim. Acta.*, **50**, 553 (2004).

227. J. Hu, H. Li and X. J. Huang, *Electrochem. Solid St.*, **8**, A66 (2005).

228. P. Poizot, S. Laruelle, S. Grugeon, L. Dupont and J. M. Tarascon, *Nature*, **407**, 496 (2000).

229. O. Delmer, P. Balaya, L. Kienle and J. Maier, *Adv. Mater.*, **20**, 501 (2008).

230. A. M. Chockla, J. T. Harris, V. A. Akhavan, T. D. Bogart, V. C. Holmberg, C. Steinhagen, C. B. Mullins, K. J. Stevenson and B. A. Korgel, *J. Am. Chem. Soc.*, **133**, 20914 (2011).

231. C. K. Chan, H. L. Peng, G. Liu, K. McIlwrath, X. F. Zhang, R. A. Huggins and Y. Cui, *Nat. Nanotechnol.*, **3**, 31 (2008).

232. T. Song, J. L. Xia, J. H. Lee, D. H. Lee, M. S. Kwon, J. M. Choi, J. Wu, S. K. Doo, H. Chang, W. Il Park, D. S. Zang, H. Kim, Y. G. Huang, K. C. Hwang, J. A. Rogers and U. Paik, *Nano Lett.*, **10**, 1710 (2010).

233. H. Kim, B. Han, J. Choo and J. Cho, *Angew. Chem. Int. Edit.*, **47**, 10151 (2008).

234. D. H. Wang, D. W. Choi, J. Li, Z. G. Yang, Z. M. Nie, R. Kou, D. H. Hu, C. M. Wang, L. V. Saraf, J. G. Zhang, I. A. Aksay and J. Liu, *ACS Nano*, **3**, 907 (2009).

235. Z. S. Wu, W. C. Ren, L. Wen, L. B. Gao, J. P. Zhao, Z. P. Chen, G. M. Zhou, F. Li and H. M. Cheng, *ACS Nano*, **4**, 3187 (2010).

236. X. L. Ji, K. T. Lee and L. F. Nazar, *Nat. Mater.*, **8**, 500 (2009).

237. G. G. Amatucci, C. N. Schmutz, A. Blyr, C. Sigala, A. S. Gozdz, D. Larcher and J. M. Tarascon, *J. Power Sources*, **69**, 11 (1997).

238. W. M. Zhang, J. S. Hu, Y. G. Guo, S. F. Zheng, L. S. Zhong, W. G. Song and L. J. Wan, *Adv. Mater.*, **20**, 1160 (2008).

239. L. F. Cui, Y. Yang, C. M. Hsu and Y. Cui, *Nano Lett.*, **9**, 3370 (2009).

240. S. H. Ng, J. Z. Wang, D. Wexler, K. Konstantinov, Z. P. Guo and H. K. Liu, *Angew. Chem. Int. Edit.*, **45**, 6896 (2006).

241. Y. S. Hu, R. Demir-Cakan, M. M. Titirici, J. O. Muller, R. Schlogl, M. Antonietti and J. Maier, *Angew. Chem. Int. Edit.*, **47**, 1645 (2008).

242. N. Dimov, S. Kugino and M. Yoshio, *Electrochim. Acta.*, **48**, 1579 (2003).

243. G. X. Wang, J. H. Ahn, J. Yao, S. Bewlay and H. K. Liu, *Electrochem. Commun.*, **6**, 689 (2004).

244. H. Kim and J. Cho, *Nano Lett.*, **8**, 3688 (2008).

245. L. F. Cui, L. B. Hu, J. W. Choi and Y. Cui, *ACS Nano*, **4**, 3671 (2010).

246. Y. S. Luo, J. S. Luo, J. Jiang, W. W. Zhou, H. P. Yang, X. Y. Qi, H. Zhang, H. J. Fan, D. Y. W. Yu, C. M. Li and T. Yu, *Energy Environ. Sci.*, **5**, 6559 (2012).

247. J.-Z. Wang, S.-L. Chou, J. Chen, S.-Y. Chew, G.-X. Wang, K. Konstantinov, J. Wu, S.-X. Dou and H. K. Liu, *Electrochem. Commun.*, **10**, 1781 (2008).

248. J. K. Kim, J. Manuel, M. H. Lee, J. Scheers, D. H. Lim, P. Johansson, J. H. Ahn, A. Matic and P. Jacobsson, *J. Mater. Chem.*, **22**, 15045 (2012).

249. X. L. Jia, C. Z. Yan, Z. Chen, R. R. Wang, Q. Zhang, L. Guo, F. Wei and Y. F. Lu, *Chem. Commun.*, **47**, 9669 (2011).

250. J. Muster, G. T. Kim, V. Krstic, J. G. Park, Y. W. Park, S. Roth and M. Burghard, *Adv. Mater.*, **12**, 420 (2000).

251. X. Y. Chen, X. Wang, Z. H. Wang, J. X. Wan, J. W. Liu and Y. T. Qian, *Nanotechnology*, **15**, 1685 (2004).

252. Z. Chen, V. Augustyn, J. Wen, Y. W. Zhang, M. Q. Shen, B. Dunn and Y. F. Lu, *Adv. Mater.*, **23**, 791 (2011).

253. S. Myung, M. Lee, G. T. Kim, J. S. Ha and S. Hong, *Adv. Mater.*, **17**, 2361 (2005).

254. J. C. P. Gabriel and P. Davidson, *Colloid Chemistry*, **226**, 119 (2003).

255. M. G. Ancona, S. E. Kooi, W. Kruppa, A. W. Snow, E. E. Foos, L. J. Whitman, D. Park and L. Shirey, *Nano Lett.*, **3**, 135 (2003).

256. X. C. Wu, Y. R. Tao, D. Lin, Z. H. Wang and H. Zheng, *Mater. Res. Bull.*, **40**, 315 (2005).

257. K. H. Seng, J. Liu, Z. P. Guo, Z. X. Chen, D. Jia and H. K. Liu, *Electrochem. Commun.*, **13**, 383 (2011).

258. T. Zhai, H. Liu, H. Li, X. Fang, M. Liao, L. Li, H. Zhou, Y. Koide, Y. Bando and D. Golberg, *Adv. Mater.*, **22**, 2547 (2010).

259. L. Noerochim, J. Z. Wang, D. Wexler, M. M. Rahman, J. Chen and H. K. Liu, *J. Mater. Chem.*, **22**, 11159 (2012).

260. G. Zhou, D.-W. Wang, F. Li, P.-X. Hou, L. Yin, C. Liu, G. Q. Lu, I. R. Gentle and H.-M. Cheng, *Energy Environ. Sci.*, **5**, 8901 (2012).

261. N. Singh, C. Galande, A. Miranda, A. Mathkar, W. Gao, A. L. M. Reddy, A. Vlad and P. M. Ajayan, *Scientific Reports*, **2**, (2012).

262. H. Nishide and K. Oyaizu, *Science*, **319**, 737 (2008).

263. P. Hiralal, S. Imaizumi, H. E. Unalan, H. Matsumoto, M. Minagawa, M. Rouvala, A. Tanioka and G. A. J. Amaratunga, *ACS Nano*, **4**, 2730 (2010).

264. D. Wei, P. Andrew, H. F. Yang, Y. Y. Jiang, F. H. Li, C. S. Shan, W. D. Ruan, D. X. Han, L. Niu, C. Bower, T. Ryhanen, M. Rouvala, G. A. J. Amaratunga and A. Ivaska, *J. Mater. Chem.*, **21**, 9762 (2011).

265. P. Knauth, *Solid State Ionics*, **180**, 911 (2009).

266. M. Koo, K.-I. Park, S. H. Lee, M. Suh, D. Y. Jeon, J. W. Choi, K. Kang and K. J. Lee, *Nano Lett.*, **12**, 4810 (2012).

267. L. B. Hu, H. Wu, F. La Mantia, Y. A. Yang and Y. Cui, *ACS Nano*, **4**, 5843 (2010).

268. R. Kotz and M. Carlen, *Electrochim. Acta*, **45**, 2483 (2000).

269. P. Simon and Y. Gogotsi, *Nat. Mater.*, **7**, 845 (2008).

270. N.-L. Wu, *Mater. Chem. Phys.*, **75**, 6 (2002).

271. T. Brousse, M. Toupin, R. Dugas, L. Athouel, O. Crosnier and D. Belanger, *J. Electrochem. Soc.*, **153**, A2171 (2006).

272. Z. Q. Niu, W. Y. Zhou, J. Chen, G. X. Feng, H. Li, W. J. Ma, J. Z. Li, H. B. Dong, Y. Ren, D. A. Zhao and S. S. Xie, *Energy Environ. Sci.*, **4**, 1440 (2011).

273. N. A. Kumar, I. Y. Jeon, G. J. Sohn, R. Jain, S. Kumar and J. B. Baek, *ACS Nano*, **5**, 2324 (2011).

274. S. Hu, R. Rajamani and X. Yu, *Appl. Phys. Lett.*, **100**, 104103 (2012).

275. A. P. Yu, I. Roes, A. Davies and Z. W. Chen, *Appl. Phys. Lett.*, **96**, 253105 (2010).

276. L. L. Zhang, X. Zhao, M. D. Stoller, Y. W. Zhu, H. X. Ji, S. Murali, Y. P. Wu, S. Perales, B. Clevenger and R. S. Ruoff, *Nano Lett.*, **12**, 1806 (2012).

277. M. Segal, *Nat. Nanotechnol.*, **4**, 611 (2009).

278. M. D. Stoller, S. Park, Y. Zhu, J. An and R. S. Ruoff, *Nano Lett.*, **8**, 3498 (2008).

279. M. F. El-Kady, V. Strong, S. Dubin and R. B. Kaner, *Science*, **335**, 1326 (2012).

280. C. Masarapu, L. P. Wang, X. Li and B. Q. Wei, *Advanced Energy Materials*, **2**, 546 (2012).

281. X. Li, J. P. Rong and B. Q. Wei, *ACS Nano*, **4**, 6039 (2010).

282. S. Chen, J. W. Zhu, X. D. Wu, Q. F. Han and X. Wang, *ACS Nano*, **4**, 2822 (2010).

283. K. W. Nam, W. S. Yoon and K. B. Kim, *Electrochim. Acta.*, **47**, 3201 (2002).

284. J. W. Lang, L. B. Kong, M. Liu, Y. C. Luo and L. Kang, *J. Solid State Electr.*, **14**, 1533 (2010).

285. Y. G. Wang, L. Yu and Y. Y. Xia, *J. Electrochem. Soc.*, **153**, A743 (2006).

286. H. K. Kim, T. Y. Seong, J. H. Lim, W. I. Cho and Y. S. Yoon, *J. Power Sources*, **102**, 167 (2001).

287. T. Zhu, J. S. Chen and X. W. Lou, *J. Mater. Chem.*, **20**, 7015 (2010).

288. D. Liu, Q. Wang, L. Qiao, F. Li, D. Wang, Z. Yang and D. He, *J. Mater. Chem.*, **22**, 483 (2012).

289. J. Liu, J. Jiang, C. Cheng, H. Li, J. Zhang, H. Gong and H. J. Fan, *Adv. Mater.*, **23**, 2076 (2011).

290. J. Liu, J. Jiang, M. Bosman and H. J. Fan, *J. Mater. Chem.*, **22**, 2419 (2012).

291. C. Guan, X. Li, Z. Wang, X. Cao, C. Soci, H. Zhang and H. J. Fan, *Adv. Mater.*, **24**, 4186 (2012).

292. J. Liu, C. Cheng, W. Zhou, H. Li and H. J. Fan, *Chem. Commun.*, **47**, 3436 (2011).

293. C. Guan, J. Liu, C. Cheng, H. Li, X. Li, W. Zhou, H. Zhang and H. J. Fan, *Energy Environ. Sci.*, **4**, 4496 (2011).

294. C. Z. Yuan, L. Yang, L. R. Hou, J. Y. Li, Y. X. Sun, X. G. Zhang, L. F. Shen, X. J. Lu, S. L. Xiong and X. W. Lou, *Adv. Funct. Mater.*, **22**, 2560 (2012).

295. S. D. Perera, B. Patel, N. Nijem, K. Roodenko, O. Seitz, J. P. Ferraris, Y. J. Chabal and K. J. Balkus, *Advanced Energy Materials*, **1**, 936 (2011).

296. V. Presser, L. F. Zhang, J. J. Niu, J. McDonough, C. Perez, H. Fong and Y. Gogotsi, *Advanced Energy Materials*, **1**, 423 (2011).

297. Z. S. Wu, W. C. Ren, D. W. Wang, F. Li, B. L. Liu and H. M. Cheng, *ACS Nano*, **4**, 5835 (2010).

298. Z. S. Wu, D. W. Wang, W. Ren, J. Zhao, G. Zhou, F. Li and H. M. Cheng, *Adv. Funct. Mater.*, **20**, 3595 (2010).

299. F. H. Li, J. F. Song, H. F. Yang, S. Y. Gan, Q. X. Zhang, D. X. Han, A. Ivaska and L. Niu, *Nanotechnology*, **20** (2009).

300. T. Brezesinski, J. Wang, S. H. Tolbert and B. Dunn, *Nat. Mater.*, **9**, 146 (2010).

301. P. C. Chen, G. Shen, S. Sukcharoenchoke and C. Zhou, *Appl. Phys. Lett.*, **94**, 043113 (2009).

302. G. L. Guo, L. Huang, Q. H. Chang, L. C. Ji, Y. Liu, Y. Q. Xie, W. Z. Shi and N. Q. Jia, *Appl. Phys. Lett.*, **99**, 083111 (2011).

303. Y. Y. Horng, Y. C. Lu, Y. K. Hsu, C. C. Chen, L. C. Chen and K. H. Chen, *J. Power Sources*, **195**, 4418 (2010).

304. W. Chen, R. B. Rakhi, L. Hu, X. Xie, Y. Cui and H. N. Alshareef, *Nano Lett.*, **11**, 5165 (2011).

305. Y. C. Chen, Y. K. Hsu, Y. G. Lin, Y. K. Lin, Y. Y. Horng, L. C. Chen and K. H. Chen, *Electrochim. Acta*, **56**, 7124 (2011).

306. L. Y. Yuan, X. H. Lu, X. Xiao, T. Zhai, J. J. Dai, F. C. Zhang, B. Hu, X. Wang, L. Gong, J. Chen, C. G. Hu, Y. X. Tong, J. Zhou and Z. L. Wang, *ACS Nano*, **6**, 656 (2012).

307. X. H. Lu, T. Zhai, X. H. Zhang, Y. Q. Shen, L. Y. Yuan, B. Hu, L. Gong, J. Chen, Y. H. Gao, J. Zhou, Y. X. Tong and Z. L. Wang, *Adv. Mater.*, **24**, 938 (2012).

308. Y. S. Luo, J. Jiang, W. W. Zhou, H. P. Yang, J. S. Luo, X. Y. Qi, H. Zhang, D. Y. W. Yu, C. M. Li and T. Yu, *J. Mater. Chem.*, **22**, 8634 (2012).

309. M. Toupin, T. Brousse and D. Bélanger, *Chem. Mater.*, **16**, 3184 (2004).

310. G. Yu, L. Hu, M. Vosgueritchian, H. Wang, X. Xie, J. R. McDonough, X. Cui, Y. Cui and Z. Bao, *Nano Lett.*, **11**, 2905 (2011).

311. L. Hu, W. Chen, X. Xie, N. Liu, Y. Yang, H. Wu, Y. Yao, M. Pasta, H. N. Alshareef and Y. Cui, *ACS Nano*, **5**, 8904 (2011).

312. C. Z. Yuan, L. R. Hou, D. K. Li, L. F. Shen, F. Zhang and X. G. Zhang, *Electrochim. Acta*, **56**, 6683 (2011).

313. K. Wang, P. Zhao, X. M. Zhou, H. P. Wu and Z. X. Wei, *J. Mater. Chem.*, **21**, 16373 (2011).

314. Y. J. Kang, S. J. Chun, S. S. Lee, B. Y. Kim, J. H. Kim, H. Chung, S. Y. Lee and W. Kim, *ACS Nano*, **6**, 6400 (2012).

315. L. B. Hu, M. Pasta, F. La Mantia, L. F. Cui, S. Jeong, H. D. Deshazer, J. W. Choi, S. M. Han and Y. Cui, *Nano Lett.*, **10**, 708 (2010).

316. L. H. Bao and X. D. Li, *Adv. Mater.*, **24**, 3246 (2012).

317. M. A. Q. Xue, F. W. Li, J. Zhu, H. Song, M. N. Zhang and T. B. Cao, *Adv. Funct. Mater.*, **22**, 1284 (2012).

318. Z. Q. Niu, P. S. Luan, Q. Shao, H. B. Dong, J. Z. Li, J. Chen, D. Zhao, L. Cai, W. Y. Zhou, X. D. Chen and S. S. Xie, *Energy Environ. Sci.*, **5**, 8726 (2012).

319. D. W. Wang, F. Li, J. P. Zhao, W. C. Ren, Z. G. Chen, J. Tan, Z. S. Wu, I. Gentle, G. Q. Lu and H. M. Cheng, *ACS Nano*, **3**, 1745 (2009).

320. J. Bae, M. K. Song, Y. J. Park, J. M. Kim, M. Liu and Z. L. Wang, *Angew. Chem.Int.Ed.*, **50**, 1683 (2011).

321. Y. Fu, X. Cai, H. Wu, Z. Lv, S. Hou, M. Peng, X. Yu and D. Zou, *Adv. Mater.*, **24**, 5713 (2012).

322. L. B. Hu, H. Wu and Y. Cui, *Appl. Phys. Lett.*, **96**, 183502 (2010).

323. X. F. Wang, B. Liu, Q. F. Wang, W. F. Song, X. J. Hou, D. Chen and G. Z. Shen, *Adv. Mater.*, **25**, 1479 (2012).

324. B. Liu, D. S. Tan, X. F. Wang, D. Chen, G. Z. Shen, *Small*, **9**, 1869 (2012).

325. K. Wang, W. J. Zou, B. G. Quan, A. F. Yu, H. P. Wu, P. Jiang and Z. X. Wei, *Adv. Energy Mater.*, **1**, 1068 (2011).

326. P. Krawiec, M. Kramer, M. Sabo, R. Kunschke, H. Fröde and S. Kaskel, *Adv. Eng. Mater.*, **8**, 293 (2006).

327. V. Bérubé, G. Radtke, M. Dresselhaus and G. Chen, *Int. J. Energy Res.*, **31**, 637 (2007).

328. A. Saito, International Journal of Refrigeration, **25**, 177 (2002).

329. S. M. Wu, G. Y. Fang and X. Liu, *Chem. Eng. Technol.*, **33**, 455 (2010).

330. M. M. Farid, A. M. Khudhair, S. A. K. Razack and S. Al-Hallaj, *Energy Conversion and Management*, **45**, 1597 (2004).

331. X. Y. Xue, S. H. Wang, W. X. Guo, Y. Zhang and Z. L. Wang, *Nano Lett.*, **12**, 5048 (2012).

332. W. X. Guo, X. Y. Xue, S. H. Wang, C. J. Lin and Z. L. Wang, *Nano Lett.*, **12**, 2520 (2012).

333. J. Bae, Y. J. Park, M. Lee, S. N. Cha, Y. J. Choi, C. S. Lee, J. M. Kim and Z. L. Wang, *Adv. Mater.*, **23**, 3446 (2011).

CHAPTER 7

NEXT GENERATION FLEXIBLE SOLAR CELLS

Wei Chen*, Wenjun Zhang, Huan Wang and Xianwei Zeng

*Wuhan National Laboratory for Optoelectronics,
Huazhong University of Science and Technology
Wuhan, Hubei 430074, P.R. China
wnlochenwei@mail.hust.edu.cn

Flexible solar cells are paving the way to low-cost electricity. Flexible fabrication of several important kinds of next generation solar cells, including dye-sensitized solar cell, organic solar cell, perovskite solar cell (PSSC), have been overviewed. The content is arranged as the general sequence of: research background/history, working principle, key components and techniques, and future research prospects.

1. Introduction

As said by International Energy Agency in 2011, solar energy can make considerable contributions to solving some of the most urgent problems the world now faces. The development of affordable, inexhaustible, and clean solar energy technologies will have huge long-term benefits. It will increase countries' energy security through reliance on an indigenous, inexhaustible and mostly import-independent resource, enhance sustainability, reduce pollution, lower the costs of mitigating climate change, and keep fossil fuel prices lower than otherwise.[1]

Though with so much benefit, solar energy up-to-date has only occupied a very small portion of global energy consumption. The photovoltaics (PV's) technology is one of the most important approaches for solar energy utilization, by converting sunlight directly into electricity. Though the worldwide installed solar PV capacity grew fast during the last 10 years (>30% per year) and reached a total of 139 GW in 2013, which is sufficient to generate 160 TWh every year, it is only about 0.85% of the electricity demand on the planet. This is far below the solar energy absorbed by the earth, which is more energy in one hour than the world used in one year.[2]

The biggest challenge for scale-up application of PV technology is the price consideration or to say how to decrease the payback time. By the end of 2012, the "best in class" crystalline Si module price has dropped to $0.50/watt, and is expected to drop to $0.36/watt by 2017.[3] Despite the falling prices it still cannot challenge the electricity price generated by fossil resources. Given the current electricity price of 0.25 euro/KWh in Germany, the PV module with installation cost of 1,700 euro/KWp requires seven years to payback by the electricity generated by the PV system.[4] Such a prolonged payback time may scare away the individual buyers, especially if the local government has no specific policy support.

In this circumstance, flexible PV technologies become a promising choice, which could further decrease the module price due to high output roll-to-roll fabrication, the as-produced lightweight, flexible modules can be integrated into, not installed on, various surfaces.[5] Especially for today's PV market dominated by rigid crystalline Si solar cells, such specific features provide flexible PVs with different advantages. Low cost, lightweight and flexible solar cells may benefit ~2 billion people lacking access to the grid in the developing world, who would be the first potential customers. The potential market of those flexible PVs could be even greater if other characteristics, such as stability and efficiency could be further improved.[5]

Shown in Fig. 1 are efficiency evolution of different kinds of solar cells, including: the first generation of crystalline Si solar cells; the second generation of conventional thin-film solar cells, such as amorphous Si, copper indium gallium selenide (CIGS), cadmium telluride (CdTe) solar cells; as well as the third generation of new concept solar cells, e.g., dye-sensitized

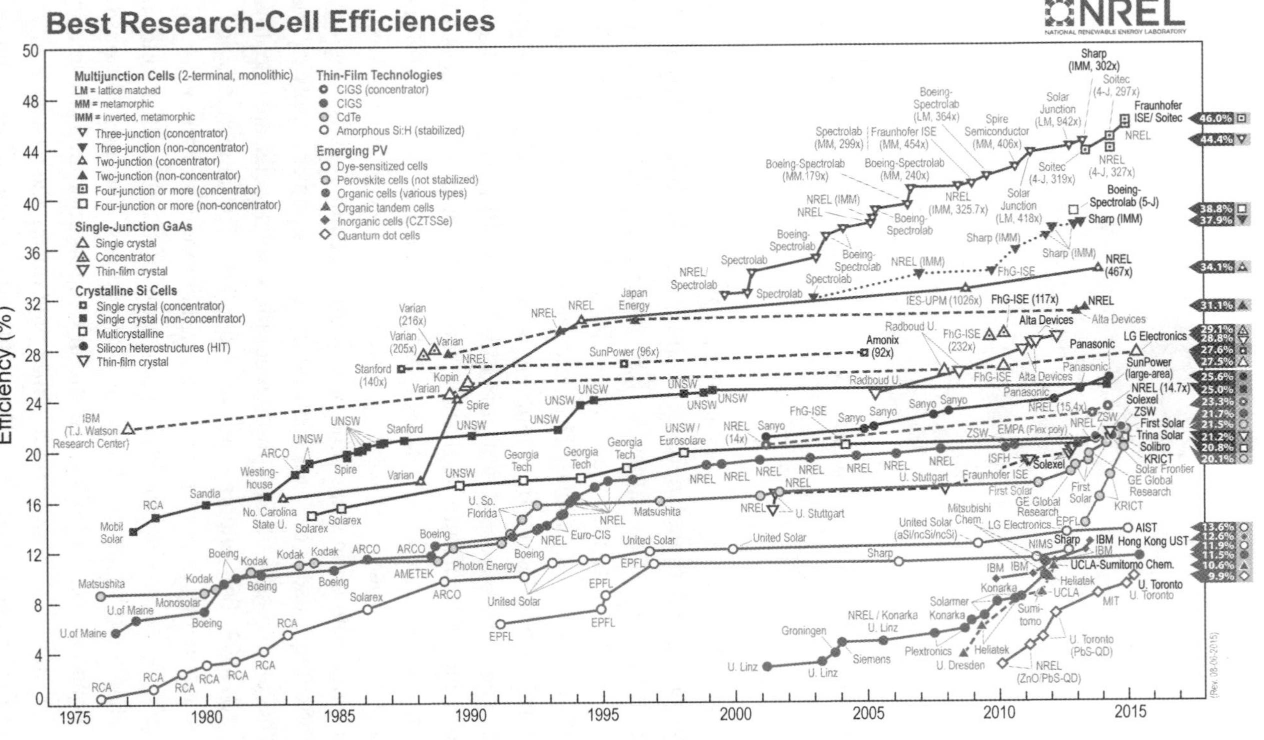

Figure 1.   The latest chart on record efficiencies of solar cells published by National Renewable Energy Laboratory (NREL).[6]

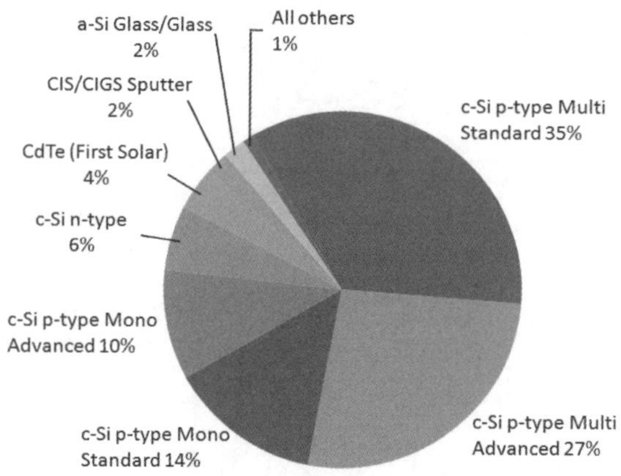

Figure 2.   2014 solar PV module production by technology.
Source: NPD solarbuzz PV Equipment Quarterly.[7]

solar cells (DSSCs), organic solar cells (OPVs), perovskites solar cells (PSSCs), etc. The data included are certified records collected by National Renewable Energy Laboratory (NREL) covering the period of past 40 years. From this figure, the different R&D levels for varying kinds of solar cells can be clearly found. The efficiencies of crystalline Si solar cells are the highest besides multiple junction solar cells and reach their plateau for many years. As one could find from Fig. 2, crystalline Si solar cells dominate more than 90% of current global PV market. The efficiencies of CdTe, CIGS thin-film solar cells are relatively inferior but also close to 20%. They, together with amorphous Si solar cell, share the other ~10% market. High efficiency multi-junction solar cells or gallium arsenide (GaAs) single junction solar cell are used for special purpose, such as power supply of spacecraft. The efficiencies of the third generation solar cells are the lowest, around 10–15%, but catching up very quickly. These technologies still remain in laboratory or pilot-plant scale, without industrialization.

The pilot-scale production and application of flexible solar cells have been implemented by various innovation companies. The most famous ones include Konarka Co., for OPVs, Uni-Solar Co. for amorphous Si solar cells in the USA, though they were both bankrupted during the past several years. However, the bad news had never stopped the investigators' steps, more and

Uni-Solar Co., the USA, offering the low cost of energy for rooftops enabled by solar

PowerFilm Solar teamed up with the U.S. Army to supply PowerFilm-equipped tents

Konarka Co., the USA, roll-to-roll fabrication of OPV

Solar bikini, a wearable technology

Ford C-Max Solar Energy Concept at the 2014 International Consumer Electronics Show

Figure 3. A quick glance on kinds of flexible solar cells, real applications and concepts: (a) Uni-Solar Co. provided, the best flexible a-Si solar cells on rooftops[8]; (b) PowerFilm Co. provided, solar tents for army[9]; (c) the roll-to-roll fabrication line of OPVs by Konarka Co[10]; (d) the concept of wearable solar cloth; (e) the solar power integrated car.[11]

more new companies emerged. From the images shown in Fig. 3, the distinct features of these companies' flexible solar cell products can be clearly identified, which can explain visually why flexible PVs are so attractive.

2. Flexible DSSCs

2.1. *The Structure of Flexible DSSCs*

DSSCs also named "Gratzel Cell" was invented by O'Regan and M. Gratzel in 1991.[12] The cell structure of conventional rigid DSSC is normally based on fluorine tin oxide (FTO) glass substrate. Flexible DSSCs structure is similar. As shown in Fig. 4, it consists of flexible substrate, conducting film, mesoporous film made of nanocrystalline metal oxides, dye sensitizers, electrolyte, and counter electrode.[13] Flexible conducting substrate and the mesoporous film on top constitute the working electrode. Flexible substrates are normally made of plastic films or metal foils. Titanium dioxide (TiO_2), zinc oxide (ZnO) are the most frequently

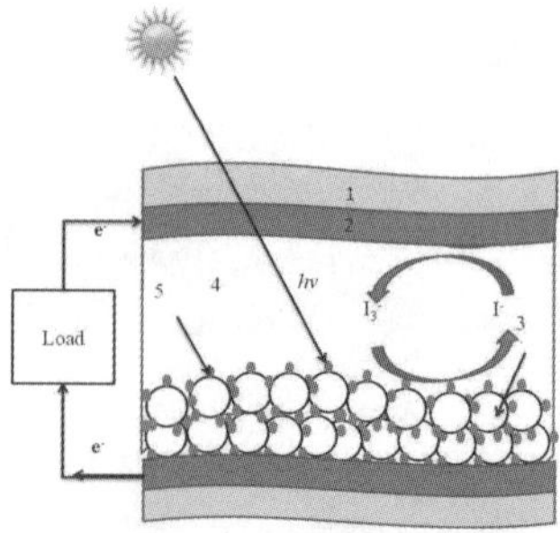

Figure 4. The typical structure of a flexible DSSC — (1) Flexible substrate, (2) Conducting film, (3) Semiconductor mesoporous film, (4) Electrolyte, and (5) dye.

used wide bandgap semiconductors in working electrodes of DSSCs.[14–16] The most frequently reported dye sensitizers are pyridine-Ru complex dyes.[17] Counter electrode is often made of platinum coated flexible conducting substrate. Electrolyte consisting of I_3^-/I^- or Co^{3+}/Co^{2+} redox couple is filled in the space between the working electrode and counter electrode.

2.2. *Working Principle of Flexible DSSCs*

In a DSSC system, a photon-to-electron conversion circle is generally accomplished through the following processes: (i) light excitation to produce electron-hole pairs at the dye monolayer, (ii) separation of the electron-hole pairs at the semiconductor/electrolyte interface by kinetic competition, (iii) free electrons or holes transport separately to the out circuit.

As illustrated in Fig. 5, the processes can be divided into the following seven steps. The first step is the absorption of a photon by the sensitizer D, leading to the excited sensitizer of D^* (Eq. 1). Second, D^* injects an electron into the conduction band of the semiconductor, leaving the sensitizer in the oxidized state of D^+ (Eq. 2). Third, the injected electron is collected through the semiconductor network and flows through the out circuit to the back contact (Eq. 3). Fourth, reduced species in the electrolyte reduce oxidized D^+ (Eq. 4). Fifth, the oxidized species in the electrolyte are reduced at the counter electrode with the help of Pt catalyst (Eq. (5)). Some important undesirable reactions can also occur, such as the recombination of injected

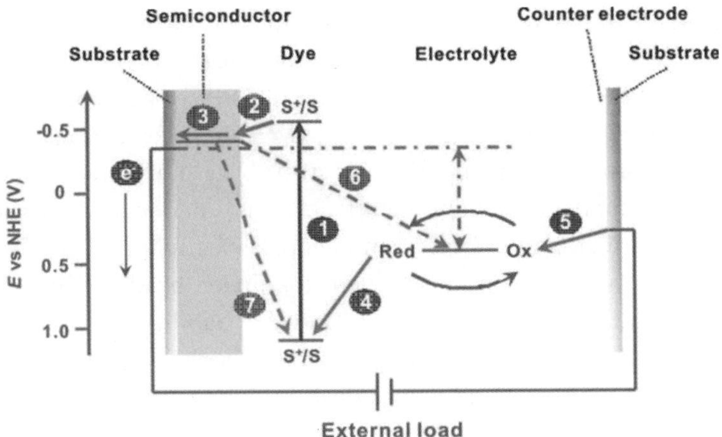

Figure 5. Schematic shown: the working principle of DSSC.

electrons with the oxidized species in electrolyte at the TiO_2 surface (Eq. (6)) or with oxidized sensitizer of D^+ (Eq. (7)).[18]

① $D + hv - D^*$, (Eq. 1)

② $D^* - D^+ + e^-$ (conduction band of TiO_2), (Eq. 2)

③ e^- (conduction band of TiO_2)
$\quad -e^-$ (conduction band of FTO), (Eq. 3)

④ $3I^- + D^+ - I_3^- + D$, (Eq. 4)

⑤ $I_3^- + 2e^-$ (counter electrode) $- 3I^-$, (Eq. 5)

⑥ $I_3^- + 2e^-$ (conduction band of TiO_2) $- 3I^-$, (Eq. 6)

⑦ $D^+ + e^-$ (conduction band of TiO_2) $- D$. (Eq. 7)

2.3. *The Research Progress on Flexible DSSCs*

2.3.1. Different Flexible Substrates

Plastic substrates Flexible plastic substrates have the advantages of light density, high flexibility, and high compatibility to the roll-to-roll techniques. At present, polyethylene terephthalate (PET) and polyethylene naphthalate (PEN) coated with indium tin oxide (ITO) are two typical conductive plastic substrates. PEN has relatively better thermal and chemical stability than PET, however, its price is five times more than that of PET.[19, 20]

Flexible DSSCs based on ITO/PET or ITO/PEN have been reported with the best 7–8% efficiency.[21–26] However, the plastic films can only withstand thermal treatment within 120–150°C. This low temperature range greatly hinders the sintering effect of TiO_2 nanoparticles. The poor physical/electrical contacts between nanoparticles within TiO_2 film, and between TiO_2 film/flexible substrate greatly hinder the effective collection of photo-induced electrons and the device stability during bending. Therefore, low temperature techniques need to be developed for the deposition of highly efficient and mechanically robust mesoporous TiO_2 film. This will be discussed in the latter section.

Metal foils

Metal foils, such as stainless steel foil, nickel foil, titanium foil, can also be flexible. Their resistance can be even lower than conductive glasses. What's more, they can endure high temperature sintering which the plastic substrates cannot. The drawback is that non-transparent metal foils need to be illuminated from the back side, which will arouse certain loss on light harvesting and charge collection.

In 2006, Grätzel *et al.*[27] used Ti foil as substrate and prepared TiO_2 film by screen printing on it. After sintering at 500°C, the flexible DSSC has achieved efficiency up to 7.2%, with platinum coated ITO/PEN as the counter electrode. Light is illustrated from the counter electrode direction. The cell structure is shown in Fig. 6.

In 2008, TiO_2 nanotube arrays prepared on Ti foil substrate by electrochemical method has been applied as photoanode for flexible DSSC

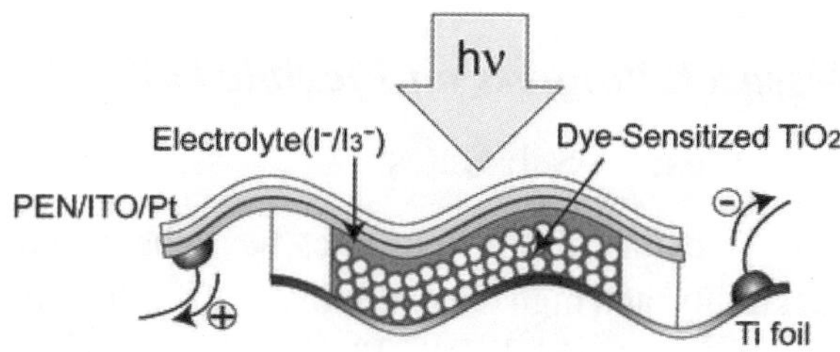

Figure 6.   Configuration of a flexible DSSC on Ti foil with solar light illuminated from counter electrode.[27]

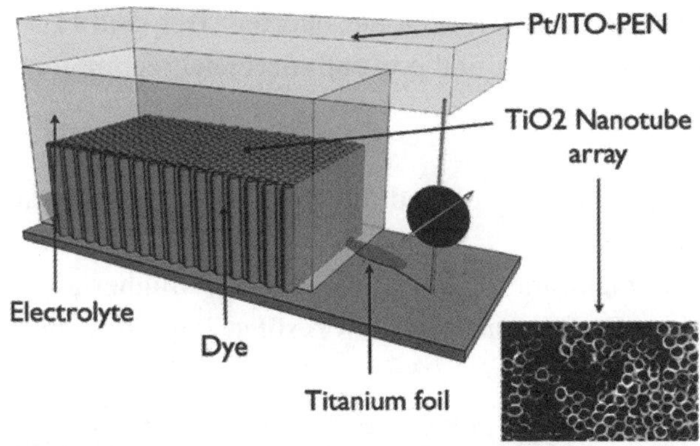

Figure 7.   Configuration of a flexible DSSC illuminated from TiO_2 nanotube array electrode prepared on Ti foil substrate.[28]

application. The platinum coated ITO/PEN was used as the counter electrode, and the ionic liquid was used as electrolyte to replace the commonly volatile organic solvent based electrolyte. The efficiency has reached 3.16%.[28] The structure is shown in Fig. 7.

Jong Hyeok Park studied the high temperature stability of different metal foils covered with an inert protection layer.[29] They used the low-cost stainless steel foil as substrate covered by a SiO_x protection layer and ITO conductive layer on top and finally got 8.16% efficiency. The cell structure is shown in Fig. 8. Inserting SiO_x layer can improve the efficiency by 17%.

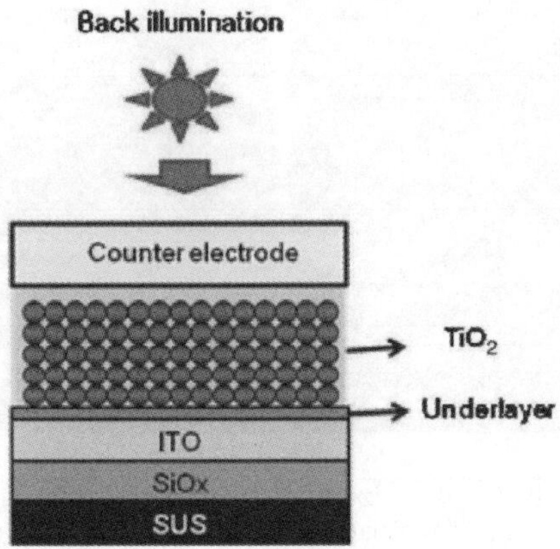

Figure 8.   Configuration of a flexible DSSC on stainless steel foil.[29]

The compact underlayer between mesoporous TiO_2 and ITO can enhance the adhesion of TiO_2 layer and prevent interfacial recombination.

Metal fibers

In 2007, Zou's group[30] employed the metal mesh as substrate in flexible DSSC, as shown in Fig. 9. Compared with conductive plastic films and metal foils, the metal mesh substrate can endure higher angle bending. It also allows complex optical way for solar light passing through the

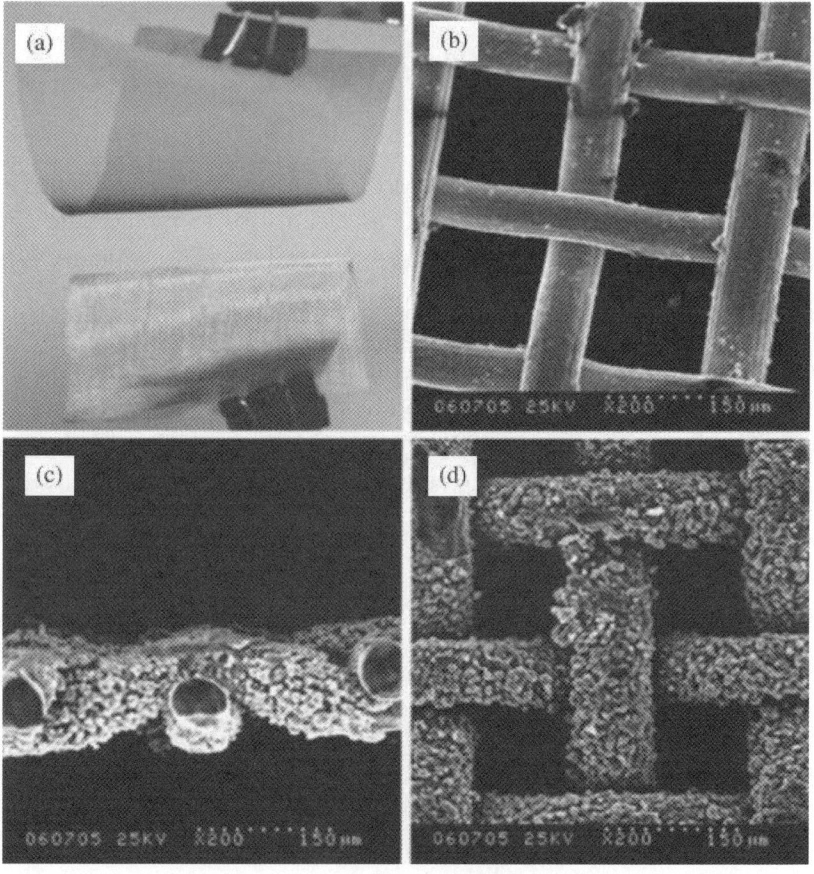

Figure 9.   (a) Optical image of TiO_2 nanoparticles coated metal mesh before and after dye loading, (b) SEM image of the bare metal mesh substrate, (c) top, and (d) sectional views of the TiO_2 nanoparticles coated metal mesh substrate.[30]

electrode, and can increase the light harvesting efficiency. The first example gave a 1.49% efficiency.

In 2008, an improved fibrous DSSC was studied by Xing Fan and co-workers (Fig. 10).[31] It consists of two spiral wound fibers as electrodes. The working electrode is made of porous TiO_2 coated stainless steel fiber. As claimed by the authors, this kind solar cell could work within small space, and the performance could not be affected by light incident angle. The fibrous DSSC obtained V_{oc} of 610 mV, J_{sc} of 0.106 mA cm^{-2} and fill

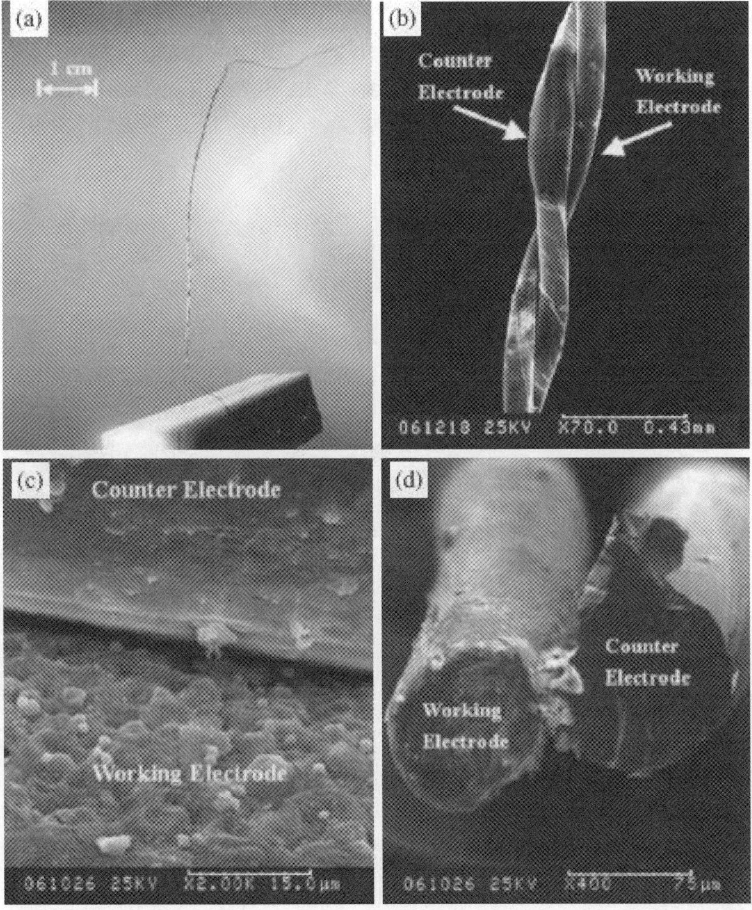

Figure 10.   (a) Optical image and (b–d) SEM images of a fibrous DSSC.[31]

factor of 0.38. It is facile to weave large area fibrous solar cell, because the metal fibers are highly conductive.

2.3.2. Low Temperature Techniques for Mesoporous TiO_2 Films Deposition

Hydrothermal method is widely used for the synthesis of nanocrystals, in which the metastable precursor is hydrolyzed in water medium under high temperature and high pressure, leading to ideal crystal formation. Zhang *et al.*[32] prepared about 10 μm thick mesoporous TiO_2 film on ITO/PET via hydrothermal method. It was obtained by firstly dispersing P25 TiO_2 nanoparticles in $TiCl_4$, $TiSO_4$ or TTIP precursors as the pastes; and after film deposition from such pastes, the films were hydrothermally treated in a Teflon autoclave at 100°C for 12 hours (Fig. 11). The hydrothermal

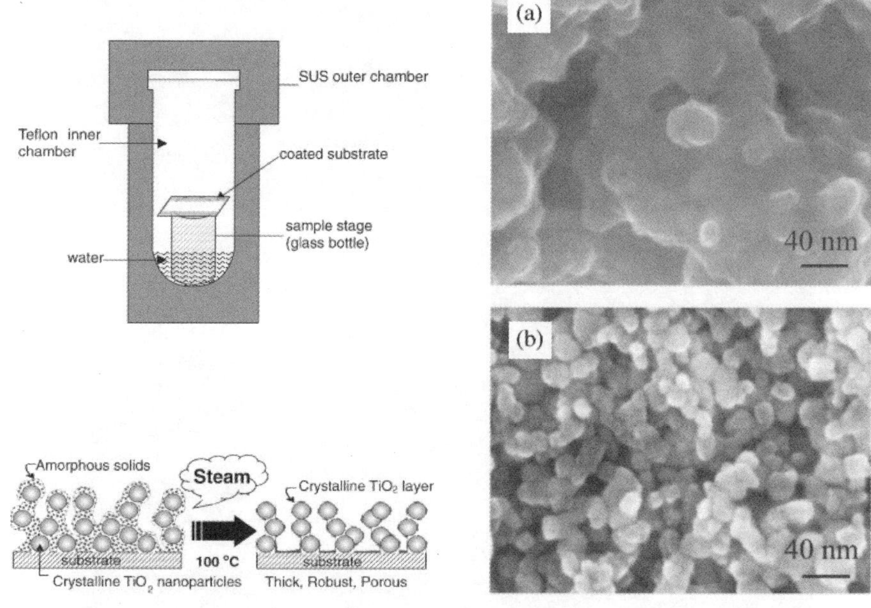

Figure 11.   Schematic shown of hydrothermal treatment on TiO_2 nanoparticles film, SEM images depict the morphology of the TiO_2 film before (a) and after (b) hydrothermal treatment.[32]

Figure 12.   A photo of TiO_2 film coated on ITO/PEN, showing good film uniformity and good adhesion to the substrate.[33]

treatment promotes crystallization of the amorphous precursors. The flexible DSSC finally got an efficiency of 2.5%.

Nanoglue strategy: Miyasaka *et al.*[33] studied the TiO_2 pastes without binder.[34] An acidic water sol containing ultrasmall TiO_2 nanocrystals (act as nanoglue) was used to promote the adhesion between larger TiO_2 particles. A high viscosity paste was obtained. The as-prepared mesoporous TiO_2 film with homogeneous thickness, tightly attached on the surface of PET/ITO flexible substrate, is shown in Fig. 12. The reported cell efficiency was up to 6.4%. Li *et al.*[35] further studied the sticking effect of TiO_2 particles with different sizes by modeling calculation, and concluded that the best ratio of P25 particles, small particles and large particles should be 5:2:2. Their flexible DSSC got 3.05% efficiency at suitable blending ratio.

Electrophoretic deposition has been employed to prepare TiO_2 film for flexible DSSC application. A typical electrophoretic deposition process is illustrated in Fig. 13. The positively charged TiO_2 nanoparticles are dispersed in organic solvent, and the flexible substrate is set as the cathode. Under >1 kV cm^{-1} static electric field, the TiO_2 nanoparticles are driven

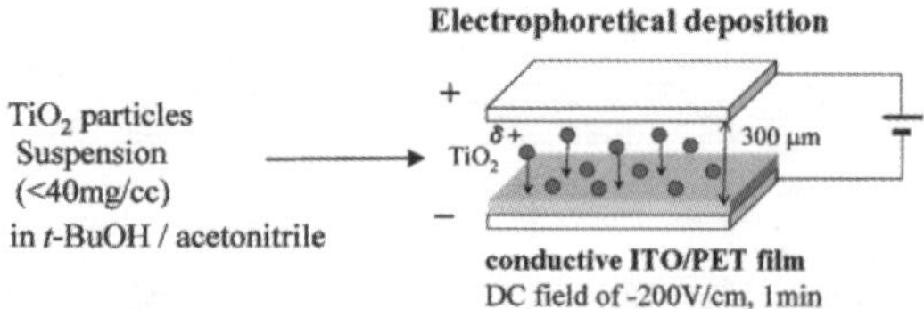

Figure 13.   Schematic diagram of electrophoretic deposition process.

Figure 14.   Flexible DSSC module based on TiO$_2$ film prepared by electrophoretic deposition.[36]

to deposit on flexible substrate with good adhesion. On the basis of this method, Miyasaka *et al.*[36] have reported a flexible DSSC with efficiency of 3–4%. TiO$_2$ nanoparticles with average size of 20 nm and 400 nm were dispersed in a mixture of tert-butyl alcohol and acetonitrile, and PET/ITO substrate was immersed in the above mixture under a direct-current electric field. Their flexible DSSC module is shown in Fig. 14.

UV-ozone treatment can be used to sinter TiO$_2$ nanoparticles because of its strong absorption within such wavelength region. In 2006, Zhang *et al.*[37] reported a facile method to prepare TiO$_2$ film on flexible plastic substrate. Firstly, a small amount of titanium isopropoxide was added to an ethanolic paste of TiO$_2$ nanoparticles, where it hydrolyzed *in situ* and connected the TiO$_2$ particles to form a homogenous and mechanically stable film of up to 10 μm thickness without crack formation. Then, the film was treated by UV-ozone to remove the residue organics deriving from titanium isopropoxide. UV-ozone treatment could extend the electron lifetime in the TiO$_2$ film. The reported efficiency was 3.27%.[38]

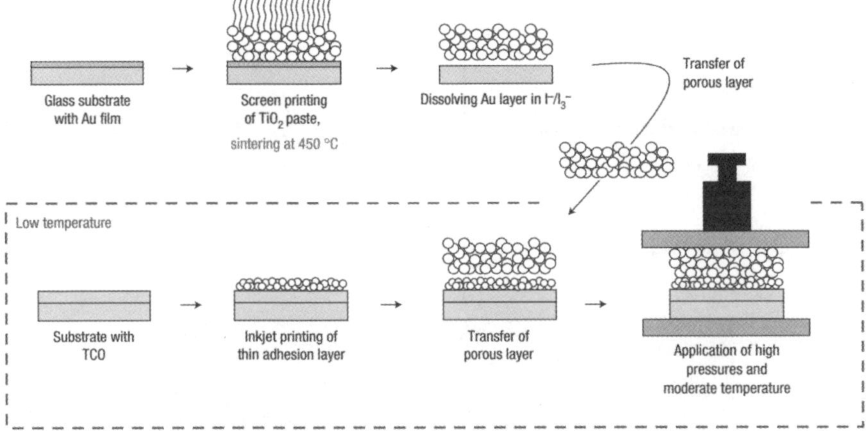

Figure 15.   Schematic shown of the film transfer/mechanical pressure strategy.[40]

Mechanical press was used to promote the inter-connection between the TiO$_2$ nanoparticles. Arakawa *et al.*[39] have used mechanical press method to prepare TiO$_2$ film, with which, they fabricated flexible DSSC and achieved a power conversion efficiency of 6.5%. After further treatement with UV-zone, the power conversion efficiency increased to 7.4%.

Dürr *et al.*[40] have integrated the high temperature sintering method and mechanical press method to fabricate TiO$_2$ film for flexible DSSC application. The process is shown in Fig. 15. The pre-sintered TiO$_2$ film exhibited good internal connection between nanoparticles, which was then transferred to another substrate by mechanical press. Using this method, they fabricated quasi-solid DSSC on ITO-PEN substrate and achieved a power conversion efficiency of 5.8%.

Cold isostatic press applies an isostatic pressure transferred by liquid oil to a powder sample in all directions. It is particularly suitable for making thin-films on plastic substrates. The as-prepared films are crack-free, highly uniform, and with good adhesion to flexible substrate (Fig. 16). Yi-Bing Cheng *et al.*[41] firstly introduced this method to prepare flexible DSSC. Based on P25 nanoparticles, they improved the efficiency of flexible DSSCs from 4.05% to 6.30% after treatment by cold isostatic press at 100 MPa.

Selective laser sintering method: In 2014, Chen *et al.* employed a fast and selective laser sintering to further treat cold isostatic pressed P25 TiO$_2$

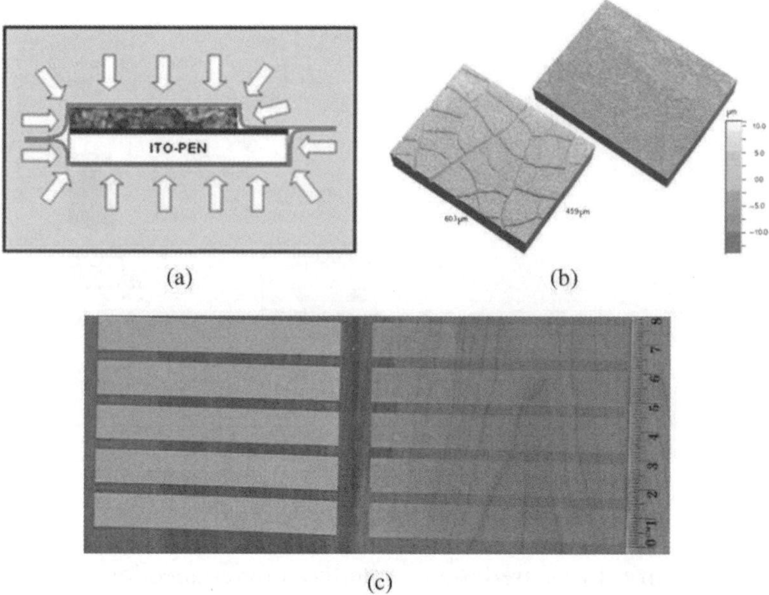

Figure 16. (a) Schematic shown: the principle of cold isostatic press technique for TiO$_2$ film deposition, (b) three-dimensional (3D) profilometry and (c) optical images of the P25 films before and after cold isostatic press.[41]

film on ITO/PEN substrate. The near-IR laser beam can be focused above the TiO$_2$ layer. After fast scan within seconds, the TiO$_2$ nanoparticles could be sintered. Phase transition from anatase to rutile has been observed. As a consequence, electron transport resistance could be reduced and recombination resistance could be increased, leading to improved charge collection efficiency. The efficiency of flexible DSSC has been improved from 4.6% to 5.7%.[42]

2.3.3. Flexible Counter Electrodes

For flexible DSSCs fabricated on plastic conductive substrates, the incident light can pass through the substrate. There is no specific requirement on transparency of the counter electrode. The catalysts fabricated on the glass substrates can all be applicable on the flexible substrates by suitable film deposition techniques. For flexible DSSCs fabricated on metal foils, the incident light can only illuminate from the counter electrodes, which

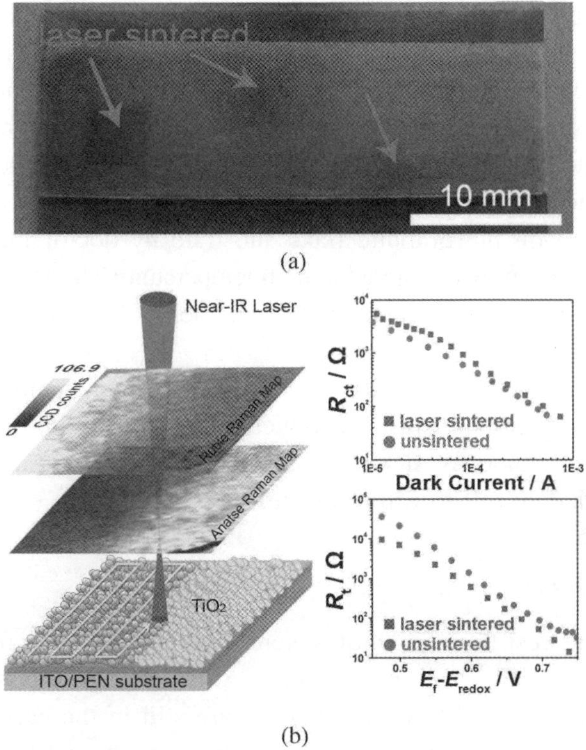

Figure 17. (a) Laser sintered patterns on cold isostatic pressed P25 TiO$_2$ film. (b) schematic shown of laser sintering induced phase transition and the effect on electron transport/recombination resistance. [42]

should be highly transparent. This increases the difficulty on flexible counter electrode preparation. Highly efficient flexible counter electrodes must have low charge transfer resistance, good chemical stability to electrolyte and good adhesion to flexible substrate.

Ma *et al.*[43] compared sputtering deposited Pt film with respect to the references prepared by pyrolysis and electroplating methods. It was found that sputtering deposited Pt film surpassed those prepared by the other two techniques.

Zou *et al.*[44] fabricated carbon-based flexible counter electrode with high porosity, consisting of 3D porous carbon fibers on a polytetrafluoroethylene film. They firstly graphitized the carbon fibers and loaded Pt nanoparticles on the fibers, and then used hot-press approach to attach the Pt modified fibers to the polytetrafluoroethylene film. The electrolyte could be stored in the 3D porous micro-structure of the counter electrode. This solved the problem, such as exfoliation of the encapsulating material and electrolyte leakage, caused by thermal expansion or contraction of solution.

Chen *et al.*[45] used pure carbon to prepare flexible counter electrode. They used flexible graphite flake as substrate. The activated carbon paste was deposited on the graphite flake substrate by doctor blade method. After dried, the film was sintered at high temperature. The DSSC fabricated with this flexible carbon counter electrode achieved a power conversion efficiency of 5%.

Lindstrm *et al.*[46] deposited a mixture of platinum nanoparticles and SnO_2 nanoparticles on plastic conductive substrate with spray coating method. The method for flexible counter electrode preparation is appropriate for scale up roll-to-roll production.

2.4. *Prospects*

Currently, the best flexible DSSCs were reported with the efficiencies of about 7–8%, relatively inferior to the DSSCs on rigid glass substrates. Practical applications of flexible DSSCs are still in the early stage, and many issues need to be resolved. Firstly, the key issue is to promote the power conversion efficiency and stability of flexible DSSC. Secondly, the manufacture process needs to be simplified to reduce cost and realize continuous roll-to-roll production. New materials, new cell structures, new techniques are important. G24i Co. in England has pioneered a production line of flexible DSSC on Ti foils. This kind of commercial investment will be a very important step, which will help convert laboratory materials/ technologies and check their effect in the real application modules.

3. Flexible Organic Solar Cells

3.1. *A Short Overview on Research History*

OPVs employ conductive organic polymers or small organic molecules to convert sunlight to electricity by the PV effect. Because of organic materials with high light extinction coefficient and their compatibility with solution process for film deposition, in principle, OPVs are very suitable for flexible fabrication. The research history of OPVs could be traced back to 1958,[47] when the PV effect of a cell with structure of Al/magnesium phthalocyanine/Ag was firstly demonstrated with a photovoltage of 200 mV. But such single layer structure of OPVs have low quantum efficiencies (<1%) and low power conversion efficiencies (<0.1%). A planar hetero-junction OPV with efficiency of 1% was then invented by Ching W. Tang of Eastman Kodak Co. in 1986.[48] This could be regarded as an important milestone during OPV research and development. In 1995, Heeger *et al.* employed phenyl-C61-butyric acid methyl ester (PCBM) as acceptor and poly[2-methoxy-5-(2-ethylhexyloxy)-1,4-phenylenevinylene] (MEH-PPV) as donor to make the first bulk-heterojunction OPV.[49] In this structure, electron donor and acceptor are blended with their interfaces controlled in a nanometer scale. Based on this structure, Konarka Technologies Co. reported an efficiency of 5.21% for plastic solar cells with an active area of 1.024 cm^2.[50] Recently, tandem structured OPVs were developed, which effectively broadened the light harvesting range. A large number of OPVs with power conversion efficiencies beyond 6% and up to 10% were reported, making this topic more than ever promising.[51,52] The new record is occupied by Heliatek GmbH, a Germany company, which employed the tandem structure to achieve a 12% efficiency.[53]

3.2. *Cell Structures, Working Principles and Key Components*

3.2.1. Cell Structures

Different structures of OPVs have been investigated during the research history. The first structure in the early stage of OPVs is with single layer of organic active material sandwiched between two conductive electrodes, shown in Fig. 18(a). The front side electrode is typically optical transparent ITO glass with high work function; the back electrode is made of low work function of metals, such as Al, Mg, Ca. The work function difference sets up an electrical filed help to split the excitons and separate the electrons and holes to different electrodes. However, in comparison to high exciton

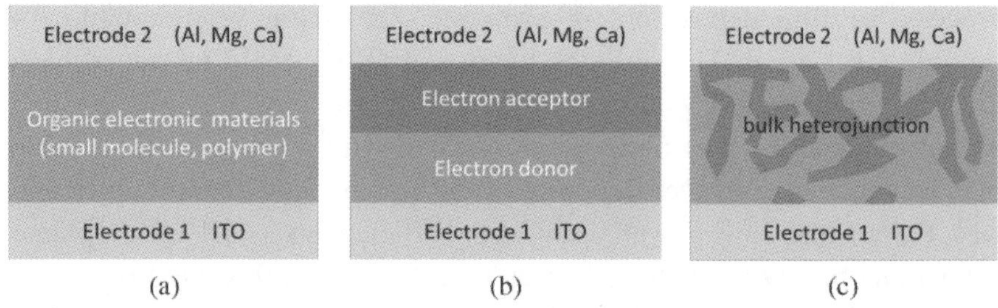

Figure 18.   Schematic shown: three typical cell structures of OPVs: (a) single layer; (b) bilayer, planar heterojunction; (c) bulk heterojunction.

binding energy of organic electronics materials (0.35–0.5 eV), the electrical filed is not sufficient to separate excitons to free carriers. As a consequence, such OPVs are normally with low quantum efficiencies (<1%) and low power conversion efficiencies (<0.1%).[54] Shown in Fig. 18(b) is a development of planar structured OPV with bilayer structure. This is called planar heterojunction OPV. The electron accepter layer with higher electron affinity and electron donor layer with lower electron affinity provide high enough electrostatic forces at their interface, which is helpful to increase excitons separation. Indeed, this bilayer structure has been proved much more efficient than the single layer structure. The short diffusion length of excitons in organic electronic materials (<10 nm) becomes the main limiting factor for such planar hetero-junction OPVs, which means a big fraction of excitons cannot reach the interface before their recombination.[55] In order to resolve this problem, bulk-heterojunction structure OPV is invented. In such a structure (Fig. 18(c)), electron donor and acceptor materials are blended to construct nanoscale interface. If the length scale of the blend is close to the exciton diffusion length, most of the excitons can reach the interface and separate efficiently.[56, 57] The state-of-art OPVs with high performance (7–12%) are all based on the bulk-heterojunction structure or the associated tandem design.

3.2.2. Working Principles

Different from inorganic solar cells, the active materials of OPVs are normally organic polymers with small dielectric constant, light absorption

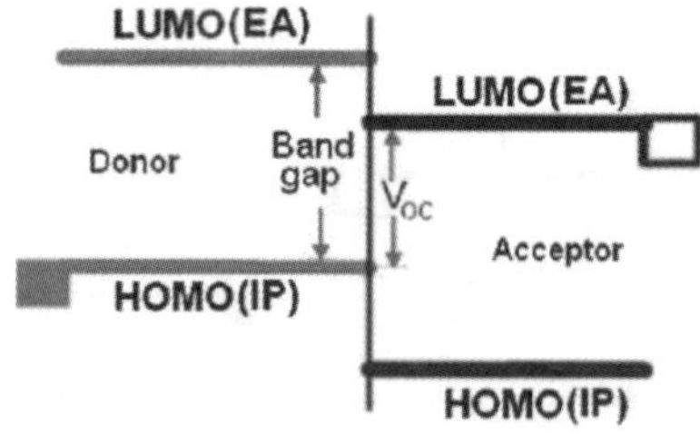

Figure 19.   Energy diagram of donoracceptor interface in OPVs.[59]

which is governed by Frenkel excitons. Their binding energy is normally with a high value of ~0.3 eV which cannot be dislocated by thermal excitation at room temperature. Therefore, an interface, as depicted in Fig. 19, between electron donor material and acceptor material with different electron affinity (about LUMO level) or ionization potential (about HOMO level) is required to generate strong electric field for efficient excitons dislocation. According to the following empirical equation[58]:

$$ \text{(8)} \quad V_{oc} = \frac{1}{e}\left(\left| E_{HOMO}^{donor} \right| - E_{LUMO}^{acceptor} \right) - 0.3, $$

V_{oc} of OPVs is determined by the difference between the HOMO of the donor and the LUMO of the acceptor. 0.3 is a typical value due to contact loss. Considering that 0.3 eV of the LUMO or HOMO offset is required for interfacial exciton dislocation, the maximum V_{oc} of OPVs typically equals to the band gap (E_g) of the light absorber minus 0.6 V.

Photocurrent generation in OPVs is determined by the four successive processes depicted in Fig. 20: (1) An exciton is generated inside the photoactive material after the absorption of a photon (Fig. 20(a)); (2) the exciton diffuses towards the donor–acceptor interface (Fig. 20(b)); (3) with the help of the interfacial electric field, the bound exciton disassociates into free carriers (Fig. 20(c)); (4) the free carriers transport towards the electrodes for final collection (Fig. 20(d)). Corresponding to these four steps, the loss mechanisms could be: (1) kinds of optical losses such as narrow E_g of photoactive materials; (2) exciton losses due to insufficient transport to the interface or inefficient dissociation **at the interface;** (3) Charge carrier collection losses due to **slow** mobility **or fast recombination.**

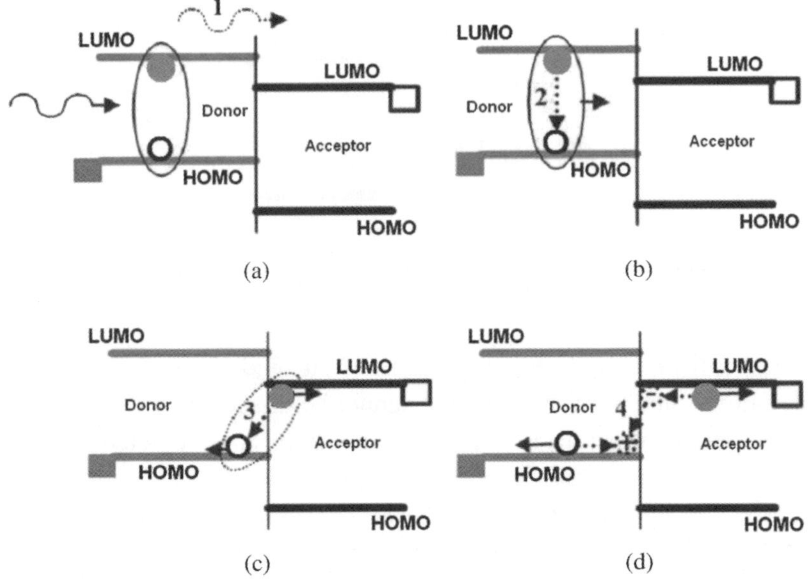

Figure 20.   Schematic shown on the four successive processes leading to photocurrent generation in OPVs.[59]

3.2.3. Key Components

The key components of bulk-heterojunction OPV include the electron acceptor and donor materials (the photoactive materials), interfacial electron or hole selective materials and electrode materials. The first couple of donor/acceptor materials in bulk-heterojunction OPVs arepoly [2-methoxy-5-(3′,7′-dimethyloctyloxy)-1,4-phenylenevinylene]/1-(3-methoxycarbonyl)-propyl-1-phenyl-[6,6]C_{61} (MDMO-PPV/PCBM). But the wide band gap of PPV materials in combination with their low mobility result in limited photocurrent, and therefore 3% efficiency at best.[60] Poly(3-hexylthiophene-2,5-diyl)/1-(3-methoxycarbonyl)-propyl-1-phenyl-[6,6]C_{61} (P3HT/PCBM) couple and their deviants were then widely studied. P3HT features as narrower band gap and better mobility than PPV materials. The efficiency of P3HT/PCBM cells quickly increased to the 5% level.[61,62] However, photovoltage of P3HT/PCBM cells was limited by the HOMO level of P3HT; photocurrent density also needs to be further improved. Therefore, deep HOMO polymers with narrower

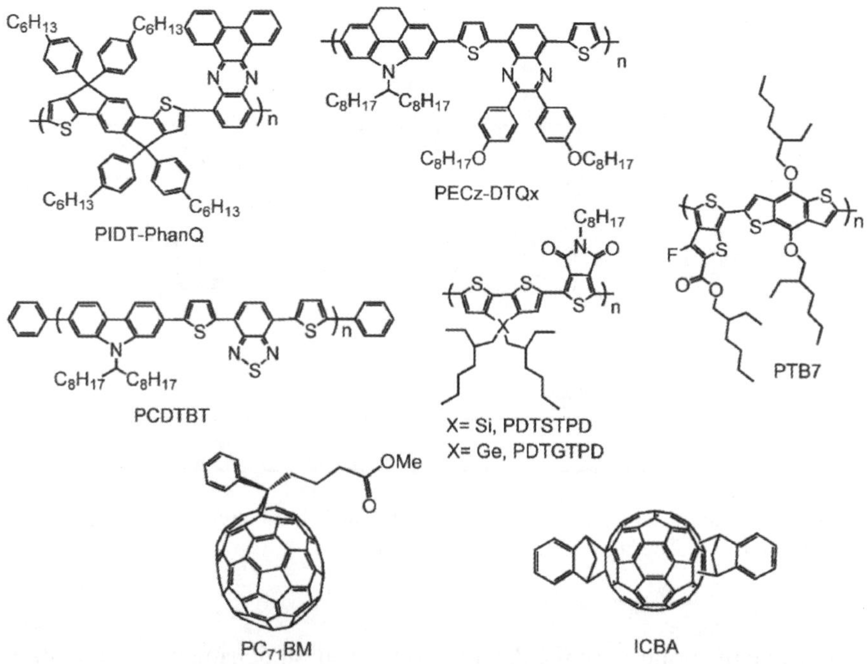

Figure 21.　Molecular structures of several typical conjugated polymer donors and fullerene acceptors successfully applied in bulk heterojunction OPVs.[65]

bandgap are highly desired, aiming to elevate the cell efficiency to be over
5%.[63] Show in Fig. 21 are the molecular structures of several donor materials with good performance records, their HOMO/LUMO levels are shown
in Fig. 22. Not like donors with such a huge diversity, acceptors other than
PCBM and related fullerenes, such as conjugated polymers, carbon nanotubes, perylenes, and inorganic semiconducting nanoparticles, have not
satisfied expectations. Among fullerene derivate, ICBA alterative to
PCBM in bulk-heterojunction OPVs was normally reported with higher
photovoltage due to its shallower LUMO level (Fig. 22).[64]

Interfacial charge selective materials are formed between the photoactive layer and the electrodes, which are essential for highly efficient and
stable OPV devices.[65,66] The main functions of interface materials are:
(1) to adjust the barrier heights between the photoactive materials and the
conductive electrodes; (2) to form selective contacts for the charges and to

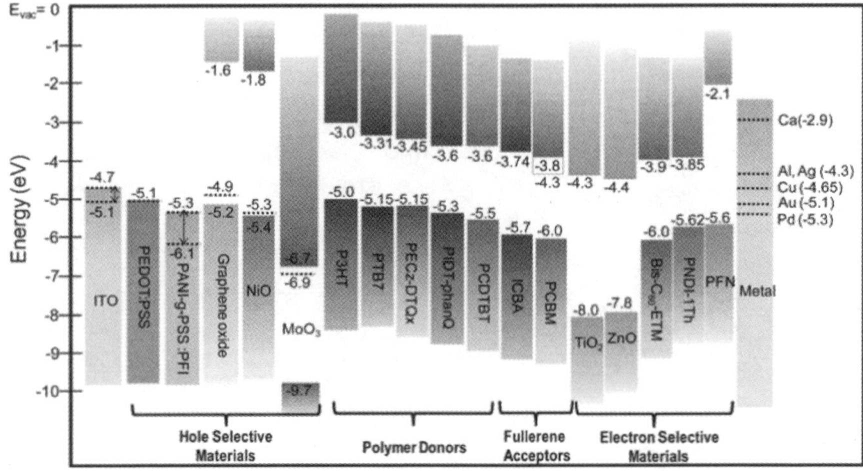

Figure 22.   Interfacial materials applied in OPVs, their band alignments with respect to the typical donor–acceptor materials are highlighted.[65]

determine the polarity of the device; (3) to prohibit detrimental interactions between the photoactive materials and electrodes. In general, they are required to (i) have energy level matching with respect to photoactive materials to promote Ohmic contact formation between electrodes and the active layer; (ii) have appropriate energy levels to improve charge selectivity for corresponding electrodes; (iii) have large bandgap to confine excitons in the active layer; (iv) possess sufficient conductivity to reduce resistive losses; (v) have chemical and physical stability to prevent undesirable reactions at the active layer/electrode interface; (vi) have the ability to be processed from solution and at low temperatures. A list of electron or hole selective materials with their band gap positions are listed in Fig. 22.[65]

3.3. *Roll-to-roll Processing Techniques*

As the key 10–10 targets of OPVs (10% power conversion efficiency and 10 years of operational stability) are promising to be achieved in the near future, the next step is about how to use efficient, scalable, and rational processing methods to realize high throughput roll-to-roll production. The general fabrication methods in the laboratory scale, such as spin

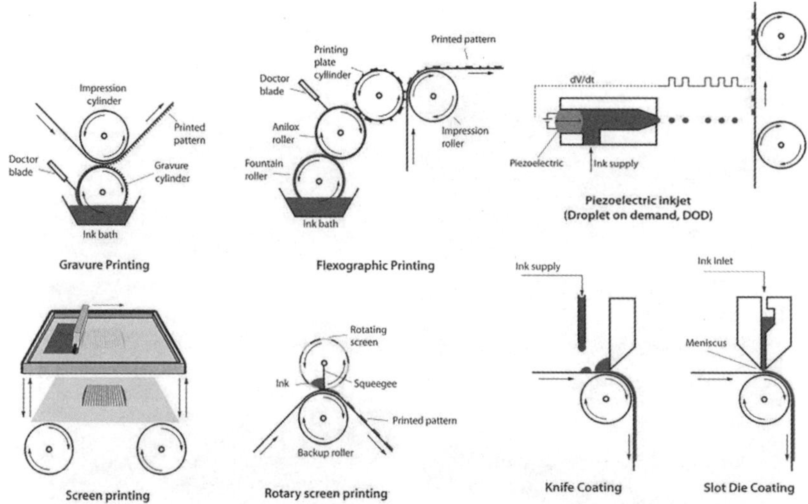

Figure 23.   Illustrations on technical features of six printing/coating techniques.[67]

coating and high vacuum thermal evaporation are not compatible in the case. Herein, several printing, coating technologies have the potential to be incorporated with the roll-to-roll production line. Their printing/coating principles are depicted in Fig. 23. Several practical examples of their applications in OPVs are shown in Fig. 24. Each of these technologies have their own advantages and disadvantages, and are shortly reviewed as follows. For more details, one could refer to the recently published review papers.[67]

3.3.1. Gravure Printing

As shown in Fig. 23, the ink was transferred from the ink tank by the gravure cylinder to the flexible substrate due to press contact by the impression cylinder. The tiny engraved cavities with designed pattern on the gravure cylinder can hold and transfer the ink. Excessive ink can be removed by the doctor blade, and therefore, the ink can only be present inside the cavities. The printed film thickness is determined by the pattern and depth of engraved cavities on the gravure cylinder. The advantage of this printing technology is about its high printing speed, which could

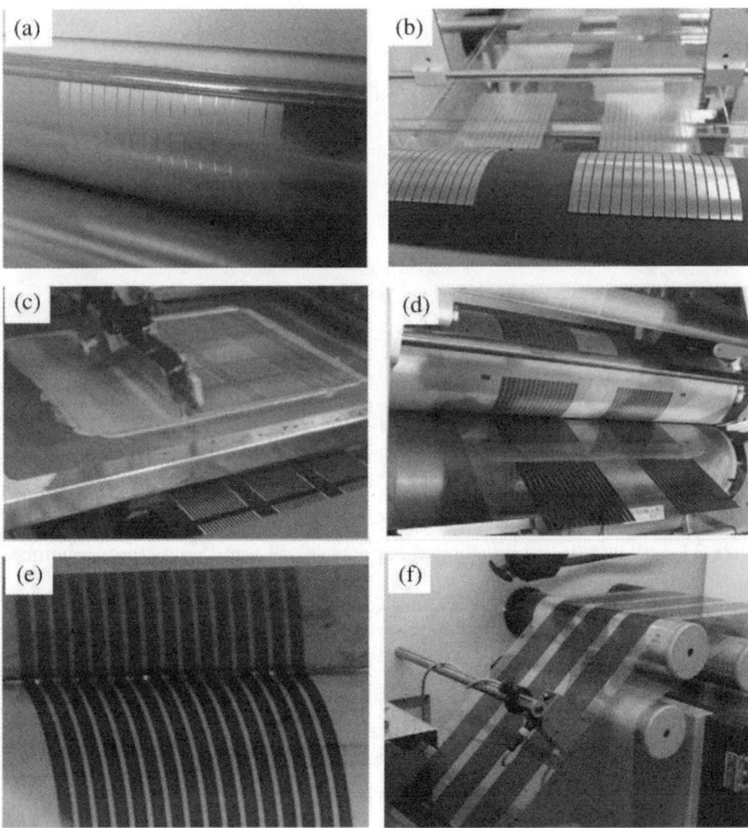

Figure 24. Several practical examples of printing different layers of OPVs (a,b) silver electrodes by flexographic printing (c) silver electrodes by flat-bed screen printing (d) conducting graphite ink, by rotary screen printing, (e,f) slot-die coating of the active layer of the OPV.[67]

reach up to 15 m/s. The disadvantage may be that the printing quality is sensitive to the ink's surface tension which needs careful optimization on the ink's composition. This technology is rarely reported in OPVs.

3.3.2. Flexographic Printing

The flexo system is a relatively new technology, which consists of two more cylinders than gravure printing (Fig. 23). The fountain roller is used for continuous transfer of ink to the ceramic anilox roller. The anilox roller with engraved cavities collects the ink and then transfers to the relief on the printing cylinder, which is normally made of rubber. The ink on the relief of the printing cylinder is finally transferred to the flexible substrate. This method has been reported more frequently than gravure printing in OPVs.

3.3.3. Screen Printing

Screen printing is superior in preparing thick films (micron meter thick) in comparison to flexographic and gravure printing. Depending on the thickness and porosity of the screen, the wet film thickness could be in the range of 10–500 μm. There are two types of screen printing: flat-bed screen printing and rotary screen printing (Fig. 23). Their principles are the same: the paste filled inside the screen mesh of the patterned mask, will be transferred to the substrate by pressure from the squeegee. The flat-bed screen printing is more suitable for small scale printing due to easier changing of the masks. The rotary screen printing is superior in high speed and continuous printing on a large scale (>10 m^2). The printing techniques are of particular effectiveness for the printing of the front and back electrodes. Screen printing of active layers have also been reported in OPVs.

3.3.4. Inkjet Printing

Inkjet printing can be easily controlled by computer that creates a digital image by propelling droplets of ink from the nozzle onto the substrate (Fig. 23). Different from other printing technologies, the printing head and the substrate is without physical contact. Therefore, it allows printing 3D multiple layers in principle. The image resolution would be controlled with high pixel density. The drawback of this method should rest with some speed limitations and restrictions on ink formulations. The inks made of chemical materials in OPVs may be corrosive or easy to be precipitated, which may greatly shorten the lifetime of expensive printing head.

3.3.5. Knife Coating and Slot-Die Coating

Knife coating is also called blade coating. The ink is supplied in front of the knife and passes through the gap between the knife and substrate

(Fig. 23). The gap height determines the as-deposited film thickness. The technique is suitable for wide areas coating without any pattern. The ink supply in slot-die coating is different from knife coating. A sharp coating head is used; the ink is pumped inside the channel of the head and deposited onto the substrate (Fig. 23). The pressure for ink supply by the pump determines the film thickness. Unlike knife coating, slot-die coating could prepare stripes of variable width, which is dependent on the width of coating head. Both knife and slot-die coating can be used for inks with large varieties in viscosity and solvents, and the coating speed could be very high (0.1–200 m/min).

3.3.6. An Example of Fully Printed OPV

In 2013, Danish researchers demonstrated a fast inline roll-to-roll printing technology for OPVs, with the six layer structure of silver-grid/ PEDOT:PSS/ZnO/P3HT:PCBM/PEDOT:PSS/silver-grid.   Vacuum deposited ITO conductive layer was replaced by silver-grid. The involved printing technologies for the whole six layers are: the first layer of hexagonal silver grid, prepared by flexographic printing at a web speed of 10 m min^{-1}; the second layer of poly(3,4-ethylenedioxy-thiophene/polystyrene sulfonate (PEDOT:PSS) prepared by rotary-screen printing; the third electron selective layer of ZnO and the fourth active layer of P3HT:PCBM were slot-die coated at the same time at a web speed of 10 m min^{-1}; the fifth layer of hole-collecting PEDOT:PSS and the sixth layer of comb-patterned silver-grid back electrode were rotary-screen printed at 2 m min^{-1}. The whole printing line setup is shown in Fig. 25. The fully roll-to-roll processed solar cells and modules finally achieved PCEs of more than 1.8% on single cells and 1.6% PCE on the active area of module.[68]

3.4. *Prospects*

Up to now, for single junction bulk-heterojunction OPVs, the best performance is over 10% efficiency, while for tandem structured bulk-heterojunction OPVs, the record data is 12% in efficiency. These rapid

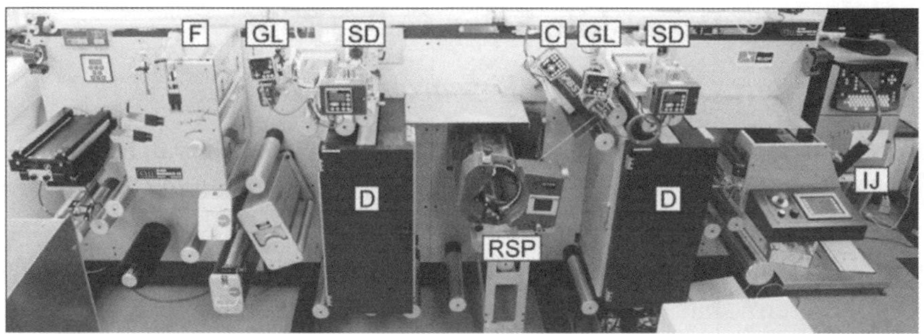

Figure 25.   A design on a completed printing line for OPVs, namely guideline detection (GL), strobe camera (C), and barcode inkjet printer (IJ) mounted on the roll-to-roll machine setup with flexo-printing (F), slot-die coating (SD), rotary-screen printing (RSP), and driers (D).[68]

progresses were mostly made within the last 10 years, reflecting the great potential of this technology in the future. Considering a typical performance ratio of 0.6 often found for thin-film technologies, a commercial OPV module with efficiency of 6–7% could be expected, which is still inferior to the commercialized amorphous Si solar cell with module efficiency of ~8%. To further increase the power conversion efficiency up to ~15% should be important, which may be reachable by optimizing energy levels, optical and electrical properties of the absorber layers and the device design. To develop narrower bandgap absorber outreaching into the 900 nm region holds the key for further efficiency improvement of the tandem structured OPVs; to improve the charge separation processes and the transport of the free charge to the electrodes, promising ways may include (1) establishment of ordered interface of donor and acceptor materials, via introducing liquid crystal semiconductor molecules facilitating phase separation; (2) increasing the dielectric constant of organic absorber. Ultimately, for $\varepsilon > 10$, binding excitons could be split into free carriers via room temperature thermal activation. Besides improving efficiency, to develop stable, low-cost materials and their facile printed technologies compatible for the roll-to-roll production line are of equal importance for successful industrialization of OPVs.

4. Flexible Perovskite Solar Cells (PSSCs)

4.1. *A Short Overview on Research History*

Perovskites represent a class of materials with the general formula of AMX_3, which is named because of the mineral component of $CaTiO_3$. For perovskite oxides, "A" site cations could be Mg^{2+}, Ca^{2+}, Sr^{2+}, Ba^{2+} and Pb^{2+}, "M" site cations could be Ti^{4+}, Si^{4+}, and "X" site anion represents O^{2-}. For perovskite halides, "A" site cations could be Li^+, Na^+, K^+, Rb^+, Cs^+, $CH_3NH_3^+$, NH_2–CH=NH^+, etc., "M" site cations could be Be^{2+}, Mg^{2+}, Ca^{2+}, Sr^{2+}, Ba^{2+}, Zn^{2+}, Ge^{2+}, Sn^{2+}, Pb^{2+}, Fe^{2+}, Co^{2+}, Ni^{2+}, etc., and "X" site anions represent F^-, Cl^-, Br^-, I^-. The typical crystal lattice is shown as Fig. 26, in which the M cation is in six-fold coordination, surrounded by an octahedron of anions (MX_6), and the A cation is in 12-fold cuboctahedral coordination.

Though the first observation of photocurrents in $BaTiO_3$ can date back to 1956, before 2009, the efficiency of PV effect in perovskites light absorbers were very low, typically much less than 1%.[70] In 2009, Miyasaka *et al.* reported their pioneer work on PSSCs based on $CH_3NH_3PbX_3$ quantum dots sensitization on mesoporous TiO_2. The reported efficiency was 3.81%.[71] Later, in 2011, Park *et al.* reported a 6.54% efficiency with similar cell configuration, also based on iodine based electrolyte.[72] Those solar cells are unstable due to the perovskite

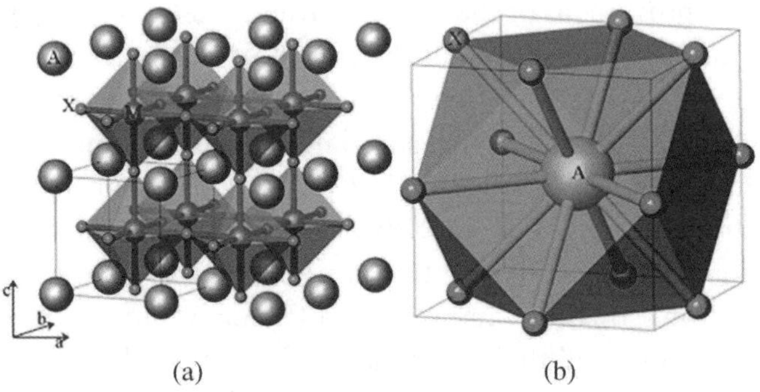

(a)   (b)

Figure 26.   (a) Crystal structure of cubic perovskite of general formula ABX_3, (b) 12-fold coordination of the A-site cation.[69]

absorbers could be dissolved or decomposed in the liquid electrolyte within several minutes. The first breakthrough in solid-state PSSCs came in late 2012, when Park collaborated with Grätzel. They used $MAPbI_3$ as a light absorber and solid state Spiro-OMeTAD as hole transfer material instead of liquid electrolyte. They got an efficiency of 9.7%.[73] Soon after that, Snaith in collaboration with Miyasaka reported their meso-superstructured PSSC based on $CH_3NH_3PbI_{3-x}Cl_x$ and inert mesoporous Al_2O_3 scaffold. They got V_{oc} of 1.13 V and an efficiency of 10.9%.[74] These pioneer works strongly encouraged the followers. The efficiency of PSSCs has been quickly elevated to be more than 15% in 2013. Grätzel *et al.* introduced a sequential deposition method for the fabrication of perovskite on mesoporous TiO_2 film, which gave an efficiency of 15% and a certified value of 14.1%.[75] Snaith *et al.* reported the planar heterojunction PSSC fabricated by vapor deposition which demonstrated 15.4% efficiency.[76] In 2014, Seok *et al.* certified a 17.9% efficiency via solvent engineering which allows the deposition of perovskite film with perfect morphology.[77] Yang *et al.* later reported an efficiency of 19.3% (not certified) via surface engineering in PSSC.[78] Up to now, to approach 20% efficiency becomes an achievable goal. Note that the efficiencies of CdTe and CIGS thin-film solar cells just reached 21% after about 40 years' development, the developing speed of PSSCs is really striking. Also noteworthy is that crystal phase formation of most of the organo-metal halides perovskites can be achieved at the temperature below 150°C, therefore, it is quite promising to fabricate flexible PSSCs with notably high performance in comparison to OPVs.

4.2. Cell Structures, Working Principles and Key Components

4.2.1. Cell Structures and Related Working Principles

Several different cell configurations have been successfully demonstrated with high performance. The first one is shown in Fig. 27(a), the concept was borrowed from solid-state dye-sensitized solar cells, where perovskite ultrathin absorber was used to replace dye sensitizers. This

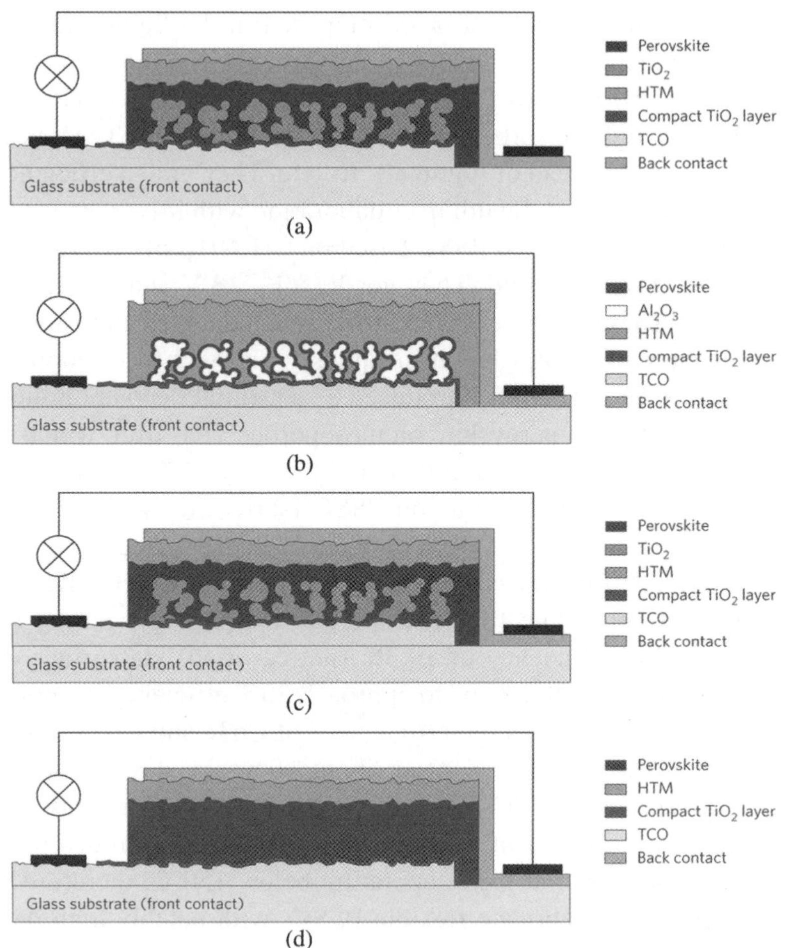

Figure 27. Typical cell structures of PSSCs which can achieve high performance (a) a mesoscopic solar cell using ultrasmall perovskite nanocrystals as light absorber, (b) nanocomposite embodiment of a PSSC where the mesoscopic TiO_2 scaffold is infiltrated by the perovskite, (c) a meso-superstructured PSSC employing a mesoporous Al_2O_3 film as the scaffold, (d) OPV like planar structured PSSC.[79]

configuration is beneficial for excitons separation at the mesoscopic interfaces with respect to TiO_2 and hole transport material, i.e., Spiro-OMeTAD. Due to perovskite light absorber with much higher light extinction coefficient than the organic dyes, the mesoporous TiO_2 film with the thickness of around 350 nm is enough for sufficient light harvesting and corresponding solar cell has been reported with 15.0% efficiency by Grätzel *et al.*[75] The certified highest efficiency of 17.9% was reported by Seok *et al.*[77] The cell configuration is shown in Fig. 27(b). Different from the mesoscopic cell configuration shown in Fig. 27(a), the perovskite film has a capping layer on top of mesoporous TiO_2 scaffold. It can work efficiently due to the excellent photoelectrical properties of perovskites, i.e., long and balanced charge diffusion length (100–1,000 nm), small binding energy of exitons (<0.030 eV).[80] Note that these features are highly desired in OPVs, which will certainly benefit exiton separation and charge collection. Therefore, the advantage of small perovskite light absorber in the mesoscopic cell configuration is not prerequisite for high external quantum efficiency of the cell", despite the fact that a thin mesoporous TiO2 layer is still retained in the champion solar cells reported by Seok et al.[77] Shown in Fig. 27(c) is the so-called meso-superstructured PSSC invented by Snaith *et al.* Instead of mesoporous TiO_2, inert mesoporous Al_2O_3 scaffold is employed, which acts as the holder for perovskite deposition.[74] Al_2O_3 scaffold cannot transport electrons. Therefore, in this cell configuration, photoinduced electrons are confined in perovskite itself, the mobility of which is much higher than that of mesoporous TiO_2. Via this cell configuration, Snaith *et al.* reported the highest efficiency of 15.9% and better than the solar cell based on mesoporous TiO_2 in their controlled experiment.[81] However, the meso-superstructured cell configuration is still under debate, as the role of the mesoscopic Al_2O_3 scaffold is not so clear. Especially, the planar cell configuration shown in Fig. 27(d) without any mesoporous scaffold has also been reported with competitive performance. The planar cell configuration becomes analogous to inverted OPVs. Compact TiO_2 and Spiro-OMeTAD act as the electron and hole selective interfacial materials, and perovskite light absorber is sandwiched between them. Not like OPVs, bulk-heterojunction is not required to enhance exciton dislocation. Snaith *et al.* firstly demonstrated a 15.4% efficiency on this configuration in which the perovskite layer was prepared by dual-source vapor deposition method. Very recently, Yang *et al.*[78] reported a 19.3% efficiency with the planar cell configuration.

4.2.2. Key Components

Perovskites light absorbers serve as the core of PSSCs, their band gaps, band edge positions, binding energy, and charge mobilities/lifetimes play the decisive roles on the devices' performance. Band gap is of primary importance. According to Shockley-Queisser limitation, the optimized band gap for single junction solar cell should be 1.41 eV; while the best band gap combination for double junction solar cell should be 1.63 eV/0.96 eV. Fig. 28 summarizes the organometal halide perovskites reported up-to-date, with their band gaps and relative band positions highlighted. According to the first-principles study, the electronic structure of halide perovskites is mostly controlled by their metal–halide (M–X) bonds, while the A site cations (A = Cs^+, $CH_3NH_3^+$, NH_2–CH=NH^+) with different size will effect on the band gaps via tuning the metal–halide bond lengths. The general rules about band gap tuning include: (1) as the electron-negativity decrease from Pb^{2+} to Sn^{2+}, a decrease of the covalent character of M–X bond is observed, the band gaps of the perovskites follow the trend of $APbX_3$ > $ASnX_3$; (2) as the electron-negativity decrease

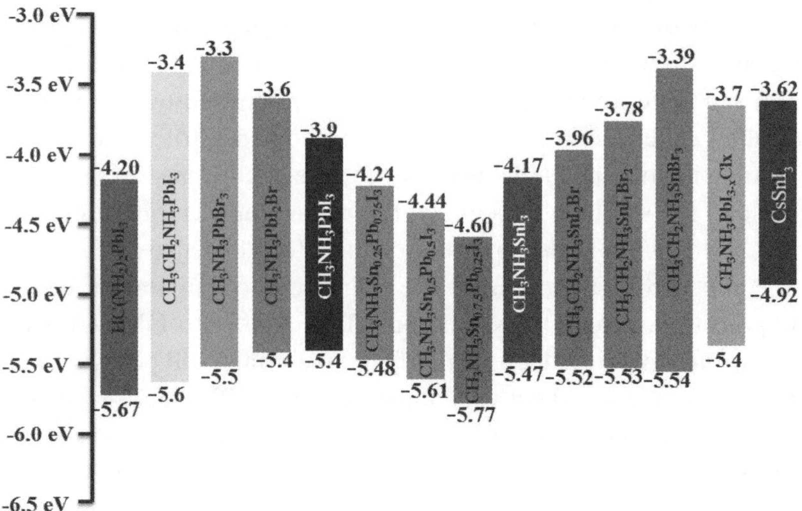

Figure 28. Summary on band gaps and band energy positions of the reported halides perovskites.

as $F^- > Cl^- > Br^- > I^-$, a decrease of the covalent character of M–X bond is observed, the band gaps of the group of perovskites follows the trend of $AMF_3 > AMCl_3 > AMBr_3 > AMI_3$; (3) the A site cation increases in ionic radius as $Cs^+ < CH_3NH_3^+ < NH_2–CH=NH^+$, and hence the lattice would be expected to expand, the band gap of lead perovskites follow the trend of $CsPbX_3 < CH_3NH_3PbX_3 < NH_2–CH=NHPbX_3$. Via mixed perovskites, the band gaps and band positions could be continuously tuned, typical examples include $CH_3NH_3PbI_{3-x}Br_x$, $CH_3NH_3Pb_{1-x}Sn_xI_3$, etc.

4.3. *Low Temperature Preparation Methods of Perovskite Films*

Up to now, there have been several different deposition methods of perovskite films which lead to achievement of high device performance. They are: (1) one-step solution deposition; (2) two-step sequential deposition; (3) dual-source vapor deposition; (4) vapor assisted solution procession; (5) improved one-step solution method with solvent extraction and (6) improved vapor assisted solution procession. The first four of them are depicted in Fig. 29.

4.3.1. One-Step Solution Deposition

One-step solution deposition is to firstly prepare the stoichiometric mixture solution of CH_3NH_3I and PbI_2 (1:1) or non-stoichiometric mixture solution of CH_3NH_3I and $PbCl_2$ (3:1) in high boiling aprotic polar solvents, e.g., DMF, DMSO, GBL, etc.; then it was spin-coated on the substrate and annealed at certain temperature (typically 70–130°C) for several minutes to several hours. The non-stoichiometric mixture solution of CH_3NH_3I and $PbCl_2$ sintering at elevated temperature (130°C) normally leads to Cl doped perovskite ($CH_3NH_3PbI_{3-x}Cl_x$) with a larger crystal size and better film coverage. Excessive CH_3NH_3I will be evaporated away. So, to stop annealing at certain time will be critical to the film's morphology and composition. The improved one-step solution method with solvent extraction is invented in order to get a much smoother surface of perovskite film, which allows the subsequent deposition of hole transport material (like Spiro-OMeTAD) or electron

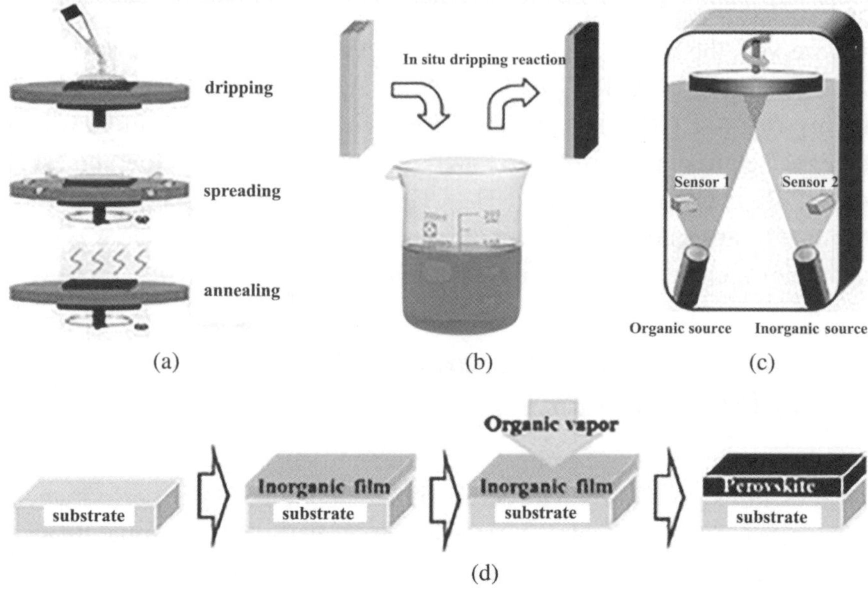

Figure 29.   Summary on the four typical methods for the preparation of perovskite films. (a) One-step solution deposition; (b) two-step sequential deposition; (c) dual-source vapor deposition; (d) vapor assisted solution procession.[69]

transport material (like PCBM) with thinner thickness and better coverage. This will result in higher shunt resistance and lower series resistance, because low mobility of these interfacial materials normally acts as a main hindering factor.[74,81]

4.3.2. Two-Step Sequential Deposition

Two-step sequential deposition is to firstly spin-coat PbI_2 film and then to immerse the film in isopronaol solution of CH_3NH_3I at certain concentration (typical 10 mg mL^{-1}) for certain time at room temperature. Increasing immersing tempeartue or prolonging the immersing time will result in larger crystals and more completed conversion of PbI_2 to perovskite. About the immersing time, in the very beginning in Gratzel *et al.*'s report, it was kept at only 20 seconds. That is because the perovksite is restricted inside the mesopores of TiO_2 scaffold without capping layer. Nanometer size of perovksite can complete the phase transition very fast. If with hundreds nanometers of perovskite capping layer on top of mesoporous scaffold, it generally requires tens minutes to accomplish full phase transition.[75] This method is suggested to be better in reproducibility, but it cannot be simply applied to other perovskites beyond $CH_3NH_3PbI_3$. Some underlayer such as PEDOT:PSS sensitive to the water-like solvents may also limit this deposition method.

4.3.3. Dual-Source Vapor Deposition

Dual-source vapor deposition is via evaporating the two sources of PbI_2 and CH_3NH_3I under vacuum and elevated temperature, the two vapors are deposited and reacted *in situ* at the substrate.[76] This method is superior for the deposition of flat, compact and uniform film with large area, and very suitable for planar structured solar cells. The only issue is about the vacuum involved process, which decreases the deposition rate and increases the cost.

4.3.4. Vapor Assisted Solution Procession

Vapor assisted solution procession is much simple in comparison to dual-source vapor deposition. This method is to firstly deposit PbI_2 films by solution spin-coating method, and then treat the film in the MAI vapor, which can be easily evaporated at 130°C without vacuum. The as-deposited perovskite film is surface smooth, pinhole-free, and with large crystal size.[82] The problem is about its reproducibility, which may be sensitive to the duration time, substrate temperature, etc. Therefore, an improved vapor assisted solution procession is invented, by directly spin-coating of MAI solution on top of PbI_2 film during high-speed rotation. This method allows the direct contact between MAI and PbI_2 which is announced with better reproducibility.

Note that, the phase formation temperature of $CH_3NH_3PbI_3$ is around 70–100°C and the evaporation temperature of CH_3NH_3I is only 130°C, therefore, all of these methods can be applied for the preparation of flexible solar cells on plastic substrates. The typical examples are simply described as follows.

4.4. *Flexible PSSCs*

In 2013, Liu *et al.* reported a PSSC with the structure of ITO/ZnO/CH$_3$NH$_3$PbI$_3$/Spiro-OMeTAD/Ag (Fig. 30).[83] The ZnO interfacial layer was prepared by nano-ink spin-coating. After optimizing the ZnO thickness ranging from 10 nm to 70 nm, they got 14.4% efficiency on ITO-glass substrate and 10.2% efficiency on ITO/PET flexible substrate.[83] Nearly at the same time, Docampo *et al.* demonstrated that a PSSC with structure of TCO/PEDOT: PSS/CH$_3$NH$_3$PbI$_{3-x}$Cl$_x$/PCBM/TiO$_x$/Al achieved power-conversion efficiency of up to 10% on glass substrates and over 6% on flexible polymer substrates.[84] The perovskite layer was prepared by one-step solution method and sandwiched between organic contacts such as PEDOT: PSS and PCBM which are widely adopted in the organic PV

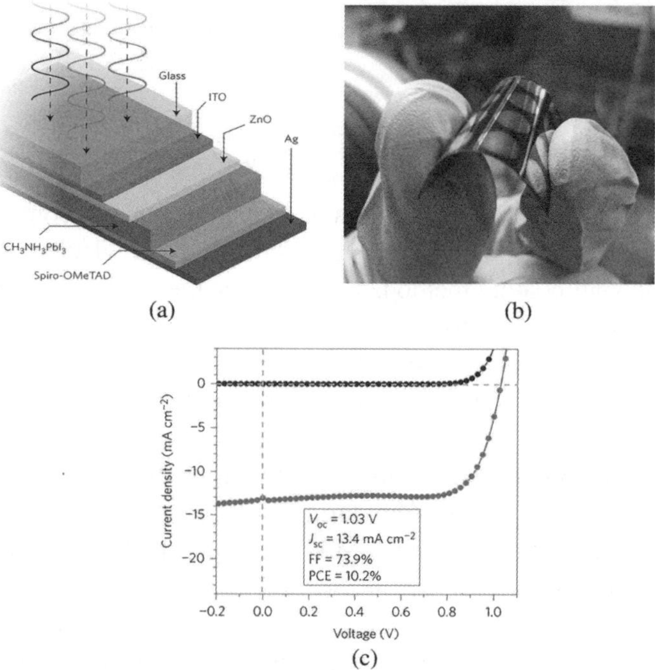

Figure 30. A flexible PSSC with the structure of ITO/ZnO/CH3NH3PbI3/Spiro-OMeTAD/Ag, (a) the cell structure, (b) an optical image of the cell, (c) the *J–V* curve.[83]

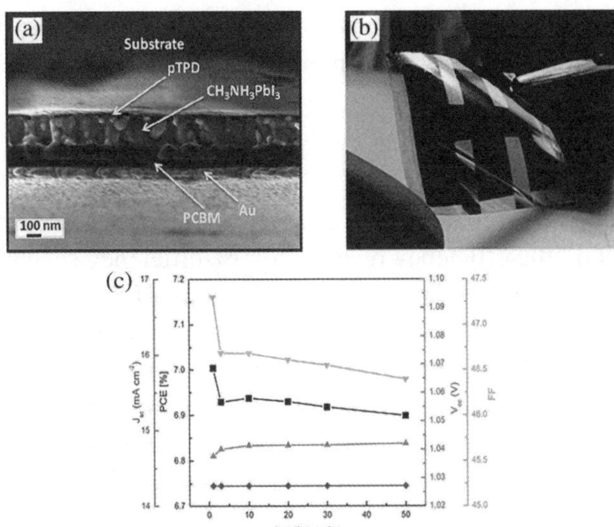

Figure 31. A flexible PSSC with the structure of ITO/ZnO/CH3NH3PbI3/Spiro-OMeTAD/Ag (a) SEM cross-section view of the cell, (b) an optical image of the cell, (c) the dependences of the performance parameters on the bending circles.[85]

community. Roldan–Carmona *et al.* employed dual-source vapor deposition to prepare a flat layer of $CH_3NH_3PbI_3$ sandwiched between PEDOT: PSS and PCBM. They first demonstrated a 12% efficiency on rigid ITO-glass substrate and later a 7% efficiency on flexible conductive substrate. The flexible PSSC can withstand extended bending without significant performance degradation (Fig. 31).[85] In 2014, Yang's group reported a cell structure of substrate/ITO/PEDOT:PSS/$CH_3NH_3PbI_{3-x}Cl_x$/PCBM/Al and demonstrated a 11.5% efficiency on ITO/glass substrate and a 9.2% efficiency on ITO/PET flexible substrate.[86] Later, based on the same cell configuration, Huang *et al.* improved the preparation of perovskite film by the improved vapor assisted solution method, and finally got a 15.3% efficiency of solar cell on ITO coated glass.[87] Nearly at the same time, Seok *et al.* improved the perovskite film quality by one-step solution method with solvent extraction and inserted a LiF layer between PCBM and Al; based on the cell configuration of ITO/ PEDOT:PSS/$CH_3NH_3PbI_3$/PCBM/LiF/Al, they got an efficiency of 14.1%.[88] Note that, these PSSCs based on PEDOT:PSS and PCBM interfacial layers can be fabricated on flexible conductive substrates. Therefore, at least 10% efficiency of flexible PSSCs is quite accessible, which is much higher than flexible OPVs and DSSCs.

4.5. *Prospects*

At present, the research on PSSCs is in a dynamic state. According to the fast progress on the efficiency records, a 20% efficiency seems now to be a realistic goal. However, large-scale application of PSSCs will depend on whether stability and toxicity issues can be solved. For example, the frequently studied $CH_3NH_3PbI_3$ will degrade in humid conditions due to dissolution and decompose at higher temperatures due to the loss of CH_3NH_3I. Lead compounds are very toxic and harmful to the environment. Especially, PbI_2 could be moderately dissolved in water. Therefore, lead free perovskites are highly desired. Possible alternatives could be Sn^{2+} based perovskites, such as $CH_3NH_3SnI_3$ and $CsSnI_3$. The biggest challenge is easy oxidization of Sn^{2+} by oxygen in the air.[89]

5. Conclusions and Outlook

This chapter has focused on flexible fabrication of several important next generation solar cells, such as DSSCs, OPVs and PSSCs. The basic concept, research history, key materials, techniques, and prospects on further development have been overviewed. Up-to-date, the best flexible DSSCs are with the efficiencies of 7–8%, while the best flexible OPVs could be approaching to 10%. The efficiency of newly emerged PSSCs quickly surpass the records of DSSCs and OPVs, to be more than 10% on flexible substrates. As mentioned above, it is primarily beneficial from the excellent photoelectric properties of perovskite light absorber. Also important is that the huge research basis on DSSCs and OPVs could be referred for the development of PSSCs. However, the limited choices on perovskite materials might be a great challenge; whether the stability and toxic issues could be solved in the future is not clear yet. Therefore, it cannot be simply judged as to which technology is superior at the current state. Perhaps, these technologies could be alive and be developed simultaneously for many years in the future.

Acknowledgments

The authors would like to express sincere thanks for the financial support by the National Natural Science Foundation (21103058), 973 Program of China (Grant No. 2011CBA00703), Basic Scientific Research Funds for Central Colleges (2013TS040).

References

1. [Online] Available: http://en.wikipedia.org/wiki/Solar_energy. (Accessed on 20th September, 2014).
2. [Online] Available: http://en.wikipedia.org/wiki/Photovoltaics. (Accessed on 20th September, 2014).
3. [Online] Available: http://www.greentechmedia.com/articles/read/solar-pv-module-costs-to-fall-to-36-cents-per-watt. (Accessed on 20th September, 2014).
4. [Online] Available: http://en.wikipedia.org/wiki/Solar_power. (Accessed on 20th September, 2014).
5. M. Pagliaro, R. Ciriminna and G. Palmisano, *Chem. Sus. Chem.*, **1**, (2008).
6. [Online] Available: http://www.nrel.gov/ncpv/. (Accessed on 20th September, 2014).
7. [Online] Available: http://www.solarbuzz.com/news/recent-findings/multi-crystalline-silicon-modules-dominate-solar-pv-industry-2014. (Accessed on 20th September, 2014).
8. [Online] Available: http://www.expo21xx.com/renewable_energy/19630_st2_solar_mountingsystems/default.htm. (Accessed on 20th September, 2014).
9. [Online] Available: http://www.powerfilmsolar.com/(Accessed on 20th September, 2014).
10. [Online] Available: http://article.wn.com/view/2014/08/18/Solar_Cell_Production_Line_Upgrade_Cycle_Driving_Module_Powe_k/. (Accessed on 20th September, 2014).
11. [Online] Available: http://edition.cnn.com/2009/TECH/01/28/solar.powered.cars/. (Accessed on 20th September, 2014).
12. B. O'regan and M. Grfitzeli, *Nature*, 353 (1991).
13. S. Ito, N-L. C. Ha, G. Rothenberger, P. Liska, P. Comte, S. M. Zakeeruddin, P. Pechy, M. K. Nazeeruddin and M. Gratzel, *Chem. Commun*, **38**, 4004 (2006).

14. K. Tennakone, G. R. R. A. Kumara, I. R. M. Kottegoda and V. P. S. Perera, *Chem. Commun.*, (1999).

15. C. Y. Jiang, X. W. Sun, K. W. Tan, G. Q. Lo, A. K. K. Kyaw and D. L. Kwong, *Appl. Phys. Lett.*, **92**, 143101 (2008).

16. K. Sayama , H. Sugihara and H. Arakawa, *Chem. Mater.*, **10**, (1998).

17. M. K Nazeeruddin, A. Kay, I. Rodicio, R. Humpbry-Baker, E. Miiller, P. Liska, N. V. lachopoulos and M. Grätzel, *J. Am. Chem. Soc.*, **115**, (1993).

18. Y. Ooyama and Y. Harima, *Eur. J. Org, Chem.*, (2009).

19. J. X. Hong, *Ind. Mater.*, 243 (2007).

20. T. Miyasaka, M. Ikegami and Y. Kijitori, *Electrochemisty*, **75**, (2007).

21. D. Zhang, T. Yoshida and H. Minoura, *Adv. Mater.*, **15**, (2003).

22. D. Zhang, T. Yoshida and K. Furuta, *J. Photoch. Photobio. A*, 164 (2004).

23. T. Yamaguchi, N. Tobe, D. Matsumoto and H. Arakawa, *Chem. Commun.*, (2007).

24. L. Yang, L. Wu, M. Wu, G. Xin, H. Lin and T. Ma, *Electrochem. Commun.*, **12**, (2010).

25. T. Miyasaka, M. Ikegami and Y. Kijitori, *J. Electrochem. Soc.*, 1545 (2007).

26. T. Miyasaka, *J. Phys. Chem. Lett.*, **2**, (2011).

27. S. Ito, G. Rothenberger, P. Liska, P. Comte, S. M. Zakeeruddin , P. Péchy, M. K. Nazeeruddin and M. Grätzel, *Chem. Commun.*, (2006).

28. D. Kuang, J. Brillet, P. Chen, M. Takata, S. Uchida, H. Miura, K. Sumioka , S. M. Zakeeruddin and M. Grätzel, *ACS Nano*, 2 (2008).

29. J. H. Park, Y. Jun, H. G. Yun, S. Y. Lee and M. G. Kang, *J. Electrochem. Soc.*, 155 (2008).

30. X. Fan, F. Wang, Z. Chu, L. Chen, C. Zhang and D. Zou, *Appl. Phy. Lett.*, **90**, (2007).

31. X. Fan, Z. Chu, F. Z. Wang, C. Zhang, L. Chen, Y. Tang and D. Zou, *Adv. Mater.*, **20**, (2008).

32. D. Zhang, T. Yoshida, K. Furuta and H. Minoura, *J. Photoch. Photobio. A*, 164 (2004).

33. N. G. Park, K. M. Kim, M. G. Kang, K. S. Ryu, S. H. Chang and Y. J. Shin, *Adv. Mater.*, **17**, (2008).

34. T. Miyasaka, M. Ikegami and Y. Kijitori, *J. Electrochem. Soc.*, 154 (2007).

35. X. Li, H. Lin, J. Li, N. Wang, C. Lin and L. Zhang, *J. Photoch. Photobio. A*, 195 (2008).

36. K. Miyoshi, M. Numao, M. Ikegami and T. Miyasaka, *Electrochemistry*, **76**, (2008).

37. S. Uchida, M. Tomiha, H. Takizawa and M. Kawaraya, *J. Photoch. Photobio. A*, 164 (2004).

38. T. N. Murakami, Y. Kijitori, N. Kawashima and T. Miyasaka, *J. Photoch. Photobio. A*, 164 (2004).

39. T. Yamaguchi, N. Tobe, D. Matsumoto and H. Arakawa, *Chem. Commun.*, (2007).

40. M. Dürr, A. Schmid, M. Obermaier, S. Rosselli, A. Yasuda and G. Nelles, *Nat. Mat.*, **4**, (2005).

41. H. C. Weerasinghe, F. Z. Huang and Y. B. Cheng, *Nano Energy*, **2**, (2013).

42. L. Q. Ming, H. Yang, W. J. Zhang, X. W. Zeng, D. H. Xiong, Z. Xu, H. Wang, W. Chen, X. B. Xiao, M. K. Wang, J. Duan, Y. B. Cheng, J. Zhang, Q. L. Bao, Z. H. Wei and S. H. Yang, *J. Mater. Chem. A*, **2**, (2014).

43. T. L. Ma, *Chem. Prog.*, **18**, (2006).

44. Z. G. Zhou, H. M. Tian and T. Tao, *CN 101140956A* (2008).

45. J. Chen, K. Li, Y. Luo, X. Guo, D. Li, M. Deng, S. Huang and Q. Meng, *Carbon*, **11**, (2009).

46. M. Ikegami, J. Suzuki, K. Teshima, M. Kawarayam and T. Miyasaka, *Solar Energy Materials & Solar Cells*, **93**, (2009).

47. D. Kearns and M. Calvin, *J. Chem. Phys.*, **29**, (1958).

48. C. W. Tang, *Appl. Phys. Lett.*, 183 (1986).

49. G. Yu, J. Gao, J. C. Hummelen, F. Wudl and A. J. Heeger, *Science-AAAS-Weekly Paper Edition*, 270 (1995).

50. C. Hoth (2007), [Online] Available: spie.org/documents/newsroom/imported/0912/0912-2007-11-09.pdf. (Accessed on 20th September, 2014).

51. T. Ameri, N. Lia and C. J. Brabecab, *Energy Environ. Sci.*, **6**, (2013).

52. J. You, L. Dou, K. Yoshimura, T. Kato, K. Ohya, T. Moriarty, K. Emery, C. C. Chen, J. Gao, G. Li and Y. Yang, *Nat. Commun.*, **4**, (2013).

53. [Online] Available: http://www.heliatek.com/. (Accessed on 20th September, 2014).

54. S. Glenis, G. Tourillon and F. Garnier, *Thin Solid Films*, 139 (1986).

55. C.W. Tang, *Appl. Phys. Lett.*, 48 (1986).

56. F. Yang, M. Shtein and S. R. Forrest. *Nat. Mater.*, **4**, (2005).

57. P. Peumans, S. Uchida and S. R. Forrest, *Nature*, **425**(6954), 158 (2003).

58. M. C. Scharber, D. Mühlbacher, M. Koppe, P. Denk, C. Waldauf, A. J. Heeger and C. J. Brabec, *Adv. Mater.*, **18**, (2006).

59. N. Yeh and P. Yeh, *Renewable and Sustainable Energy Reviews*, 21 (2013).

60. M. M. Wienk, J. M. Kroon, W. J. H. Verhees, J. Knol, J. C. Hummelen, P. A. van Hal and R. A. J. Janssen, *Angew. Chem. Int. Ed.*, **42**, (2003).

61. J. Y. Kim, S. H. Kim, H.-H. Lee, K. Lee, W. Ma, X. Gong and A. J. Heeger, *Adv. Mater.,* **18**, (2006).

62. M. D. Irwin, D. B. Buchholz, A. W. Hains, R. P. H. Chang and T. J. Marks, *PNAS*, 105 (2008).

63. Y. Liang, Y. Wu, D. Feng, S.-T. Tsai, H.-J. Son, G. Li and L. Yu, *J. Am. Chem. Soc.*, 131 (2009).

64. Y. He, H-Y. Chen, J. Hou and Y. Li, *J. Am. Chem. Soc.*, 132 (2010).

65. H-L. Yip and A. K-Y. Jen, *Energy Environ. Sci.*, **5**, (2012).

66. R. Steim, F. R. Koglera and C. J. Brabeccd *J. Mater. Chem.*, **20**, (2010).

67. R. Søndergaard, M. Hösel, D. Angmo, T. T. Larsen-Olsen and F. C. Krebs, *Materials Today*, **15**, (2012).

68. M. Hçsel, R. R. Søndergaard, M. Jørgensen and F. C. Krebs, *Energy Technol.*, **1**, (2013).

69. P. Gao, M. Grätzel and M. K. Nazeeruddin, *Energy Environ. Sci.*, **7**, (2014).

70. D. Cao, C. Wang, F. Zheng, W. Dong, L. Fang and M. Shen, *Nano Lett.*, **12**, (2012).

71. A. Kojima, K. Teshima, Y. Shirai and T. Miyasaka, *J. Am. Chem. Soc.*, 131 (2009).

72. H. Im, C. R. Lee, J. W. Lee, S. W. Park and N. G. Park, *Nanoscale*, **3**, (2011).

73. H. S. Kim, C. R. Lee, J. H. Im, K. B. Lee, T. Moehl, A. Marchioro, S-J. Moon, R. Humphry-Baker, J-H. Yum, J. E. Moser, M. Grätzel and N. G. Park, *Sci. Rep.*, 591 (2012).

74. M. M. Lee1, J. Teuscher, T. Miyasaka, T. N. Murakami and H. J. Snaith, *Science*, 338 (2012).

75. J. Burschka, N. Pellet, S. J. Moon, R. Humphry-Baker, P. Gao, M. K. Nazeeruddin and M. Gratzel, *Nature*, 499 (2013).

76. M. Liu, M. B. Johnston and H. J. Snaith, *Nature*, 501 (2013).

77. N. J. Jeon, J.H. Noh, Y. C. Kim, W.S. Yang, S. Ryu and S. Seok, *Nat. Mater.*, **13**, (2014).

78. H. Zhou, Q. Chen, G. Li, S. Luo, T. Song, H. S. Duan, Z, Hong, J.i You, Y. Liu and Y. Yang, *Science*, 345 (2014).

79. M. Grätzel, *Nat. Mater.*, **13**, (2014).

80. S. D. Stranks, G. E. Eperon, G. Grancini, C. Menelaou, M. J. P. Alcoce, T. Leijtens, L. M. Herz, A. Petrozza and H. J. Snaith, *Science*, 342 (2013).

81. K. Wojciechowski, M. Saliba, T. Leijtens, A. Abate and H. J. Snaith, *Energy Environ. Sci.*, **7**, (2014).

82. Q. Chen, H. Zhou, Z. Hong, S. Luo, H. S. Duan, H. H. Wang, Y. Liu, G. Li and Y. Yang, *J. Am. Chem. Soc.*, 136 (2014).

83. D. Liu and T. L. Kelly, *Nat. Photonics*, **8**, (2013).

84. P. Docampo, J.M. Ball, M. Darwich, G.E. Eperon and H. J. Snaith, *Nat. Commun.*, **4**, (2013).

85. C. Roldán-Carmona, O. Malinkiewicz, A. Soriano, G. M. Espallargas, A. Garcia, P. Reinecke, T. Kroyer, M. I. Dar, M. K. Nazeeruddine and H. J. Bolink, *Energy Environ. Sci.*, **7**, (2014).

86. J. You, Z. Hong, Y. Yang, Q. Chen, M. Cai, T. B. Song, C. C. Chen, S. Lu, Y. Liu, H. Zhou and Y. Yang, *ACS Nano*, **8**, (2014).

87. Q. Wang ,Y. Shao, Q. Dong, Z. Xiao, Y. Yuan and J. Huang, *Energy Environ. Sci.*, **7**, (2014).

88. J. Seo, S. Park, Y. C. Kim, N. J. Jeon, J. H. Noh, S. C. Yoon and S. Seok, *Energy Environ. Sci.*, **7**, (2014).

89. F. Hao, C. C. Stoumpos, D. H. Cao, R. P. H. Chang and M. G. Kanatzidis, *Nature Photonics*, **8**, (2014).

CHAPTER 8

FLEXIBLE SOLAR CELLS

Dongdong Li,*,‡ Dongliang Yu,* Zhiyong Fan,†,§ Linfeng Lu*
and Xiaoyuan Chen*

*Shanghai Advanced Research Institute
Chinese Academy of Sciences, 99 Haike Road
Zhangjiang Hi-Tech Park, Pudong, Shanghai 201210, P. R. China
†Department of Electronic and Computer Engineering
The Hong Kong University of Science and Technology
Clear Water Bay, Kowloon, Hong Kong, P. R. China
‡lidd@sari.ac.cn
§eezfan@ust.hk

Flexible solar cells have attracted steadily increasing interest as a convenient and alternative energy source. Their light-weight and foldable features open up new opportunities for portable and building integrated applications. In this chapter, recent progress on organic, inorganic and organic–inorganic flexible solar cells constructed on various flexible substrates will be described. And the principles behind the mainstream flexible technologies, and the future challenges in this area will be discussed.

1. Introduction

Photovoltaic (PV) devices, i.e., solar cells, are capable of converting clean and abundant solar irradiation into electricity directly. They are important renewable power sources as alternatives to other sources of energy such as fossil fuels and nuclear energy, and have attracted enormous attention.[1–4] Through decades of research and development, PV technologies have achieved great improvement. The PV market has grown at a remarkable rate and solar energy is expected to be a major source of power generation for the world. Cumulative installed solar PV capacity has increased rapidly from ~1.28 GW in 2,000 to ~138 GW in 2013.[5] So far, the Si wafer-based technology has been well developed by industry, but several other emerging thin-film technologies based on inorganic[6] and organic materials[7,8] have drawn the attention of researchers.

Compared with Si wafer-based PV technologies, thin film PV holds great promise in the quest for clean renewable energy. Especially, the light-weight, mechanically flexible thin-film technologies, i.e., flexible solar cells, open up new opportunities for PV devices in terms of building-integrated, wearable and portable applications. Manufacturing solar cells on flexible substrates is highly attractive because it enables the implementation of high-throughput roll-to-roll (R2R) deposition methods, which could afford lightweight, economic solar modules as compared to conventional batch-to-batch or in-line methods used for rigid glass substrates. Furthermore, flexible PV panels can be easily integrated with infrastructures of various shapes and sizes, consequently stimulating the design of innovative energy-generating products. Therefore, flexible solar cells offer a convenient and alternative energy source for indoor and outdoor applications.[2,3,9,10]

In this chapter, we summarize the recent progress that has been made in the field of flexible solar cells, including PV materials, substrate materials, and efficient light management. Different types of active layers endow the flexible solar cells different properties, and the principles behind the flexible technologies are discussed. In addition, a perspective on the future development of flexible solar cells and new opportunities offered by these devices is provided.

2. Flexible (PVs)

Flexible PVs lead to light-weight, low-cost solar modules that can be integrated into various surfaces. However, full exploration of these PV devices requires further development and improvement of cell performances through better understanding of cell behavior, material usage, and optimized processes. Until now, a variety of flexible PV technologies has been developed based on various materials with diverse deposition methods. Here some (inorganic, organic and organic–inorganic) typical flexible solar cells are discussed briefly.

2.1. *Inorganic Flexible Solar Cells*

2.1.1. Flexible Thin-Film c-Si Solar Cells

Generally, crystalline silicon (c-Si) solar cells are rigid and fragile, although they hold a dominant market share due to the advantages of material abundance, broad spectral absorption range, high carrier mobilities, and mature technology.[11] Wafer-based c-Si solar cells usually have a thickness of around 200 μm, and can gain effciencies as high as 24.7% in the research lab.[12] When the thickness decreases to ~47 μm, a 21.5% efficient thin c-Si solar cell can still be obtained.[13] More importantly, thin film c-Si solar cells have a potentially higher open circuit voltage (V_{oc}) and effciency limits imposed by Auger recombination compared with bulk wafers, makes thin film c-Si a good candidate for high performance flexible solar cells.[14] Meanwhile further reducing the c-Si thickness down to a few micrometers can significantly reduce material usage and cost. Figure 1(a) shows a 3 μm thick Si film that was wrapped around a plastic rod with a diameter of 7 mm.[15]

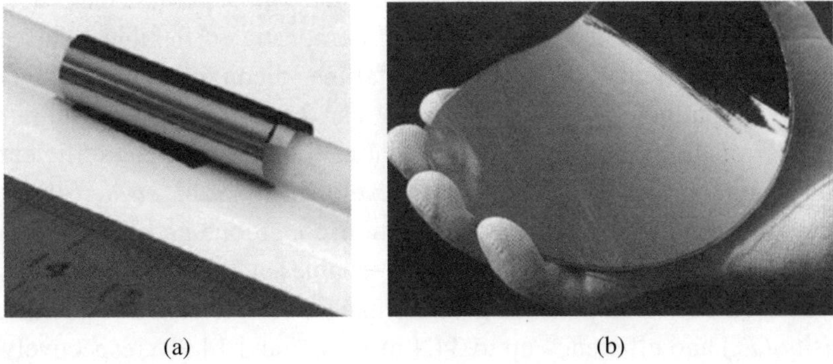

(a)   (b)

Figure 1.   (a) A 3 μm thick Si film was wrapped around a plastic rod with diameter of 7 mm. Reprinted from Ref. 15. (b) A flexible, 150 mm diameter, 25 μm thick c-Si sheet laminated between two, 200 μm thick protective foils. Reprinted from Ref. 2.

Because the wire diameter (150–200 μm), used in the conventional wire-saw technique, is much bigger than the wafer thickness, the kerf loss as a fraction of the total ingot will dramatically increase. In order to fabricate low-cost and flexible c-Si PVs, several methods have been developed. One of the most famous approaches is the so-called epitaxial layer transfer (ELTRAN) process, which enables a high quality mono-c-Si thin-film transfer on foreign substrates.[16] Most popular, the epitaxial growth occurs on the topmost porous silicon layer with enclosed voids of several 10 nm in diameter, which is termed as quasi-monocrystalline silicon (QMS) and serves as seeding layer. After *in situ* device fabrication, the solar cell is attached to a transparent superstrate and detached from the host wafer directly by mechanical force. By employing this method, Schubert *et al.*[2,17] successfully demonstrated flexible c-Si solar cells with 150 mm diameter sandwiched in two plastic sheets (Fig. 1(b)), and an effciency up to 14.6% was achieved. Besides, the preparation of c-Si thin film can also be achieved by etch release process. Yoon *et al.*[18] demonstrated ultrathin flexible c-Si solar microcells. These microcells were composed of well-organized interconnected microbar solar cells, with the microbars fabricated through the anisotropic undercut etching of silicon wafers. The individual microcells are 15–20 μm thick and show 6–8% efficiencies. Moreover, the efficiency of the flexible module remains unchanged at a bending radius of 4.9 mm and only shows little degradation with bending up to 200 cycles, which suggests good mechanical robustness for flexible applications. Another effective method to prepare c-Si thin film is simple exfoliation from bulk silicon wafers. Initiated by a pre-determined crack, the c-Si thin film is finally exfoliated. Bedell *et al.*[19] demonstrated flexible solar cells with an efficiency of 4.3% on monocrystalline silicon foils prepared by the controlled spalling. This low efficiency can be ascribed to the ultrathin thickness of the silicon foil (~3 μm) that cannot absorb light sufficiently. Thicker silicon foils (~25 μm) were developed by Saha *et al.*[20] through optimizing the metal thickness, deposition conditions and the pre-determined crack parameters. This process enables a double side texturing process and delivered promising performances with a short circuit current density (J_{sc}) and efficiency up to 34.4 mA cm^{-2} and 14.9% respectively.

So far it is still challenging for mass production of thin-film c-Si down to several tens of micrometer. However, except for the demand of delicate substrate fabrication, the flexible solar cell fabrication process is similar to the conventional wafer-based c-Si solar cells that includes the doping of the back surface field, diffusion of the selective emitter, deposition of passivation, anti-reflection layer and the contacts. Other state-of-the-art technologies, such as Heterojunction with Intrinsic Thin layer (HIT) and Passivated Emitter Rear Cell (PERC) solar cells, can also potentially be adopted on the flexible c-Si wafers.

2.1.2. Flexible Thin-Film a-Si:H/μc-Si:H Solar Cells

Hydrogenated amorphous (a-Si:H)/microcrystalline (μc-Si:H) silicon possess higher absorption coefficient and better weak light performance than that of c-Si due to the presence of amorphous component, that requires only a thickness of a few hundred nanometers for a-Si:H and a few micrometers for μc-Si:H solar cells.[21] The other difference is the much reduced processing temperature (< 200°C) of a-Si:H/μc-Si:H silicon thin-film solar cells[21] than that of c-Si ones (>1000°C). Owing to the above advantages, a-Si:H/μc-Si:H solar cells are quite suitable for the large scale deployment of low-cost PVs on flexible and low temperature tolerance substrates.

The active layer of a-Si:H/μc-Si:H single junction solar cell is typically composed of three silicon layers, i.e., the n- and p-type doped layers and an intrinsic (i-) layer between them. The n- and p-layers create an internal field across the i-layer, while the i-layer is responsible for absorbing the most incident light. Generally, a-Si:H/μc-Si:H films are deposited by plasma-enhanced chemical vapor deposition (PECVD) at below 200°C using H_2 diluted SiH_4 as the source precursor, and PH_3 and B_2H_6 as the n- and p-type dopant sources. With regard to mass production, flexible a-Si:H and μc-Si:H solar cells are preferred to be fabricated via R2R processing techniques, which improves the production efficiency substantially, and in turn reduces the cost greatly.[22,23]

In practice, however, thin-film a-Si:H solar cells have not been able to hold a significant share of the global PV market. It can be mainly ascribed to their low stable average efficiency of 6% or less of large-area singlejunction PV modules.[24] Research is continuing into finding ways to improve the efficiency. Excellent light management on the PVs and multi-junction designs can effectively help to increase the solar cell performance by

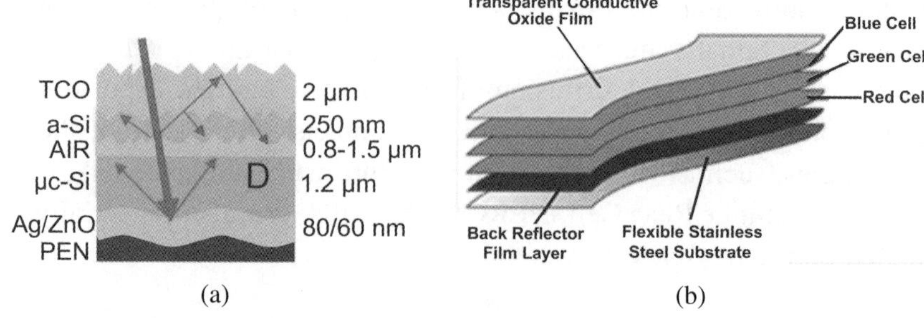

Figure 2.   (a) Tandem micromorph a-Si:H/μc-Si:H with AIR deposited on textured substrates covered with thin Ag/ZnO. Reprinted from Ref. 28. (b) Schematic of a triple-junction structure containing amorphous silicon. Reprinted from Ref. 3.

maximizing the light absorption. Micromorph tandem solar cell is a typical design, pioneered by University of Neuchatel in the 1990s.[25,26] The performance of μc-Si:H solar cells improves significantly by combining with thin a-Si:H top cells. Given that the bandgaps of a-Si:H and μc-Si:H in the tandem devices are ~1.7 eV and ~1.2 eV, respectively, a much better utilization of the solar spectrum and hence a higher power conversion efficiency is achieved.[27] And flexible tandem micromorph cells on plastic substrate with a stabilized efficiency of 9.8% have been fabricated.[28] The configuration is shown in Fig. 2(a), where an asymmetric intermediate reflector (AIR) design uncouples the growth and light scattering issues of the tandem micromorph solar cells. Meanwhile, triple-junction cells, for example a-Si:H/a-SiGe:H/a-SiGe:H, are also demonstrated by tuning the bandgap through controlling the silicon-to-germanium ratio to more efficiently harvest the incident light. The schematic is shown in Fig. 2(b). For example, United Solar, a US PV manufacturer, fabricated triple-junction amorphous silicon alloy solar cells on stainless steel (SS) with an initial efficiency of 14.6% and a stable efficiency of 13%.[10,29]

2.1.3. Flexible Thin-Film Cadmium Telluride (CdTe) Solar Cells

Up to now, CdTe has been one of the leading thin-film PV materials due to its proximity to the ideal band gap (1.45 eV) for PV conversion efficiency and its high optical absorption coefficient. Generally CdTe

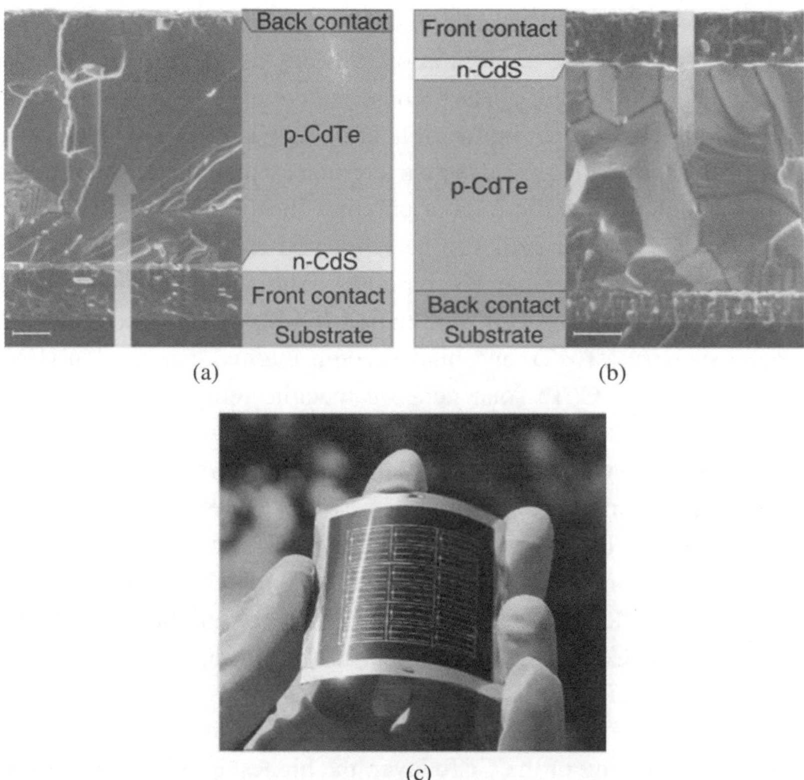

Figure 3. Scanning electron micrograph and schematic of the cross-section of a CdTe solar cell in the conventional superstrate configuration (a) and the substrate configuration. (b) which allows the use of opaque substrates like metal foils. In substrate configuration Mo/MoO$_x$ and i-ZnO/ZnO:Al are used as electrical back and front contact, respectively. The scale bars correspond to 1 μm. The arrows show the direction of illumination. (c) Photograph of a sample with several CdTe solar cells on flexible metal foil. Reprinted from Ref. 30.

thin-film solar cells are made by heterojunction of p-type CdTe with n-type CdS partner window layer. Figures 3(a) and 3(b) show the cross-sections of such CdS/CdTe thin-film solar cells in superstrate and substrate configuration, separately.[30] The superstrate configuration means that the light enters the solar cells through the substrate which has to be a transparent material for optical photons. For substrate configuration however, the light passes through the opposite side rather than from the base.

In conventional CdTe solar cells based on transparent conductive oxide (TCO) glass superstrate configuration, 98% of the device thickness and weight is taken up by the glass.[31] Glass is known to be rigid and brittle, which is unfavorable to fabrication and installation. The weight and mechanical issues with glass can be overcome by the utilization of flexible substrates, such as metallic foils or polymer sheets.

Thin-film CdTe solar cells can be fabricated on thin foils of metals or polymers, obtaining flexible PV modules. And the common deposition technologies are close spaced sublimation (CSS), electrodeposition (ED), magnetic sputtering (MS), and high vacuum thermal evaporation (HVE), etc.[32] Since 1990s, CdTe solar cells on metallic molybdenum foils have been developed; with the thermally evaporated CdTe as absorber layer a cell efficiency of 5.3% was achieved.[33] While employing radio-frequency (RF) magnetron sputtering for deposition of CdS and CdTe and an interlayer of nitrogen-doped ZnTe between CdTe and the back contact, an efficiency of 7.8% was demonstrated for CdTe solar cells on molybdenum substrates.[34] The improved efficiency can be attributed to the efficient ohmic back contact between CdTe and metallic substrates through the interlayer of nitrogen-doped ZnTe. Considerable efforts have been spared to improve the performance of flexible CdTe solar cells. And so far, through precisely controlling Cu doping of the CdTe layer, the highest efficiency in substrate configuration on metallic foils was 11.5%.[30] The flexibility of the CdTe solar cell on molybdenum substrates is demonstrated in Fig. 3(c). Meanwhile, the highest reported efficiency in flexible CdTe solar cells of 14.05% was achieved with superstrate configuration using ultra-thin glass.[35]

Such foil-based solar cells can be manufactured in a roll-to-roll process that offers many manufacturing advantages such as lower equipment size, thus lower costs, higher material utilization, increased fabrication scalability, and high speed of deposition.[36] Further, flexible CdTe solar cells have high specific power of more than 2 kW/kg and the highest stability to particle irradiation, making them potentially better suited for space applications.[37,38]

2.1.4. Flexible Thin-Film Cu(In,Ga)Se$_2$ (CIGS) Solar Cells

The quaternary compound CIGS is cadmium-free and thus a more environment friendly material for thin-film solar cells. The first chalcopyrite

I-III-VI$_2$ solar cells were developed in the 1970s, and they were based on Ga-free absorbers, resulting in relatively low cell efficiencies.[39] After 1990, alloying of Ga into the compound was found to significantly improve the device efficiency, because the addition of Ga can engineer the bandgap of CIGS continuously from around 1.0 eV for pure CIS (Ga-free) to 1.7 eV for pure CGS (In-free).[40] State-of-the-art CIGS absorber generally has an average bandgap energy of around 1.2 eV, which corresponds to a relative Ga content of $x = [Ga]/([Ga] + [In]) = 0.3–0.4$.[41] And highest efficiency of 20.4% was achieved on rigid glass substrate by elemental co-evaporation process [EMPA, 2013], which is the highest among the thin-film solar cell technologies.

Meanwhile, deposition of CIGS films onto flexible substrates has also attracted extensive research interests. Flexible CIGS solar cells could adapt to many new field of applications and significantly decrease production costs by employing roll-to-roll manufacturing. Development of lightweight flexible CIGS solar cells on flexible metal foils or plastic films have the potential for achieving high specific power, which would be good candidate for space power applications because of the near optimum bandgap for AM0 solar radiation spectrum in space.[42] Chirilă *et al.*[41] obtained solar cells on flexible polymer films with 18.7% efficiency by adjusting the composition gradient in the CIGS absorber layer. The micrograph and photograph of the flexible CIGS solar cell are shown in Fig. 4.

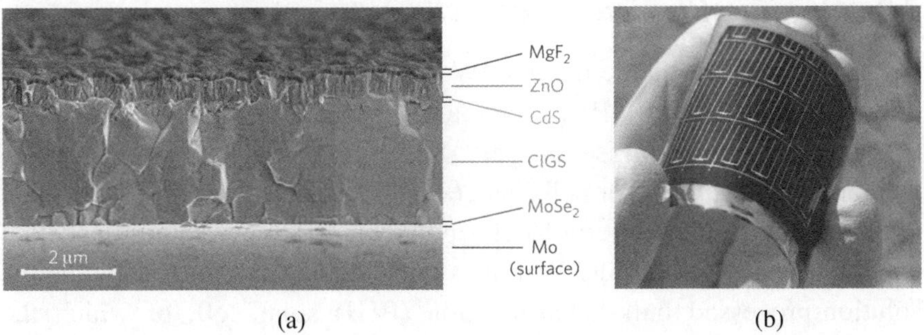

(a)   (b)

Figure 4.   (a) Scanning electron micrograph of a cross-section obtained from a complete CIGS device on PI with an efficiency greater than 18%. To minimize reflection losses a top layer of MgF$_2$ is used. (b) Photograph of a sample on PI with single cells. Reprinted from Ref. 41.

Recently, they deposited CIGS solar cells on polyimide (PI) film and achieved a remarkable efficiency of 20.4%, which even surpassed the cell efficiency on rigid glass, and it could match the performance of the best polycrystalline Si wafer-based device.[43]

For high-efficiency CIGS solar cells, the absorber layers are generally grown by vacuum-based methods such as thermal coevaporation at high substrate temperature or sputtering of precursor layers followed by selenization in Se vapor or H_2Se gas, which encounters a high manufacturing cost and poor device yield. In contrast, non-vacuum methods promise significantly lower cost and reduced material losses. Non-vacuum deposition of absorber has become an intensively pursued goal in PV research and a variety of solution-based approaches have been demonstrated.[44] Typically, chemically synthesized nanocrystals with controlled element stoichiometry and crystal phase disperse in appropriate solvents, resulting in a paint or ink. And then the paint or ink is coated on flexible substrates followed by the removal of the solvents, thus forming absorber layers. The non-vacuum methods have been used to produce CIGS devices with efficiencies of up to 14%.[45] In parallel, CdTe devices fabricated by solution-based processing achieved an efficiency as high as 12.3%.[46] It can be noted that the combination of non-vacuum processing with roll-to-roll manufacturing technology would offer the potential for drastic cost reduction through improved materials utilization and high device throughput.

2.2. *Organic Flexible Solar Cells*

Compared with inorganic flexible PV technologies, the core advantage of organic photovoltaic (OPV) technology is the ease of the fabrication, which offers the promise of very low-cost manufacturing process.[47] Research on organic solar cells generally focuses on solution-processable organic semiconducting molecules/polymers or on vacuum-deposited small-molecular materials. A typical and successful technique is the solution-processed bulk-heterojunction (BHJ) solar cell, of which the active layers comprise a donor (p-type semiconductor) and an acceptor (n-type semiconductor).[48] Important representatives of donor semiconducting molecules/polymers are pentacene, anthracene, rubrene, oligothiophenes or polypyrrole, polyacetylene, poly(3-hexylthiophene)

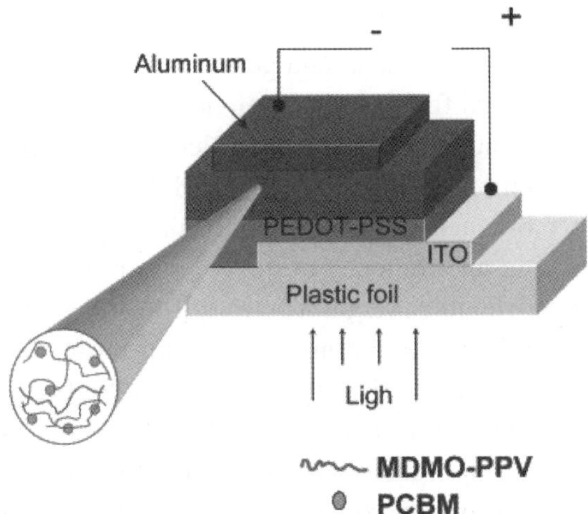

Figure 5. Bulk heterojunction configuration in organic solar cells. Reprinted from Ref. 51.

(P3HT), poly(p-phenylene vinylene) (PPV).[49] A soluble derivative of C_{60}, PCBM (1-(3-methoxycarbonyl) propyl-1-phenyl[6,6]C_{61}) is widely used as the electron acceptor.[50] A configuration of BHJ solar cell is shown in Fig. 5.[51] Both the donor and acceptor interpenetrate one another, so that the interface between them is spatially distributed. Therefore, the bulk heterojunction ensures a high interfacial area and thus an optimal donor acceptor contact.

Up to now, BHJ solar cells based on the polymer blends of PTB7 with $PC_{71}BM$ exhibited an efficiency up to 7.4%,[47] while BHJ solar cells based on DTS(PTTh$_2$)$_2$: $PC_{70}BM$ molecular composites gained an efficiency as high as 6.7%.[52] Besides single junction OPVs, a 10.6% efficiency solution-processed polymer tandem[53] and a 10.7% efficiency vacuum-processed small molecule organic tandem (Heliatek in Germany) OPVs have been demonstrated, which holds promises for the future OPV technologies. However, these OPVs were fabricated on rigid glass substrates coated with indium tin oxide (ITO) as the transparent conductor. In order to take full advantage of OPVs for vehicles and buildings, flexible and even stretchable substrates are preferred. And given the brittleness of ITO that prevents OPVs from being flexible, intensive efforts have been devoted to develop highly conductive and transparent ITO-free electrodes in order to fabricate OPVs on flexible substrates.[54] These alternatives belong to four material groups: polymers; metal and polymer composites; metal nanowires and ultra-thin metal films; and carbon nanotubes and graphene.[54] The use of highly conductive PEDOT:PSS electrode instead of the ITO showed a good demonstration. P3HT:PCBM bulk heterojunction solar cells were then fabricated on ultrathin polyethylene terephthalate (PET) substrates, achieving an efficiency of 4.2%.[55] Currently, the efficiency of OPVs obtained on flexible, vacuum-free, roll-to-roll processed modules processed under ambient conditions still remain unsatisfactory (<2%).[54] Compared with the inorganic flexible solar cells (c-Si, CdTe or CIGS), the maximum power conversion that organic flexible solar cells can achieve is clearly low. However, organic flexible solar cells have a lot of potential in view of manufacturing by roll-to-roll processing, low cost, flexibility and light weight.

2.3. *Organic–Inorganic Flexible Solar Cells*

2.3.1. Flexible Dye-Sensitized Solar Cells (DSSCs)

Since DSSCs were first reported by Grätzel in 1991,[56] this novel solar cell has been intensively investigated due to its relatively high conversion efficiency and low cost. The typical structure and working mechanism is roughly depicted in Fig. 6(a).[57] A DSSC generates photocurrent through ultra-fast injection of electrons from photo-excited dye molecules into the conduction band of a semiconductor such as TiO_2[57] or ZnO.[58] This process is followed by dye regeneration and hole transportation to the counter electrode. DSSCs have advantages in collecting energy of low intensity light (diffused light, indoor illumination, etc.) without lowering voltage.[59,60] As they are less sensitive to partial shadowing and low-level illumination, DSSCs work well over a wide range of lighting conditions and orientation, making DSSC-based modules particularly well suited for building integrating applications. Figure 6(b) demonstrates an example of a large belt-shaped DSSC on ITO-coated polyethylene naphthalate (PEN) films. The high voltage of the cell, >0.7 V, is maintained under exposure to indoor illumination.[61]

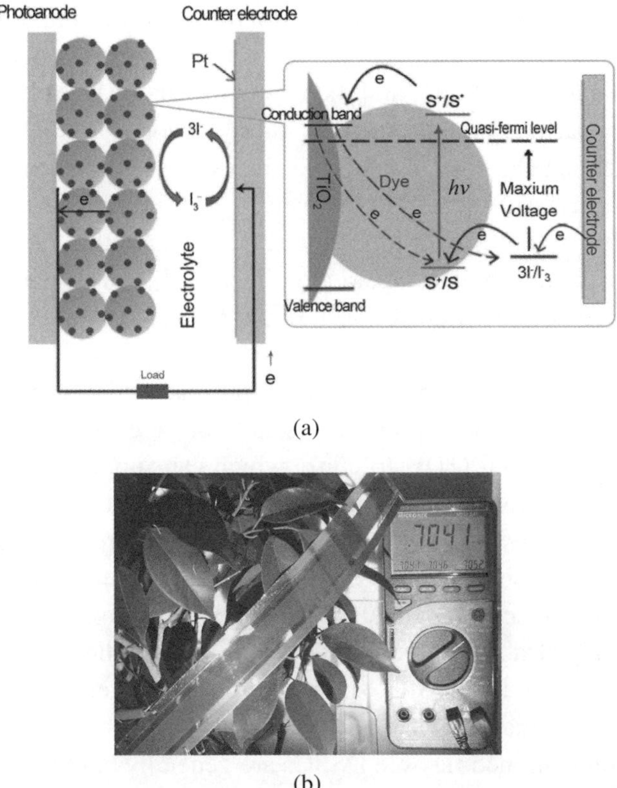

Figure 6. (a) Sketch map of the structure and working mechanism of a DSSC device. Reprinted from Ref. 57. (b) A flexible film-type DSSC made on a belt-shaped plastic ITO-PEN. Exposure of the cell to indoor illumination generates a DC voltage exceeding 0.7 V. Reprinted from Ref. 61.

Although the efficiency of DSSCs based on conductive glass substrates has been reported as over 12%,[62] there is a great deal of interest in cheaper, flexible DSSCs using transparent ITO coated plastic electrodes in place of the glass. However, as the plastic-based electrode cannot undertake the high sintering temperature (450–550°C) generally necessary to enhance the TiO_2 particle interconnection, different low temperature coating and deposition techniques have been investigated for the preparation of TiO_2 electrodes on flexible substrates. Mechanical compression is one of the widely used techniques for fabricating DSSCs on flexible plastics, and this method can be compatible to roll-to-roll manufacturing process.[63,64] This method involves the deposition of semiconductor particle layer on conductive plastic, and then compression of the particle layer to enhance the particle-particle and film-substrate connectivity. Through this method, Yamaguchi *et al.*[64] achieved the highest efficiency of 8.1% on ITO coated PEN substrate. Alternatively, metal substrates such as SS and Ti foil have also attracted a lot of interest for fabricating flexible DSSCs as they can sustain high temperature sintering.[65–67] Through controlling the dark current, the efficiency of flexible DSSCs on stainless steel substrate can reach as high as 8.6%.[65]

Apart from using TiO_2 nanoparticles as photoactive layer in DSSCs, self-organized TiO_2 nanotube (TNT) layers also can be good candidates due to their one-dimensional (1D) structure, which can supply a direct path for electron transfer from TiO_2 to substrate.[68] Vomiero *et al.*[69] directly fabricated TNT arrays by an anodization of Ti thick film pre-deposited on PET and Kapton HN substrates. And the obtained flexible DSSCs with 6 μm length TNTs film achieved an efficiency of 3.5%. Meanwhile, the flexible DSSCs with TNT arrays directly on the Ti foil, after filled with TiO_2 nanoparticles and coated with SrO, showed a remarkable efficiency of 5.39% under back-illumination.[70]

The aforementioned flexible DSSCs are generally based on liquid electrolyte, which leads to the poor stability of DSSCs due to the leakage and deterioration of the electrolyte. To overcome this issue, solid-state or quasi-solid-state flexible DSSCs have been developed by introducing polymer gels into the electrolytes or by replacing the liquid electrolyte with a solid-state polymer hole conductor.[71–74] In 2003, flexible solid-state solar cells based on dye-sensitised nanocrystalline Al_2O_3 coated TiO_2 films and an I_2/NaI doped solid-state polymer electrolyte were demonstrated, achieveing an efficiency of ~5.3% under 10 mW cm^{-2} AM1.5 illumination.[73] Ting *et al.*[74] fabricated all-plastic substrate flexible DSSCs by using a gel electrolyte, and an efficiency of 4.57% was reached under 100 mW cm^{-2} AM1.5 illumination. Solid-state flexible DSSCs have advantages for the robust sealing and encapsulation technique. When the solid-state electrolytes were applied in the flexible device, it is expected that the long-term performance and stability would be improved further, and accelerate practical application.

2.3.2. Flexible Perovskite Solar Cells

The rise of metal halide perovskites as light harvesters has stunned the PV community, and they are even addressed as a game changer in PVs.[75] The $CH_3NH_3PbX_3$ (X = Br, I) perovskite nanocrystals were firstly used as sensitizer in liquid-electrolyte based DSSCs in 2009, but poor performance were observed.[76,77] And then in 2012, a key advance was made by using the metal halide perovskites in solid-state DSSCs. Not only did the conversion efficiency double, but the cell stability also improved greatly.[78] Subsequently, perovskites attracted great interest in PV devices due to their panchromatic light absorption and ambipolar behavior, that is, they can act as both light harvester and hole and electron transport layers.[78–82] So far, a new class of solar cell based on mixed organic-inorganic halide perovskites has achieved rapid emergence. Recently the group of Sang II Seok has obtained a certified efficiency of 17.9%, and an uncertified efficiency of 19.3%.[83]

Although great improvements have been made in perovskite-based solar cells, flexible ones are not much investigated. Bolink and coworkers [84] recently hold the view that the perovskite based solar cells are very well suited for roll-to-roll processing on flexible substrates. They have successfully prepared flexible perovskite based solar cells on PET substrate by thermal evaporation of $CH_3NH_3PbI_3$, achieving a power conversion efficiency of 7%. And the flexible solar cells were bent over a roll with a diameter of 5.5 cm. No reduction of the key performance indicators was observed after 50 bending cycles. The layout and photograph of the flexible perovskite solar cell are shown in Fig. 7. Besides the thermal evaporation method, the solution-processed method is most commonly used to fabricate perovskite compounds. Through this method, Snaith *et al.*[85] prepared $CH_3NH_3PbI_{3-x}Cl_x$ as perovskite absorber and fabricated inverted flexible PV devices on ITO-coated PET substrates with an efficiency of 6.4%. Similarly, Hong *et al.*[86] fabricated perovskite-based solar cells via a low-temperature (<120°C) solution-processed approach. They gained efficiency as high as 9.2% on ITO-coated PET flexible substrate.

During the rapid development of perovskite solar cells, the present perovskites share a common disadvantage with CdTe, namely reliance on

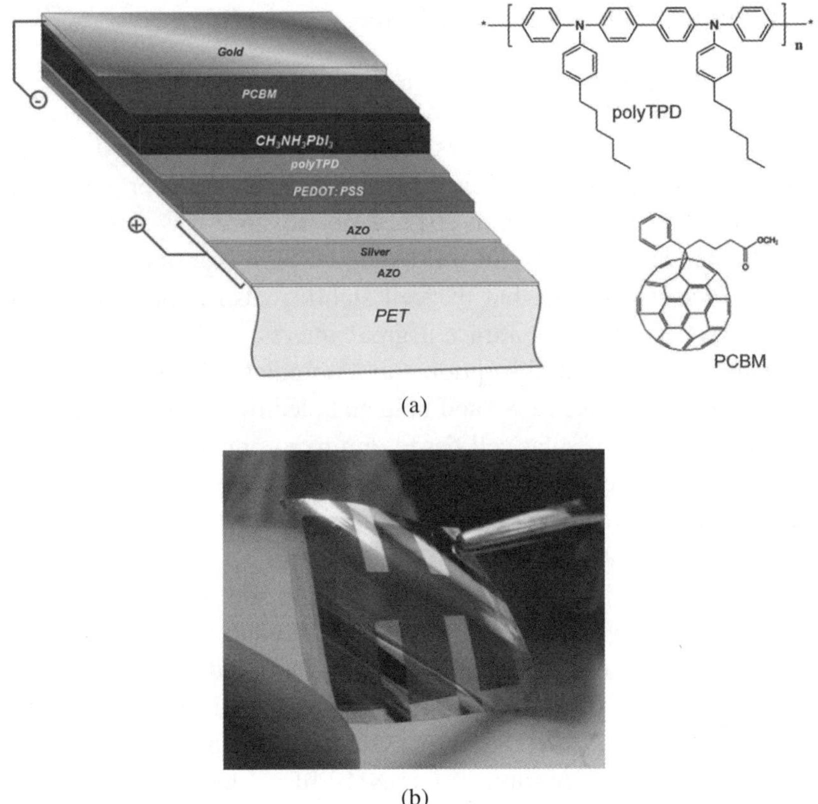

Figure 7.   (a) Schematic layout of the flexible perovskite solar cell and chemical structure of the materials used as the electron and hole blocking layer. (b) Photograph of the flexible solar cell. Reprinted from Ref. 84.

an environmentally hazardous heavy metal (Cd or Pb). However, it is believed that research into the present perovskites might allow more precise determination of the features, encouraging identification and investigation of non-toxic material systems with similar properties.[82,87] Advantages over existing PV technologies, such as material properties that simplify the manufacture of high-performance devices, may give rise to low processing costs and simple implementation of attractive products.

2.3.3. Flexible Hybrid BHJ Solar Cells

Hybrid BHJ solar cells which incorporate both organic and inorganic semiconducting materials aim to combine the advantages of both material classes. As an alternative to fully organic solar cells, hybrid BHJ solar cells have been demonstrated by combining semiconducting polymers including P3HT and PCBM, as the donor, with different inorganic materials, such as TiO_2 and ZnO, as the acceptor.[88,89] The fundamental operating principal in a fully organic or hybrid BHJ solar cell is identical.[90] Hybrid BHJ solar cells, which commonly consists of well-ordered 1D nanostructures for carriers to the collecting electrodes, and filled with polymer as an electron donor, has drawn much attention. Recently, Olson *et al.*[91] have demonstrated first hybrid BHJ solar cells by combination of a semiconducting polymer P3HT and a ZnO nanorod array, gaining a power conversion efficiency of 0.53%. the introduction of PCBM into these hybrid devices significantly improves the efficiency up to 2.03%.

Flexible hybrid BHJ solar cells have also been demonstrated based on well-ordered ZnO nanowires and P3HT. With transparent and conducting single walled carbon nanotube (SWNT) thin-films acting as current collector on PET substrates, an efficiency of 0.6% was achieved.[92] In addition, the use of ZnO nanowires instead of PCBM provides a cost effective alternative for BHJ solar cells. Most interestingly, Wang and coworkers[93] first demonstrated a flexible hybrid energy cell that was capable of simultaneously or individually harvesting thermal, mechanical, and solar energies to power some electronic devices. Well-aligned ZnO nanowire arrays grown on the flexible PET substrate (Fig. 8(a)) and P3HT were combined for harvesting solar energy, while a polarized poly(vinylidene fluoride) film-based nanogenerator (NG) was used to harvest thermal and mechanical energies. The schematic diagram of the flexible hybrid energy cell is shown in Fig. 8(b).

Even though the efficiencies of hybrid BHJ solar cells are still limited, it is expected that their efficiency will continue to increase by structure optimization, for example, control of the structure down to dimensions of ~10 nm, comparable to the exciton diffusion length in polymers.[90]

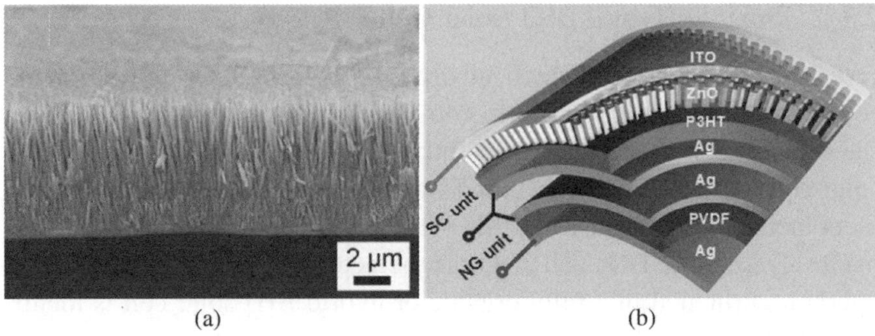

Figure 8. (a) Cross-section SEM image of the ZnO nanowire array. (b) Schematic diagram of the fabricated hybrid energy cell. Reprinted from Ref. 93.

3. Solar Cells on Flexible Substrates

Except for flexible c-Si solar cells that do not need extra substrate to support the cells, most of the flexible thin-film solar cells require flexible substrates to deposit the component layers on. The flexible substrates are diversified from polymers to metals. They can roughly be divided into two types, transparent and opaque flexible substrates. Different configurations and designs in flexible solar cells are dependent on different types of substrates. And to find suitable flexible substrates for depositing thin-film solar cells, different physical and chemical properties of the substrates have to be taken into account. Demands such as thermal stability, suitable coefficient of thermal expansion, chemical inertness, and surface smoothness have to be fulfilled. Low cost and light weight are also necessary for an optimal substrate material. Furthermore, given the flexibility of the solar cells, large-area high-throughput roll-to-roll printing technologies would help these PV modules to achieve mass production with low cost and decent efficiencies.

3.1. *Transparent Substrates for Flexible Solar Cells*

Generally, thin-film solar cells can be developed in two configurations, i.e., superstrate configuration and substrate configuration, depending on the direction through which light enters the cell. And as mentioned in

Table 1.   Commonly used flexible transparent substrates for solar cells.

Flexible substrate	Solar cell	Active layer preparation method	Temperature (°C)	PCE (%)	Refs.
PET	OPV	Solution-processed method	110	3.8	94
PET	OPV	Solution-processed method	150	4.2	55
PET	DSSC	Transfer and Compression method	Room temperature	5.8	95
PET	Perovskite	Thermal evaporation	250	7	96
PET	Perovskite	Solution-processed method	<120	9.2	86
PEN	DSSC	Compression method	Room temperature	8.1	64
PI (superstrate)	CdTe	Thermal evaporation	420	13.8	97
PI (superstrate)	OPV	Solution-processed method	110	2.62	98
Flexible Glass	CdTe	CSS	600	14.05	35

Section 2.1.3, the superstrate configuration requires the substrate to be reasonably transparent, as to allow enough light to pass into the cell. The most commonly used flexible substrates for superstrate configuration are polymers, such as PET, PEN, and PI. Actually, polymers can be employed in both configurations depending on their transparency. Commonly used flexible transparent substrates for solar cells are shown in Table 1.

Because of their flexibility, high transparency, light weight and low cost properties, PET and PEN are the mostly wide used flexible substrates. However, when processing temperature is above 150°C these plastic substrates will be deformed, which brings challenges to the fabrication of active layers in solar cells. Therefore, some preparation methods at low temperature have to be developed for flexible PVs to overcome this issue. Among the various flexible solar cells, organic solar cells show outstanding flexiblity and even stretchability on PET substrate owing to their inherent properties. Peumans *et al.*[94] fabricated flexible OPVs by employing ITO coated or Ag nanowires/PEDOT:PSS composite coated PET film as electrodes, achieving power conversion efficiencies (PCE) of 3.4% and 3.8%, respectively. Furthermore, stretchable OPVs have been successfully fabricated on ultrathin PET substrate. The thickness of the PET substrate was only 1.4 μm and that of the total device was less than 2 μm. The stretchable OPVs could obtain nearly identical efficiency to those constructed on ITO-coated glass substrates with an efficiency of 4.2%.[55]

When fabricating flexible plastic-based DSSCs, the ITO coated PET and PEN substrates are prevailing. Through a modified compression method, Durr *et al.*[95] employed a lift-off technique to transfer the pre-sintered porous TiO_2 nanoparticle film onto ITO coated PET. After combing with high mechanical pressures, the fabricated PET-based DSSCs showed an efficiency of 5.8%. Chu *et al.*[58] presented peel-off process that enables the transfer of vertically aligned high crystalline quality ZnO nanowire onto ITO coated PET substrates without post annealing and high mechanical pressure process. Meanwhile, DSSCs on ITO coated PEN also achieved relatively high efficiencies at room temperature by compression method.[64,99] Similarly, on PET substrate flexible perovskite-based solar cells have also been investigated by two different deposition methods, solution-processed and vapor-deposited methods, and efficiencies of 9.2% and 7% were obtained, respectively.[86,96]

Since the limitation of processing temperature, many flexible solar cells, especially the inorganic ones, are rarely fabricated on PET or PEN due to their poor stability at high processing temperatures. Thus, finding a polymer film that could withstand high processing temperatures was a challenge initially. In this regard, PI seems to be the best candidate. The PI films such as *Kapton* and *Upilex*[TM] can tolerate high temperatures up to 450°C. However, the PI films are usually dark yellow and strongly absorb visible radiation, and hence superstrate cells fabricated on such PI films will yield limited current due to optical absorption loss.[35,100] One of the strategies employed to minimize this problem is by reducing the PI thickness, thereby reducing the optical absorption by the PI. Instead of using the 50–125 μm commercial PI, typical PI films of 7.5–12 μm thickness have been reported to give higher cell performance. By using a DuPont clear PI foil with the thickness of 7.5 μm a very high efficiency of 13.8% was reported for a flexible CdTe superstrate solar cell.[97] Besides the yellow-colored PI films, recently, *Dupont* introduced the new *Kapton* colorless PI film having both high temperature tolerance needed and higher transparency, which helped in producing the new record efficiency of polymer-based solar cells (EMPA, 2011). Also, based on the colorless PI, flexible ITO-free OPVs with good mechanical integrity were obtained, achieving an efficiency of ~2.62%.[98]

Even though great improvements have been made in fabricating flexible solar cells on PI substrates, the maximum processing temperature (~450°C) and optical absorption loss still compromise their potential for good cell performances. Recently Corning Incorporated launched flexible glass, such as 100 μm-thick *Corning® Willow® Glass*. And Willow Glass has already been demonstrated as the backplane glass in active-matrix displays.[101,102] Figure 9 shows the transmittance of Willow flexible glass, common rigid glass substrates (1.1 mm Corning 7059 and 3.8-mm soda-lime) and a 7.5 μm-thick Kapton polyimide foil.[35] The PI foil has no transmission below 400 nm due to absorption. In contrast, the ultrathin glass is highly transparent, and more significantly amenable to higher processing temperature. With the flexible glass substrate held at

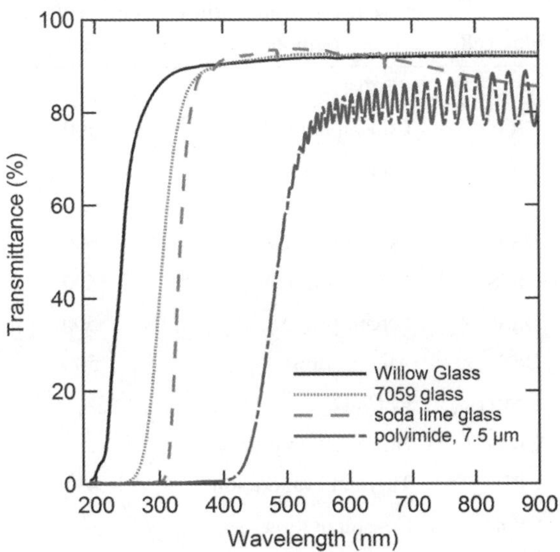

Figure 9.   Ultraviolet-visible transmittance of *Willow Glass* compared to *Corning 7059* glass, soda-lime glass, and 7.5 μm *Kapton* polyimide. Reprinted from Ref. 35.

600°C the CdTe layers were deposited by CSS, and a certified record conversion efficiency of 14.05% for a flexible CdTe solar cell was achieved.[35]

3.2. *Opaque Substrates for Flexible Solar Cells*

Opaque substrates for flexible solar cells mainly include metal foils and dark color polymer sheets, such as SS, enameled mild steel, Ti, Al, Mo, and PI. Compared with transparent substrates, opaque substrates can only be used in the substrate configuration due to their optical opacity or poor transparency. However, metal foils can endure comparatively high processing temperatures; thus, they are preferred by many inorganic flexible solar cells. Some commonly used opaque substrates for flexible solar cells are shown in Table 2.

SS is the most popular choice owing to its advantages of low cost, high mechanical strength and ease of preparation.[29,110] United Solar

Table 2.   Commonly used flexible non-transparent substrates for solar cells.

Flexible substrate	Solar cell	Active layer preparation method	Temperature (°C)	PCE (%)	Refs.
SS	a-Si:H	PECVD	—	13	10, 29
SS	CIGS	Co-evaporation	<500	17.7	103
Enameled mild steel	CIGS	Co-evaporation	600	17.6	104
SS	DSSC	Doctor blading	600	8.6	65
Ti	CIGS	Co-evaporation	500	17.9	105
Ti	DSSC	Screen-printing	500	8.46	67
Ti	DSSC	Anodization	500	5.39	66
Al	a-Si:H	PECVD	—	7.92	106
Al	CIGS	Non-vacuum printing	—	17.1	107
Mo	CdTe	Thermal evaporation	435	11.5	30
Mo	CIGS	Co-evaporation	—	14.6	108
PI (substrate)	a-Si:H	PECVD	—	9.7	109
PI (substrate)	CIGS	Co-evaporation	—	18.7	41

demonstrated a good example of triple-junction amorphous silicon alloy solar cells on SS, achieving a stable efficiency of 13% in 1997.[29] SS foil is also a promising substrate for flexible CIGS solar cells because of its stability at the high temperatures involved during CIGS processing. In 2012, the highest efficiency of 17.7% for flexible CIGS cells on SS was reported.[103] Furthermore, the substrate temperature during CIGS growth was well below 500°C, which effectively reduced impurity diffusion into the CIGS layer without using an additional diffusion barrier such as SiO_x, Al_2O_3 or Cr.[103] Whereas, compared to SS, mild steel as a flexible substrate can introduce impurity diffusion in the CIGS layer more easily. By depositing enamel layers onto mild steel foil as a diffusion barrier layer, as an intrinsic sodium source, and also as the insulating layer for monolithic, CIGS cells with a PCE of 17.6% (certified) were obtained.[104] Besides inorganic flexible solar cells, flexible DSSCs also achieved good performances on SS. By introducing a thin TiO_x layer through sol-gel spin coating to the ITO- and SiO_x-sputtered SS substrate, record metal-based DSSCs with an efficiency of 8.6% were achieved.[65]

Besides SS many other commonly used metal foil substrates have been studied, such as Ti, Al and Mo. Takeshi *et al.* obtained 17.9% efficiency flexible CIGS cells on Ti foil by deploying a three state co-evaporation of CIGS deposition and a ZnS(O,OH) buffer layer.[105] Ti foil is a promising substrate for flexible DSSCs given that the Ti substrate as support of the nanosized TiO_2 can not only reduce the cost, but also improve the cell performance with low internal resistance. Ma *et al.*[67] improved the PCE to 8.46% by introducing TiO_2 nanoforest underlayer between TiO_2 film and Ti-foil for flexible DSSCs. Al foils have also been studied as flexible substrates for CIGS and a-Si:H solar cells. When Al foils were used as flexible substrates for CIGS solar cells, the highest efficiency of 17.1% was obtained.[107] And meanwhile, Li and coworkers obtained flexible single junction a-Si:H solar cells on patterned Al foils,[106,111] achieving improved efficiency up to 7.92%. Figure 10(a) displays a large scale flexible thin-film solar cell with an active area of 12 cm^2. The efficiencies only experienced a marginal drop even at bending angles of up to 120°. Moreover, the efficiency remains 82% of the initial efficiency after 1,000 cycles of bending (Fig. 10(b)).

Among the metal substrates, Mo is the most attractive for flexible CdTe solar cells due to its matched thermal expansivity, given as

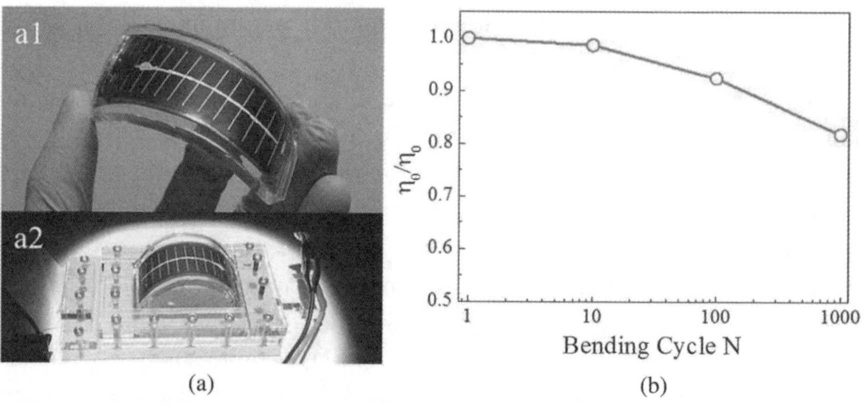

(a)   (b)

Figure 10. Optical images of (a1) flexible a-Si:H solar cell and (a2) bending angle depending PV performance measurement setup. (b) Relative efficiency of the solar cell after 10, 100, and 1,000 bending cycles. Reprinted from Ref. 106.

4.8×10^{-6} K^{-1} (CdTe) and 5.0×10^{-6} K^{-1} (Mo). Thus, even annealed at temperatures up to 550°C for 2 h, CdTe films neither peel off nor form blisters or bubbles.[33,112] By precisely controlling Cu doping of the CdTe layer, the PCE of flexible CdTe solar cells on Mo foils can reach as high as 11.5%.[30] Also, flexible CIGS solar cells on Mo foils were developed by the Institute for Advanced Industrial Science and Technology (AIST, Japan), and an efficiency of 14.6% was obtained.[108] In addition, Mo is the preferred electrical back contact because, during the growth of the CIGS absorber, a thin interface layer of $MoSe_2$ is formed, resulting in a quasi-ohmic contact.[113] Chirilă *et al.*[41] deposited 600 nm thick Mo as back contact on PI by DC sputtering, and developed flexible CIGS solar cells on 25 μm thick PI substrate, achieving an efficiency of 18.7%. On the other hand, the PI sheets endow flexible a-Si:H solar cells good potential for light weight and even wearable PV devices, because the PI sheets can endure the elevated temperature during PECVD and sputtering processes.[114] Although the mechanical strength and thermal stability of polymers are not comparable with those of metal foils, high efficiency devices have also been realized.[23,109] For example, an efficiency of 9.7% was harvested from the triple-junction a-Si:H solar cells on PI substrate by roll-to-roll technology.[109]

As flexible substrates, metals and special polymer substrates have been used. Metals have the advantage of being able to withstand high deposition temperature, but they generally have rather high density, roughness, and coefficient of thermal expansion (especially Cu and Al). Furthermore, most of them, especially steel, contain metallic impurities (e.g., Fe) that are detrimental for solar cell performance.[115] Polymer substrates have a much lower density and roughness than that of metals, but they cannot sustain high processing temperatures. Therefore, use of a low-temperature process is necessary, but this generally leads to significantly inferior absorber quality.[116] On the other hand, one of the major advantages of polymer substrates is their electrically insulating property. It allows for direct monolithic integration of solar cells, developing modules through successive patterning of layers in-between the different deposition steps.[117] Whereas, solar cells deposited on metal substrates have to be stringed and tabbed or connected in a shingling-type configuration. Monolithic integration on conductive substrates should be developed by depositing intermediate dielectric barrier layers on metallic substrates between the substrate and electrical back contact, and the layers also serve as diffusion barrier against impurities from the substrate.[32,117,118]

3.3. *R2R Technologies*

The decent efficiencies demonstrated on various flexible substrates promote researchers to realize low-cost, high-throughput and large-areal flexible PV devices by integrating roll-to-roll R2R technologies. R2R technologies provide a route to large scale production of flexible solar cells and can be suitable for many kinds of solar cells, such as a-Si:H solar cells, CIGS solar cells, and OPVs.

R2R processing is believed to be a prerequisite for the successful manufacture of OPVs at sufficiently low cost for the technology to be competitive with other thin-film solar cell technologies such as a-Si:H, CdTe, and CIGS.[119] In 2009, OPVs became commercially available from Konarka by R2R techniques. The most commonly employed transparent conductive substrate is ITO-coated PET foils. Figure 11 shows both the roll and device layout along with a zoom on the device stack. All the processing was completed in ambient air using solution processing.[119] This

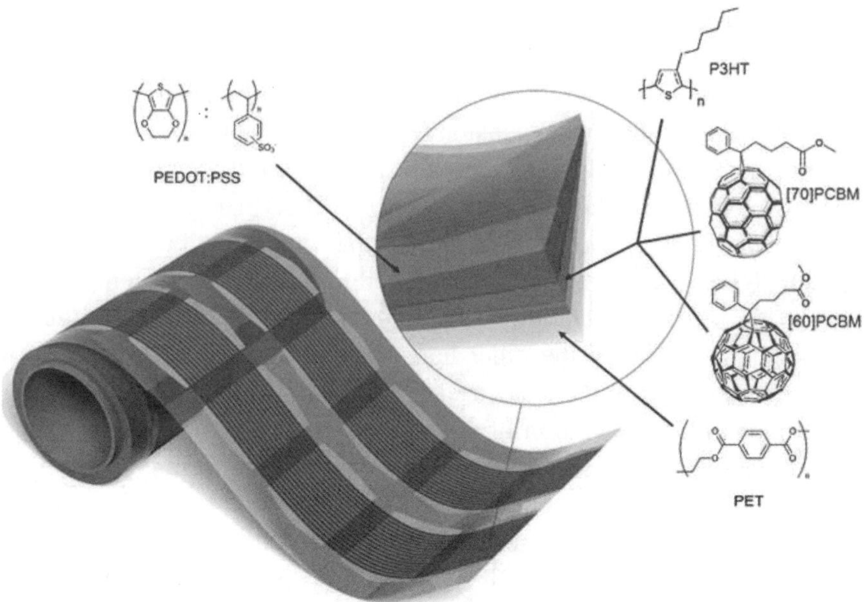

Figure 11.   A roll of the printed modules showing how two parallel sets of modules were prepared simultaneously. The zoom-in shows the layer stack which is PET-ITO-ZnO-P3HT:PCBM-PEDOT:PSS-printed silver. The device stack was encapsulated using roll-to-roll lamination post production. Reprinted from Ref. 119.

roll had a web width of 305 mm, and the simultaneous slot-die coating of two 16-stripe modules was carried out. The performance reached when using[70] PCBM was marginally better than[60] PCBM and the best performance reached for a single module was 2.75%.[119] An inline coating and printing machine from Grafisk Maskinfabrik A/S is shown in Fig. 12, which comprises winders, unwinders, double sided web cleaning, double pass ovens, tension zones and coating and printing apparatuses etc.[119] Such instruments enable the fabrication of OPVs through full R2R processing. The R2R processed OPVs in ambient atmosphere still show low efficiencies,[54,120] but the driving force for research within the field is the huge potential to enable high throughput production of cheap solar cells. Life cycle analysis and financial analysis have demonstrated that R2R processing of OPVs could yield very short energy pay-back times (EPBT)

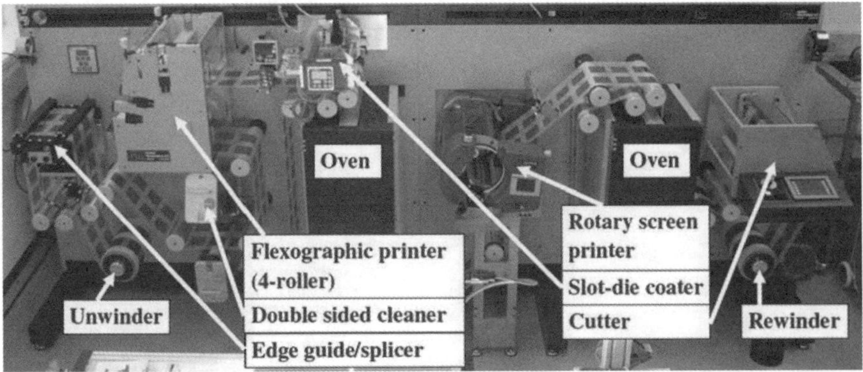

Figure 12. Inline printing machine with two ovens and three different coating and printing methods. The system has three independent web tension zones (lower). Reprinted from Ref. 119.

by using simple approaches.[121] It is reported that by avoiding scarce elements such as indium, avoiding the use of vacuum processing, and using only solar heat and solar electricity, an EPBT of the order of one day is possible.[122]

In 1980, Energy Conversion Devices, Inc. (ECD) began the development of a continuous R2R manufacturing technology for the production of a-Si:H solar cells. And its joint venture United Solar shows good demonstration in the fabrication of a-Si:H solar cells using R2R technologies (Fig. 13(a)). In a high-volume manufacturing plant operated by United Solar (Michigan, USA), solar cells are deposited on rolls of SS. Rolls of SS (2,500 m long, 36 cm wide, and 125 μm thick) move in a continuous manner in four machines to complete the fabrication of the solar cell. A wash machine washes the web one roll at a time; a back-reflector machine deposits the back reflector by sputtering Al and ZnO on the washed rolls; an amorphous silicon alloy processor deposits the layers of a-Si:H and a-SiGe:H alloy; and finally an anti-reflection coating machine deposits ITO on top of the rolls.[3,22] Beside SS, the R2R process is also doable on flexible polymer substrates. Fuji Electric developed a-Si:H/a-SiGe:H-based tandem solar cells on PI substrate and achieved a stabilized efficiency of 9% in a 40 cm × 80 cm cell by employing R2R

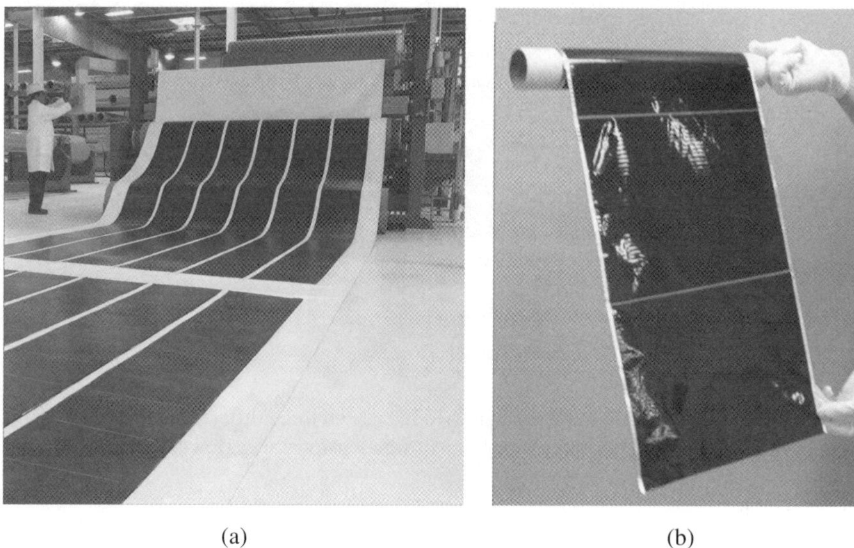

(a)   (b)

Figure 13.   (a) Roll of laminate roofing product integrating Unites Solar Ovonic triple-junction a-Si:H on stainless steel with a membrane at Solar Integration Technology's plant in CA, US. Reprinted from Ref. 3. (b) A string of the triple-junction cells on ~25 μm polymer substrates. Reprinted from Ref. 109.

technologies.[23] And as shown in Fig. 13(b), a string of triple-junction a-Si:H/a-SiGe:H/a-SiGe:H cells on polymer substrates were fabricated.[109] And then, United Solar demonstrated triple-junction ultra-light-weight solar cells on ~25 μm plastic substrate using R2R deposition. This material gives an initial aperture-area efficiency of 9.84% and specific power of ~1200 W/kg.[109,110]

Due to the development of CIGS "nano ink", the US company Nanosolar commercialized the CIGS thin-film solar cells by R2R printing the corresponding ink on aluminum foil in 2007.[3] Besides, CIGS layer can be deposited in the same R2R setup with only one winding and unwinding unit at ZSW (Centre for Solar Energy and Hydrogen Research, Germany). Mo sputtering as well as CIGS and In_2S_3 evaporation can be performed without breaking the vacuum. Singlestage and multistage CIGS growth process on PI were implemented in the R2R setup, and cell efficiencies over 10% can be achieved.[117]

4. Light Management on Flexible Solar Cells

Performances of PV devices largely rely on their optical absorption and carrier collection dynamics. Different light-trapping strategies have been implemented in order to improve the optical absorption of the solar cell, thus enhancing the PCE. One of the conventional methods was applying anti-reflective coating (ARC) on solar cells, and typically Si_3N_4, Al_2O_3 or MgF_2 thin-films were used. An MgF_2 ARC normally increases the absolute flexible CIGS cell efficiency by approximately 1.0–1.2%.[117] However, such an ARC works the best for individual wavelengths and normal incident direction. Therefore, conventional ARC might not be the best solution for PV devices as solar light is broadband in nature and its incident angle varies throughout the day.[123] Recently, Fan and coworkers[124] have utilized a facile molding process to fabricate flexible plastic AR films with 3D nanocone arrays on the front surface. This AR film can significantly reduce the reflectance of the superstrate above the CdTe light-absorbing layer, resulting in appreciable device performance improvement. Furthermore, it was found that the improved AR effect can be observed with oblique incident angle, which is highly beneficial for practical deployment of PV panels. On the other hand, while increasing the absorber material thickness of the PV device could lead to higher optical absorption; it will unavoidably increase the manufacturing cost. More importantly, a thicker absorber implied a longer minority charge carrier diffusion path length, which put a constraint of device performance. Thus, employing a nanostructure for light trapping is an appealing strategy to accommodate both optical absorption and carrier collection by utilizing a thinner absorber material to harvest more light and simultaneously facilitate carrier collection.[125–129]

In the case of flexible c-Si solar cells, the reduced thickness would compromise their performances greatly due to the weak absorption property. In this regard, Wang *et al.*[15] fabricated ultrathin c-Si by KOH etching the Si wafer to a uniform thickness from 10 μm to thinner than 2 μm. Then they patterned double-side nanotextures on the free-standing ultrathin c-Si films for efficient light absorption (Fig. 14). It was found that the top-side texturing enhanced the photocurrent largely due to the anti-reflection

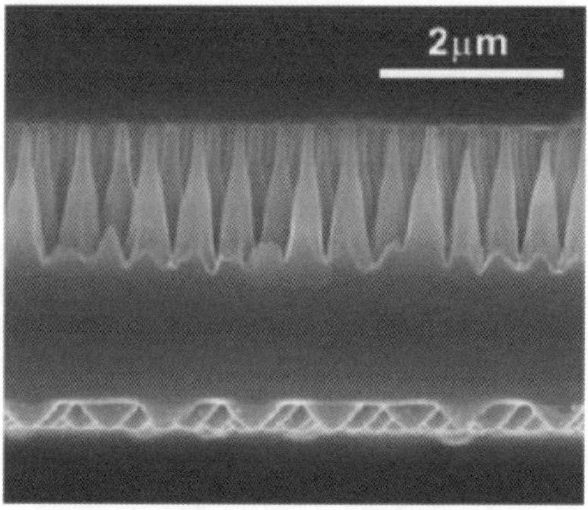

Figure 14.   Cross-sectional SEM image of a double-sided nanotextured c-Si film. Reprinted from Ref. 15.

effect caused by the gradual effective refractive index, while the back-side patterning increased the photocurrent further by the light scattering effect. Following the nanotexturing processing, a 6.8 μm thick front-side nano-textured solar cell with a PCE of 6.2% was fabricated without supporting substrates.

Meanwhile, Fan *et al.*[127,130] fabricated highly ordered nanopillar arrays, as shown in Figs. 15(a) and 15(b). These nanopillar arrays can be tuned through shape and geometry control to achieve the optimal absorption efficiency. The obtained single crystalline absorber material with a thickness of only 2 μm exhibits an impressive absorbance of ~99% over wavelengths, λ = 300–900 nm. As an example, a PV structure that incorporated three-dimensional (3D) singlecrystalline *n*-CdS nanopillars, embedded in polycrystalline thin films of *p*-CdTe was demonstrated, as depicted in Fig. 16(a). From Fig. 16(b), we can see that the conversion efficiency drastically and monotonically increases with the embedded nanopillar height (*H*) in CdTe. By increasing *H*, the space charge region area is effectively increased accompanied with much improved carrier collection efficiency. It was demonstrated these structures enabled highly versatile solar modules on both rigid and flexible substrates with enhanced carrier

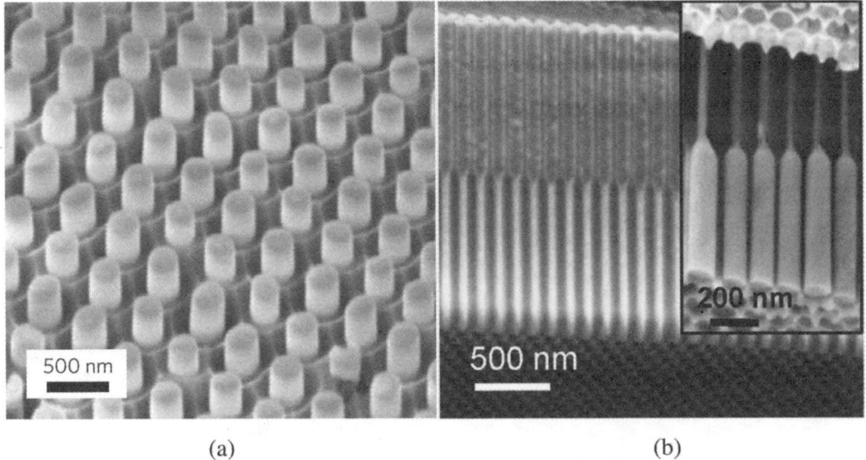

(a)   (b)

Figure 15. (a) A CdS nanopillar array after partial etching of the anodic aluminum membrane. Reprinted from Ref. 130. (b) Cross-sectional SEM images of a blank anodic aluminum membrane with dual-diameter pores and the Ge dual-diameter nanopillars (inset) after the growth. Reprinted from Ref. 127.

collection efficiency arising from the geometric configuration of the nanopillars.

Plasmonic light trapping and patterned back reflector are also two commonly used light management approaches to maximize absorption within a thin absorber. The localized plasmonic resonance can lead to a highly intensified local electric field. Thus, if there is a PV material in the proximity of the nanoparticles, significant energy absorption can be observed. Meanwhile, patterned back reflector allows the incident photon to experience multiple scattering before being reflected out of the device, which implies a higher chance of being absorbed by the active layer. And sometimes, the two approaches can be integrated in PVs to boost higher light absorption and efficiencies.

Plasmonic light trapping is a distinctive way of light management as compared with the aforementioned structures. Plasmon enhancement typically involves metal structures at the nanometer scale. Chen *et al.*[131] demonstrated a great advantage of plasmonic Au nanoparticles for efficient enhancement of flexible CIGS PV devices. In Fig. 17(a), the *J–V* characteristics show that the PCE can be enhanced from 8.31% to 10.36% with the incorporation of Au nanoparticles to harvest more incident light

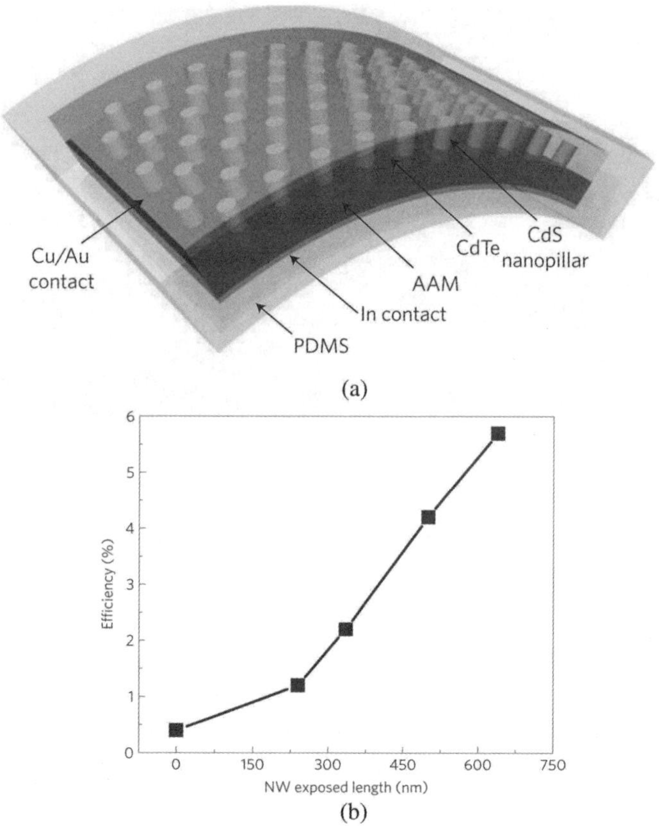

(a)

(b)

Figure 16.   (a) Schematic diagram of a bendable solar nanopillar cell module embedded in PDMS. (b) Efficiency of solar nanopillar cells as a function of the embedded nanopillar height, *H*. NW: nanowire. Reprinted from Ref. 130.

energy at the plasmonic resonance regions. And as shown in Fig. 17(b), a strong elimination of light reflectance can be found in the spectra of the plasmonic CIGS solar cell in the visible range of 500–700 nm. The incorporation of Au nanoparticles can improve the absorption at wavelengths in the high intensity region of the solar spectrum, especially at a large incident angle of solar irradiation, which will facilitate developing ink-printing flexible CIGS PVs. Similarly, Jankovic *et al.*[132] reported light management in an OPV by incorporating a plasmonic nanoparticle into

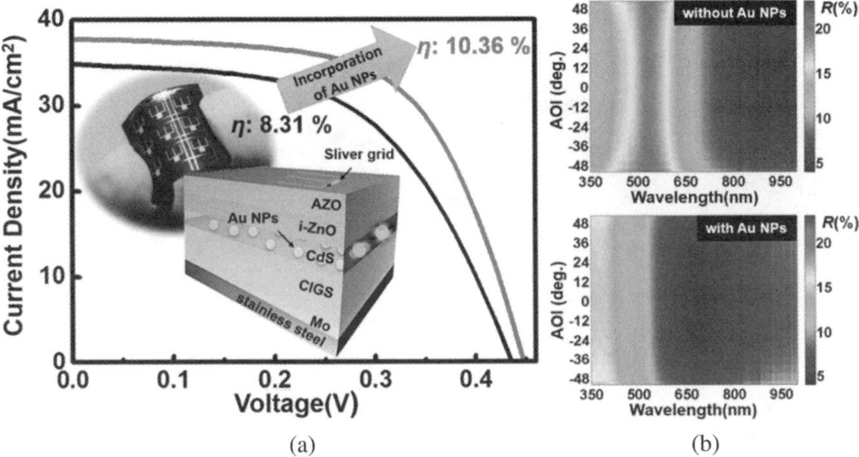

(a)   (b)

Figure 17. (a) Comparison of the *J-V* characteristics between conventional and plasmonic flexible CIGS PV solar cell. The insets are the photograph and schematic layer structure of flexible plasmonic CIGS solar cell respectively. (b) Angle-resolved reflectance spectra of the CIGS PVs with and without Au nanoparticles. Reprinted from Ref. 131.

the active layer in order to capitalize on the light scattering effect as well as the near-field enhanced local surface plasmon resonance. Figure 18(a) presents the schematic of the OPV device on flexible PEN with a P3HT/PC$_{60}$BM active layer incorporated with Au/SiO$_2$ nanorods. It can be seen from Fig. 18(b) that Au nanospheres/nanorods with different aspect ratios had different effect on the device performacnes. Further, this work revealed that the spectrally matched extinction spectra of Au/SiO$_2$ nanoparticles with device external quantum efficiency were of paramount importance to performance improvement.

On the other hand, besides depositing textured TCO layers, there have been few available strategies to directly fabricate textured structures for efficient light management on flexible substrates. Recently, Li and Fan *et al.*[111,133] demonstrated a-Si:H thin film solar cells with enhanced light absorption and performance on low-cost nanodent array aluminium foils, as shown in Figs. 19(a) and 19(b). These flexible aluminum foils with ordered nanodents were fabricated by large-scale aluminum anodization.

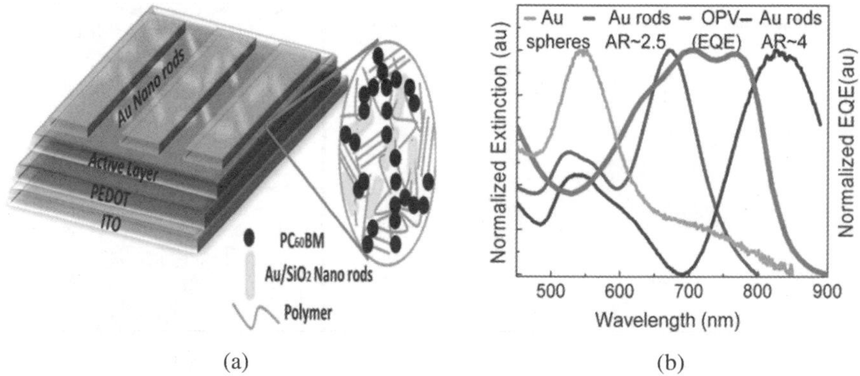

Figure 18. (a) Schematic of an OPV device with the active layer incorporated with plasmonic Au/SiO$_2$ nanorods. (b) Normalized EQE spectrum of the OPV and normalized extinction spectra of corresponding colloidal solutions with different aspect ratios (AR). Reprinted from Ref. 132.

Such light trapping structures lead to scattering and gradient of effective refractive index in the short wavelength region. And further, two more mechanisms of surface plasmon resonances and waveguide modes can be excited, which resulted in the enhancement in the long wavelength region. As a result, J_{sc} and PCE of solar cells on patterned Al were improved by 31% and 27% respectively compared with those on flat Al, and even better than those on commercial Asahi ANS14 glass (Fig. 19(c)). In addition, through combination with a R2R anodization process,[134] this novel substrate holds attractive potential to be integrated with the present R2R processing system of flexible solar cells. They also reported a-Si:H PV devices fabricated on periodic and random nanospike arrays with controlled geometrical factor including the spike pitch and height.[106,135] The typical periodic and random nanospike arrays and the deposited solar cells are shown in Figs. 20 and 21, respectively. These PV devices demonstrated improved performances owning to the improved photon capturing capability. In an optimal case, the efficiency of a flexible a-Si:H PV device fabricated on nanospikes can be improved by 43% as compared to its planar counterpart.[135]

As the 3D nanostructures have demonstrated enticing potency to boost performance of PV devices, Li and Fan *et al.*[136] achieved perfectly ordered

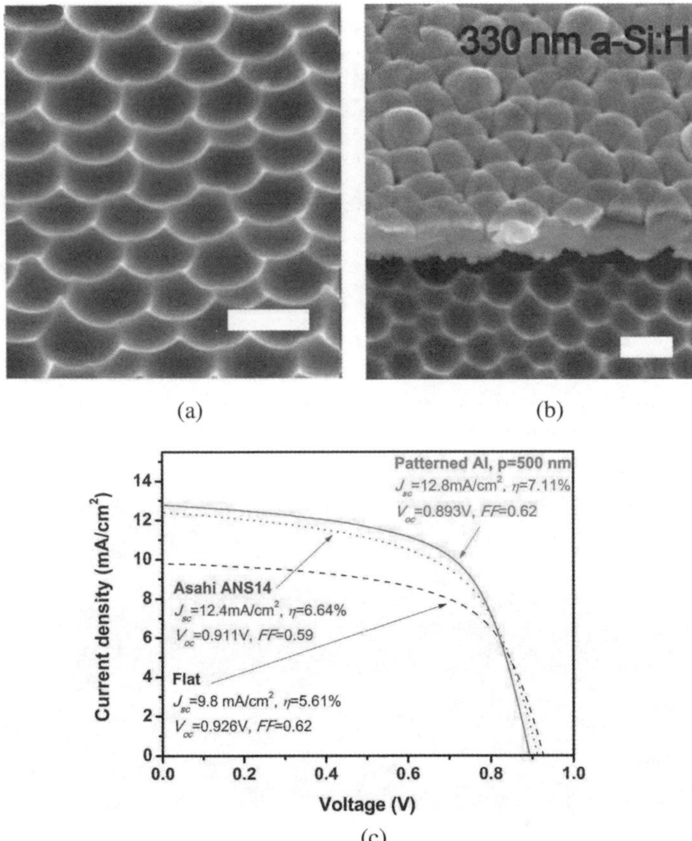

(a)   (b)

(c)

Figure 19.   (a) Tilted view (~45°) of the patterned Al substrate. (b) 330 nm silicon. The scale bars in all SEM images represent 500 nm. (c) *J–V* curves under AM 1.5 irradiation of a-Si:H solar cells built on a patterned Al substrate (p = 500 nm), Asahi ANS14 FTO glass and flat glass. Reprinted from Ref. 111.

inverted cone (i-cone) arrays with different aspect ratios and pitches, on a-Si:H solar cells were fabricated. The typical morphologies of the i-cone arrays before and after a-Si:H device fabrication are shown in Figs. 22(a) and 22(b), respectively. It was found that the 0.5-aspect-ratio i-cone-based device performed the best on both light absorption capability and energy conversion efficiency, which is 34% higher than that of the flat counterpart. Moreover, as shown in Fig. 22(c), the i-cone-based device enhanced

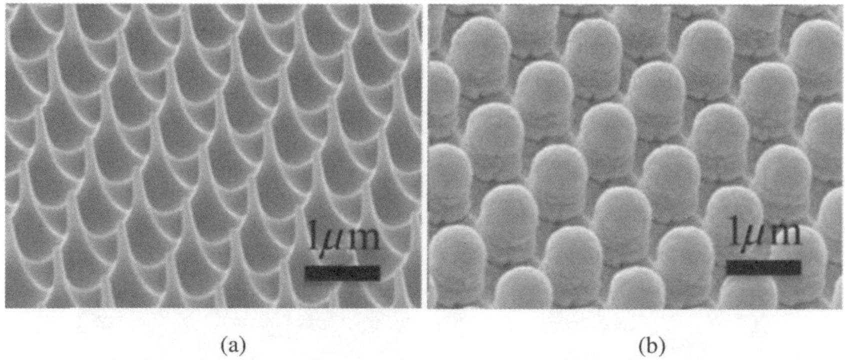

(a)   (b)

Figure 20.   SEM of (a) periodic nanospike arrays. (b) thin film a-Si:H solar cell based on them. Reprinted from Ref. 135.

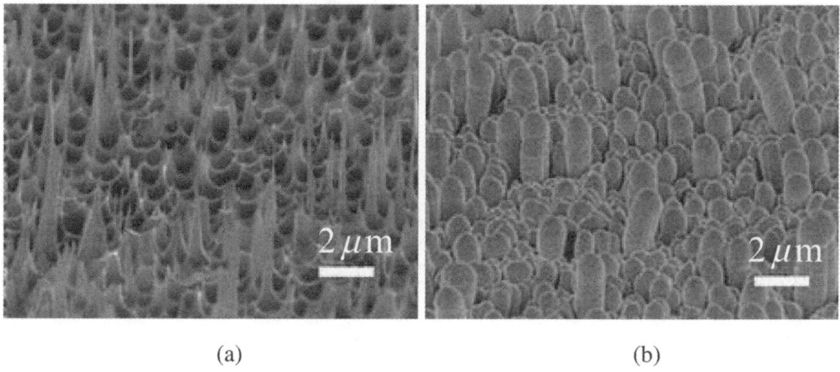

(a)   (b)

Figure 21.   SEM of (a) random nanospike arrays. (b) thin-film a-Si:H solar cell based on them. Reprinted from Ref. 106.

the light absorption and device performance over the flat reference device omnidirectionally.

5. Conclusions and Outlook

As an important PV technology, flexible thin-film solar cells offer promise as low cost, light weight, and flexible power sources that can be used for a wide range of next generation portable applications. This chapter has summarized recent progress on organic, inorganic and organic-inorganic

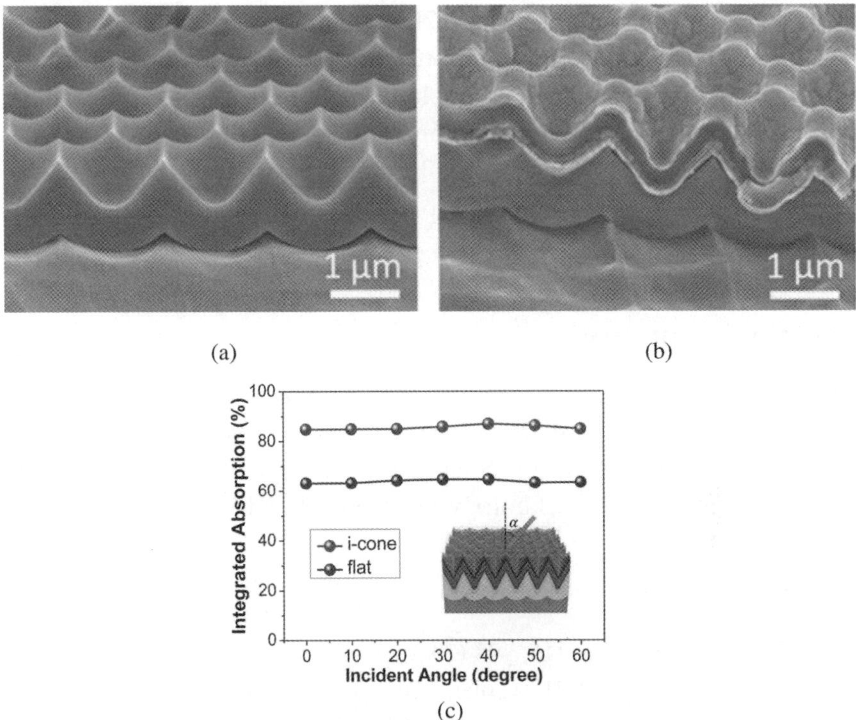

Figure 22.   SEM images of 1.5 µm pitch i-cone arrays with aspect ratios of 0.5 before (a) and after (b) a-Si:H device fabrication. (c) Integrated absorption of (b) and the flat reference device. Reprinted from Ref. 136.

hybrid flexible solar cells, which are deposited on various flexible substrates. These flexible solar cells have achieved decent power convention efficiencies. Proper light management on PVs can efficiently facilitate the efficiency improvement. Although bulk crystalline Si PVs have been dominating the PV market, thin-film PVs have multiple advantages over thick crystalline Si for flexible applications. With the fast evolving portable and personal electronic devices, such as smart phones, embedded sensors, and portable display devices, the lightweight and flexibility of the components/devices have become more important and attractive. Some flexible solar cells have been fabricated with R2R technologies successfully, such as a-Si:H and CIGS solar cells and organic solar cells, while some have demonstrated promising potential to be compatible with R2R technologies. In conjunction with R2R techniques, flexible PV technologies will provide lightweight and affordable solar modules that can be integrated into various surfaces (including buildings). And furthermore, with future materials optimization and structure design innovation, it is expected that high performance flexible solar cells will bring revolutionary advances in technology and play an important role in flexible electronics.

Acknowledgments

This work was supported by the National Natural Science Foundation of China (Grant Nos. 61474128, 51102271, 11174308), Science & Technology Commission of Shanghai Municipality (14JC1492900), Hong Kong Innovation Technology Fund (ITS/117/13).

References

1. N. S. Lewis, *Science*, **315**, 798 (2007).
2. M. B. Schubert and J. H. Werner, *Mater. Today*, **9**, 42 (2006).
3. M. Pagliaro, R. Ciriminna and G. Palmisano, *Chem. Sus. Chem.*, **1**, 880 (2008).
4. N. S. Lewis, *Chem. Sus. Chem.*, **2**, 383 (2009).
5. EPIA, (June 2014), Global Market Outlook for Photovoltaics 2014–2018. http://www.epia.org/home.
6. K. L. Chopra, P. D. Paulson and V. Dutta, *Prog. Photovoltaics Res. Appl.*, **12**, 69 (2004).
7. M. Gratzel, *Nature*, **414**, 338 (2001).
8. B. Kippelen and J. L. Bredas, *Energ. Environ. Sci.*, **2**, 251 (2009).
9. Z. Fan and A. Javey, *Nat. Mater.*, **7**, 835 (2008).
10. S. Hegedus, *Prog. Photovoltaics*, **14**, 393 (2006).
11. R. M. Swanson, *Prog. Photovoltaics Res. Appl.*, **14**, 443 (2006).
12. J. Zhao, A. Wang, M. A. Green and F. Ferrazza, *Appl. Phys. Lett.*, **73**, 1991 (1998).
13. A. Wang, J. Zhao, S. R. Wenham and M. A. Green, *Prog. Photovoltaics Res. Appl.*, **4**, 55 (1996).
14. M. A. Green, *IEEE T. Electron Dev.*, **31**, 671 (1984).

15. S. Wang, B. D. Weil, Y. Li, K. X. Wang, E. Garnett, S. Fan and Y. Cui, *Nano Lett.*, **13**, 4393 (2013).

16. T. Yonehara, K. Sakaguchi and N. Sato, *Appl. Phys. Lett.*, **64**, 2108 (1994).

17. C. Berge, M. Zhu, W. Brendle, M. B. Schubert and J. H. Werner, *Sol. Energy Mater. Sol. Cells*, **90**, 3102 (2006).

18. J. Yoon, A. J. Baca, S. I. Park, P. Elvikis, J. B. Geddes, L. F. Li, R. H. Kim, J. L. Xiao, S. D. Wang, T. H. Kim, M. J. Motala, B. Y. Ahn, E. B. Duoss, J. A. Lewis, R. G. Nuzzo, P. M. Ferreira, Y. G. Huang, A. Rockett and J. A. Rogers, *Nat. Mater.*, **7**, 907 (2008).

19. S. W. Bedell, D. Shahrjerdi, B. Hekmatshoar, K. Fogel, P. A. Lauro, J. A. Ott, N. Sosa and D. Sadana, *IEEE J. Photovolt.*, **2**, 141 (2012).

20. S. Saha, M. M. Hilali, E. U. Onyegam, D. Sarkar, D. Jawarani, R. A. Rao, L. Mathew, R. S. Smith, D. W. Xu, U. K. Das, B. Sopori and S. K. Banerjee, *Appl. Phys. Lett.*, **102**, (2013).

21. S. Pizzini, *Advanced Silicon Materials for Photovoltaic Applications* (John Wiley & Sons, 2012), p. 311.

22. M. Izu and T. Ellison, *Sol. Energy Mater. Sol. Cells*, **78**, 613 (2003).

23. Y. Ichikawa, T. Yoshida, T. Hama, H. Sakai and K. Harashima, *Sol. Energy Mater. Sol. Cells*, **66**, 107 (2001).

24. P. Lechner and H. Schade, *Prog. Photovoltaics Res. Appl.*, **10**, 85 (2002).

25. J. Meier, S. Dubail, R. Fluckiger, D. Fischer, H. Keppner and A. Shah, paper presented at the *Proc. 1st World Conf. Photovoltaic Energy Conversion*, Waikoloa, HI (1994).

26. S. Guha and J. Yang, *IEEE T. Electron Dev.*, **46**, 2080 (1999).

27. S. Guha, J. Yang, A. Pawlikiewicz, T. Glatfelter, R. Ross and S. R. Ovshinsky, *Appl. Phys. Lett.*, **54**, 2330 (1989).

28. T. Söderström, F. J. Haug, V. Terrazzoni-Daudrix and C. Ballif, *J. Appl. Phys.*, **107**, 014507 (2010).

29. J. Yang, A. Banerjee and S. Guha, *Appl. Phys. Lett.*, **70**, 2975 (1997).

30. L. Kranz, C. Gretener, J. Perrenoud, R. Schmitt, F. Pianezzi, F. La Mattina, P. Blosch, E. Cheah, A. Chirila, C. M. Fella, H. Hagendorfer, T. Jager, S. Nishiwaki, A. R. Uhl, S. Buecheler and A. N. Tiwari, *Nat. Commun.*, **4**, 2306 (2013).

31. A. Seth, G. B. Lush, J. C. McClure, V. P. Singh and D. Flood, *Sol. Energy Mater. Sol. Cells*, **59**, 35 (1999).

32. M. M. Aliyu, M. A. Islam, N. R. Hamzah, M. R. Karim, M. A. Matin, K. Sopian and N. Amin, *Int. J. Photoenergy*, (2012).

33. V. P. Singh, J. C. McClure, G. B. Lush, W. Wang, X. Wang, G. W. Thompson and E. Clark, *Sol. Energy Mater. Sol. Cells*, **59**, 145 (1999).

34. I. Matulionis, S. Han, J. A. Drayton, K. J. Price and A. D. Compaan, in *2001 MRS Spring Meeting* D. L. R. W. Birkmire, R. Noufi, H. W. Schock, Ed. (MRS Proceedings, 2001), Vol. 668(H8), p. 23.

35. W. L. Rance, J. M. Burst, D. M. Meysing, C. A. Wolden, M. O. Reese, T. A. Gessert, W. K. Metzger, S. Garner, P. Cimo and T. M. Barnes, *Appl. Phys. Lett.*, **104**, 143903 (2014).

36. J. Perrenoud, B. Schaffner, S. Buecheler and A. N. Tiwari, *Sol. Energy Mater. Sol. Cells*, **95**, S8 (2011).

37. A. N. Tiwari, A. Romeo, D. Baetzner and H. Zogg, *Prog. Photovoltaics Res. Appl.*, **9**, 211 (2001).

38. S. Chandramohan, R. Sathyamoorthy, P. Sudhagar, D. Kanjilal, D. Kabiraj, K. Asokan and V. Ganesan, *J. Mater. Sci. Mater. Electronics*, **18**, 1093 (2007).

39. J. L. Shay, S. Wagner and H. M. Kasper, *Appl. Phys. Lett.*, **27**, 89 (1975).

40. S. H. Wei, S. B. Zhang and A. Zunger, *Appl. Phys. Lett.*, **72**, 3199 (1998).

41. A. Chirilă, S. Buecheler, F. Pianezzi, P. Bloesch, C. Gretener, A. R. Uhl, C. Fella, L. Kranz, J. Perrenoud, S. Seyrling, R. Verma, S. Nishiwaki, Y. E. Romanyuk, G. Bilger and A. N. Tiwari, *Nat. Mater.*, **10**, 857 (2011).

42. Q. Lin, H. Huang, Y. Jing, H. Fu, P. Chang, D. Li, Y. Yao and Z. Fan, *J. Mater. Chem.*, *C* **2**, 1233 (2014).

43. A. Chirilă, P. Reinhard, F. Pianezzi, P. Bloesch, A. R. Uhl, C. Fella, L. Kranz, D. Keller, C. Gretener, H. Hagendorfer, D. Jaeger, R. Erni, S. Nishiwaki, S. Buecheler and A. N. Tiwari, *Nat. Mater.*, **12**, 1107 (2013).

44. M. Kaelin, D. Rudmann and A. N. Tiwari, *Sol. Energy*, **77**, 749 (2004).

45. C. J. Hibberd, E. Chassaing, W. Liu, D. B. Mitzi, D. Lincot and A. N. Tiwari, *Prog. Photovoltaics Res. Appl.*, **18**, 434 (2010).

46. M. G. Panthani, J. M. Kurley, R. W. Crisp, T. C. Dietz, T. Ezzyat, J. M. Luther and D. V. Talapin, *Nano Lett.*, **14**, 670 (2014).

47. Y. Liang, Z. Xu, J. Xia, S. T. Tsai, Y. Wu, G. Li, C. Ray and L. Yu, *Adv. Mater.*, **22**, E135 (2010).

48. G. Yu, J. Gao, J. C. Hummelen, F. Wudl and A. J. Heeger, *Science*, **270**, 1789 (1995).

49. A. Mishra and P. Bauerle, *Angew. Chem. Int. Ed.*, **51**, 2020 (2012).

50. F. Wudl, *Acc. Chem. Res.*, **25**, 157 (1992).

51. S. Gunes, H. Neugebauer and N. S. Sariciftci, *Chem. Rev.*, **107**, 1324 (2007).

52. Y. M. Sun, G. C. Welch, W. L. Leong, C. J. Takacs, G. C. Bazan and A. J. Heeger, *Nat. Mater.*, **11**, 44 (2012).

53. J. You, L. Dou, K. Yoshimura, T. Kato, K. Ohya, T. Moriarty, K. Emery, C. C. Chen, J. Gao, G. Li and Y. Yang, *Nat. Commun.*, **4**, 1446 (2013).

54. D. Angmo and F. C. Krebs, *J. Appl. Polym. Sci.*, **129**, 1 (2013).

55. M. Kaltenbrunner, M. S. White, E. D. Glowacki, T. Sekitani, T. Someya, N. S. Sariciftci and S. Bauer, *Nat. Commun.*, **3**, 770 (2012).

56. B. Oregan and M. Gratzel, *Nature*, **353**, 737 (1991).

57. H. Y. Chen, D. B. Kuang and C. Y. Su, *J. Mater. Chem.*, **22**, 15475 (2012).

58. S. Chu, D. D. Li, P. C. Chang and J. G. Lu, *Nanoscale Res. Lett.*, **6**, 38 (2011).

59. T. Miyasaka, Y. Kijitori and M. Ikegami, *Electrochemistry*, **75**, 2 (2007).

60. M. Ikegami, J. Suzuki, K. Teshima, M. Kawaraya and T. Miyasaka, *Sol. Energy Mater. Sol. Cells*, **93**, 836 (2009).

61. T. Miyasaka, *J. Phys. Chem. Lett.*, **2**, 262 (2011).

62. A. Yella, H. W. Lee, H. N. Tsao, C. Y. Yi and A. K. Chandiran, *Science*, **334**, 1203 (2011).

63. H. Lindstrom, A. Holmberg, E. Magnusson, L. Malmqvist and A. Hagfeldt, *J. Photochem. Photobiol. A*, **145**, 107 (2001).

64. T. Yamaguchi, N. Tobe, D. Matsumoto, T. Nagai and H. Arakawa, *Sol. Energy Mater. Sol. Cells*, **94**, 812 (2010).

65. J. H. Park, Y. Jun, H. G. Yun, S. Y. Lee and M. G. Kang, *J. Electrochem. Soc.*, **155**, F145 (2008).

66. J. G. Chen, C. Y. Chen, C. G. Wu, C. Y. Lin, Y. H. Lai, C. C. Wang, H. W. Chen, R. Vittal and K. C. Ho, *J. Mater. Chem.*, **20**, 7201 (2010).

67. J. An, W. Guo and T. L. Ma, *Small*, **8**, 3427 (2012).

68. D. D. Li, P. C. Chang, C. J. Chien and J. G. Lu, *Chem. Mater.*, **22**, 5707 (2010).

69. A. Vomiero, V. Galstyan, A. Braga, I. Concina, M. Brisotto, E. Bontempi and G. Sberveglieri, *Energ. Environ. Sci.*, **4**, 3408 (2011).

70. J.-G. Chen, C.-Y. Chen, C.-G. Wu, C.-Y. Lin, Y.-H. Lai, C.-C. Wang, H.-W. Chen, R. Vittal and K.-C. Ho, *J. Mater. Chem.*, **20**, 7201 (2010).

71. A. Du Pasquier, *Electrochimca. Acta.*, **52**, 7469 (2007).

72. R. H. Lee, T. F. Cheng, J. W. Chang and J. H. Ho, *Colloid Polym. Sci.*, **289**, 817 (2011).

73. S. A. Haque, E. Palomares, H. M. Upadhyaya, L. Otley, R. J. Potter, A. B. Holmes and J. R. Durrant, *Chem. Commun.*, 3008 (2003).

74. L.-C. Chen, J.-M. Ting, Y.-L. Lee and M.-H. Hon, *J. Mater. Chem.*, **22**, 5596 (2012).

75. S. Kazim, M. K. Nazeeruddin, M. Gratzel and S. Ahmad, *Angew. Chem. Int. Ed.*, **53**, 2812 (2014).

76. A. Kojima, K. Teshima, Y. Shirai and T. Miyasaka, *J. Am. Chem. Soc.*, **131**, 6050 (2009).

77. J. H. Im, C. R. Lee, J. W. Lee, S. W. Park and N. G. Park, *Nanoscale*, **3**, 4088 (2011).

78. M. M. Lee, J. Teuscher, T. Miyasaka, T. N. Murakami and H. J. Snaith, *Science*, **338**, 643 (2012).

79. I. Grinberg, D. V. West, M. Torres, G. Gou, D. M. Stein, L. Wu, G. Chen, E. M. Gallo, A. R. Akbashev, P. K. Davies, J. E. Spanier and A. M. Rappe, *Nature*, **503**, 509 (2013).

80. A. Mei, X. Li, L. Liu, Z. Ku, T. Liu, Y. Rong, M. Xu, M. Hu, J. Chen, Y. Yang, M. Gratzel and H. Han, *Science*, **345**, 295 (2014).

81. M. Z. Liu, M. B. Johnston and H. J. Snaith, *Nature*, **501**, 395 (2013).

82. M. Gratzel, *Nat. Mater.*, **13**, 838 (2014).

83. R. F. Service, *Science*, **344**, 458 (2014).

84. C. Roldan-Carmona, O. Malinkiewicz, A. Soriano, G. M. Espallargas, A. Garcia, P. Reinecke, T. Kroyer, M. I. Dar, M. K. Nazeeruddin and H. J. Bolink, *Energ. Environ. Sci.*, **7**, 994 (2014).

85. P. Docampo, J. M. Ball, M. Darwich, G. E. Eperon and H. J. Snaith, *Nat. Commun.*, **4**, 2761 (2013).

86. J. B. You, Z. R. Hong, Y. Yang, Q. Chen, M. Cai, T. B. Song, C. C. Chen, S. R. Lu, Y. S. Liu, H. P. Zhou and Y. Yang, *ACS Nano*, **8**, 1674 (2014).

87. M. A. Green, A. Ho-Baillie and H. J. Snaith, *Nat. Photonics*, **8**, 506 (2014).

88. P. A. van Hal, M. M. Wienk, J. M. Kroon, W. J. H. Verhees, L. H. Slooff, W. J. H. van Gennip, P. Jonkheijm and R. A. J. Janssen, *Adv. Mater.*, **15**, 118 (2003).

89. W. J. E. Beek, M. M. Wienk and R. A. J. Janssen, *Adv. Mater.*, **16**, 1009 (2004).

90. J. Weickert, R. B. Dunbar, H. C. Hesse, W. Wiedemann and L. Schmidt-Mende, *Adv. Mater.*, **23**, 1810 (2011).

91. D. C. Olson, J. Piris, R. T. Collins, S. E. Shaheen and D. S. Ginley, *Thin Solid Films*, **496**, 26 (2006).

92. H. E. Unalan, P. Hiralal, D. Kuo, B. Parekh, G. Amaratunga and M. Chhowalla, *J. Mater. Chem.*, **18**, 5909 (2008).

93. Y. Yang, H. L. Zhang, G. Zhu, S. Lee, Z. H. Lin and Z. L. Wang, *ACS Nano*, **7**, 785 (2013).

94. W. Gaynor, G. F. Burkhard, M. D. McGehee and P. Peumans, *Adv Mater.*, **23**, 2905 (2011).

95. M. Durr, A. Schmid, M. Obermaier, S. Rosselli, A. Yasuda and G. Nelles, *Nat. Mater.*, **4**, 607 (2005).

96. C. Roldán-Carmona, O. Malinkiewicz, A. Soriano, G. Mínguez Espallargas, A. Garcia, P. Reinecke, T. Kroyer, M. I. Dar, M. K. Nazeeruddin and H. J. Bolink, *Energ. Environ. Sci.*, **7**, 994 (2014).

97. J. C. Perrenoud, Ph.D. thesis, ETH (2012).

98. J. W. Lim, D. Y. Cho, K. T. Eun, S. H. Choa, S. I. Na, J. H. Kim and H. K. Kim, *Sol. Energy Mater. Sol. Cells*, **105**, 69 (2012).

99. T. Yamaguchi, N. Tobe, D. Matsumoto and H. Arakawa, *Chem. Commun.*, 4767 (2007).

100. X. Mathew, J. P. Enriquez, A. Romeo and A. N. Tiwari, *Sol. Energy*, **77**, 831 (2004).

101. S. M. Garner, H. Mingqian, L. Po-Yuan, S. Chao-Feng, L. Chueh-Wen, H. Yen-Min, R. Hsu, D. Jau-Min, H. Je-Ping, C. Yi-Jen, J. J. ChiehLin, L. Xinghua, M. Sorensen, L. Jianfeng, P. Cimo and C. Kuo, *J. Disp. Technol.*, **8**, 590 (2012).

102. S. Hoehla, S. Garner, M. Hohmann, O. Kuhls, L. Xinghua, A. Schindler and N. Fruehauf, *J. Disp. Technol.*, **8**, 309 (2012).

103. F. Pianezzi, A. Chirilă, P. Blösch, S. Seyrling, S. Buecheler, L. Kranz, C. Fella and A. N. Tiwari, *Prog. Photovoltaics Res. Appl.*, **20**, 253 (2012).

104. R. Wuerz, A. Eicke, F. Kessler, S. Paetel, S. Efimenko and C. Schlegel, *Sol. Energy Mater. Sol. Cells*, **100**, 132 (2012).

105. T. Yagioka and T. Nakada, *Applied Physics Express*, **2**, (2009).

106. S.-F. Leung, K.-H. Tsui, Q. Lin, H. Huang, L. Lu, J.-M. Shieh, C.-H. Shen, C.-H. Hsu, Q. Zhang, D. Li and Z. Fan, *Energ. Environ. Sci.*, (2014).

107. G. Brown, P. Stone, J. Woodruff, B. Cardozo and D. Jackrel, in *Photovoltaic Specialists Conference (PVSC), 2012 38th IEEE.* (2012), pp. 003230–003233.

108. S. Niki, M. Contreras, I. Repins, M. Powalla, K. Kushiya, S. Ishizuka and K. Matsubara, *Prog. Photovoltaics Res. Appl.*, **18**, 453 (2010).

109. X. Xu, K. Lord, G. Pietka, F. Liu, K. Beemink, B. Yan, C. Worrel, G. DeMaggio, A. Banerjee, J. Yang and S. Guha. (IEEE, San Diego, CA, USA, 2008), pp. 1–6.

110. X. Xu, T. Su, S. Ehlert, G. Pietka, D. Bobela, D. Beglau, J. Zhang, Y. Li, G. DeMaggio, C. Worrel, K. Lord, G. Yue, B. Yan, K. Beernink, F. Liu, A. Banerjee, J. Yang and S. Guha, in *Photovoltaic Specialists Conference (PVSC), 2010 35th IEEE.*, (2010), pp. 001141–001146.

111. H. Huang, L. Lu, J. Wang, J. Yang, S.-F. Leung, Y. Wang, D. Chen, X. Chen, G. Shen, D. Li and Z. Fan, *Energ. Environ. Sci.*, **6**, 2965 (2013).

112. V. P. Singh and J. C. McClure, *Sol. Energy Mater. Sol. Cells*, **76**, 369 (2003).

113. M. Bar, S. Nishiwaki, L. Weinhardt, S. Pookpanratana, W. N. Shafarman and C. Heske, *Appl. Phys. Lett.*, **93**, 042110 (2008).

114. A. R. M. Yusoff, M. N. Syahrul and K. Henkel, *B. Mater. Sci.*, **30**, 329 (2007).

115. P. Reinhard, A. Chirila, P. Blosch, F. Pianezzi, S. Nishiwaki, S. Buechelers and A. N. Tiwari, *IEEE J. Photovolt.*, **3**, 572 (2013).

116. R. Caballero, C. A. Kaufmann, T. Eisenbarth, T. Unold, R. Klenk and H.-W. Schock, *Prog. Photovoltaics Res. Appl.*, **19**, 547 (2011).

117. M. Powalla, W. Witte, P. Jackson, S. Paetel, E. Lotter, R. Wuerz, F. Kessler, C. Tschamber, W. Hempel, D. Hariskos, R. Menner, A. Bauer, S. Spiering, E. Ahlswede, T. M. Friedlmeier, D. Blazquez-Sanchez, I. Klugius and W. Wischmann, *IEEE J. Photovolt.*, **4**, 440 (2014).

118. K. Herz, E. Kessler, R. Wachter, M. Powalla, J. Schneider, A. Schulz and U. Schumacher, *Thin Solid Films*, **403**, 384 (2002).

119. F. C. Krebs, J. Fyenbo and M. Jorgensen, *J. Mater. Chem.*, **20**, 8994 (2010).

120. F. C. Krebs, T. Tromholt and M. Jørgensen, *Nanoscale*, **2**, 873 (2010).

121. R. Sondergaard, M. Hosel, D. Angmo, T. T. Larsen-Olsen and F. C. Krebs, *Mater. Today*, **15**, 36 (2012).

122. N. Espinosa, M. Hosel, D. Angmo and F. C. Krebs, *Energ. Environ. Sci.*, **5**, 5117 (2012).

123. S. F. Leung, Q. P. Zhang, F. Xiu, D. L. Yu, J. C. Ho, D. D. Li and Z. Y. Fan, *J. Phys. Chem. Lett.*, **5**, 1479 (2014).

124. K. H. Tsui, Q. F. Lin, H. T. Chou, Q. P. Zhang, H. Y. Fu, P. F. Qi and Z. Y. Fan, *Adv. Mater.*, **26**, 2805 (2014).

125. R. Yu, Q. F. Lin, S. F. Leung and Z. Y. Fan, *Nano Energy*, **1**, 57 (2012).

126. B. Hua, Q. F. Lin, Q. P. Zhang and Z. Y. Fan, *Nanoscale*, **5**, 6627 (2013).

127. Z. Y. Fan, R. Kapadia, P. W. Leu, X. B. Zhang, Y. L. Chueh, K. Takei, K. Yu, A. Jamshidi, A. A. Rathore, D. J. Ruebusch, M. Wu and A. Javey, *Nano Lett.*, **10**, 3823 (2010).

128. Q. F. Lin, B. Hua, S. F. Leung, X. C. Duan and Z. Y. Fan, *ACS Nano*, **7**, 2725 (2013).

129. S. F. Leung, M. Yu, Q. F. Lin, K. Kwon, K. L. Ching, L. L. Gu, K. Yu and Z. Y. Fan, *Nano Lett.*, **12**, 3682 (2012).

130. Z. Y. Fan, H. Razavi, J. W. Do, A. Moriwaki, O. Ergen, Y. L. Chueh, P. W. Leu, J. C. Ho, T. Takahashi, L. A. Reichertz, S. Neale, K. Yu, M. Wu, J. W. Ager and A. Javey, *Nat. Mater.*, **8**, 648 (2009).

131. S.-C. Chen, Y.-J. Chen, W. T. Chen, Y.-T. Yen, T. S. Kao, T.-Y. Chuang, Y.-K. Liao, K.-H. Wu, A. Yabushita, T.-P. Hsieh, M. D. B. Charlton, D. P. Tsai, H.-C. Kuo and Y.-L. Chueh, *ACS Nano*, **8**, 9341 (2014).

132. V. Jankovic, Y. Yang, J. B. You, L. T. Dou, Y. S. Liu, P. Cheung, J. P. Chang and Y. Yang, *ACS Nano*, **7**, 3815 (2013).

133. R. Guo, H. Huang, P. Chang, L. Lu, X. Chen, X. Yang, Z. Fan, B. Zhu and D. Li, *Nano Energy*, **8**, 141 (2014).

134. M. H. Lee, N. Lim, D. J. Ruebusch, A. Jamshidi, R. Kapadia, R. Lee, T. J. Seok, K. Takei, K. Y. Cho, Z. Fan, H. Jang, M. Wu, G. Cho and A. Javey, *Nano Lett.*, **11**, 3425 (2011).

135. S. F. Leung, L. L. Gu, Q. P. Zhang, K. H. Tsui, J. M. Shieh, C. H. Shen, T. H. Hsiao, C. H. Hsu, L. F. Lu, D. D. Li, Q. F. Lin and Z. Y. Fan, *Sci. Rep.*, **4**, (2014).

136. Q. F. Lin, S. F. Leung, L. F. Lu, X. Y. Chen, Z. Chen, H. N. Tang, W. J. Su, D. D. Li and Z. Y. Fan, *ACS Nano*, **8**, 6484 (2014).

CHAPTER 9

RECENT ADVANCES IN FIBER SUPERCAPACITORS

Lingxia Wu,*,¶ Jinping Liu*,‡,¶ and Yuanyuan Li[†,§]

*School of Chemistry, Chemical Engineering and Life Science
and State Key Laboratory of Advanced Technology
for Materials Synthesis and Processing,
Wuhan University of Technology
Wuhan 430070, P. R. China
[†]School of Optical and Electronic Information,
Huazhong University of Science and Technology
Wuhan 430074, P.R. China
[¶]Institute of Nanoscience and Nanotechnology,
Central China Normal University
Wuhan 430079, P.R. China
[‡]liujp@mail.ccnu.edu.cn
[§]liyynano@hust.edu.cn

Fiber supercapacitor is one kind of promising flexible supercapacitor, which can complement or even replace microbatteries in wearable/portable miniaturized electronics and micro-/nano-electromechanical systems. This chapter provides an overview on the recent developments in the evaluation methods/metrics, electrode materials, device structures and preliminary applications of fiber supercapacitors (SCs). Some perspectives on future electrode/device design trends and directions are also proposed.

1. Introduction

The increasing demand for energy has triggered tremendous research efforts for energy harvesting, conversion and storage from clean and renewable energy sources.[1–3] The availability of high-performance energy storage devices plays a critical role in the effective utilization of the harvested and converted new energy. Batteries and supercapacitors (SCs) are such kinds of energy storage devices, which have received numerous interests from both industry and academia.[4–7] In batteries, energy is stored in a chemical form in the active materials on the basis of bulk redox reactions.[8] The charge/discharge rate and the power performance of batteries are determined by the kinetics of the electrochemical reactions as well as mass transfer. Therefore, batteries in general provide high energy density, but have low power density and very limited cycling stability. SCs, as a new type of charge storage device, can provide higher power density and longer cycle life than batteries, but are with apparently lower energy density.[9–11] This is because of the fact that in SCs energy is stored by means of ion adsorption at the electrode/electrolyte interfaces or via reversible redox reaction on the electrode surface/near surface. In addition to providing high power and long lifetime, SCs show superior performance to batteries in case of unstable energy supply situations, since their charge/discharge curves are linear and they can work effectively at any potential within their potential window.[11–12] In contrast, batteries supply energy mainly by the utilization of the discharge plateau at a specific potential.

Based on the energy storage mechanism, SCs can be classified into two kinds: one is electric double layer capacitor (EDLC) where charges are physically stored; the other is pseudocapacitor where charges are stored chemically through surface/near-surface Faradaic reactions. For EDLCs, the active electrode materials are typically various carbon-based materials including activated carbon,[13] onion-like carbon,[14] carbide-derived carbon,[15] singlewalled/multiwalled carbon nanotubes,[16] and graphene/reduced graphene oxide (rGO),[17] etc. due to their high specific surface area and good conductivity. In pseudocapacitor, transition-metal oxides/hydroxides (such as MnO_2, RuO_2, NiO, CoO_x, FeO_x, etc.) and conducting polymers — polyaniline (PANI), polypyrrole (PPy), poly(3,4-ethylenedioxythiophene)

(PEDOT), are generally employed as the electrode materials.[18–20] These materials can provide fruitful surface redox reactions and thus provide high charge storage capacitance.

With the rapid development of a wide spectrum of flexible electronics such as roll-up displays, hand-held portable devices, and even wearable electronic devices, it becomes critically important to develop flexible, lightweight, and miniaturized SCs as the power sources.[21–23] To this end, SCs are expected to be designed into a fully solid state to avoid the possible electrolyte leaking that not only results in environmental pollution but also deteriorates the device stability. As compared to the SCs using liquid electrolyte, solid-state flexible SCs have better flexibility, higher safety and superior stability, which are more suitable for practical applications.[23–29] Among various flexible SCs, fiber SCs are a new class of flexible device architecture, which have the following distinctive advantages.[30–32] Firstly, the one-dimensional fiber structure renders the device fabrication with more flexibility, opening up great opportunities for design innovation.[30] Fiber SCs have already been constructed into several types of geometries such as parallel, biscrolled and coaxial structures. Secondly, fiber SCs have outstanding mechanical flexibility, so that they could even be stretchable, this fulfills the essential requirements of power sources for wearable electronics.[31] Thirdly, fiber SCs are typically with micro-/submicrosized diameter, they are the basic units of wearable energy storage devices and can be woven or knitted into different areas and thicknesses without any restrictions. In particular, they can be applied at different environments and in various places (e.g., in miniaturized devices and on human body).

In recent years, there are significant advances in the fundamental research of fiber SCs.[24,25,30–58] In order to provide useful guidance on how to achieve high comprehensive performance of fiber SCs, it is imperative and urgent to update the progress in this field. This chapter review firstly discusses the evaluation methods and metrics on the electrochemical performance of fiber SCs. Then, the recent development of electrode materials, device structures as well as the potential applications of fiber SCs will be summarized in sequence. Lastly, we will give a brief outlook on electrode material and device design trends for fiber SCs.

2. Evaluation Methods/Metrics of a Fiber SC Performance

2.1. *Specific Capacitance, Energy/Power Density, and Equivalent Series Resistance (ESR)*

Similar to the methods used for traditional SC characterization, cyclic voltammetry (CV) and galvanostatic charge/discharge (GCD)[2,8,11,18] are most commonly utilized to evaluate the electrochemical performance of a fiber SC. The CV curve of the fiber EDLC generally demonstrates a rectangular shape since there is no Faradaic reaction between the electroactive material and the electrolyte. By contrast, for fiber pseudocapacitor, there are usually redox peaks with a deviation from the rectangular shape in CV. In both cases, however, the average specific capacitance can be calculated by using the voltammetric charge integrated from a CV curve, according to Eq. (1)[8,18]:

$$C = \frac{Q}{2AV} = \frac{1}{2AVv} \int_{V-}^{V+} I(V)\,dV, \tag{1}$$

where Q is the total charge obtained by the integration of the positive and negative sweeps in a CV curve, v is the sweep rate, and $V = V_+ - V_-$, representing the potential window of the fiber SC. Considering the special geometry and small size of fiber SC, length capacitance (C_L), areal capacitance (C_A) and volumetric capacitance (C_V) are usually taken into account by scientists and engineers. Thus, A in the above formula can represent fiber SC's length, surface area or volume.

In GCD technique, a constant-current charge/discharge mode is used. In such a case, the cell potential is in general linear with respect to the time, so that the specific capacitance of a fiber SC device can be calculated as follows[59–61]:

$$C = \frac{It}{AV}, \tag{2}$$

where I, t, and V correspond to the discharge current, discharge time, and potential window of the cell (excluding the IR drop), respectively. A in most cases represents the total volume of fiber SC.

One of the earliest reports on fiber SCs was from Wang's research group at Georgia Institute of Technology.[45] In the work, a fiber SC was fabricated by twisting two fiber electrodes around each other. The electrode fiber was the Kevlar fiber coated with gold and then grown with ZnO nanowires and MnO_2 layer. C_L of 0.2 mF cm^{-1} and C_A of 2.4 mF cm^{-2} per fiber SC device were reported. The highest areal capacitance C_A of 116.3 mF cm^{-2} was demonstrated very recently for a fiber SC consisting of hierarchically structured carbon nanotube-graphene (CNT at graphene) fiber electrodes and tested in a PVA-H_3PO_4 gel electrolyte.[36]

Power density and energy density are two important parameters for evaluating the electrochemical performance of SCs. Specifically for fiber SC, volumetric energy (E_V) and power densities (P_V) are highly required as these two can unambiguously reflect the energy storage capability of the micro-SC in limited space. The volumetric energy (E_V) and power densities (P_V) of a fiber SC can be determined from Eqs. (3) and (4)[62,63]:

$$E = \frac{1}{2}CV^2, \tag{3}$$

$$P = \frac{E}{t_{\text{discharge}}}, \tag{4}$$

in which C is the capacitance from the Eq. (1) or (2), V is the potential window of the fiber SC, and $t_{\text{discharge}}$ is the discharge time. It should be noted that in some cases areal energy (E_A) and power densities (P_A) were also reported for fiber SCs in the literature.

Ragone plot of volumetric energy density versus average volumetric power density is commonly used to demonstrate the electrochemical performance of a fiber SC. The plot can not only reveal the energy and power density values, but also display the variation of energy density with the increase of power density. Figure 1(a) shows the comparison of Ragone plots for several typical fiber SCs reported in literature (Ni(OH)$_2$//C, pen ink//pen ink, MnO_2//MnO_2). The Ragone plots in Fig. 1(b) compare the volumetric performance of a CNT at graphene-based fiber SC[36] with commercially available state-of-the-art energy storage systems. The fiber SC has a high volumetric energy density of ~6.3 mWh cm^{-3}, which is about ten times higher than the energy densities of commercially available SCs

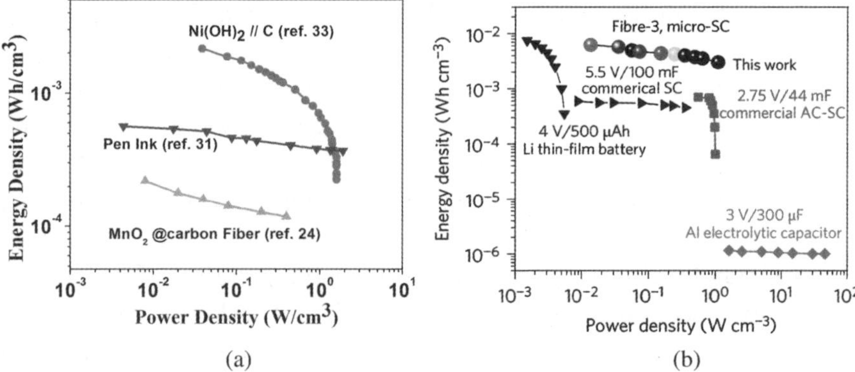

Figure 1. (a) Volumetric energy and power density comparison of three recently reported fiber SCs. (b) Energy and power densities of the CNT at graphene fiber SC[36] compared with commercially available state-of-the-art energy storage systems.

(2.75 V/44 mF and 5.5 V/100 mF, the energy density of which is <1 mWh cm^{-3}) and even comparable to the 4 V/500 μAh thin-film lithium battery (0.3–10 mWh cm^{-3}). The maximum volumetric power density for the fiber SC is ~1,085 mW cm^{-3}, comparable to the commercially available SCs and more than two orders of magnitude higher than that of lithium thin-film batteries. The volumetric energy density is the highest among all solid-state fiber SCs reported to date.

Table 1 summarizes the main parameters (C_L, C_A, C_V, E_A, E_V, P_A, P_V, etc.) of electrochemical performance of various fiber micro-SCs investigated so far in the literature. From Table 1, typical performance metrics of different types of fiber SCs can be obtained and compared directly. However, in order to ensure a correct and fair comparison for different fiber SCs, it is highly recommended that the following points should be considered seriously: (1) Specify the scan rate for CV or the current density for GCD when reporting the specific capacitance. Point out clearly the capacitance reported for single electrode or for the whole device. This will make the comparison more convenient; (2) provide the mass of the active materials if reporting gravimetric capacitance, provide the dimensions of fiber electrodes (diameter, length), the thickness of solid-state electrolyte, and the area/volume of the fiber SC device. Also specify how the area and volume of the cell are calculated.

ESR is well known as the internal resistance, which is frequency and temperature dependent. It reflects the electrical conductivity of the whole

Table 1. Summary of the performance of typical fiber SCs reported in the literature.

Device & Materials	C_L [mF cm^{-1}]	C_A [mF cm^{-2}]	C_V [F cm^{-3}]	E_A [mWh cm^{-2}]	E_V [mWh cm^{-3}]	P_A [mW cm^{-2}]	P_V [mW cm^{-3}]	Cycle	Voltage [V]	Refs.
Ni(OH)$_2$//C	6.67	35.67	2.4	1×10^{-2}	2.16	7.3	1,600	10^4	1.5	33
GCF-N//GCF-MnO$_2$	/	/	11.1	/	5	/	929	10^4	1.8	34
Co$_3$O$_4$//Graphene	—	—	2.1	—	0.62	—	1,470	10^3	1.5	35
CNT at Graphene (SYM)	—	116.3	45	—	6.3	—	1,085	10^4	1	36
RGO at CNT(SYM)	5.3	177	158	3.84×10^{-3}	3.5	—	40,000	2×10^3	0.8	37
PEDOT at CNT(SYM)	0.47	73	179	—	1.4	—	—	10^4	0.8	38
CNT at MnO$_2$(SYM)	/	/	25.4	/	3.52	/	127	/	0.9	39
PANI(SYM)	/	19	/	9.5×10^{-4}	/	4.2	/	10^4	0.6	40
CNT(SYM)	0.029	8.66	32.09	/	/	/	/	10^4	1	41
MWCNT at OMC (SYM)	1.91	39.7	/	1.77×10^{-3}	/	0.043	/	5×10^2	1	42
CNT fiber(SYM)	/	4.99	13.5	2.26×10^{-4}	0.6	0.493	1,320	10^4	0.8	43
Pen Ink(SYM)	/	11.9–19.5	/	2.7×10^{-3}	/	0.042–9.1	/	1.5×10^4	1	31
MWCNT at MnO$_2$ (SYM)	0.019	3.53	/	/	1.73	/	790	10^3	2	44
ZnO at MnO$_2$(SYM)	0.2	2.4	/	2.7×10^{-5}	/	0.014	/	/	0.5	45
GF at 3D-G(SYM)	0.018	1.2–1.7	/	1.7×10^{-4}	/	0.1	/	/	0.8	46
MnO$_2$ at carbon Fiber (SYM)	/	/	2.5	—	0.22	—	400	10^4	0.8	24
PPy at MnO$_2$ at CNT(SYM)	/	1,490	/	3.3×10^{-2}	/	13	/	3×10^3	0.8	32

"SYM" represents a symmetric device configuration. "/" means "not available". C_L and C_A and C_V mean the length, areal and volumetric capacitance, respectively. E_A and E_V are the areal and volumetric energy density. P_A and P_V are the areal and volumetric power density.

fiber SC including the active materials, current collectors and electrolyte, etc. Electrochemical impedance spectroscopy (EIS)[8] is usually used to determine the value of the ESR. The measurement of ESR is necessary for determining the precise *IR* drop in the GCD curves, which is used for *V* and capacitance calculation according to Eq. (2).

2.2. *Cycling Stability and Mechanical Flexibility*

One of the most important requirements for fiber SCs is the long term cycling stability. Once the SC is integrated into wearable or miniaturized electronics, it is generally impractical and impossible to replace or repair it during its lifetime. Ideally, fiber SCs are rated for hundreds of thousands of cycles and even more, and often outlast the lifetime of the device they power. Although some applications require higher energy densities, cycling stability will be one of the most critical challenges to the development of fiber SCs. In practice, long term cycling with good capacitance retention should be demonstrated. Generally, the cycling stability can be evaluated by using repeated CV or GCD techniques as described in Section 2.1. Figure 2(a) displays one example which used GCD method to check the cycleability of fiber SC.[36] One can also compare the CV curves at different cycles to show the cycle performance, as in Fig. 2(b).[31] In both cases, the plot of capacitance versus cycle number typically needs to be presented, which can illustrate clearly the capacitance fading rate. It is noted that the cycling testing must be conducted at reasonable rates, as cycling too fast will not realistically show the ability of the device to perform over the long term. In Fig. 2(c), another stability testing method is presented, which mainly evaluates the stability of the fiber SC upon continuous rate variation.[41] For fiber SC, cycling stability of the device under bending conditions is also highly required. Figures 2(d) and 2(e) show two examples of evaluating the cycle performance of fiber SCs under bending conditions.[36,42] The cycle times of so far reported fiber SCs are summarized in Table 1 for comparison. From the data, most of the reported fiber SCs has good cycle performance.

Besides cycling stability, mechanical flexibility is also an important parameter for evaluating the electrochemical performance of fiber SCs. Flexibility testing is used to determine the full range a fiber SC can be bent, twisted or stretched before destruction occurs. It thus provides

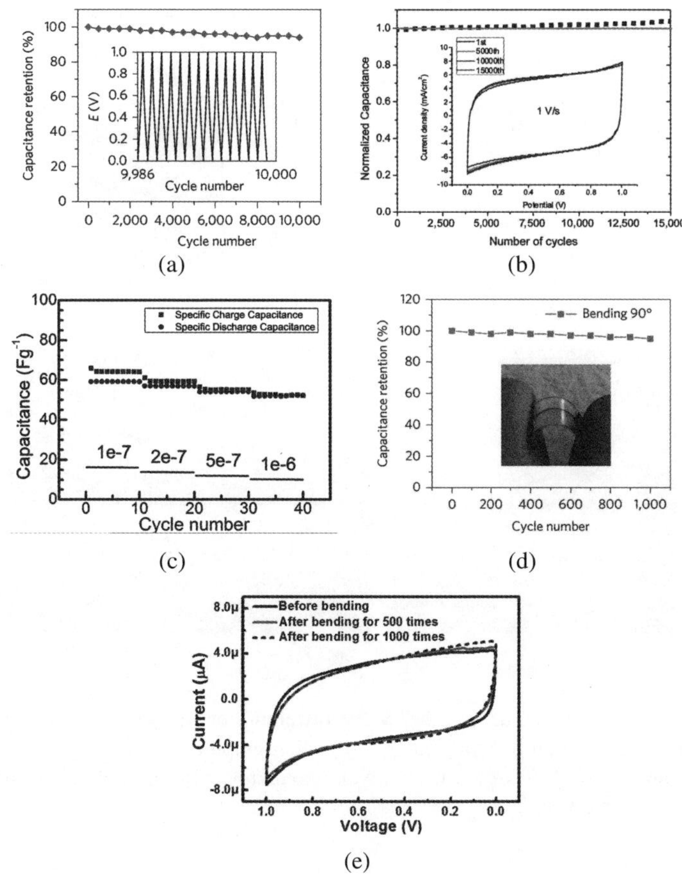

Figure 2.   (a) Cycle life testing by GCD technique.[36] (b) Cycle life testing by CV technique.[31] (c) Cycle life testing at continuously varied rates.[41] (d) Cycle life testing of the fiber SC under bending state.[36] (e) Cycle life testing by comparing CV curves before and after bending of fiber SC.[42]

insights into a capacitor's strength and elasticity. Recently, many papers reported the fiber SC device's flexibility by different ways. Shi's group at Tsinghua University reported a fiber SC that can be bent into different shapes without capacitance decay (Fig. 3(a)).[51] They demonstrated the mechanical flexibility of their device by testing the CVs at different bending states. In 2014, Xia's group from Fudan University reported the flexible characteristic of a wire-shaped micro-SC (Fig. 3(b)) and confirmed the good flexibility of device via GCD technique.[33] Mechanical flexibility

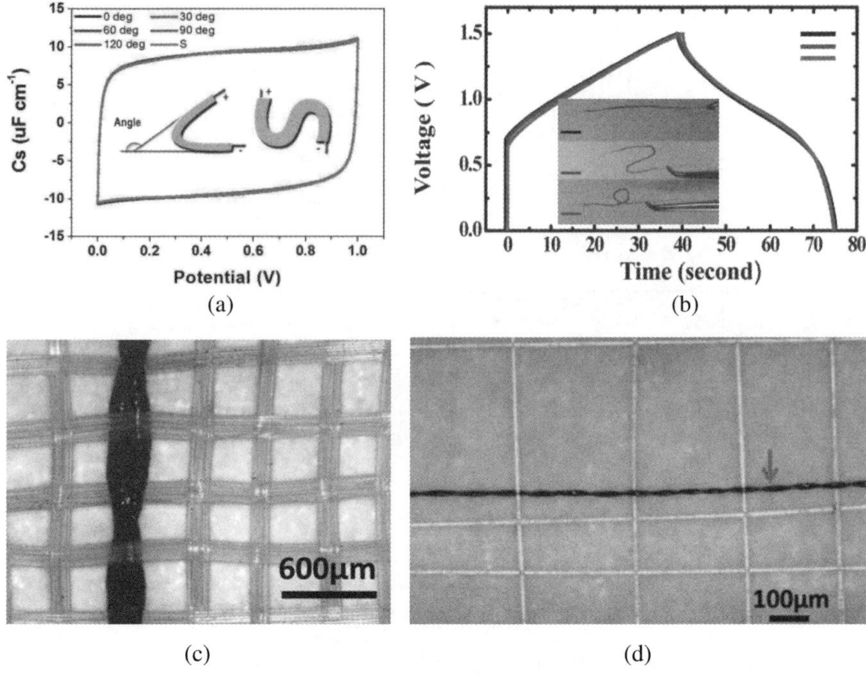

Figure 3. (a) CV curves of the fiber SC at different bending angles.[51] (b) Charge/discharge curves at different bending states.[33] (c) Micrograph of a fiber SC wire woven into a polyurethane textile.[42] (d) A SC wire woven into textile composed of aramid fibers. The red arrows show the wire.[52]

directly determines if the SC device can be woven or knitted, which is especially important for integrating fiber SCs into wearable electronic products.[42,52] Peng's group from Fudan University reported a fiber SC with extremely high mechanical property.[42] In their work, the SC wires could even be woven into textile structures (Fig. 3(c)). In addition, no obvious decrease in the electrochemical performance had been detected for the SC when the textile was bent. Another work from the same group developed a new family of polyaniline composite fibers incorporated with aligned multi-walled CNTs, which was used to construct a fiber SC with very high flexibility.[52] Due to the high flexibility, the fiber SCs can be easily woven into clothes, as shown in Fig. 3(d).

3. Materials for the Fiber Electrode

Various wired/yarn/fiber electrode materials have been explored for fiber SCs, those include carbon-based materials, metal oxides/hydroxides and conducting polymers. After analyzing the data available in literatures (also listed in Table 1), we can see clearly that carbon-based materials were the most widely used.[24,31–33,39,41–44,46,54,55] Several papers discussed about using metal oxides/hydroxides,[33,45] and in most cases metal oxides and polymers are likely be integrated in carbonaceous matrix for fiber electrodes. The purpose of combining oxides/polymers with carbon materials is to increase the capacitance of fiber electrode while still maintaining good electrical conductivity and mechanical flexibility.

To the best of our knowledge, Wang's group reported one of the first fiber electrodes. They used the Kevlar fiber modified with gold and then grown with ZnO nanowires and MnO_2 layer.[45] Kevlar fiber is known for good flexibility but is insulating, Au was coated on the surface in order to improve the conductivity but it is expensive. ZnO itself is not capacitive and served mainly as the backbone to anchor the pseudocapacitive MnO_2 (Fig. 4(a)). MnO_2 is a promising active material for fiber SCs. Later, Zhou *et al.* also reported a core-shell fiber electrode by growing MnO_2 directly on carbon fiber.[24] The carbon fiber served as a flexible scaffold and current collector (Fig. 4(b)). The as-fabricated SC device showed good rate capability with a scan rate as high as $20\ Vs^{-1}$, a volumetric capacitance of $2.5\ F\ cm^{-3}$, and a volumetric energy density of $2.2 \times 10^{-4}\ Wh\ cm^{-3}$. Moreover, the fiber SC integrated with a triboelectric generator (TG) could power a commercial liquid crystal display or a light-emitting-diode, demonstrating potential applications for self-powered nanotechnology. Peng *et al.* reported an aligned multi-walled carbon nanotube (MWCNT)/ MnO_2 composite fiber electrode by depositing MnO_2 on the MWCNT. Interestingly, the amount of MnO_2 can be controlled by the electrodeposition cycles.[44] A length capacitance of $0.019\ mF\ cm^{-1}$ and an areal capacitance of $3.53\ mF\ cm^{-2}$ were demonstrated for the fiber SC. In 2013, a CNT-modified cotton yarn was synthesized via a simple "dip-coating" method and was further used for loading MnO_2 and PPy (Fig. 4(c)).[32] The resulting composite yarn electrode was assembled into a plastic tube with

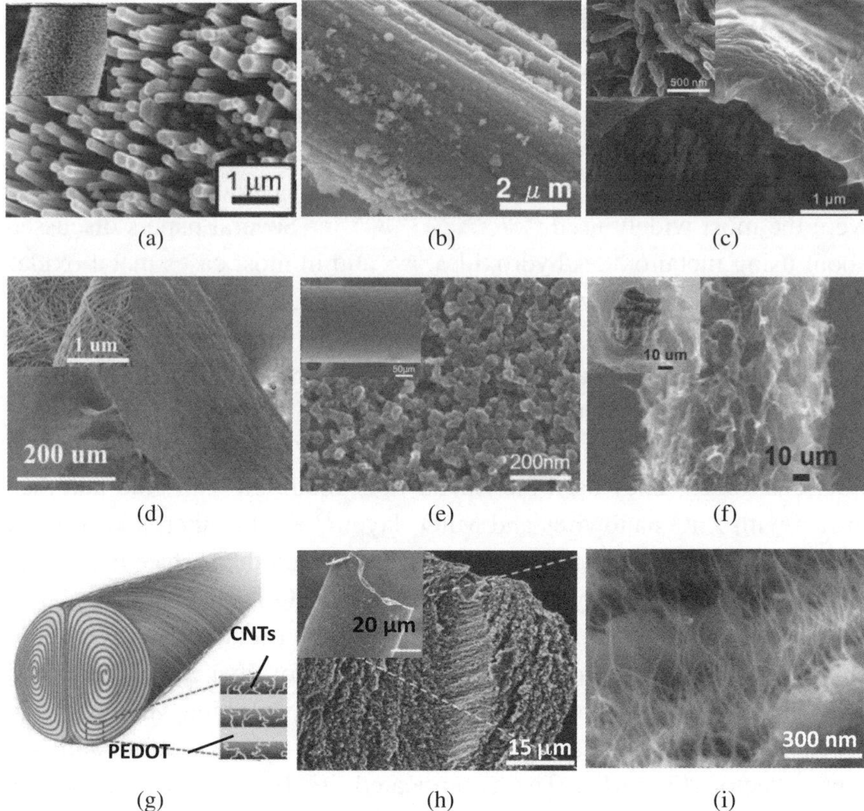

Figure 4. (a) ZnO nanowire arrays grown on Au coated Kevlar fiber. Inset shows the low magnification.[45] (b) SEM image of MnO_2 on carbon fiber.[24] (c) SEM image of the fiber surface. Inset is the three dimensional (3D) PPy-MnO_2-CNT-cotton thread multi-grade nanostructures.[32] (d) SEM images of $Ni(OH)_2$ nanowires on Ni wire.[33] (e) SEM image of graphite ink nanoparticles. Inset show SEM image of the plastic fiber electrode coated with pen ink film.[31] (f) SEM image of the graphene hierarchical fiber cross-section.[46] (g) Biscrolled CNT-PEDOT fiber.[38] (h) Fracture end area of rGO fiber.[36] (i) SWNT bundles are shown attached to the edges and surfaces of rGO.[36]

PVA-H_3PO_4 as the electrolyte, forming a fiber SC. An impressive capacitance of ~1 F cm^{-2} was achieved at the scan rate of 10 mVs^{-1} for the yarn electrode. Such a high value in general could not be obtained for MnO_2 thin-film based electrodes and devices.

In addition to MnO_2, other high-capacitance pseudocapacitive materials such as $Ni(OH)_2$ and Co_3O_4 were also grown on fiber/wire current collectors for fiber SCs. A recent example was demonstrated by Xia and Wang *et al.* from Fudan University.[33] They first prepared $Ni(OH)_2$ nanowire film on a metal wire (Fig. 4(d)), and then assembled with an ordered mesoporous carbon (OMC) fiber electrode and a polymer electrolyte to establish a fiber SC. The as-prepared SC device displayed a length/areal capacitance of 6.67 mF cm^{-1} (or 35.67 mF cm^{-2}) and a high areal/volumetric energy density of 0.01 mWh cm^{-2} (or 2.16 mWh cm^{-3}), which are about 10–100 times that of previous reported data.[20,31,41–42,44–45] Shen *et al.* used Ni wire electrode grown with Co_3O_4 nanowire to construct a fiber SC, which demonstrated a volumetric capacitance of 2.1 F cm^{-3} and a volumetric energy density of 0.62 mWh cm^{-3}.[35]

Carbon-based active materials can also be deposited on fiber current collector for fiber SC. Zou *et al.* reported a fiber electrode utilizing commercial graphitic pen ink coated metal/plastic wire (Fig. 4(e)). The resulting SC can be scanned up to 1Vs^{-1}, demonstrating excellent rate capability.[31] A length capacitance of ~0.5 mF cm^{-1} of the fiber electrode was reported. This work presented a very simple way to fabricate fiber electrode architecture with cheap active materials. The use of inactive wire as the backbone of the fiber electrode, however, will lead to low volumetric energy density of the entire device when the active materials' capacitance is low.

In the above examples, electrode active materials were typically deposited onto the 1D current collector to form the fiber electrode. The current collector in most cases only facilitated the electron transport without providing capacitance. There is some recent progress in designing "active material-current collector" integrated fiber electrodes. Such kind of fiber electrode makes the active materials come into intimate contact with current collector, and in some cases the fiber electrode itself serves both as active material and current collector. This avoids the presence of "dead volume" in the fiber electrode and can result in high volumetric capacitance. For example, Meng *et al.* designed and fabricated a unique all graphene hierarchical fiber electrode, in which the high electronic conduction of the core graphene and the highly exposed surface areas of 3D graphene sheets were combined into a single yarn, thus offering great advantages for electron transport and ion diffusion (Fig. 4(f)).[46] Another good example was from Baughman's group, which demonstrated a very unique PEDOT-CNT-biscrolled yarn made entirely of active material, and scrolled with electrolyte (Fig. 4(g)).[38] The resulting fiber SC showed excellent electrochemical performance at a fast scan rate of $1V\ s^{-1}$ (even bent). Impressively, this electrode yarn had an extremely high volumetric capacitance of 179 F cm^{-3}, much superior to most fiber electrodes reported so far.

Very recently, Chen *et al.* reported a hierarchically structured carbon microfiber electrode made of an interconnected network of aligned single-walled carbon nanotubes (SWNTs) with interposed nitrogen-doped rGO sheets (Figs. 4(h) and 4(i)).[36] The authors developed a scalable method to continuously produce the fibers using a silica capillary column functioning as a hydrothermal microreactor. The fiber electrode had large specific surface area and high electrical conductivity ($102\ S\ cm^{-1}$), showing a specific volumetric capacity as high as 305 F cm^{-3} in H_2SO_4 electrolyte (three-electrode cell, at 73.5 mA cm^{-3}). The resultant full fiber SC with PVA/H_3PO_4 gel electrolyte, free from binder, current collector and separator, had a high volumetric capacitance of ~45 F cm^{-3} and a outstanding volumetric energy density of ~6.3 mWh cm^{-3}.

In general, carbon-based fiber electrodes have good rate capability. By contrast, pseudocapacitive material-dominant fiber electrodes (metal oxides and conducting polymers, etc.) are kinetically unfavorable at high rates although they can provide large capacitance. In this regard, hybrid fiber electrodes by integrating carbon-based and pseudocapacitive materials together at the nanoscale should be very promising. There were only a few reports on such hybrid fiber electrodes and MnO_2 was mostly investigated. It is still a huge challenge to develop hybrid fiber electrodes with good conductivity, flexibility, and high rate performance from other attractive pseudocapacitive materials (such as FeO_x, MoO_3).

4. Geometrical Structure Types of the Fiber SC

Starting from 1D fiber electrode, different types of fiber SCs can be established by using different assembling methods. Figure 5 shows the five typical geometrical structures of fiber SCs reported in the literature.

The first one is the "parallel" structure encapsulated into a plastic tube, as shown in Fig. 5(a). Zou's group from Peking University designed such a SC structure, which in general consists of two parallelly aligned fiber electrodes, a helical spacer wire, and an electrolyte.[31] Inside the tube, the

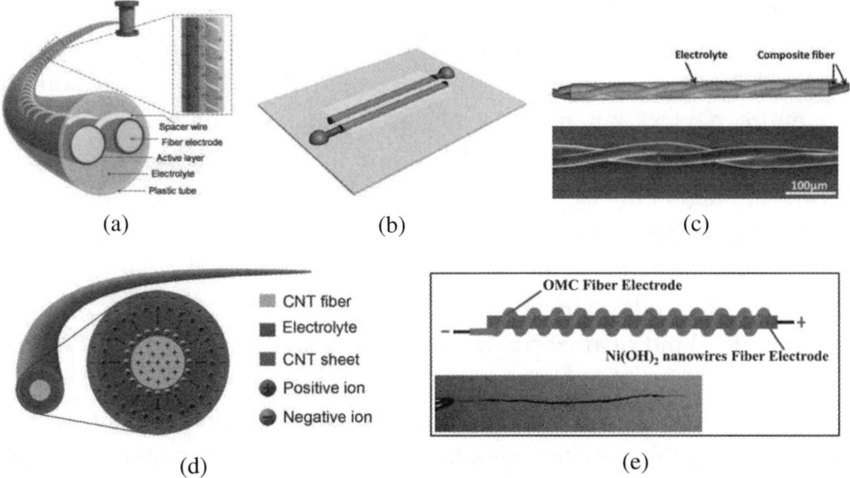

Figure 5.   (a) "Parallel" structure encapsulated into a plastic tube.[31] (b) "Parallel" structure fixed on a planar plastic.[24] (c) Biscrolled structure.[52] (d) Coaxial structure.[41] (e) "Single-twisting" structure.[33]

spacer wire is evenly twisted onto the surface of one fiber electrode with a specific pitch. Although the entire SC device is flexible and the authors demonstrated extensive flexibility testing for the first time, there are some obvious drawbacks of such structure. Firstly, the use of helical spacer wire to avoid short circuiting during bending makes the SC fabrication process complicated. Secondly, the plastic tube in the fiber SC accommodates much space so that increases the "dead volume" in the device. In addition, if liquid electrolyte such as H_2SO_4 and Na_2SO_4 is used, it would leak easily from the tube. All these make the large-scale industrial production impossible.

It is known that most liquid battery electrolytes ($LiPF_6$, etc.) are toxic.[44] Thus, they are not recommended for fiber SC. Solid/gel electrolytes should be the most realistic type of electrolyte for real wearable/miniaturized electronic applications. In particular, if solid or gel electrolytes are chosen, it is not necessary to introduce spacer wire and plastic tube any more. From the literature, gel electrolyte was commonly utilized.

The second fiber SC structure is the "parallel" structure fixed on a planar plastic. One example was demonstrated by Zhou *et al.*[24] In their work, gel electrolyte PVA-H_3PO_4 was utilized, and two MnO_2-carbon fiber electrodes were placed onto a polyester (PET) substrate in parallel to assemble the SC device (Fig. 5(b)). This structure employs gel electrolyte to solidify the entire device, but needs a planar flexible substrate as a support. It might be considered for powering future flexible miniaturized electronics and applied in micro-electromechanical systems, but is not suitable for wearable or textile energy storage.

The third structure is the biscrolled structure. In such a structure, the two fiber electrodes are bi-twisted with each other. Peng *et al.* once reported this kind of fiber SC by assembling two polyaniline-CNT composite fibers (Fig. 5(c)).[52] This structure is relatively compact and easy to be fabricated. The gel electrolyte serves as the separator without wasting volume. Thus, this structure should be one of the best to achieve fiber SC with high volumetric energy storage capability. By using this structure, the entire SC device can maintain most flexibility of electrode fibers, so that it can be easily woven into clothes and other portable devices by conventional textile technology.

The fourth structure is the coaxial structure, in which one electrode is planar and wrapped around the other one. A typical example also came from Peng *et al.* who reported a coaxial fiber EDLC from the aligned CNT fiber and sheet, with a polymer gel sandwiched between them (Fig. 5(d)).[41] During the device fabrication, the PVA/H_3PO_4 electrolyte can be both infiltrated into the fiber and coated on the surface. On further considering the unique coaxial SC structure, the reduction of the contact resistance between the two electrodes can be envisioned. This type of fiber SC is the most convenient to be woven into various shapes and dimensions.

The fifth structure is the "single-twisting" structure. The difference of this structure from the biscrolled one is that only one fiber electrode needs to be twisted while the other one is kept straight. As shown in Fig. 5(e), Xia *et al.* reported such a structured fiber SC by twisting the OMC fiber negative electrode around the Ni(OH)$_2$-nanowire fiber positive electrode.[33] Gel electrolyte of PVA-NaOH was employed. This SC structure is very useful for pairing two fiber electrodes with different capacitances, and in general the low-capacitance fiber electrode should be designed longer and needs to be scrolled. However, the flexibility of this structure may be inferior to that of the

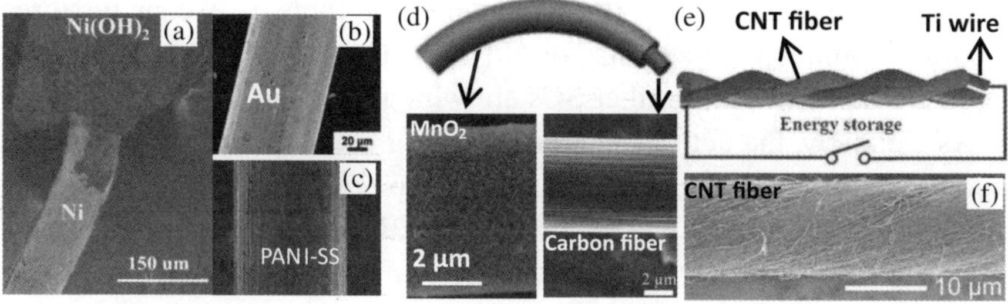

Figure 6. Different current collectors used in fiber SCs: (a) Nickel metal.[33] (b) Au wire.[51] (c) Stainless steel wire.[40] (d) Carbon microfiber.[58] (e) Ti wire and CNT fiber.[50] (f) Aligned CNT fiber.[41]

biscrolled and coaxial ones because the two electrodes have risk of shifting from each other.

Current collector is an essential part in the structure of fiber SCs. It not only is responsible for the electron collection, but also renders the entire device with mechanical flexibility. At the end of this section, we summarized various kinds of current collectors used in previous reports, such as nickel metal wire,[31,33,35] Au wire,[51] stainless steel wire,[40] carbon microfiber,[35,43,64] CNT/graphene fiber,[32,36,39,41–44,50–52] Kevlar fiber coated with Au,[45] etc. as shown in Fig. 6. Among them, the carbonaceous fibers were widely used.

5. Symmetric/Asymmetric Fiber SC

In addition to construct fiber SCs into different geometrical architectures, they can also be assembled with symmetric and asymmetric configurations. Symmetric SC is typically utilizing the same electrode as both the positive and negative electrodes, and the electrode materials can either be EDLC materials or pseudocapacitive materials. The working voltage of symmetric SC is generally lower than 1 V. Asymmetric SC, however, has one EDLC-type electrode as the power source and the other battery-type or pseudocapacitive electrode as the energy source. This renders different electrochemical windows of two kinds of electrodes to achieve a high working voltage. On the basis of the energy density Eq. (3), it is obvious that asymmetric SCs typically have higher energy density than symmetric ones.[33–35] Despite this, most of the publications have reported on the symmetric fiber SCs.[24,39–41,43–46] As shown in Table 1, except for three reported asymmetric fiber SCs of $Ni(OH)_2$//C, GCF-N//GCF-MnO_2 and Co_3O_4//Graphene, all the fiber SCs are with a symmetric structure.

As we know, the active material's mass in each microsized fiber electrode is very low, thus it is indeed difficult to reach the charge balance of the two fiber electrodes by simply adjusting the mass. In addition, as also mentioned in Section 3, pseudocapacitive material-dominant fiber electrodes with good electrical conductivity and mechanical flexibility are very lacking. The above two reasons make the construction of asymmetric fiber SC much more difficult than symmetric one.

6. Preliminary Applications

Fiber SCs are expected to have many potential applications. In this section, we will mainly summarize some preliminary applications reported in literature. (1) Powering small electronic products. For example, Xia's group demonstrated that their fiber SCs connected in series can successfully operate a red LED light (Fig. 7(a)).[33] The fiber SCs connected in series and in parallel from Gao's group were used to drive a LED segment display (Fig. 7(b)).[32] (2) Integrating into self-powered systems. In 2013, Zhou *et al.* reported that the fiber SC integrated with a TG could be charged and used to power commercial electronic devices, such as a liquid crystal display and a LED (Fig. 7(c)).[24] In a work from Chen's group, the workers constructed a self-powered nanosystem, in which three fully charged microfiber SCs are assembled in series to power an ultraviolet photodetector based on TiO_2 nanorod arrays grown on a fluorine-doped tin oxide glass substrate (Fig. 7(d)).[36] (3) Textile energy storage. For instance, Peng's group reported that the fiber SC wires could be woven into a commercial textile (Fig. 7(e)).[39] The other papers from the same group also reported that their SC wires can be easily woven into clothes, which have been shown in Figs. 3(c) and 3(d). Baughman *et al.* from University of Texas-Dallas and collaborators reported that their biscrolled yarn SC can be woven into a glove (Fig. 7(f)).[38] The fiber SC from Chao Gao's group was also woven into cloth (Fig. 7(g)).[37] (4) Integrating with microenergy conversion devices. One early example of this application is from

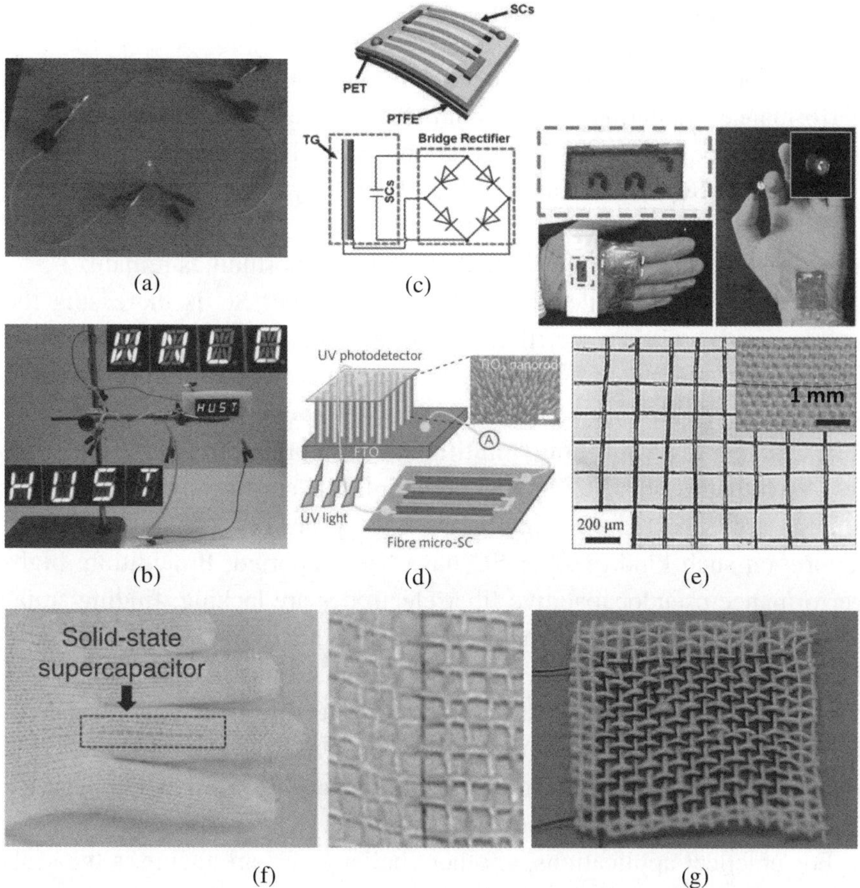

Figure 7. (a) Digital photo that shows two microfiber SCs in series to light a LED.[33] (b) Photograph of fiber SC used to drive a LED segment display.[32] (c) Digital photograph of the energy modulus that is powering commercialized LCD and LED.[24] (d) Schematic of a self-powered nanosystem.[36] (e) Optical images of hand-made fiber SC textile consisting of 15 years. Inset: A fiber electrode woven into a commercial textile.[39] (f) The fiber SC woven into a glove.[38] (g) Two fiber SCs woven into a cloth.[37]

Zou *et al.* who combined a pseudocapacitive fiber and highly flexible solar cell fiber together. In detail, anodically deposited polyaniline was jointly used as the electrode of the fiber DSSC and fiber SC, so that solar energy conversion and storage were achieved in one device.[65]

7. Conclusion and Outlook

This article has focused on recent developments of the fiber SCs. The performance evaluation methods and metrics, the key electrode materials, and the device structures/configurations as well as preliminary applications of fiber SCs are discussed. It has been only two to three years since the first fiber SC was reported. Thus, the research on such a topic is still at its infancy stage and many challenges and opportunities remain.

One of the key challenges to develop the fiber SC is increasing the energy density without sacrificing the power density. It is well known that EDLC electrode material-based symmetric fiber SC has high power density; the energy density of such device is however generally low, making it insufficient to power future multifunctional, portable/wearable electronics. Asymmetric fiber SCs having both high capacitance and high working voltage are expected to enhance the energy density. Nevertheless, few reports on such kind of fiber SC have been reported. In addition, high-performance pseudocapacitive fiber electrodes are lacking; finding suitable fiber electrodes to match the pseudocapacitive one is more difficult. Therefore, one promising research direction should be fabricating new pseudocapacitive materials/structures and selecting matching fiber electrode pairs to construct asymmetric fiber SCs. The development of simple and facile methods to achieve the charge balance of positive and negative fiber microelectrodes is particularly important.

For practical applications, another challenge we are facing is the scalable construction of fiber SCs. To this end, the materials synthesis methods should be easily scaled-up and be compatible with industrial techniques. Common coating techniques such as dip coating and screen printing are most simple and promising for scale-up fabrication. However, recent progress in small-scale fiber SCs involved some lab techniques such as hydrothermal and electro-deposition. There is still an open question: are these techniques scalable? In addition, the reproducibility of materials synthesis as well as device fabrication is critically important for commercialization of fiber SCs. Last but not the least, if applied in future wearable energy storage, the fiber SCs should not only be highly mechanically flexible, but must also be resistant to washing and other external influences such as sunshine irradiation, etc.

Research in fiber SCs is extremely exciting and full of great opportunities. It is believed that with the joint efforts of both scientists and engineers, fiber SCs will be practically used for wearable, portable miniaturized electronics in the near future.

Acknowledgments

This work was supported by grants from the national natural science foundation of China (Grant Nos. 51102105, 11104088), the science fund for distinguished young scholars of Hubei province (Grant No. 2013cfa023), the youth Chenguang project of science and technology of Wuhan city (Grant No. 2014070404010206), self-determined research funds of CCNU from the colleges basic research and operation of MOE (ccnu14a02001), and the self-determined innovation foundation of huazhong university of science and technology (Grant No. 2013027).

References

1. J. R. Miller and P. Simon, *Science*, **321**, 651 (2008).
2. P. Simon and Y. Gogotsi, *Nat. Mater.*, **7**, 845 (2008).
3. C. Zhou, Y. W. Zhang, Y. Y. Li and J. P. Liu, *Nano Lett.*, **13**, 2078 (2013).
4. D. S. Su and R. Schlogl, *Chem. Sus. Chem.*, **3**, 136 (2010).
5. C. Liu, F. Li, L.-P. Ma and H.-M. Cheng, *Adv. Mater.*, **22**, E28 (2010).
6. M. S. Whittingham, *Chem. Rev.*, **104**, 4271 (2004).
7. M. Winter and R. J. Brodd, *Chem. Rev.*, **104**, 4245 (2004).
8. A. Burke, *J. Power Sources*, **91**, 37 (2000).
9. J. P. Liu, J. Jiang, C. W. Cheng, H. X. Li, J. X. Zhang, H. Gong and H. J. Fan, *Adv. Mater.*, **23**, 2076 (2011).
10. L. Y. Yuan, X. Xiao, T. P. Ding, J. W. Zhong, X. H. Zhang, Y. Shen, B. Hu, Y. H. Huang, J. Zhou and Z. L. Wang, *Angew. Chem. Int. Ed.*, **51**, 4934 (2012).
11. L. L. Zhang and X. S. Zhao, *Chem. Soc. Rev.*, **38**, 2520 (2009).
12. A. S. Arico, P. Bruce, B. Scrosati, J. M. Tarascon and W. V. Schalkwijk, *Nat. Mater.*, **4**, 366 (2005).
13. J. Gamby, P. L. Taberna, P. Simon, J. F. Fauvarque and M. Chesneau, *J. Power Sources*, **101**, 109 (2001).

14. D. Pech, M. Brunet, H. Durou, P. H. Huang, V. Mochalin, Y. Gogotsi, P. L. Taberna and P. Simon, *Nat. Nanotechnol.*, **5**, 651 (2010).

15. J. Chmiola, C. Largeot, P. L. Taberna, P. Simon and Y. Gogotsi, *Science*, **328**, 5977 (2010).

16. L. B. Hu, J. W. Choi, Y. Yang, S. Jeong, F. L. Mantia, L. F. Cui and Y. Cui, *P. Natl. Acad. Sci. USA*, **106**, 21490 (2009).

17. M. D. Stoller, S. J. Park, Y. W. Zhu, J. H. An and R. S. Ruoff, *Nano Lett.*, **8**, 3498 (2008).

18. M. Toupin, T. Brousse and D. Belanger, *Chem. Mater.*, **16**, 3184 (2004).

19. J. Zhang, W. Chu, J. W. Jiang and X. S. Zhao, *Nanotechnology*, **22**, 125703 (2011).

20. P. Jampani, A. Manivannan and P. N. Kumta, *Electrochem. Soc. Interface.*, **19**, 57 (2010).

21. M. F. El-Kady, V. Strong, S. Dubin and R. B. Kaner, *Science*, **335**, 1326 (2012).

22. X. H. Lu, T. Zhai, X. H. Zhang, Y. Q. Shen, L. Yuan, B. Hu, L. Gong, J. Chen, Y. H. Gao, J. Zhou, Y. X. Tong and Z. L. Wang, *Adv. Mater.*, **24**, 938 (2012).

23. C. Z. Meng, C. H. Liu, L. Z. Chen, C. H. Hu and S. S. Fan, *Nano Lett.*, **10**, 4025 (2010).

24. X. Xiao, T. Li, P. Yang, Y. Gao, H. Jin, W. Ni, W. Zhan, X. Zhang, Y. Cao, J. Zhong, L. Gong, W. C. Yen, W. Mai, J. Chen, K. Huo, Y. L. Chueh, Z. L. Wang and J. Zhou, *ACS Nano*, **6**, 9200 (2012).

25. A. B. Dalton, S. Colling, E. Munoz, J. M. Razal, V. H. Ebron, J. P. Ferraris, J. N. Coleman, B. G. Kim and R. H. Baughman, *Nature*, **423**, 703 (2003).

26. M. Kaemgen, C. K. Chan, J. Ma, Y. Cui and G. Gruner, *Nano Lett.*, **9**, 1872 (2009).

27. Z. Weng, Y. Su, D. W. Wang, F. Li, J. Du and H. M. Cheng, *Adv. Energ. Mater.*, **1**, 917 (2011).

28. L. X. Wu, R. Z. Li, J. L. Guo, C. Zhou, W. P. Zhang and J. P. Liu, *AIP Adv.*, **3**, 082129 (2013).

29. W. P. Si, C. L. Yan, Y. Chen, S. Oswald, L. Y. Han and O. G. Schmidt, *Energy Environ. Sci.*, **6**, 3218 (2013).

30. Y. H. Kwon, S. W. Woo, H. R. Jung, H. K. Yu, K. Kim, B. H. Oh, S. Ahn, S. Y. Lee, S. W. Song, J. Cho, H. C. Shin and J. Y. Kim, *Adv. Mater.*, **24**, 5192 (2012).

31. Y. P. Fu, X. Cai, H. W. Wu, Z. B. Lv, S. C. Hou, M. Peng, X. Yu and D. C. Zou, *Adv. Mater.*, **24**, 5713 (2012).

32. N. S. Liu, W. Z. Ma, J. Y. Tao, X. H. Zhang, J. Su, L. Y. Li, C. X. Yang, Y. H. Gao, D. Golberg and Y. Bando, *Adv. Mater.*, **25**, 4925 (2013).

33. X. L. Dong, Z. Y. Guo, Y. F. Song, M. Y. Hou, J. G. Wang, Y. G. Wang and Y. Y. Xia, *Adv. Funct. Mater.*, **24**, 3405 (2014).

34. D. S. Yu, K. L. Goh, Q. Zhang, L. Wei, H. Wang, W. C. Jiang and Y. Chen, *Adv. Mater.*, **3**, 61 (2014).

35. X. F. Wang, B. Liu, R. Liu, Q. F. Wang, X. J. Hou, D. Chen, R. M. Wang and G. Z. Shen, *Angew. Chem. Int. Ed.*, **53**, 1849 (2014).

36. D. S. Yu, K. L. Goh, H. Wang, L. Wei, W. C. Jiang, Q. Zhang, L. Dai and Y. Chen, *Nat. Nanotechnol.*, **9**, 555 (2014).

37. L. Kou, T. Huang, B. Zheng, Y. Han, X. L. Zhao, K. Gopalsamy, H. Y. Sun and C. Gao, *Nat. Commun.*, **5**, 3754 (2014).

38. J. A. Lee, M. K. Shin, S. H. Kim, H. U. Cho, G. M. Spinks, G. G. Wallace, M. D. Lima, X. Lepro, M. E. Kozlov, R. H. Baughman and S. J. Kim, *Nat. Commun.*, **4**, 1970 (2013).

39. C. S. Choi, J. A. Lee, A. Y. Choi, Y. T. Kim, X. Lepro, M. D. Lima, R. H. Baughman and S. J. Kim, *Adv. Mater.*, **26**, 2059 (2013).

40. Y. P. Fu, H. W. Wu, S. Y. Ye, X. Cai, X. Yu, S. C. Hou, H. Kafafy and D. C. Zou, *Energy Environ. Sci.*, **6**, 805 (2013).

41. X. L. Chen, L. B. Qiu, J. Ren, G. Z. Guan, H. J. Lin, Z. T. Zhang, P. N. Chen, Y. G. Wang and H. S. Peng, *Adv. Mater.*, **25**, 6436 (2013).

42. J. Ren, W. Y. Bai, G. Z. Guan, Y. Zhang and H. S. Peng, *Adv. Mater.*, **25**, 5965 (2013).

43. P. Xu, T. L. Gu, Z. Y. Cao, B. Q. Wei, J. Y. Yu, F. X. Li, J. H. Byun, W. B. Lu, Q. W. Li and T. W. Chou, *Adv. Energy Mater.*, **4**, (2014).

44. J. Ren, L. Li, C. Chen, X. L. Chen, Z. B. Cai, L. B. Qiu, Y. G. Wang, X. R. Zhu and H. S. Peng, *Adv. Mater.*, **25**, 1155 (2013).

45. J. Bae, M. K. Song, Y. J. Park, J. M. Kim, M. Liu and Z. L. Wang, *Angew. Chem. Int. Ed.*, **50**, 1683 (2011).

46. Y. N. Meng, Y. Zhao, C. G. Hu, H. H. Cheng, Y. Hu, Z. P. Zhang, G. Q. Shi and L. T. Qu, *Adv. Mater.*, **25**, 2326 (2013).

47. T. Chen, Z. B. Yang and H. S. Peng, *ChemPhysChem.*, **14**, 1777 (2013).

48. K. Jost, G. Dion and Y. Gogotsi, *J. Mater. Chem. A*, **2**, 10776 (2014).

49. K. Wang, Q. H. Meng, Y. J. Zhang, Z. X. Wei and M. H. Miao, *Adv. Mater.*, **25**, 1494 (2013).

50. T. Chen, L. B. Qiu, Z. B. Yang, Z. B. Cai, J. Ren, H. P. Li, H. J. Lin, X. M. Sun and H. S. Peng, *Angew. Chem. Int. Ed.*, **51**, 11977 (2012).

51. Y. R. Li, K. X. Sheng, W. J. Yuan and G. Q. Shi, *Chem. Commun.*, **49**, 291 (2013).

52. Z. B. Cai, L. Li, J. Ren, L. B. Qiu, H. J. Lin and H. S. Peng, *J. Mater. Chem. A.*, **1**, 258 (2013).

53. B. Liu, D. S. Tan, X. F. Wang, D. Chen and G. Z. Shen, *Small*, **9**, 1998 (2012).

54. X. Li, X. B. Zang, Z. Li, X. M. Li, P. Z. Sun, X. Lee, R. J. Zhang, Z. H. Huang, K. L. Wang, D. H. Wu, F. Y. Kang and H. W. Zhu, *Adv. Funct. Mater.*, **23**, 4862 (2013).

55. Q. H. Meng, H. P. Wu, Y. N. Meng, K. Xie, Z. X. Wei and Z. X. Guo, *Adv. Mater.*, **26**, 4100 (2014).

56. V. T. Le, H. Kim, A. Ghosh, J. Kim, J. Chang, Q. A. Vu, D. T. Pham, J. H. Lee, S. W. Kim and Y. H. Lee, *ACS Nano*, **7**, 5940 (2013).

57. Z. B. Yang, J. Deng, X. L. Chen, J. Ren and H. S. Peng, *Angew. Chem., Int. Ed.*, **52**, 13453 (2013).

58. J. Y. Tao, N. S. Liu, W. Z. Ma, L. W. Ding, L. Y. Li, J. Su and Y. H. Gao, *Scientific Reports*, 3 (2013).

59. A. K. Shukla, S. Sampath and K. Vijayamohanan, *Curr. Sci.*, **79**, 1656 (2000).

60. J. R. Miller and P. Simon, *Electrochem. Soc. Interface.*, **17**, 53 (2008).

61. H. D. AbruÇa, Y. Kiya and J. C. Henderson, *Phys. Today.*, **61**, 43 (2008).

62. H. J. Liu, J. Wang, C. X. Wang and Y. Y. Liang, *Adv. Energy Mater.*, **1**, 1101 (2011).

63. L. B. Hu, M. Pasta, F. L. Mantia, L. F. Cui, S. Jeong, H. D. Deshazer, J. W. Choi, S. M. Han and Y. Cui, *Nano Lett.*, **10**, 708 (2010).

64. B. S. Shim, W. Chen, C. Doty, C. L. Xu and N. A. Kotov, *Nano Lett.*, **8**, 4151 (2008).

65. Y. P. Fu, H. W. Wu, S. Y. Ye, X. Cai, X. Yu, S. C. Hou, H. Kafafy and D. C. Zou, *Energy Environ. Sci.*, **6**, 805 (2013).

CHAPTER 10

FLEXIBLE ELECTRONIC DEVICES BASED ON ELECTROSPUN MICRO-/NANOFIBERS

Bin Sun,* Miao Yu,*,† Yun-Ze Long*,‡,§ and Wen-Peng Han*

*Collaborative Innovation Center for Low-Dimensional
Nanomaterials & Optoelectronic Devices,
College of Physics, Qingdao University, Qingdao 266071, P.R. China
†Department of Mechanical Engineering,
Columbia University, New York 10027, USA
‡State Key Laboratory Cultivation Base of New Fiber Materials
and Modern Textile, Qingdao University,
Qingdao 266071, P.R. China
§yunze.long@163.com; yunze.long@qdu.edu.cn

Demonstrating various outstanding characters, such as large surface area, high length-to-diameter ratio, tunable surface morphologies, flexible surface functionality, etc., electrospun micro-/nanofibers with excellent structural and mechanical properties are adaptive for the fabrication of flexible electronic devices, which have been extensively explored in recent years. In this chapter, recent advances in two types of flexible electronic devices (bendable electronic devices and stretchable electronic devices) based on the electrospun ultrafine fibers have been summarized and reviewed. It is found that not only the materials but also the architectures of the electrospun fibers can affect the function and performance of the devices. In addition, some challenges such as electrospinnability of the organic precursors and the mechanical performance of the electrospun inorganic fibers in this area also exist, although some positive results have been achieved.

1. Introduction

It is believed that an important future of devices is with systems that avoid the rigid, brittle and planar nature of existing classes, to enable new applications. Since 1960s, the first report of flexible solar cells which was fabricated by thinning single crystal silicon wafer cells to about 100 μm and then assembling them on a plastic substrate to provide flexibility, micro-/nanoscale devices with outstanding flexibility have aroused much more interest in the last couple of decades.[1,2] These devices demonstrate fascinating properties such as biocompatibility, flexibility, light weight, shock resistance, softness, and transparency, which make them potential application in some areas such as paper displays, wearable devices, and energy-storage devices, and photodetectors that are impossible to achieve with hard, planar integrated circuits that exist today.[3–9] Sekitani *et al.*[7] have fabricated non-volatile memory transistors on flexible plastic substrates. The small thickness of the dielectrics allows very small program and erase voltages ($\leq$6 volts) to produce a large, non-volatile, reversible threshold-voltage shift. The transistors endure more than 1,000 programs and erase cycles, which is within two orders of magnitude of silicon-based floating-gate transistors widely employed in flash memory.

Flexible electronic devices, which are usually built on bendable/ elastic substrates, involve a wide range of device and materials technologies. Achieving bendability is a first step to stretchability. In general, any materials which have enough thinness will reveal bendability, by virtue of bending strains that are directly proportional to thickness. For example, a sheet of single crystalline silicon with 100 nm thickness on a 20 mm-thick plastic substrate experiences peak strains of only 0.1% on bending to a radius of curvature of 1 cm, which is well below the fracture limits for silicon ($\sim$1%).[10] However, simple bendability only allows roll-to-roll production of some types of materials, but does not guarantee conformal bonding of these devices to non-planar substrates apart from cylinders and cones.[11] Namely, simple bendability often has insufficient stretchability, which will limit the flexible devices application in some fields that have large curvatures or undergo a large mechanical deformation[11–13] although they demonstrate outstanding performances compared with their planar counterparts. Thus, manufacturing devices with high stretchability that can survive under large deformations is of equal importance.

To date, flexible electronic devices have been fabricated using a variety of techniques, which can be divided into two main strategies: (i) the whole structures prepared on a rigid substrate (e.g., a Si wafer or a glass plate) in advance, and then transferred onto flexible substrates *via* some methods such as transfer/contact printing, blown bubble films,[14–24] and so on. Although with some advantages such as precision and repeatability, it is becoming obvious that both physical and economic factors in this strategy will restrict their further advances. Therefore, many researchers focus on strategy (ii) the circuits fabricated directly in scalable ways onto the flexible substrates such as metal foils, thin sheets of plastic, and even paper, using techniques including low-temperature deposition, solution processing, nano-/micromolding, and electro-spinning,[25–42] etc. Figure 1 shows some methods of strategy (ii) for fabricating flexible devices.

Electrospinning is a simple and versatile approach for fabricating uniform micro-/nanofibers in a continuous process and at long length scales. The as-spun fibers exhibit several fascinating properties, such as large surface area, high length/diameter ratio, flexible surface functionality, and tunable surface morphologies, which make them the underlying application in optoelectronics, sensors, catalysis, textiles, filters, fiber reinforcement, tissue engineering, drug delivery, wound healing, and so forth.[43–49] Especially, compared with other one-dimensional (1D) nanostructures (e.g., nanorods and nanotubes), the feature of long continuity provide nanofibers with high axial strength combined with extreme flexibility. Namely, the electrospun fibrous membranes have excellent structural mechanical properties, which are adaptive for flexible devices.

In this chapter, based on the electrospinning technique backgrounds in general, we put emphasis on the recent development of flexible (bendable and stretchable) electronic devices based on electrospinning and electrospun products, specially the long and continuous micro-/nanofibers. Furthermore, some challenges in this area have been discussed in depth.

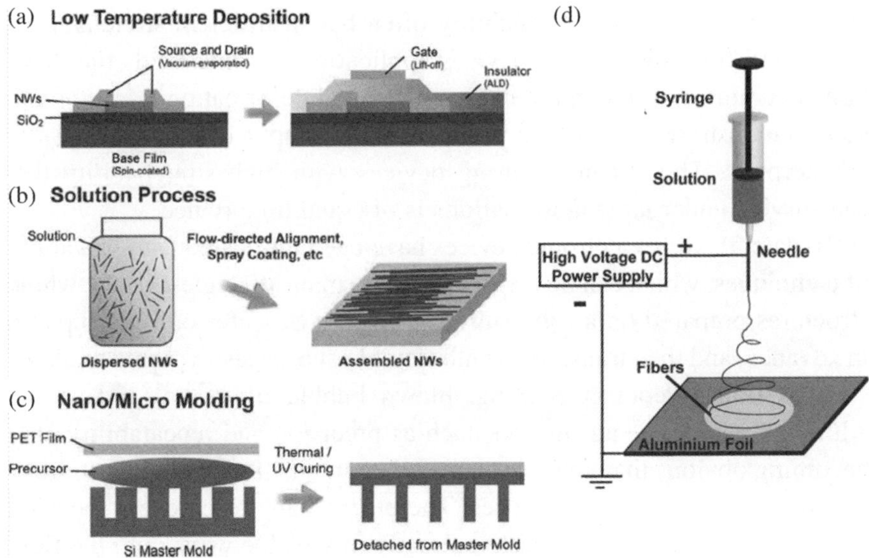

Figure 1. Some methods of strategy (ii) for fabricating flexible devices: (a) low-temperature deposition method, (b) solution-assisted assembling process, (c) nano-/micromolding, and (d) electrospinning. Reproduced from Ref. 25. (Copyright © 2014, the Royal Society of Chemistry.)

2. Electrospinning Backgrounds

Electrospinning was studied by Formhals in the 1930s[50] and has regained tremendous interest in the 1990s.[51] A typical electrospinning setup is composed of three basic elements: a high voltage DC (or AC) power supply, a capillary (including a solution container and a spinneret) and a grounded metal collector (usually aluminum foil). One electrode of the high voltage power supply is connected with the spinneret containing spinning solution/melt and the other attached to the collector, which is usually grounded, as shown in Fig. 1(d). During electrospinning process, the precursor solution is extruded from a spinneret to form a small droplet in the presence of an electric field, and then the charged solution jets are extruded from the cone. Generally the fluid extension occurs first in uniform, and then the straight flow-lines undergo vigorous whipping and/or splitting motion due to fluid instability and electrically driven bending instability. Finally, the continuous as-spun fibers are deposited commonly as a two-dimensional (2D) non-woven mat on the collector. However, the random orientation has limited further applications of the electrospun products, not only in the area of tissue engineering such as muscle, nerve, blood vessel, and connective tissue, which require a well-controlled cellular and extracellular matrix (ECM) organization for proper tissue function, but also in other areas such as electronics, photonics, and actuators which also need direct, fast charge transfer or regular, uniform structures. In order to tackle this shortcoming, it is necessary to generate structure-controlled micro-/nanofibers. To date, numerous groups have devoted themselves to this subject, and it is delightful to observe that besides fiber membranes without orientation, other fibrous structures and organization (e.g., parallel and crossed fiber arrays, helical or wavy fibers, twisted fiber yarns, patterned fibrous web, and three-dimensional (3D) stacks) (Fig. 2) based on not only polymers of synthetic or biological nature, but also metals, metal oxides, ceramics, organic/organic, organic/inorganic as well as inorganic/inorganic composite systems have been electrospun successfully *via* modified electrospinning process or collectors, which will extend further application of as-spun fibers in many fields. Anyhow, it should emphasis that the morphology and diameters of the electrospun fibers are deeply influenced

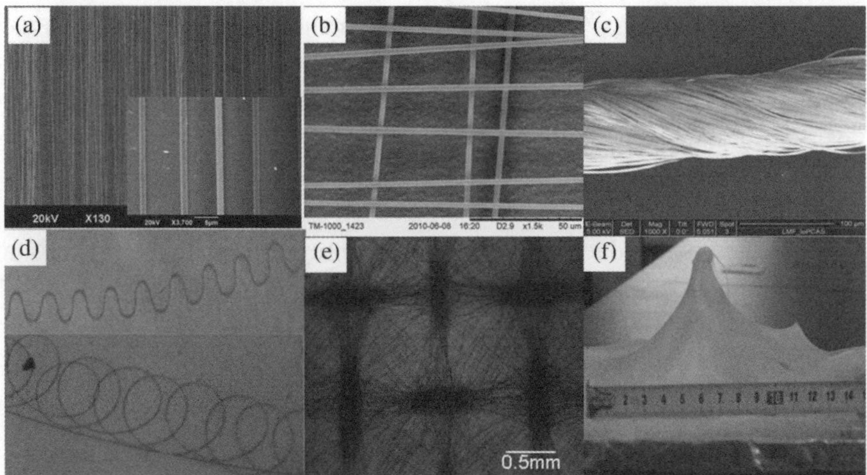

Figure 2. Electrospun (a) aligned and (b) crossed fibrous arrays, (c) twisted fibrous yarns, (d) helical and wavy fibers, (e) fibrous pattern, (f) 3D stacks.

by a series of parameters: (i) the intrinsic properties of the solution such as the type of polymer, the conformation of polymer chain, viscosity/concentration, elasticity, electrical conductivity, and the polarity and surface tension of the solvent; (ii) the operational conditions such as the strength of applied electric field, the distance between spinneret and the collector, and the feeding rate for the polymer solution; (iii) the environment parameters including the humidity and temperature of the surroundings.[52]

Numerous systems of materials have been eletrospun successfully for flexible electronic devices. According to our early analysis, there are three main strategies.[25] (i) Direct electrospinning: this approach is the easiest and simplest. Here, the as-spun nano-/microfibers based on conductive or piezoelectric polymers are deposited onto the flexible substrates for direct use, and in order to improve the electrical or mechanical properties, sometimes additives such as metal nanoparticles, carbon black, carbon nanotubes, and graphene are selected to be added into the precursor previously.[39,40,53–58] (ii) Electrospinning precursor nanofibers: in this case, the precursors of a monomer or mixture of organic/organic, organic/inorganic are electrospun firstly, and then a post-process such as calcinating or carbonizing is adopted to generate pure inorganic fibers (e.g., carbon, metal and semiconducting fibrous membrane), which is needed for flexible electronic devices.[59–65] (iii) Electrospun nanofibers as template: as we know, due to the limitations on molecular weight and solubility unsuitable for electrospinning, not all the polymer solutions with appreciating viscosity can be electrospun directly up to now. Therefore, some non-conductive polymers can be electrospun into nanofibers and used as a template, and then some other materials such as metals (e.g., Au and Ag), semiconductors, conductive polymers (e.g., polyaniline), and even carbon nanotubes, are deposited on the surfaces of the electrospun nanofibers by sputtering, evaporation, *in situ* polymerization, infiltration, and so forth (sometimes, the polymer fiber cores have to be removed subsequently according to requirement). Finally the as-prepared samples are transferred to the flexible substrates to fabricate flexible electronic devices.[38,66–69]

Here it should be noted that although the electrospinnable polymers may assist with formation of conducting polymer fibers, which can be used directly, they thus decrease the conductivity of the as-spun fibers remarkably.[70] Many inorganic semiconductive nanofibers show excellent electrical properties, but their mechanical properties need to be further improved. And inorganic semiconductive fibers usually form polycrystals after calcination, which may limit their applications in some cases, because of the lower carrier mobilities and transconductances compared with their singlecrystal counterparts.[25] Moreover, for all these devices, no matter what kind of materials are applied and what kind of functional mechanism is working, a flexible substrate is always needed. Thus the characteristics of the substrates often impose limitations on materials choice and fabrication processes. As an example, most low-cost polymer substrates degrade at temperatures above 300°C, making them incompatible with conventional techniques for the deposition and doping of most established classes of inorganic semiconductors.[10]

3. Flexible Electronic Devices *via* Electrospinning

Due to outstanding features of the electrospun nano/microfibers (e.g., flexible surface functionality, tunable surface morphologies and superior mechanical performance), development of flexible electronic devices based on the electrospun fibers have been driven rapidly in recent years. In this section, we mainly introduce some recent progresses relating to electrospun fibers for flexible electronic devices, including bendable devices and stretchable devices

3.1. *Electrospun Bendable Electronic Devices*

3.1.1. Sensors

An emerging development trajectory in electronics focuses on flexible sensors which are gaining increasing interest in a number of applications, including biomedical, food control, domotics and robotics, which have very light weight, robustness, and low cost. Developments of electrospinning have offered opportunities in this field to construct more efficient interfaces with electronic components whose size is comparable to that of molecules.[71] These electrospun products with huge specific surface area, high porosity, and superior mechanical performance, meet ideally the desired requirements for ultrasensitive sensors, especially pressure sensors,[40,55,56,72–74] gas sensors,[75,76] acoustic emission sensors,[77] Surface-Enhanced Raman Scattering (SERS),[78] chemiresistive sensors,[79] optical sensors,[80,81] and so on. Here the conductive or piezoelectric polymers are dominated because they can exploit deformations induced by small forces, through pressure, mechanical vibration, elongation/compression, bending or twisting,[82] and fibrous arrays are able to demonstrate better performance than non-woven mats or isolated strands in this area. For example, Persano *et al.*[55] have electrospun a flexible pressure sensor based on aligned nanofibrous arrays (average diameter of 260 nm) of poly(vinylidene-fluoride-co-trifluoroethylene) (P(VDF-TrFe)) (Fig. 3(a)), which have particularly attractive piezoelectric properties, revealing large response to even mini applied pressures (~60–80 dB) (Fig. 3(b)). Meanwhile, a preferential alignment of the main molecular chains along the fiber longitudinal axis and an enhancement of the orientation of piezoelectric active dipoles in the direction perpendicular to this axis are observed. This pressure sensor can work in any orientation and can be mounted on any surface by means of transparent, skin-conformal plastic sheets. Here the conductive and piezoelectric polymers are dominated because they can exploit deformations induced by small forces, through pressure, mechanical vibration, elongation/compression, bending or twisting,[72] and fibrous arrays are able to demonstrate better performance than the non-woven mats or isolated strands in this area.

Metal nanoparticles impregnated flexible and free-standing polymer fibers construct new substrates that have received increased attention in the context of SERS devices.[83] And these nanofibers can be easily made *via* electrospinning in large quantities. Zhang *et al.*[78] introduced a method to deposit Ag nanoparticles on electrospun amidoxime surface-functionalized polyacrylonitrile (ASFPAN) nanofibrous membranes by *in situ* seeding and seed-mediated growth (Fig. 3(c)). High SERS activities/sensitivities were observed from the ASFPAN-Ag nanoparticles nanofibrous membranes, as shown in Fig. 3(d). And it is found the density and size of Ag nanoparticles had an important impact on the SERS activity/sensitivity. Their results confirmed that the enhancement of Raman signals is due to the presence of hot spots between/among Ag nanoparticles on the nanofiber surfaces. Electrospun nanofibers with assembled

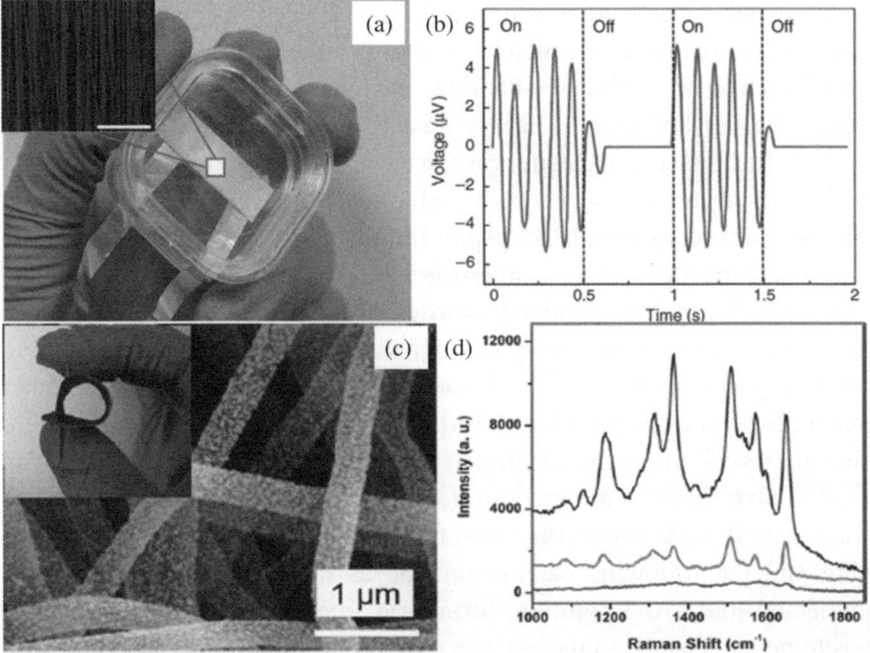

Figure 3.   (a) Photograph of a P(VDF-TrFe) nanofiber-based pressure sensor (inset is the SEM micrograph of fibrous arrays (scale bar = 10 μm)). (b) Output voltage collected from this device exposed to 70 dB sound intensity. Reproduced from Ref. 55. (Copyright © 2013, Nature Publishing Group.) (c) SEM micrographs of ASFPAN-Ag NPs nanofibrous membranes (insert is photograph depicts the excellent mechanical flexibility/resilience of a ASFPAN-Ag NPs nanofibrous membrane). (d) Typical SERS spectra of the ASFPAN nanofibrous membranes after the deposition of Ag NPs for three minutes and treated with R6G solutions with different concentrations (bottom to top: 10, 100, and 1,000 ppb). Reproduced from Ref. 78. (Copyright © 2012, American Chemical Society.)

nanoparticles offer a promising tool for the fabrication of flexible devices, but several aspects such as mechanical strength, stability of the fibers in different solvents and reproducibility need to be thoroughly investigated.[83]

3.1.2. Supercapacitors

In order to power the flexible electronic devices which maintain functionality while they are on a non-planar surface or undergoing a deformation, rechargeable and flexible energy storage devices become a necessity. Among supercapacitors which store energy in double layers on the electrode/electrolyte interface with opposite charges for providing a high power density and modest energy density in complementing the battery system, flexible supercapacitors[84–88] have aroused broad interest due to its possibility to create new design opportunities for developing portable, tiny and flexible electronics. Electrospun fibrous mats have potential for flexible supercapacitors because of the large interconnected voids between the fibers which provide abundantly porous structures for the mobile ions in the electrolyte to move freely through the elastomeric separator during bending and releasing.[62–65,89–91] Sometimes, combined with some other materials such as polyaniline (PANI)[64] and grapheme,[65] the products are able to reveal improved electrical and electrochemical properties. Yan *et al.*[64] have reported a method to fabricate carbon nanofiber-polyaniline composite flexible paper (Fig. 4(a)) for supercapacitors. After electrospinning and the following calcination, the as-spun carbon nanofibers are immersed into PANI solution, forming a freestanding CNF-PANI composite paper. Thanks to the existence of PANI, the product has improved electrical conductivity and electrochemical performances compared with the CNF paper, as shown in Fig. 4(b). The same result is observed when graphene nanosheet was added into the CNF paper, the composite exhibits the largest specific capacitance of 197 F g^{-1}, about 24% higher than that of pure CNF paper.[65]

Furthermore, Xue *et al.*[62] use electrospun MnO_2 aligned fibrous mats as active materials for flexible supercapacitors due to the smaller crystalline size and the higher proton diffusion of MnO_2 nanomaterials which offer the potential for further improvement in capacitance. The as-prepared supercapacitor has H_3PO_4-PVA gel acting as both electrolyte and substrate, demonstrating good transparency and flexibility (Fig. 4(c)), and the transparency of the film varies with the thickness of the MnO_2 layer. The electrospun MnO_2-based supercapacitor can preserve more than 95% of its specific capacitance (from 341.4 – 328.2 F g^{-1}) as the current density increases from 0.1–0.2 mA cm^{-2}, as shown in Fig. 4(d). And the specific capacitance of these electrospun MnO_2 nanofibers is higher than that of the electrodeposited MnO_2.

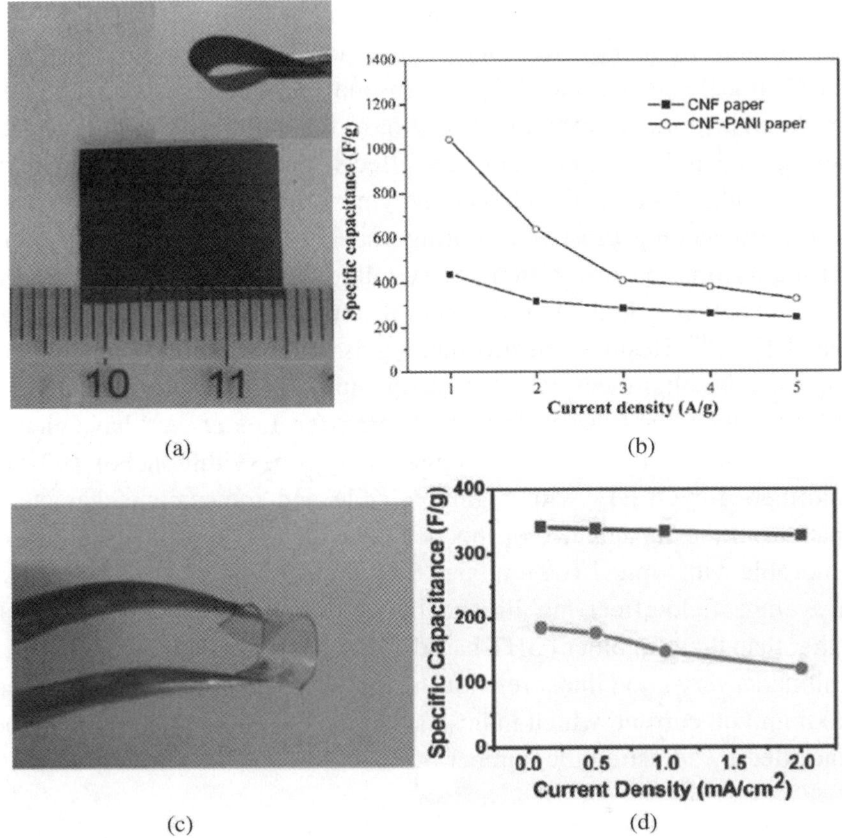

Figure 4.   (a) Digital camera images of the CNF-PANI composite paper (the inset images indicate the flexibility of the samples). (b) Specific capacitance of the CNF paper and the CNF-PANI composite paper as a function of discharge current. Reproduced from Ref. 64. (Copyright © 2011, the Royal Society of Chemistry.) (c) Optical images of a folded capacitor. (d) Specific capacitance of electrospun MnO_2-based supercapacitor (red) and directly calcined MnO_2-based supercapacitor (green) at difference charge/discharge current densities. Reproduced from Ref. 62. (Copyright © 2011, the Royal Society of Chemistry.)

3.1.3. Organic Field-Effect Transistors (OFETs)

Due to the low fabrication temperature which allows a wide range of substrate (e.g., polymeric, paper) possibilities, and the potentially reduced cost compared to hydrogenated amorphous silicon thin-film transistors (TFTs), OFETs are attractive for a variety of electronics applications, particularly those which may require or benefit from flexible substrates such as radio-frequency (RF) identification tags, next-generation displays, and chemical sensors, etc.[92–97] Electrospun one-dimensional nanofibers can do well in this field because they have the ability to promote charge transfer and reduce grain boundary effects, providing increasing field-effect mobility compared with other polymer OFETs.[98–102] This is because the electrospinning process can bring enhanced π–π molecular stacking and highly ordered orientation of crystallites in the as-spun nanofibers, which result in a high mobility significantly higher than that of spin-coated film.[101] Besides, electrospinning is able to fabricate nanofibers with uniaxial alignment, which is a very important factor especially for OFETs, which need direct, fast charge transfer. Lee *et al.*[58] have eletro-spun organic TFTs based on aligned poly(3-hexylthiophene) (P3HT) nanofibers for OFETs with highly flexible and transparent characters. Apart from the capacitance of this device, which is far higher than those achievable with typical 100 nm thick SiO_2 dielectric layers, the device has an average field-effect mobility to be ~2 cm^2 V^{-1} s^{-1}, which is much higher than those in other P3HT-based TFTs. Most interesting, the devices exhibited a very good linear relationship between the number of fibers and maximum on current, which indicates that the electrical properties can be controlled by adjusting the number of fibers.

3.1.4. Electrodes

As a key component with favorable mechanical strength, large capacitance, and higher energy density, flexible electrode is important for flexible electronic devices. There are several approaches that have been adapted for the fabrication of flexible electrodes, such as electropolymerization,[103,104] photolithography,[105] chemical vapor deposition,[106] flow-directed assembly,[107] electrospinning,[108–112] etc. Particularly, with the application of some new materials (e.g., graphene), the methods become more flexible accordingly. For example, Liu *et al.*[111] reported a novel approach to fabricate flexible electrodes. Herein, the precursors were first electrospun and then followed by calcinations to obtain Cu–Ni alloy nanofibers. After that, the sample was put into a two-temperature-zone CVD furnace to grow graphene on the surface of Cu–Ni nanofibers by using solid polystyrene (PS) as the carbon source. It is found that the thin-films based on grapheme coated Cu–Ni nanofibers have excellent flexibility, and the resistance changed little even after the films bent to a curvature with the radius of 3 mm, much less than that of the graphene coated Ni nanofibers, as shown in Fig. 5(a).

Moreover, some of the flexible devices, such as touch screens, interactive electronics, and flexible photovoltaics (PVs), need the electrode to not only provide advantages mentioned above, but also transparency that enable light pass through as much as possible. Thus flexible transparent electrodes are faced with increasingly large demands. However, most of the flexible transparent electrodes are made from indium tin oxide (ITO) films, which are brittle, high cost, and have low infrared transmittance, despite the excellent electronic performances. Till now various materials

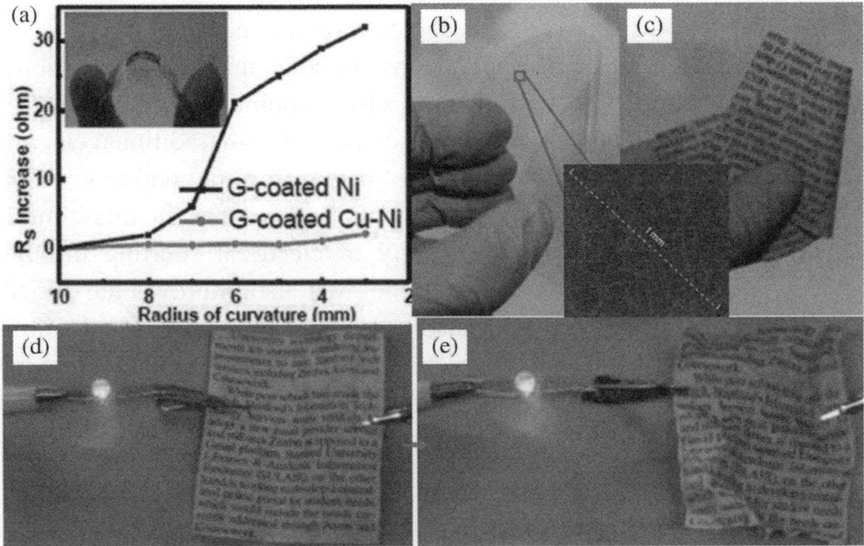

Figure 5.   (a) The resistance increase of G-coated Cu–Ni and G-coated Ni NF films after bending to different radii of curvature. Reproduced from Ref. 111. (Copyright © 2014, the Royal Society of Chemistry.) Gold nanotrough networks transferred onto a (b) PET plastic and (c) paper substrate, the insert is top-view SEM images of gold nanotrough networks. (d–e) Photographs of "conducting paper", fabricated by transferring gold nanotrough networks onto paper. After crushing and unfolding, the paper remained conducting. Reproduced from Ref. 38. (Copyright © 2013, Nature Publishing Group.)

such as metal, polymer and composites have been fabricated directly or with other methods, which may be capable for displacing ITO films as flexible transparent electrodes.[38,59–61,66,67,113,114] Cui' s team[38] have fabricated flexible transparent electrodes based on a metal nanotrough network. Here the continuous polymer nanofibrous mats as random networks or oriented arrays are electrospun as template, and coated with a thin layer of material using standard thin-film deposition techniques. After transferring, the electrodes with transparency and flexibility (Figs. 5(b) and 5(c)) have perfect sheet resistance (Rs) and optical transmittance (T), which are two key parameters that determine the applications of transparent electrodes. Most interesting, even after crushing and then unfolding the electrode on paper, the electrode remained conductive (Figs. 5(d) to 5(e)), with only a limited change in the resistance, which is ascribed that the nanotroughs remain continuous during folding, undergoing nanoscale deformation to relax the applied stress.

However, the surface roughness of electrospun transparent electrodes is a little high for the practical applications. To solve this problem, in some cases an annealing process following electrospinning is required.[59,113] This treatment results in less roughness compared with traditional electrospun mats by melting the as-spun nanofibers to form a continuous film. And adjacent nanofibers fuse together at the junctions in this course, which not only reduces the probability of electrical shorting through devices fabricated on top of the electrode[59] but also improves the optical transmittance.[113] Cui's group[114] also synthesized interconnected, ultralong junction-free metal nanowire networks by combining electrospinning and electroless deposition at ambient temperature and pressure. The results show that the Rs–T performance has been enhanced.

3.2. *Electrospun Stretchable Electronic Devices*

Mechanical flexibility (bendability) is a useful characteristic that can be achieved with macrodevices and substrates, and/or clever engineering approaches such as neutral mechanical plane designs.[10] However, stretchability is a different and much more challenging characteristic than bendability. Interest in stretchable electronics derives not only from the extreme levels of bendability that can be achieved, but also from the ability to integrate electronics with complex curvilinear surfaces, such as hemispheres for electronic eye imagers, wings of an aircraft for structural health monitoring, and biological systems for implants, sensors, or wearable electronics. None of these examples is possible with electronics that offer only bendability.[10] Nowadays, devices with good stretchability have been prepared mainly in two different but complementary strategies: (i) assembling a device directly from stretchable materials[115,116] and (ii) engineering new structural constructs from established materials.[11,117] Therefore, we in this section put emphasis on recent achievements relating to electrospinning for stretchable electronic devices following these two strategies, respectively.

3.2.1. Devices Based on Stretchable Materials

As we know, the electrospun polymer nanofibers have superior flexible surface functionality and mechanical performance, which enable the fibrous membranes experience tensile and compressive strains. Herein, the materials for stretchable electronic devices could be conductive polymers such as poly(3,4-ethylenedioxythiophene: poly(styrenesulfonate) (PEDOT:PSS),[54] poly(3-hexylthiophene) (P3HT),[118,119] elastomertic polymer such as Polyurethane (PU),[120] and in order to improve the relative properties, some other materials, such as carbon black nanoparticles,[54] metal nanoparticles,[121] carbon-nanotubes [69] were chosen as required additives. The typical purpose is to fabricate strain sensors,[53,54,121] electrodes,[69,118] transistors,[119] and supercapacitors.[120] For example, Li *et al.*[120] have introduced a dynamically stretchable supercapacitor using elastomeric electrospun (PU) membrane as the separator, as well as buckled singlewalled carbon nanotube(SWNT) macrofilms as the electrodes, and tetraethylammonium tetrafluoroborate in propylene carbonate as organic electrolyte (Fig. 6(a)). It is observed that the specific capacitance of the supercapacitor is improved at the maximum strain (31.5% strain) applied than that of the original unstretched state at 0% strain. In addition, cyclic voltammograms (CVs) of the supercapacitors at a scan rate of 100 mV s^{-1} at static 0% strain and static 31.5% strain show a perfectly symmetrical rectangular shape (Fig. 6(b)), and it is found this device can be reversibly charged and discharged under the dynamic stretching and releasing at an

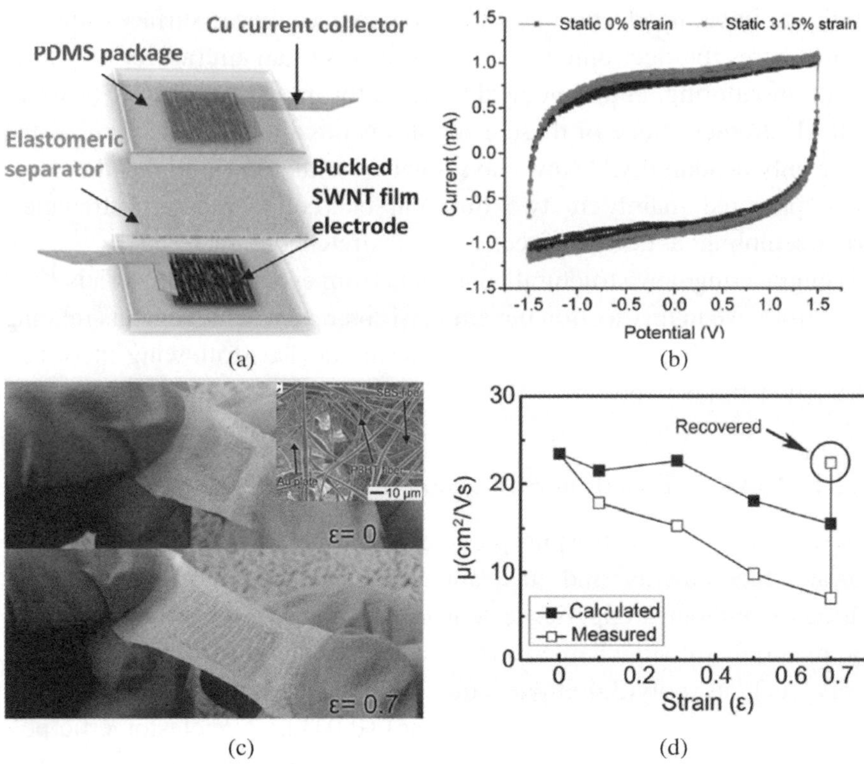

Figure 6. (a) Schematic of the main components of the dynamically stretchable super-capacitor. (b) cyclic voltammogram (CV) curves under static 0% strain and 31.5% strain at a scan rate of 100 mV s^{-1}. The device (under the static 31.5% strain) gives approximately 4% larger specific capacitance than that of the cell under the static 0% strain. Reproduced from Ref. 120. (Copyright © 2012, American Chemical Society.) (c) Digital image exhibiting the dimensional change of the device array observed by stretching at $\varepsilon =$ 0.7 (insert is SEM image of the SBS fiber mat, Au plate electrode, and P3HT fibers). (d) Mobility change as a function of strain. The measured mobilities (hollow) were converted into corrected values (solid) by considering the dimensional change of the channel layer. Reproduced from Ref. 119. (Copyright © 2014, Wiley-VCH.)

extremely high strain rate. After 1,000 galvanostatic charge-discharge cycles at a current density of 10 A g^{-1}, the device achieved excellent 98.5% capacitance retention, under a low strain rate of 1.11% strain s^{-1}, after finishing ~677 stretching/releasing cycles. At an extremely high strain rate of 4.46% s^{-1}, the device still achieved 94.6% capacitance retention, after finishing ~2,521 cycles. Shin *et al.*[119] have fabricated stretchable polymer transistors entirely of stretchable components (Fig. 6(c)). Here stretchable Au nanosheet electrodes, polyelectrolyte gel for the gate dielectric, the electrospun poly(3-hexylthiophene) (P3HT) nanofibers for the active channel materials, and poly(styrene-*b*-butadiene-b-styrene) (SBS) electrospun elastomer fiber mat as a substrate to enhance the mechanical stability of the transistors fabricated on the substrate are used. The combination of the four stretchable components provided a high hole mobility ($\mu = 18$ cm^2 V^{-1} s^{-1}) (Fig. 6(d)) at the strain $\varepsilon = 0.7$ and showed excellent electrical stability over 1,500 stretching cycles.

3.2.2. Devices Based on Stretchable Structures

Electrospun fibrous products have unique superiority for the fabrication of stretchable devices. Nonetheless, there are still problems to be solved. For instance, the isotropy of non-woven mats from traditional electrospinning restricts the direct, fast charge transfer along a certain orientation. The aligned fibrous arrays with anisotropy can do it well, but single nanofibers are too fragile to fracture, thus it is of significance to seek new structures for the experience of stretchability. One typical approach to achieve stretchability of the electronic devices is to construct the materials into "wavy", "helical", or "curled" architectures. Here, the systems can be stretched reversibly to large levels of strain without fracturing the materials because these strains can be accommodated through changes in the amplitudes and wavelengths of the curled or wavy structures.[11,122–125] Namely, these unusual structures can realize a prestrain control during stretch and release.

The fabrication of micro/nanofibers with curled or helical structures *via* electrospinning has been well studied recent years.[126–134] Some electrospun fibers with curled structures from two polymers with different properties, such as conductivity or elasticity, are electrospun, either as a blend or as a core-shell or side-by-side system. Here it is suggested that the helical construction were formed spontaneously upon contact with a conducting substrate due to the viscoelastic contraction of the fiber upon partial charge neutralization.[127,128] Other such structures fabricated on pure single-components can be attributed to the physical forces caused by the electrically driven bending instability of the electrospinning jet.[47,126,131] According to Reneker's model,[47,126] when bending perturbations began, the straight electrospinning jet grew rapidly into a coil in response to the repulsive electrical forces between the positive charges carried with the jet, and the straight electrospinning jet spontaneously forms a coil in response to the positive changes carried by the jet upon impingement on the collector.[132] Anyhow, most stretchable electronic devices demand not only perfect stretchability but also excellent electrical properties (e.g., rapid charge mobility, low resistance), thus it is very significant to combine curled structures with aligned fibers. Sun *et al.*[57] have fabricated stretchable strain sensors based on aligned microfibrous arrays of poly(3,4-ethylenedioxy-thiophene):poly(styrene sulfonate)-poly(vinyl pyrrolidone) (PEDOT: PSS-PVP) with curled architectures by a novel reciprocating-type electrospinning setup with a spinneret in straightforward simple harmonic motion. Besides a high room-temperature conductivity of the composite fibers, and a high repeatability and durability, the sensors can be stretched reversibly with a linear elastic response to strain up to 4% owing to the curled architectures of the as-spun fibrous polymer arrays, which is three times higher than that from electrospun non-woven mats. These results may be helpful for the fabrication of stretchable devices which have potential applications in some fields such as soft robotics, elastic semiconductors, and elastic solar cells.

Lately, researchers are focusing on using near field electrospinning approach to fabricate curled, wavy or helical patterns.[132,135] Because of the short electrode to-collector distance (500 μm–10 mm) in near-field electrospinning, bending instability and splitting of the charged jet in electrospinning are avoided, thus long and uniform fibers with precise-controlled morphologies can be easily generated. Duan *et al.*[73] have fabricated highly stretchable piezoelectric devices using this approach. Here the straight and uniform nanofibers of polyvinylidene fluoride (PVDF) were electrospun onto a prestrained poly(dimethylsiloxane) (PDMS) substrate using a modified near-field electrospinning termed mechano-electrospinning (MES). After the prestrained substrate released, the PVDF fibers buckle under compressive strain, forming aligned fibrous arrays, as shown in Fig. 7(a). The piezoelectric device showed a stable pressure measurement not only under the pressure perpendicular to the substrate surface but also tensile

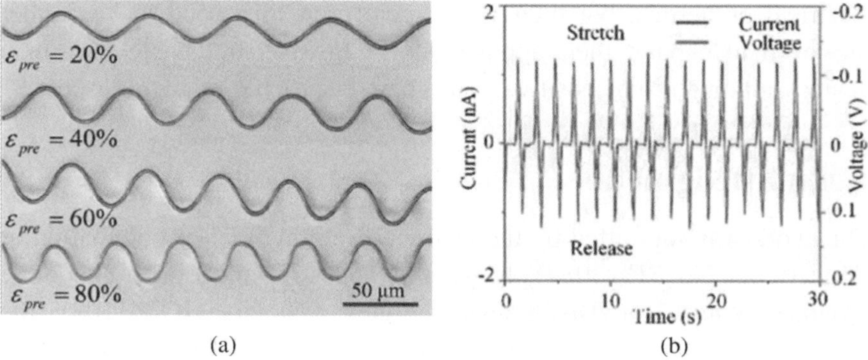

(a)   (b)

Figure 7.   (a) Images of in-surface buckled fibers for various prestrain levels. (b) Output current and voltage of the device consisting of 120 PVDF fibers measured with respect to time under an applied strain of 30% at 0.5 Hz. Reproduced from Ref. 73. (Copyright © 2014, the Royal Society of Chemistry.)

stress along the fibrous axis (Fig. 7(b)), with potential to be integrated into stretchable electronics such as human monitoring and artificial skin.

4. Summary and Outlook

Electrospinning is an effective approach for fabricating uniform ultrafine fibers with diameters ranging from several micrometers down to a few nanometers. Superior properties such as large surface area, high length/ diameter ratio, flexible surface functionality, tunable surface morphologies, and excellent mechanical performance of the electrospun micro/ nanostructures quite qualified for flexible electronic devices. In this chapter, based on the electrospinning backgrounds in general, some recent results of the electrospun products for bendable and stretchable electronic devices are presented. With achievements, some challenges in this field have been met and many practical problems have to be solved. For instance, due to the limitations on molecular weight and solubility unsuitable for electrospinning, only a few conductive polymer solutions with appreciating viscosity have been electrospun directly up to now, it is still a challenge to electrospin much more conducting polymers in proper ways for the need of manufacturing flexible electronic devices. In addition, most inorganic semiconductive nanofibers *via* electrospinning show excellent electrical properties, but their mechanical properties need to be further improved to expand their applications in flexible/stretchable electronic devices, although some positive results have been achieved in this aspect.[136]

Acknowledgments

This work was supported by the National Natural Science Foundation of China (Grant No. 51373082), the Natural Science Foundation of Shandong Province, China for Distinguished Young Scholars (JQ201103), the Taishan Scholars Program of Shandong Province (ts20120528), the National Key Basic Research Development Program of China (973 special preliminary study plan, 2012CB722705), the natural science foundation of Shandong Province (ZR2014EMM010) the Project of Shandong Province Higher Educational Science and Technology Program (J13LJ07), and the Program for Scientific Research Innovation Team in Colleges and Universities of Shandong Province, China.

References

1. R. L. Crabb and F. C. Treble, *Nature*, **213**, 1223 (1967).
2. K. A. Ray, *IEEE Trans. Aerosp. Electron. Syst.*, **3**(1), 107 (1967).
3. R. F. Service, *Science*, **312**, 1593 (2006).
4. X. F. Duan, *MRS Bull.*, **32**, 134 (2007).
5. P. C. Chen, G. Z. Shen, H. Chen, Y. Shi and C. W. Zhou, *ACS Nonlin.*, **4**, 4403 (2010).
6. Z. R. Wang, H. Wang, B. Liu, W. Z. Qiu, J. Zhang, S. H. Ran, H. T. Huang, J. Xu, H. W. Han, D. Chen and G. Z. Shen, *ACS Nano.*, **5**, 8412 (2011).
7. T. Sekitani, T. Yokota, U. Zschieschang, H. Klauk, S. Bauer, K. Takeuchi, M. Takamiya, T. Sakurai and T. Someya, *Science*, **326**, 1516 (2009).
8. X. F. Wang, W. F. Song, B. Liu, G. Chen, D. Chen, C. W. Zhou and G. Z. Shen, *Adv. Funct. Mater.*, **23**, 1202 (2013).
9. T. Y. Kim, H. Kim, S. W. Kwon, Y. Kim, W. K. Park, D. H. Yoon, A. R. Jang, H. S. Shin, K. S. Suh and W. S. Yang, *Nano Lett.*, **12**, 743 (2012).
10. A. J. Baca, J. H. Ahn, Y. Sun M. A. Meitl, E. Menard, H. S. Kim, W. M. Choi, D. H. Kim, Y. Huang and J. A. Rogers, *Angew. Chem. Int. Ed.*, **47**, 5524 (2008).

11. D. Y. Khang, H. Q. Jiang, Y. Huang and J. A. Rogers, *Science*, **311**, 208 (2006).

12. H. Yu, L. Ai, M. Rouhanizadeh, D. Patel, E. S. Kim and T. K. Hsiai, *J. Microelectromech. Syst.*, **17**, 1178 (2008).

13. R. Buchner, K. Froehner, C. Sosna, W. Benecke and W. Lang, *J. Microelectromech. Syst.*, **17**, 1114 (2008).

14. S. Inoue, S. Utsunomiya, T. Saeki and T. Shimoda, *IEEE Trans. Electron Devices*, **49**, 1353 (2002).

15. Y. Lee, H. Li and S. J. Fonash, *IEEE Electron Device Lett.*, **24**, 19 (2003).

16. A. Asano and T. Kinoshita, *SID Symp. Digest Tech. Papers*, **33**, 1196 (2002).

17. C. Berge, T. Wagner, W. Brendle, C. Craff-Castillo, M. B. Schubert and J. H. Werner, *Mat. Res. Soc. Symp. Proc.*, **769**, 53 (2003).

18. R. Stewart, A. Chiang, A. Hermanns, F. Vicentini, J. Jacobsen, J. Atherton, E. Boiling, F. Cuomo, P. Drzaic and S. Pearson, *Proc SPIE — Int. Soc. Opt. Eng.*, **4712**, 350 (2002).

19. I. C. Cheng and S. Wagner, *Springer, CA*, **1**, (2009).

20. J. H. Ahn, H. S. Kim, K. J. Lee, S. Jeon, S. J. Kang, Y. G. Sun, R. G. Nuzzo and J. A. Rogers, *Science*, **314**, 1754 (2006).

21. M. Madsen, K. Takei, R. Kapadia, H. Fang, H. Ko, T. Takahashi, A. C. Ford, M. H. Lee and A. Javey, *Adv. Mater.*, **23**, 3115 (2011).

22. A. Carlson, A. M. Bowen, Y. Huang, R. G. Nuzzo and J. A. Rogers, *Adv. Mater.*, **24**, 5284 (2012).

23. Y. G. Sun and J. A. Rogers, *Adv. Mater.*, **19**, 1897 (2007).

24. G. H. Yu, A. Y. Cao and C. M. Lieber, *Nat. Nanotechnol.*, **2**, 372 (2007).

25. B. Sun, Y. Z. Long, Z. J. Chen, S. L. Liu, H. D. Zhang, J. C. Zhang and W. P. Han, *J. Mater. Chem. C*, **2**, 1209 (2014).

26. M. Hasan, J. Rho, S. Y. Kang and J. H. Ahn, *Jpn. J. Appl. Phys.*, **49**, 05EA01 (2010).

27. X. F. Duan, C. M. Niu, V. Sahi, J. Chen, J. W. Parce, S. Empedocles and J. L. Goldman, *Nature*, **425**, 274 (2003).

28. M. C. McAlpine, R. S. Friedman, S. Jin, K. H. Lin, W. U. Wang and C. M. Lieber, *Nano Lett.*, **3**, 1531 (2003).

29. J. Y. Kim, K. Lee, N. E. Coates, D. Moses, T. Q. Nguyen, M. Dante and A. J. Heeger, *Science*, **317**, 222 (2007).

30. A. Lund, C. Jonasson, D. Haagensen and B. J. Hagstrom, *Appl. Polym. Sci.*, **126**, 490 (2012).

31. G. H. Gelinck, H. E. A. Huitema, E. Van Veenendaal, E. Cantatore, L. Schrijnemakers, J. B. P. H. Van der Putten, T. C. T. Geuns, M. Beenhakkers, J. B. Giesbers, B. H. Huisman, E. J. Meijer, E. M. Benito, F. J. Touwslager, A. W. Marsman, B. J. E. Van Rens and D. M. De Leeuw, *Nat. Mater.*, **3**, 106 (2004).

32. T. Someya, T. Sekitani, S. Iba, Y. Kato, H. Kawaguchi and T. Sakurai, *Proc. Natl. Acad. Sci. USA*, **101**, 9966 (2004).

33. C. Pang, G. Y. Lee, T. I. Kim, S. M. Kim, H. N. Kim, S. H. Ahn and K. Y. Suh, *Nat. Mater.*, **11**, 795 (2012).

34. M. K. Kwak, H. E. Jeong and K. Y. Suh, *Adv. Mater.*, **23**, 3949.

35. W. G. Bae, D. Kim, M. K. Kwak, L. Ha, S. M. Kang and K. Y. Suh, *Adv. Healthcare Mater.*, **2**, 109 (2013).

36. S. C. B. Mannsfeld, B. C. K. Tee, R. M. Stoltenberg, C. V. H. H. Chen, S. Barman, B. V. O. Muir, A. N. Sokolov, C. Reese and Z. N. Bao, *Nat. Mater.*, **9**, 859 (2010).

37. C. Pang, C. Lee and K. Y. Suh, *J. Appl. Polym. Sci.*, **130**, 1429 (2013).

38. H. Wu, D. Kong, Z. C. Ruan, P. C. Hsu, S. Wang, Z. Yu, T. J. Carney, L. Hu, S. Fan and Y. Cui, *Nat. Nanotechnol.*, **8**, 421 (2013).

39. H. C. Chang, C. L. Liu and W. C. Chen, *Adv. Funct. Mater.*, **23**, 4960 (2013).

40. G. Ren, F. Cai, B. Li, J. Zheng and C. Xu, *Macromol. Mater. Eng.*, **298**, 541 (2012).

41. J. K. Kim, J. Manuel, M. H. Lee, J. Scheers, D. H. Lim, P. Johansson, J. H. Ahn, A. Matica and P. Jacobsson, *J. Mater. Chem.*, **22**, 15045 (2012).

42. S. Cavaliere, S. Subianto, I. Savych, D. J. Jones and J. Rozière, *Energy Environ. Sci.*, **4**, 4761 (2011).

43. D. J. Lipomi and Z. Bao, *Energy Environ. Sci.*, **4**, 3314 (2011).

44. D. Li and Y. Xia, *Adv. Mater.*, **16**, 1151 (2004).

45. X. Lu, W. Zhang, C. Wang, T. C. Wen and Y. Wei, *Prog. Polym. Sci.*, **36**, 671 (2011).

46. A. Greiner and J. H. Wendorff, *Adv. Polym. Sci.*, **219**, 107 (2008).

47. D. H. Reneker and A. L. Yarin, *Polymer*, **49**, 2387 (2008).

48. S. Agarwal, J. H. Wendorff and A. Greiner, *Polymer*, **49**, 5603 (2008).

49. J. D. Schiffman and C. L. Schauer, *Polym. Rev.*, **48**, 317 (2008).

50. A. Formhals, US Patent No. 1975504, (1934).

51. D. H. Reneker and I. Chun, *Nanotechnology*, **7**, 216 (1996).

52. B. Sun, Y. Z. Long, H. D. Zhang, M. M. Li, J. L. Duvail, X. Y. Jiang and H. L. Yin, *Prog. Polym. Sci.*, **39**, 862 (2014).

53. M. K. Tiwari, A. L. Yarin and C. M. Megaridis, *J. Appl. Phys.*, **103**, 044305 (2008).

54. N. Liu, G. Fang, J. Wan, H. Zhou, H. Long and X. Zhao, *J. Mater. Chem.*, **21**, 18962 (2011).

55. L. Persano, C. Dagdeviren, Y. Su, Y. Zhang, S. Girardo, D. Pisignano, Y. Huang and J. A. Rogers, *Nat. Commun.*, **4**, 1633 (2013).

56. Q. Gao, H. Meguro, S. Okamoto and M. Kimura, *Langmuir*, **28**, 17593 (2012).

57. B. Sun, Y. Z. Long, S. L. Liu, Y. Y. Huang, J. Ma, H. D. Zhang, G. Shen and S. Xu, *Nanoscale*, **5**, 7041 (2013).

58. S. W. Lee, H. J. Lee, J. H. Choi, W. G. Koh, J. M. Myoung, J. H. Hur, J. J. Park, J. H. Cho and U. Jeong, *Nano Lett.*, **10**, 347 (2010).

59. H. Wu, L. Hu, M. W. Rowell, D. Kong, J. J. Cha, J. R. McDonough, J. Zhu, Y. Yang, M. D. McGehee and Y. Cui, *Nano Lett.*, **10**, 4242 (2010).

60. P. C. Hsu, H. Wu, T. J. Carney, M. T. McDowell, Y. Yang, E. C. Garnett, M. Li, L. Hu and Y. Cui, *ACS Nano.*, **6**, 5150 (2012).

61. H. Li, W. Pan, W. Zhang, S. Huang and H. Wu, *Adv. Funct. Mater.*, **23**, 209 (2013).

62. M. Xue, Z. Xie, L. Zhang, X. Ma, X. Wu, Y. Guo, W. Song, Z. Li and T. Cao, *Nanoscale*, **3**, 2703 (2011).

63. V. Presser, L. Zhang, J. J. Niu, J. McDonough, C. Perez, H. Fong and Y. Gogotsi, *Adv. Energy Mater.*, **1**, 423 (2011).

64. X. Yan, Z. Tai, J. Chen and Q. Xue, *Nanoscale*, **3**, 212 (2011).

65. Z. Tai, X. Yan, J. Lang and Q. Xue, *J. Power Sources*, **199**, 373 (2012).

66. Y. K. Fuh and L. C. Lien, *Nanotechnology*, **24**, 055301 (2013).

67. S. Soltanian, R. Rahmanian, B. Gholamkhass, N. M. Kiasari, F. Ko and P. Servati, *Adv. Energy Mater.*, **3**, 1332 (2013).

68. Y. Z. Long, R. Huang, C. C. Tang, H. D. Zhang, B. Sun, S. Xu, W. X. Wang and Q. Chang, Chinese Patent, CN201210196840.0 (2012).

69. T. A. Kim, S. S. Lee, H. Kim and M. Park, *RSC Adv.*, **2**, 10717 (2012).

70. Y. Z. Long, M. M. Li, C. Gu, M. Wan, J. L. Duvail, Z. Liu and Z. Y. Fan, *Prog. Polym. Sci.*, **36**, 1415 (2011).

71. B. Ding, M. Wang, X. Wang, J. Yu and G. Sun, *Mater. Today*, **13**, 16 (2011).

72. C. Hou, T. Huang, H. Wang, H. Yu, Q. Zhang and Y. Li. *Sci. Rep.*, **3**, 3138 (2013).

73. Y. Q. Duan, Y. A. Huang, Z. P. Yin, N. B. Bu and W. T. Dong, *Nanoscale*, **6**, 3289 (2014).

74. D. Mandal, S. Yoon and K. J. Kim. *Macromol. Rapid Commun.*, **32**, 831 (2011).

75. O. S. Kwon, E. Park, O. Y. Kweon, S. J. Park and J. Jang, *Talanta*, **82**, 1338 (2010).

76. L. Zhang, X. Wang, Y. Zhao, Z. Zhu and H. Fong, *Mater. Lett.*, **68**, 133 (2012).

77. X. Chen, S. Guo, J. Li, G. Zhang, M. Lu and Y. Shi, *Sens. Actuat. A*, **199**, 372 (2013).

78. L. Zhang, X. Gong, Y. Bao, Y. Zhao, M. Xi, C. Jiang and H. Fong, *Langmuir*, **28**, 14433 (2012).

79. K. Saetia, J. M. Schnorr, M. M. Mannarino, S. Y. Kim, G. C. Rutledge, T. M. Swager and P. T. Hammond, *Adv. Funct. Mater.*, **24**, 492 (2014).

80. X. Liu, L. Gu, Q. Zhang, J. Wu, Y. Long and Z. Fan, *Nat. Commun.*, **5**, 4007 (2014).

81. M. Xi, X. Wang, Y. Zhao, Z. Zhu and H. Fong, *Appl. Phys. Lett.*, **104**, 133102 (2014).

82. J. Sirohi and I. Chopra, *J. Intell. Mater. Syst. Struct.*, **11**, 246 (2000).

83. L. Polavarapua and L. M. Liz-Marzán, *Phys. Chem. Chem. Phys.*, **15**, 5288 (2013).

84. Q. Wu, Y. Xu, Z. Yao, A. Liu and G. Shi, *ACS Nano.*, **4**, 1963 (2010).

85. Y. Y. Horng, Y. C. Lu, Y. K. Hsu, C. C. Chen, L. C. Chen and K. H. Chen, *J. Power Sources*, **195**, 4418 (2010).

86. M. F. El-Kady and R. B. Kaner, *Nat. Commun.*, **4**, 1475 (2013).

87. X. Lu, T. Zhai, X. Zhang, Y. Shen, L. Yuan, B. Hu, L. Gong, J. Chen, Y. Gao, J. Zhou, Y. Tong and Z. L. Wang, *Adv. Mater.*, **24**, 938 (2012).

88. Y. J. Kang, S. J. Chun, S. S. Lee, B. Y. Kim, J. H. Kim, H. Chung, S. Y. Lee and W. Kim, *ACS Nano.*, **6**, 6400 (2012).

89. A. Laforgue, *J. Power Sources*, **196**, 559 (2010).

90. Y. S. Wang, S. M. Li, S. T. Hsiao, W. H. Liao, P. H. Chen, S. Y. Yang, H. W. Tien, C. C. M. Ma and C. C. Hu. *Carbon*, **73**, 87 (2014).

91. H. Klauk (Ed.), *Organic Electronics: Materials, Manufacturing and Applications*, Wiley-VCH Verlag GmbH & Co. KGaA, Weinheim, Germany (2006).

92. Z. Bao and J. Locklin (Ed.), *Organic Field Effect Transistors*, CRC Press, New York, USA (2007).

93. M. Heeney and I. McCulloch, Semiconducting polythiophenes for field-effect transistor devices in flexible electronics: synthesis and structure property relationships, in *Flexible Electronics: Materials and Applications*, Eds. W. S. Wong and A. Salleo, Springer, New York, USA, p. 261 (2009).

94. S. R. Forrest, *Nature*, **428**, 911 (2004).

95. J. T. Mabeck and G. G. Malliaras, *Anal. Bioanal. Chem.*, **384**, 343 (2006).

96. C. J. Brabec, N. S. Sariciftci and J. C. Hummelen, *Adv. Func. Mater.*, **11**, 15 (2001).

97. B. Lim, K. J. Baeg, H. G. Jeong, J. Jo, H. S. Kim, J. W. Park, Y. Y. Noh, D. J. Vak, J. H. Park, J. W. Park and D. Y. Kim, *Adv. Mater.*, **21**, 2808 (2009).

98. W. Wang, Z. Li, X. Xu, B. Dong, H. Zhang, Z. Wang, C. Wang, R. H. Baughman and S. Fang, *Small*, **7**, 597 (2011).

99. W. Wang, X. Lu, Z. Li, J. Lei, X. Liu, Z. Wang, H. Zhang and C. Wang, *Adv. Mater.*, **23**, 5109 (2011).

100. A. L. Briseno, S. C. B. Mannsfeld, X. M. Lu, Y. J. Xiong, S. A. Jenekhe, Z. Bao and Y. Xia, *Nano Lett.*, **7**, 668 (2007).

101. C. J. Lin, J. C. Hsu, J. H. Tsai, C. C. Kuo, W. Y. Lee and W. C. Chen, *Chem. Phys.*, **212**, 2452 (2011).

102. S. Y. Min, T. S. Kim, B. J. Kim, H. Cho, Y. Y. Noh, H. Yang, J. H. Cho and T. W. Lee, *Nat. Commun.*, **4**, 1773 (2013).

103. D. W. Wang, F. Li, J. Zhao, W. Ren, Z. G. Chen, J. Tan, Z. S. Wu, I. Gentle, G. Q. Lu and H. M. Cheng, *ACS Nano.*, **3**, 1745 (2009).

104. Z. L. Wang, R. Guo, G. R. Li, H. L. Lu, Z. Q. Liu, F. M. Xiao, M. Zhang and Y. X. Tong, *J. Mater. Chem.*, **22**, 2401 (2012).

105. B. A. Hollenberg, C. D. Richards, R. Richards, D. F. Bahr and D. M. Rector, *J. Neurosci. Methods*, **153**, 147 (2006).

106. X. Jia, Z. Chen, A. Suwarnasarn, L. Rice, X. Wang, H. Sohn, Q. Zhang, B. M. Wu, F. Wei and Y. Lu, *Energy Environ. Sci.*, **5**, 6845 (2012).

107. F. Liu, S. Song, D. Xue and H. Zhang, *Adv. Mater.*, **24**, 1089 (2012).

108. Y. Gao, V. Presser, L. Zhang, J. J. Niu, J. K. McDonough, C. R. Pérez, H. Lin, H. Fong and Y. Gogotsi, *J. Power Sources*, **201**, 368 (2012).

109. S. Liu, Z. Wang, C. Yu, H. B. Wu, G. Wang, Q. Dong, J. Qiu, A. Eychmüller and X. W. Lou, *Adv. Mater.*, **25**, 3462 (2013).

110. W. Li, Z. Yang, J. Cheng, X. Zhong, L. Gu and Y. Yu, *Nanoscale*, **6**, 4532 (2014).

111. Z. D. Liu, Z. Y. Yin, Z. H. Du, Y. Yang, M. M. Zhu, L. H. Xie and W. Huang, *Nanoscale*, **6**, 5110 (2014).

112. F. Xiang, J. Zhong, N. Gu, R. Mukherjee, I. K. Oh, N. Koratkar and Z. Yang, *B*, 6, 201 (2014).

113. Y. L. Huang, A. Baji, H. W. Tien, Y. K. Yang, S. Y. Yang, S. Y. Wu, C. C. M. Ma, H. Y. Liu, Y. W. Mai and N. H. Wang, *Carbon*, **75**, 3473 (2012).

114. P. C. Hsu, D. Kong, S. Wang, H. Wang, A. J. Welch, H. Wu and Y. Cui, *J. Am. Chem. Soc.*, 136, 10593 (2014).

115. T. Yamada, Y. Hayamizu, Y. Yamamoto, Y. Yomogida, A. Izadi-Jajafabadi, D. N. Futaba and K. Hata, *Nat. Nanotechnol.*, **6**, 296 (2011).

116. T. Sekitani and T. Someya, *Adv. Mater.*, **22**, 2228 (2010).

117. D. H. Kim, J. Xiao, J. Song, Y. Huang and J. A. Rogers, *Adv. Mater.*, **22**, 2108 (2010).

118. G. D. Moon, G. H. Lim, J. H. Song, M. Shin, T. Yu and B. Lim, U. Jeong, *Adv. Mater.*, **25**, 2707 (2013).

119. M. Shin, J. H. Song, G. H. Lim, B. Lim, J. J. Park and U. Jeong, *Adv. Mater.*, **26**, 3706 (2014).

120. X. Li, T. Gu and B. Wei, *Nano Lett.*, **12**, 6366 (2012).

121. M. Park, J. Im, M. Shin, Y. Min, J. Park, H. Cho, S. Park, M. B. Shim, S. Jeon, D. Y. Chung, J. Bae, J. Park, U. Jeong and K. Kim, *Nat. Nanotechnol.*, **7**, 803 (2012).

122. Y. G. Sun, W. M. Choi, H. Jiang, Y. G. Huang and J. A. Rogers, *Nat. Nanotechnol.*, 1, 201 (2006).

123. Y. G. Sun, V. Kumar, I. Adesida and J. A. Rogers, *Adv. Mater.*, **18**, 2857 (2006).

124. S. P. Lacour, J. Jones, S. Wagner, T. Li and Z. Suo, *Proc. IEEE*, **93**, 1459 (2005).

125. S. P. Lacour, S. Wagner, Z. Y. Huang and Z. Suo, *Appl. Phys. Lett.*, **82**, 2404 (2003).

126. D. H. Reneker, A. L. Yarin, H. Fong and S. Koombhongse, *J. Appl. Phys.*, **87**, 4531 (2000).

127. R. Kessick and G. Tepper, *Appl. Phys. Lett.*, **84**, 4807 (2004).

128. Y. Xin, Z. H. Huang, E. Y. Yan, W. Zhang and Q. Zhao, *Appl. Phys. Lett.*, **89**, 053101 (2006).

129. T. Lin, H. X. Wang and X. G. Wang, *Adv. Mater.*, **17**, 2699 (2005).

130. B. F. Zhang, C. J. Li and M. Chang, *Polym. J.*, **41**, 252 (2009).

131. M. K. Shin, S. I. Kim and S. J. Kim, *Appl. Phys. Lett.*, **88**, 223109 (2006).

132. T. Han, D. H. Reneker and A. L. Yarin, *Polymer*, 48, 6064 (2007).

133. J. Yu, Y. J. Qiu, X. X. Zha, M. Yu, J. L. Yu, J. Rafique and J. Yin, *Eur. Polym. J.*, **44**, 2838 (2008).

134. C. C. Tang, J. C. Chen, Y. Z. long, H. X. Yin, B. Sun and H. D. Zhang, *Chinese Phys. Lett.*, **28**, 056801 (2011).

135. Y. A. Huang, N. Bu, Y. Duan, Y. Pan, H. Liu, Z. Yin and Y. Xiong, *Nanoscale*, **5**, 12007 (2013).

136. S. Huang, H. Wu, M. Zhou, C. Zhao, Z. Yu, Z. Ruan and W. Pan, *NPG ASIA Mater.*, **6**, 86 (2014).

INDEX